Fundamentals of Inorganic Glasses

Fundamentals of Inorganic Glasses

Arun K. Varshneya

New York State College of Ceramics
Alfred University
Alfred, New York

ACADEMIC PRESS, INC.
Harcourt Brace & Company, Publishers

Boston San Diego New York
London Sydney Tokyo Toronto

This book is printed on acid-free paper. ♾

Academic Press, Inc.
1250 Sixth Avenue, San Diego, CA 92101-4311

United Kingdom Edition published by
Academic Press Limited
24–28 Oval Road, London NW1 7DX

Library of Congress Cataloging-in-Publication Data

Varshneya, Arun K.
Fundamentals of inorganic glasses / Arun K. Varshneya.
p. cm.
Includes bibliographical references and index.
ISBN 0-12-714970-8 (acid-free paper)
1. Glass. I. Title.
TP857.V37 1993
620.1′44—dc20 93-16591
CIP

Printed in the United States of America
93 94 95 96 97 BB 9 8 7 6 5 4 3 2 1

In memory of
my father
Nathi Lal Varshneya

Contents

Preface

"What, a glass scientist?" "What's that?" These have often been the typical responses to my indicated profession in social circles. Clearly, steel's impact on society has been powerful enough for the term "metallurgist" to be recognizable as a profession. Glass has yet to graduate to this level of recognition despite the fact that indulgence in drinking fluids out of a glass vessel, and looking at the world through a pair of eyeglasses and through a room window have been around for quite some time. Presumably, such aberrations will be corrected in the now long-overdue *materials age* when, along with crystalline ceramics such as ceramic superconductors, glass fiber for communication links will be a part of the common household vocabulary. As it happens, my father never had any confusion between a "metallurgist" and a "glass scientist." He was a laboratory glass supplier in India and knew some 30 years ago that a future for glass professionals existed. And so, there I was ... headed toward becoming a glass scientist. (Thanks, Dad, for that remarkable foresight!) As such, one of my primary purposes in writing this book is to convey that feeling of "identity" to the young that a glass professional (scientist, engineer, or technologist) does belong to a reputable caste. The day is not far, probably, when some education about glass will find its way through every college-level engineering and science curriculum.

A second purpose is to bring together a host of fine quality books on glass into a single book which has the flavor of being a textbook for an undergraduate student—comprehensive, yet confining itself to a general understanding of the topics. Trying to strike a balance between the depth and the breadth has always been my aim. Unfortunately, I did have to set limits on the coverage. This book is about *inorganic glasses*, and mostly about their *science*. Glasses based upon the carbon chains and macromolecules have not been included. Also, details of the technology and engineering of glass and glass product manufacture have been spared for a later date.

The book is intended to be a textbook on glass science suitable for teaching at a junior/senior level in a materials curriculum. Emphasis has been placed upon developing the fundamental concepts, whether they were ultimately proven wrong or not. As such, the book may also be useful to industrial scientists and engineers who are attempting to acquire a basic knowledge in glass. While all efforts have been made to avoid deep scientific discussions and heavy mathematics, there are places where such was unavoidable. Because of the size of Chapter 13, a summary has been written at the end. Some topics in phase separation (Chapter 4), much of the glass transformation range behavior (Chapter 13), and some topics in dielectric properties (Chapter 15), electronic conduction (Chapter 16) and optical properties (Chapter 19) could be spared for a second-time reading or, perhaps, for graduate-level instructions.

In writing the book, I have taken a teacher's point of view. The organization of the chapters is almost the way I like to teach "Introduction to Glass Science" to our students with one exception; Chapter 20 (Fundamentals of Inorganic Glassmaking) is taught after Chapter 5, primarily because the students get a bit "itchy" to learn some technology after a load of structures. Several key ideas have been set in italics: many key words are set in bold lettering. Occasionally, it may appear as if I am leading the reader by the hand—please forgive me for this audacity on my part.

I strongly recommend that students practice the drawing of glass networks. (One picture is worth a thousand words.) Likewise, I urge them to attack at least some of the questions posed at the end of most chapters. Answers to a few are provided. Further consultation of "Suggested Reading" is always encouraged.

I am sure that many errors have slipped by in this first attempt. Please drop me a note if you can help bring even the smallest of corrections or improvements to this book.

June 30, 1993

Arun K. Varshneya
New York State College of Ceramics
Alfred University
Alfred, NY 14802

Acknowledgments

I am forever grateful to my own teacher, Professor Alfred R. Cooper, Jr., of Case Western Reserve University, for several wonderful years of association. His knowledge, insight, and objective thinking about glass problems were a model for me.

I am indebted to Harold Rawson of Sheffield (U.K.), Prabhat Gupta of Ohio State University, George Scherer of Du Pont Company, Joe Simmons of the University of Florida, and Alastair Cormack of Alfred University, who read parts of this book (voluntarily). Their constructive criticism helped the content of this book immensely.

I would like to express my sincere appreciation to several of my colleagues and members of administration at the New York State College of Ceramics for their sustained colleagueship, comradery, and constant encouragement. Of these, I owe special thanks to Bill LaCourse: besides all the stimulating technical discussions in the hallways and the no-charge book loans, he almost always had a medicine for the various computer hiccups such that my floppies rarely needed to see the trash.

Frequent technical discussions with Oleg Mazurin of the Russian Academy of Sciences, St. Petersburg (Russia), were quite useful. Thanks are also due to Tony DiGaudio of Williamsville, New York, for help with computer-graphics.

The patience, understanding, encouragement, and continuing love expressed by my wife, Darshana, and daughters, Pooja, Kajal, and Rupal, helped me endure the pains of writing this book.

June 30, 1993

Arun K. Varshneya
Alfred, New York

Chapter 1

Introduction

1.1. Brief History

The word *glass* is derived from a late-Latin term *glæsum*, used to refer to a lustrous and transparent material. Another word often used to refer to glassy substances is *vitreous*, originating from the Latin word *vitrum*. Luster, or shine, and in particular its durability when exposed to the elements of nature, were probably the most significant properties of glass recognized by early civilizations. Glazed stone beads from Egypt date back to 12,000 B.C. Several of the artifacts unearthed from the tombs of the pharaohs exhibit excellent glass inlay work in a variety of colors. As independent objects, glassware perhaps existed roughly five to six thousand years ago. The technology of the glass window exploiting the property of transparency had developed around the birth of Christ and was developed to new heights of artistry by the Christian Church during the Middle Ages. Many of these beautifully stained windows, which can still be viewed in a number of churches over the European continent, show the deep commitment of the church to preserve the history of mankind and religious teachings through the medium of glass.

Many of the uses of glass in the modern world continue to exploit the transparency, luster, and durability properties of glass. Containers, windows, lighting, insulation, fiber, stemware, and other hand-crafted art objects are typical of these traditional uses. At this point, it is worth noting that for a material to be used in a product it must have certain desirable properties that determine its use. Later on in our discussion, it will become clear that *the properties of transparency, luster, and durability are neither sufficient nor necessary to describe "glass."* Through the application of basic sciences to the study of glass, newer properties of glasses have been developed, and hence, newer products have been conceived.

As may be expected, much of the glass science developed on the basis of the major commercial uses of glass. More than 99% of the commercial tonnage consists of glass compositions that are oxides. A large percentage of these are silica-based. This includes even the highly specialized application of glass to microelectronic packaging where the annual volume of sale may be low but glass is the "value-adding" component, i.e., the application of glass enhances the value of the assembly after the incorporating process. It is not surprising that when the term "glass" is used in scientific conversation, oxide glasses are usually implied. Over the past two to three decades, however, the possibility of some exotic uses of glass such as repeaterless transoceanic or transcontinental telecommunication lines and the delivery of CO_2 laser power to perform microsurgery has triggered a great many studies of non-oxide glasses. It is well, therefore, to review our thoughts on the various families of glasses, their compositions, and their uses before we delve into the science of glass.

1.2. Glass Families of Interest

Table 1-1 presents a summary of the various inorganic glass families that are of commercial interest. All the glasses listed here are silica-based.

One may note that, besides silica, other major constituents usually are the alkalis, the alkaline earths, alumina, boric oxide, and lead oxide. Compounds such as arsenic and sulfur are added as traces (minor constituents) intentionally. Many of the reasons for the major component additions should become clear as we continue. The various glass families are discussed in what follows.

1.2.1. Vitreous Silica

Vitreous silica is the most refractory glass in commercial use. In addition to its refractoriness, it has a high chemical resistance to corrosion (particularly to acids), a very low electrical conductivity, a near-zero ($\sim 5.5 \times 10^{-7}/°C$) coefficient of thermal expansion, and good UV transparency. Because of the high cost of manufacture, the uses of vitreous silica are mostly limited to

Table 1-1. Commercial Glass Compositions in Oxides Per Cent by Weight

	Vitreous silica	Vycor	Plate	Window	Bottle or container	Bulb	Tubing	Lime tableware	Pyrex type	Ther-mometer	Borosilicate crown	Lead tableware	Glass halogen lamp	E glass	S glass	Optical flint
SiO_2	100.0	94.0	72.7	72.0	74.0	73.6	72.1	74.0	81.0	72.9	69.6	67.0	60.0	52.9	65.0	49.8
Al_2O_3			0.5	0.6	1.0	1.0	1.6	0.5	2.0	6.2		0.4	14.3	14.5	25.0	0.1
B_2O_3		5.0							12.0	10.4	9.9			9.2		
SO_3			0.5	0.7	Tr.								0.3			
CaO			13.0	10.0	5.4	5.2	5.6	7.5		0.4			6.5	17.4		
MgO				2.5	3.7	3.6	3.4			0.2				4.4	10.0	
BaO					Tr.						2.5		18.3			13.4
PbO												17.0				18.7
Na_2O		1.0	13.2	14.2	15.3	16.0	16.3	18.0	4.5	9.8	8.4	6.0	0.01			1.2
K_2O					0.6	0.6	1.0			0.1	8.4	9.6	Tr.	1.0		8.2
ZnO																8.0
As_2O_5			Tr.	Tr.	Tr.	Tr.		Tr.		Tr.	0.3	Tr				0.4

astronomical mirrors, optical fibers, crucibles for melting high-purity silicon, and high-efficacy lamp envelopes. In one technique, the glass is obtained by melting high-purity quartz crystals or beneficiated sand at temperatures in excess of 2,000°C. In a second technique, $SiCl_4$ is sprayed into an oxy–hydrogen flame or water-vapour-free oxygen plasma. Silica vapors deposit on a substrate and are consolidated subsequently at ~1,800°C.

1.2.2. Soda-Lime Glass

Soda-lime glass or soda-lime-silicate glass is perhaps the least expensive and the most widely used of all the glasses made commercially. Most of the beverage containers, glass windows, and incandescent and fluorescent lamp envelopes are made from soda-lime glass. It has good chemical durability, high electrical resistivity, and good spectral transmission in the visible region. Because of its relatively high coefficient of thermal expansion ($\sim 100 \times 10^{-7}/°C$), it is prone to thermal shock failure, and this prevents its use in a number of applications. Large-scale continuous melting of inexpensive batch materials such as soda ash (Na_2CO_3), limestone ($CaCO_3$), and sand at 1,400–1,500°C makes it possible to form the products at high speeds inexpensively.

1.2.3. Borosilicate Glass

Small amounts of alkali added to silica and boron oxide make a family of glasses which are utilized for their low thermal expansion coefficient ($\sim 30\text{–}60 \times 10^{-7}/°C$) and a high resistance to chemical attack. Laboratory glassware, household cooking utensils, and automobile headlamps are prime examples of their usage. Glasses can be made commercially in a manner similar to the soda-lime glasses, but require slightly higher temperatures (~1,550–1,600°C). The high cost of B_2O_3 makes them much less competitive compared to the soda-lime glasses for common products.

1.2.4. Lead Silicate Glass

This family of glasses contains PbO and SiO_2 as the principal components with small amounts of soda or potash. These glasses are utilized for their high degree of brilliance (as stemware or "crystal"), large working range (useful to make art objects and intricate shapes without frequently reheating the glass), and high electrical resistivity (e.g., for electrical feedthrough components). PbO additions increase the fluidity of glass and its wettability to oxide ceramics. Hence, high lead borosilicate glasses (generally without any alkali additions) are used extensively in microelectronics (e.g., for conductor, resistor, and dielectric pastes).

1.2.5. Aluminosilicate Glass

Commercial glasses of this family have varying amounts of alkalis and a moderate amount of alkaline earths such as MgO and CaO. The low-alkali-containing aluminosilicates have high values of elastic moduli and a high resistance to chemical corrosion. Compositions commonly called, "E-glass" and "S-glass" are used as the load-bearing fiber component in fiber-reinforced plastics. The electrical resistance of the alkali-free alkaline earth aluminosilicates is comparable to (and may, in some cases, even exceed) that of vitreous silica. These glasses are intermediates between the soda-lime and the vitreous silica for refractoriness and thermal expansion ($\sim 50 \times 10^{-7}/°C$). A major commercial use of this family is in the high-efficacy lamps involving the tungsten–halogen cycle (for instance, the lamp inside an automobile halogen headlamp), where the glass can be sealed directly to molybdenum electrical leads.

1.2.6. Other Silica-Based Oxide Glasses

The primary application in this class is the optical glasses used in optical components and devices, including ophthalmics. The inclusion of BaO, ZnO, La_2O_3, Nd_2O_3 and others in soda limes, and borosilicates yields glasses which have useful refractive index and optical dispersion properties.

1.2.7. Other Non-Silica-Based Oxide Glasses

Oxide glasses that do not have silica as a principal component do not have much commercial use. B_2O_3-based and P_2O_5-based glasses are readily attacked by water. However, their study has been extremely important towards enhancing our understanding of glass structure. About the only non-silica oxide glasses that have some commercial interest are the boro-aluminates (e.g., "Cabal" glasses with electrical resistivities exceeding that of silica), alkaline earth aluminates (as a high-temperature sealant and IR-transmitting glass), and V_2O_5-based glasses. Many glasses, where large amounts of V_2O_5, TeO_2, Bi_2O_3, or Sb_2O_3 are present in addition to some silica, have very low liquidus temperatures. Such glasses can be quite fluid at ordinary temperatures. Hence, some of these glasses have found use in glass sealing of electronic components. Tellurite glasses have very high refractive indices (in excess of 2.0), and hence have found use in optical systems.

1.2.8. Halide Glasses

Of the known halide glasses, those based entirely on heavy metal inorganic fluorides attracted the most attention during the 1980s. It has been shown that the theoretical attenuation in these heavy metal fluoride glasses (HMFGs) can be as little as 10^{-3} dB/km at around 3.5 μm wavelength. (Note: 1 dB = $-4.343 \ln(I/I_0)$, where I_0 and I are the incident and the exiting intensities.) For a 6,000 km distance of travel, a light signal would suffer a 6 dB loss, and so a quarter of the original intensity would still be transmitted. This clearly indicates the possibility of utilizing HMFGs for repeaterless, continuous glass fiber cable to carry transcontinental or transoceanic telecommunications. Such repeaterless communications would not be possible in oxide glasses, where, for instance, in silica glass the attenuation minimum is shown to be not less than about 0.16 dB/km (at 1.6 μm wavelength).

The most studied composition in the HMFG family is the "ZBLA" glass, which comprises $57ZrF_4{\cdot}36BaF_2{\cdot}4LaF_3{\cdot}3AlF_3$ (mol. %). Improvements have been sought by changing various constituents or by substituting, for instance LaF_3 by YF_3 (the "ZBYA" glass), or by adding varying amounts of other constituents such as fluorides, chlorides, and even small amounts of oxides. In almost all cases, the starting raw materials (generally fluorides) have to be better than 5 to 6N (99.999–99.9999%) pure. They are melted in nonreactive crucibles typically made of platinum or vitreous carbon at 800–1,000°C. A reactive atmosphere, such as CCl_4, SF_6, and NF_3, is maintained during the melting to remove oxygen and –OH impurities, which otherwise would degrade the transmission properties to the point of rendering the glass useless.

HMF glasses, in general, are extremely prone to crystallization. As a result, despite 15 years of intense research, glasses better than ~0.2 dB/km loss are yet to be reported. Because these glasses can also be readily attacked by water, their practicality in telecommunication technology relative to that of the vitreous silica fibers is being questioned. It is likely that the application of HMFGs will be limited to mid-IR transmitting short-haul sensors.

1.2.9. Amorphous Semiconductors

Elements such as Si, Ge, P, As, compounds such as $CdGe_xAs_2$ ($x = 0$ to 1.2, a family $A^{II}B^{IV}C_2^{V}$ called **tetrahedral glasses**) and $Si_{1-x}H_x$ (where $x = 0.1$ to 0.2), and mixtures of Si, Ge, Sb, etc., with S, Se, and Te (chalcogen-based) retain their semiconducting behavior (observed in the crystalline state) even in the glassy form. (*A useful abbreviation is "a-" and "c-" for the amorphous and the crystalline states, respectively.*) It was discovered in the 1960s that these glasses displayed switching between high and low electrical conductivity states (while remaining semiconducting), potentially making them the best possible choice for computer memory [1]. The enthusiasm decayed considerably after it was realized that the switching occurred as a result of transition between localized melting and onset of crystallization — a phenomenon that could not be fully controlled. The switching could be carried out optically using short laser pulses. Devices based upon electrical and optical pulse switching behavior are in commercial production for computer memories. Amorphous semiconductors are also photovoltaics, which makes them excellent candidates for solar cell technology. Although *c*-Si is used for such purposes in space probes, the *a*-Si can be produced more inexpensively in the form of thin films by vapor deposition on cold substrates. The hydrogenation of *a*-Si (by glow discharge plasma decomposition of SiH_4 silane) apparently cleans up the "poisoning defect sites" in *a*-Si and, thus, produces $a\text{-}Si_{1-x}H_x$ ($x = 0.1$ to 0.2), which is a more efficient photoreceptor for use in the solar cells.

The chalcogen-based compounds are discussed next.

1.2.10. Chalcogenide and Chalcohalide Glasses

Glasses obtained by melting chalcogen elements (group VI: S, Se, and Te) with one or more of groups V and IV elements are called chalcogenide glasses. Compositions modified by adding halogens are called "chalcohalides." The primary interest in these glasses comes from their semiconducting (switching) behavior, photoconductivity, and IR-transmitting properties. The photoconductivity property is utilized in xerography (photocopying). Interest has also been high in more recent years upon the recognition that, although these glasses generally appear grey/black and are opaque in ordinary visible light, they begin to transmit in the near IR and continue to wavelengths as long as 18 to 20 μm. This then makes them a potential fiber material candidate for transporting the CO_2 laser wavelength (10.6 μm) in such applications as laser-assisted microsurgery.

Like the halides, the chalcogenides also require O_2- and OH-free melting conditions to assure good transmission behavior. Raw materials (pure elements) have to be distilled or sublimed to increase the purity level to 6N. Glasses are generally melted in evacuated silica glass ampoules at 800–1,000°C inside a rocking furnace to achieve a high degree of homogeneity. Chalcogenides are much more resistant to water attack than are the halides. For this reason, it is possible that chalcogenides, rather than halides, may be the material of choice for mid-IR sensor applications.

1.2.11. Glassy Metals

Glassy metals (sometimes also referred to as amorphous metals or metallic glasses) are noncrystalline materials composed of either pure metals or combinations of metals and metalloids. They are metals in the sense that their electrical, magnetic, and optical properties are typical of metals. (However, their electrical resistivity often decreases with temperature.) They are not semiconductors. An example of a glassy metal is $Fe_{40}Ni_{40}P_{14}B_6$, sold under the trademark Metglas.

These glasses are made in the form of thin tapes or fibers using very high-speed quenching techniques — for instance, by passing liquid streams between high-thermal-diffusivity rollers or by melt-spinning. The cooling rates are generally in the order of 10^5–10^8 °C/s. Because they are readily produced as thin ribbons, the primary commercial use of amorphous Fe-based ferromagnets is in flexible magnetic shielding and power transformer core laminations. These materials have very low *B–H* hysteresis curve losses. In addition, they have about three times higher electrical resistivity than their crystalline counterparts. As a result, their use as power transformer core laminations can lead to as much as 30% power savings. Glassy metals have extremely high mechanical strengths approaching theoretical values. Hence, as fiber-reinforced composites, they could be candidate materials for high strength-to-weight ratio applications (such as high-speed aircraft, space vehicles, etc.). There has been some interest in their superconducting properties for application in superconducting magnetic torus used inside thermonuclear reactors, primarily because their amorphous nature makes them less susceptible to radiation damage. However, the superconducting transition temperatures are quite low (about 9 K).

1.2.12. Oxyhalide, Oxynitride and Oxycarbide Glasses

As the name implies, these are hybrid families where the oxygen is substituted in part by halogen (usually F and Cl), N or C. Many are potentially high performance materials, but remain in the research laboratories so far. Stable glasses containing small amounts of the substituent (generally ~1–4%) may be prepared by melting together halides, nitrides or carbides with oxides over narrow range of compositions. The melt viscosities of the oxyhalides are generally lower than those of the oxides, hence, some of the Pb-oxyhalide glasses may have use as ultralow-temperature sealing glasses. Because of their very high anionic electrical conductivity (section 14.8.4), glasses of this family are also potential candidates for use as solid electrolytes [2]. Bulk oxynitride glasses containing silicates, phosphates and 5–10% nitrogen can also be prepared by ammoniating the oxide melts [3, 4]. Oxycarbide glasses with large (20–40%) carbon content can only be prepared currently in thin specimen forms by a sol–gel route (section 20.3.3) [5]. These glasses may, in fact, be nanoscale composites of carbon and oxide glasses giving them a black appearance. Nitriding and carbiding of glasses enhances the high temperature mechanical and rheological properties greatly; carbiding is superior to nitriding [6], hence, they are of considerable commercial interest potentially.

1.3 A Brief Note on Glasses Found in Nature

Suggested reading: L. D. Pye, J. G. O'Keefe, and V. D. Frechette, eds., *Natural Glasses*, North-Holland, Amsterdam, 1984 (*J. Non-cryst. Sol.*, Vol. 67). J. G. O'Keefe, *Tektites*, The University of Chicago Press, 1963.

Despite the various erosion mechanisms, there is quite an abundance of natural glasses on the earth. Obsidian, which is volcanic in origin, is perhaps the most familiar example. The chemical composition of a typical California obsidian is

$$74SiO_2 \cdot 13.5Al_2O_3 \cdot 1.6FeO/Fe_2O_3 \cdot 1.4CaO \cdot 4.3Na_2O \cdot 4.5K_2O \cdot 0.7MnO \text{ (wt \%)},$$

which makes the glass just another good member of the alkali aluminosilicate family. Most obsidians are generally less than 65 million years old; it is alleged that these would tend to devitrify over longer periods. Obsidians are not dry glasses: they may contain as much as $\sim 1\%$ water. This may be the reason for their foaming behavior when heated.

Examples of other glasses found on the earth are fulgarites, which are made by fusion when lightning strikes soil; glasses of meteoritic origin; impact glasses (impactites), which might have been formed either during a meteoritic impact event by "shock transformation" (diaplectic glasses) or by fusion of local minerals and rocks due to absorbed heat; and tektites. Not to be mixed up with these are the man-made glasses from industrial wastes (fragments of glass containers, sheets, highway building materials, and reflector beads from road signs, etc.) and from nuclear explosions.

Scientifically, the most intriguing and hotly debated natural glass is the tektite. There are many tektite-strewn fields: the more studied ones are the Australasian tektites on land and the associated microtektites (in deep-sea deposits in the Indian Ocean, Philippine Sea, and western-equatorial Pacific Ocean), the moldavites of central Europe, and the Libyan Desert glass (from western Egypt). The Australasian tektites are black to dark brown in color, typically

$$75SiO_2 \cdot 13Al_2O_3 \cdot 4FeO/Fe_2O_3 \cdot 3.5MgO/CaO \cdot 4Na_2O/K_2O \cdot 0 \cdot 7TiO_2 \text{ (wt \%)},$$

and have been dated to be roughly 700,000 years old using the ^{40}K–^{40}Ar method. Specimens are liquid splash forms — spheres, teardrops, dumbbells and buttons — and are generally 1–2 cm in dimension, although a specimen as large as 10–20 cm and weighing 12.8 kg has been found. Estimates of the total quantities of microtektites alone are as high as about 10^{10} metric tons. The microtektites appear clear to light yellow-brown in color and are generally less than 1 mm in diameter. These glasses are remarkably very homogeneous [7]. O'Keefe has suggested that many of the characteristics of the tektites, in particular the homogeneity, indicate that these glasses were molten for long times in space. After calculating possible trajectories, O'Keefe concluded that the Australasian tektites had to be of lunar volcanic origin, as opposed to being the result of a terrestrial meteoritic impact. The moldavites are pale yellow to yellowish green, typically having a composition about 75–80% SiO_2·9–12% Al_2O_3·1–3% FeO/Fe_2O_3·0.3% Na_2O·3.5% K_2O·2–3% CaO. They are estimated to be 15 million years old and are acknowledged to be fused ejecta associated with a meteorite impact that formed the Ries crater in Germany. The Libyan Desert glass specimens are pale yellow to yellowish green, contain $\geqslant$97% SiO_2, include nearly all of their 0.5% iron as Fe_2O_3, and are dated to be about 28 million years old. Estimates of the total mass as it exists today range up to about 1,500 tons, and specimens as large as 7.5 kg have been recovered. There have been suggestions that the glass has been formed by a sol–gel process (Section 20.3.3), since an associated meteoritic impact crater site has yet to be found.

Also studied and analyzed extensively (by reputable laboratories around the globe) are the lunar glasses returned from the Apollo expeditions. Irregular fragments, spherules, teardrops, and dumbbells of various sizes, as small as a few microns, have been found. Glasses are known to be of both lunar volcanic and meteoritic impact origin. The compositions of some of the particles returned from the Apollo 12 expedition have been found [8] to be

$$38\text{–}50SiO_2 \cdot 10\text{–}27Al_2O_3 \cdot 4\text{–}20FeO \cdot 0.4Na_2O/K_2O \cdot 9\text{–}14CaO \cdot 6\text{–}13MgO \cdot 2\text{–}4TiO_2 \text{ (wt \%)},$$

although there were regions of inhomogeneities having as much as 89% SiO_2. Note the high FeO and the low alkali. A specimen of known lunar origin has been discovered [9] in Antarctica.

Natural glasses are relics of the past; they bear the marks of time. They (naturally) present the opportunities for studying chemical durability as a function of glass composition. Several of the compositions have shown remarkable resistance to weathering. Perhaps, aside from looking for the origin of the earth and the universe, some important lessons learned from the study of natural glasses are in the area of waste immobilization to help us clean up our precious planet.

Exercises

1. Name four of the major technological achievements in glass which have had the most profound impact on mankind.

(Answer: The number one development has to be the glass window, which enabled sunlight to come into dwelling units while keeping the harsh elements of nature out. Lenses (ophthalmics for improved vision; microscope and telescope optics to see microorganisms and distant stars), the light bulb envelope, and the semiconducting glasses (for computer memory, solar cells, and photocopiers) rank after the window. Glass containers and glass fibers for telecommunication probably do not rank as high.)

References

1. S. R. Ovshinsky, *Phys. Rev. Lett.* **21**, 1450 (1968).
2. C. A. Gressler and J. E. Shelby, *J. Appl. Phys.*, **66**(3), 1127–1131 (1989).
3. M. R. Reidmeyer and D. E. Day, Proc. of 12th Univ. Conf. on Glass, Alfred, NY, 1993. To be published.
4. M. Heuberger and L. D. Pye, *Glastech. Ber.*, **62**(7), 223–230 (1989).
5. H. Zhang and C. G. Pantano, *J. Am. Ceram. Soc.*, **73**(4), 958–963 (1990).
6. J. Homeny, G. G. Nelson, S. W. Paulik and S. H. Risbud, *J. Am. Ceram. Soc.*, **70**(5), C114–C116 (1987).
7. A. K. Varshneya and A. R. Cooper, *J. Geophys. Res.* **74**(27), 6845–6852 (1969).
8. C. H. Greene, L. D. Pye, H. J. Stevens, D. E. Rase, and H. F. Kay, *Proc. 2nd Lunar Sci. Conf.*, Vol. 3, pp. 2049–2055. MIT Press, Cambridge, Massachusetts, 1971.
9. R. F. Fudali, M. Kreutzberger, G. Kurat, and F. Brandstaetter, in *Natural Glasses* (L. D. Pye, J. G. O'Keefe, and V. D. Frechette, eds.), pp. 383–396. North Holland, Amsterdam, 1984. (Also, *J. Non-cryst. Sol.*, Vol. 67.)

Chapter 2

Fundamentals of the Glassy State

2.1 What is Glass?

Our impressions of the nature of glass are affected by everyday experience with the material. To the layman, glass is a transparent solid that breaks easily. Yet a number of glass types, in particular the chalcogenides and metallic glasses, are opaque in the visible spectrum. Also, high-strength glasses can be made by a variety of techniques, some being used in bulletproof security glazing. The American Society for Testing Materials (ASTM) at one time defined glass as "*an inorganic product of fusion which has been cooled to a rigid condition without crystallizing.*" Even this definition is too restrictive, as many organic glass systems are known and fusion is not the only means of making a glass. The sol–gel process (see Section 20.3.3) of making a glass avoids the normally high temperatures employed for the fusion of glass. Chemical vapor deposition is yet another technique which completely avoids fusion of constituent materials.

The outward appearance of glasses is essentially solid-like. The density, the mechanical properties, and the thermal properties of glasses are similar to those of the corresponding crystals. However, unlike crystals, glasses do not have a sharp, well-defined melting point. Unlike most crystals, glasses do not cleave in preferred directions. In the absence of applied forces and internal stresses, glasses are essentially isotropic. The isotropicity of physical properties makes glasses resemble liquids. It follows that the atomic arrangements in glass must display the long-range disorder typical of liquids. In order not to be overly restrictive, we are left to *define glass as a "solid with liquid like structure," "a noncrystalline solid," or simply as "an amorphous solid,"* with the understanding that the amorphous characteristic here is intended to describe atomic disorder as evidenced by an x-ray diffraction (XRD) analysis; it excludes substances such as "amorphous" powders that may simply be "micro-crystals" and that display more or less sharp peaks in XRD analysis. Many physicists restrict the word "glass" to mean an amorphous solid that undergoes a "glass transition" when heated. Glass transition, which we shall discuss in detail in Chapter 13, involves a continuity of volume and a steep change in physical properties such as the specific heat, thermal expansion, etc., from a "supercooled" liquid state. Substances such as Si and Ge have been brought to the disordered solid state only by means of thin-film techniques (thermal evaporation, ion implantation, etc.) not involving melt-quenching. Apparently, this disordered state for both materials is thought to have a volume discontinuity from the liquid state. As a result, *a*-Si and *a*-Ge are often excluded from the class of substances called "glass" by the physicists. In our description of the word "glass" we will include these and other amorphous solids where the volume–temperature relationships are yet to be established. To get a clearer picture of the fundamentals of glass, the balance of this chapter is devoted to a consideration of the volume–temperature relationship (the *V–T* diagram) of the glass with respect to a liquid and a crystal, and pair correlation/radial distribution functions to describe atomic disorder in terms of the spatial relationships between the locations of various atoms in a substance.

2.2. The *V–T* Diagram

Consider a small volume of material at a high temperature in liquid form. Its state is given by the point "a" on the *V–T* diagram (Fig. 2-1). On cooling, the volume gradually decreases along the path "abc." Point "b" corresponds to T_m, the melting point of the corresponding crystal, which may be defined as the temperature at which the solid and the liquid have the same vapor pressure or have the same Gibbs free energy. At this temperature, an infinitely small amount of crystals is in thermodynamic equilibrium with the liquid. However, for a perceptible level of crystallization, some finite amount of undercooling of the liquid to a point "c" below T_m is required. Crystallization occurs if, and only if, (i) there are a sufficiently large number of nuclei present in the mass, *and* (ii) a large enough crystal growth rate exists. The location of the point "c" below T_m varies depending upon when the thermodynamic

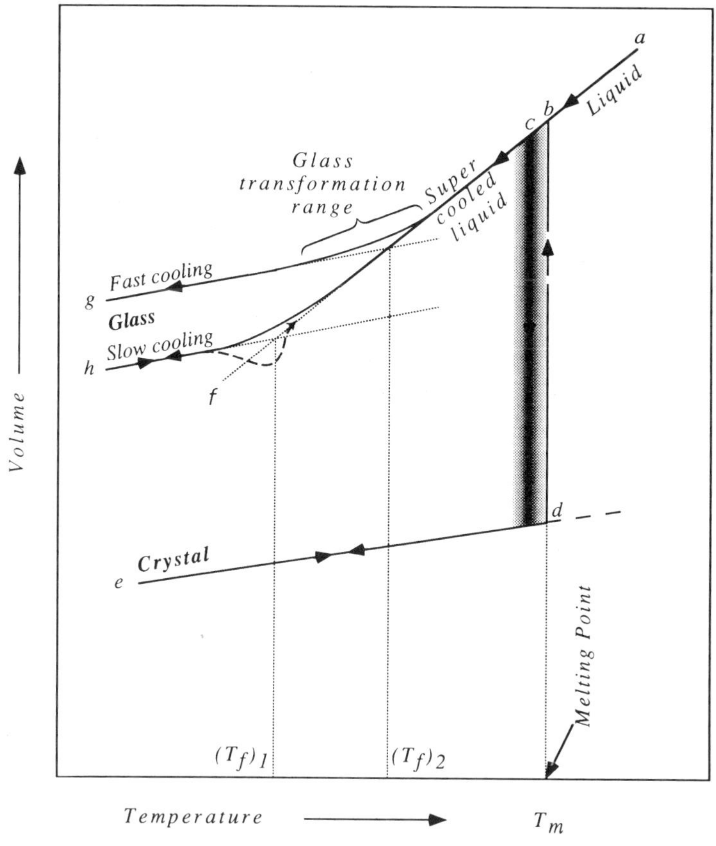

Figure 2-1. The volume–temperature diagram for a glass-forming liquid.

driving force created by the undercooling causes a particular group of atoms to transform from the liquid state to the crystal state, and upon the velocity at which the atoms from the liquid can be transported to the crystal–liquid interface. For these reasons, we have shown a wide, shaded region with varying probability representing the crystallization path. (These thermodynamic and kinetic concepts are explained further in Section 3.3.) A volume shrinkage generally accompanies the crystallization. Upon further cooling, the crystals so formed shrink along the crystal line to the point "e."

If crystallization does not occur below T_m (mostly because the cooling rate is high), the liquid mass moves into the **supercooled liquid** state along the line "bcf," which is an extrapolation of the line "abc." No discontinuities in the V–T are observed. The volume, however, shrinks continuously, i.e., the structure of the liquid rearranges itself into a lower volume along the line "bcf" required by the lower energy corresponding to a lowered temperature. As cooling continues, the molecules become less and less mobile, i.e., the viscosity of the system rapidly increases. At sufficiently low temperatures, the molecular groups cannot rearrange themselves fast enough to reach the volume characteristic of that temperature. The state line then starts a smooth departure from "bcf" and soon becomes a near-straight line (often roughly parallel to "de"), ending at point "g," when cooled fast, or at "h," when cooled slowly. The material in the near-straight, low-temperature part of the curve behaves essentially as a solid. *This is the glassy state.*

The smooth curve between the onset of departure from the supercooled liquid line and the completion to a seemingly rigid condition is termed the **glass transition region**, or the **glass transformation range**. It must be emphasized that the transition to the glassy state does not occur at a single, sharp value of the temperature. In the upper regions, glass has a viscosity of $\sim 10^8$ Pa·s ($= 10^9$ poise) or less, whereas in the glassy state the viscosity exceeds $\sim 10^{15}$ Pa·s or more to qualify for appearance as a solid. The intersection of the extrapolated glass line and the supercooled liquid line is termed the **fictive temperature** (T_f). One may imagine that T_f *is the temperature at which the structure of the supercooled liquid is instantly frozen into the glass.* (It will be stated in Chapter 13, that this concept is only approximately correct.)

The departure from the supercooled liquid line is dependent upon the rate of cooling. Slower cooling allows the structure to rearrange itself to stay on "bcf" somewhat longer, and hence the more slowly cooled glass at "h" would be expected to have a lower volume (higher density) and a lower fictive temperature than a more quickly cooled glass at point "g." For now, it suffices to suggest that the volume of the corresponding crystal is expected to be less than that of the most slowly cooled glass. (There are some interesting anomalies and questions. Some are discussed here. Some others will be discussed in greater detail in Chapter 13.)

When the glass at "h" is reheated, the state smoothly moves through the transition region along the dashed curve to the supercooled liquid state "fcb" and ultimately to the liquid state. The *V–T curve NEVER retraces its path in the transition region* (for reasons discussed in Chapter 13). Other than the gradually increasing fluidity and, perhaps, a color change due to radiative emission, in general there are no changes in the physical appearance of glass with the reheating.

If the crystals at "e" are heated, the state will move along the crystal line up to "d" (= the melting temperature, T_m) past the shaded region, melt at T_m to reach point "b," and subsequently follow the liquid path "ba." One may question whether it is possible to superheat a crystal or not. Melting produces atomic disorder in a substance and proceeds from the surface inwards. Hence, crystals may be superheated to temperatures beyond T_m if the heat is provided internally or if melting produces a very high-viscosity liquid phase such that atoms within the interior of the crystal do not have "room or time to disorder." Ordinarily, the melting of a crystal is much sharper (on the temperature scale) than the freezing of a liquid.

Note that, with regard to the slight separation "b–c" between the crystallization and the melting paths near T_m, the V–T diagram presented here differs slightly from those usually presented by other authors.

To understand the structural relationship between the glassy state and other states of matter, we need to understand the pair distribution function and the radial distribution function used commonly to describe the distribution of atoms in amorphous solids.

2.3. Pair Correlation Function and Radial Distribution Function

In a material (gas, liquid, or solid) where the average number density of atoms is $\rho_0 = n/v$, n being the total number of atoms and v the total volume, the number density of atoms at a distance r from an origin atom is given by $\rho_0 g(r)$. The function $g(r)$ is the ratio of the number density of atoms at r to the homogeneous density, ρ_0, and is called the **pair distribution function.** The term $\rho_0 g(r)dr$ then gives the probability, dp, of finding an atom center in dr that is at a distance r from the origin atom. (The reader should note that $g(r)$ is a probability density function.) In a disordered system the number density approaches ρ_0 at very large distances from any given origin atom, and hence

$$\lim_{r \to \infty} g(r) = 1.$$

The increasing repulsion between overlapping electron orbitals (Pauli's exclusion principle) discourages the close approach of two atoms, i.e.,

$$\lim_{r \to 0} g(r) = 0.$$

A schematic variation of $g(r)$ with r is sketched in Fig. 2-2a. Note that the function may rise from zero and oscillate about 1 until asymptotically approaching 1. The magnitude of the oscillations directly gives the higher or lower density at r.

Some authors define the **pair correlation function**, $h(r)$, by

$$h(r) = g(r) - 1.$$

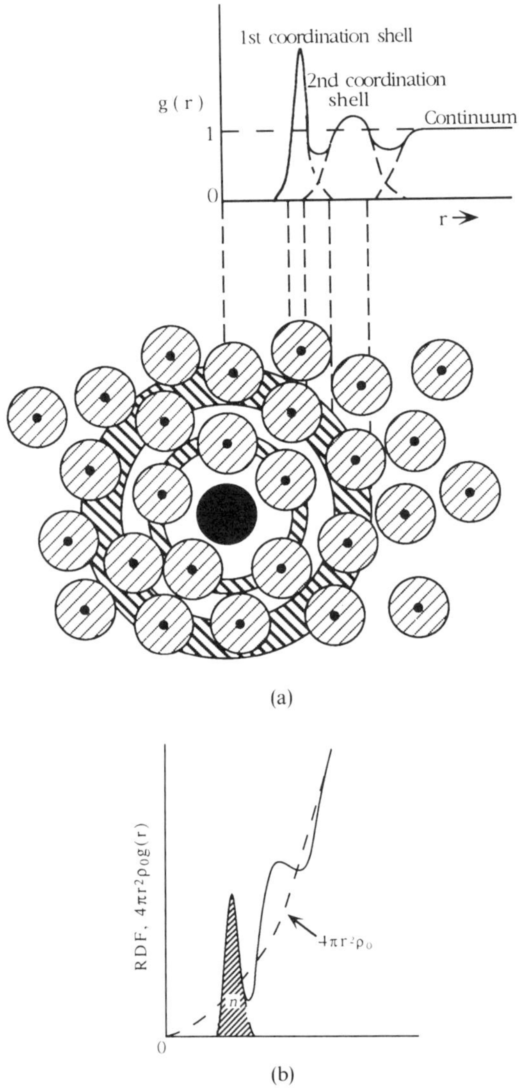

Figure 2-2. (a) Pair distribution function, $g(r)$, versus radial distance, r. (b) Radial distribution function versus r. For uniform probability, the RDF is simply a parabola.

This definition has the advantage that, in disordered systems, $h(r)$ approaches zero at large distances, implying no correlation (which agrees with our intuition), and approaches -1 at $r = 0$, implying perfect negative correlation corresponding to the non-overlap of the atoms (again agreeing with our intuition). Both the $g(r)$ and $h(r)$ provide essentially the same information.

Instead of the pair distribution function or the pair correlation function, sometimes the **radial distribution function** (the RDF) is also used. The RDF is defined by $4\pi r^2 \rho_0 g(r)$, which is the radial density of atoms at distance r from the origin atom A. The number of atoms in a spherical shell at a distance r and $r + dr$ from A is given by $4\pi r^2 \rho_0 g(r) dr$. Clearly, at large values of r, the RDF tends to become a smooth parabola (Fig. 2-2b). If the RDF is integrated between r_1 and r_2, where the limits describe the lower and the upper end of a "peak," then by definition

$$\int_{r_1}^{r_2} 4\pi r^2 \rho_0 g(r) dr = \text{the area under the peak,}$$

[which is equal to the number of neighbors to the origin atom A residing within r_1 and r_2].

If the peak is the first one from origin, then the preceding expression gives the first coordination number for A (Fig. 2-2b). In polyatomic systems, it is usual to find atoms of a different type B around the origin atom A. The expression then gives the coordination number of B about A.

Occasionally the RDFs are plotted as $[4\pi r^2 \rho_0 g(r) - 4\pi r^2 \rho_0]$ versus r. In such a plot, the parabola has been subtracted out, and the distribution function oscillates about the abscissa. Other ways of plotting the data are the reduced RDF, sometimes called the total correlation function $t(r)$, defined by

$$t(r) = 4\pi r \rho_0 g(r),$$

and the differential correlation function $d(r)$, defined by

$$d(r) = [4\pi r \rho_0 g(r) - 4\pi r \rho_0].$$

(Note: The reader may find that various authors use different names for many of these distribution functions.)

It is easy to note that the reduced functions have simply been obtained by dividing $g(r)$ and the RDF by r. The division by r makes the peaks more symmetrical, and hence the peaks can be defined with more precision, allowing better integration of the area to determine the coordination numbers. Typical plots of the reduced functions are shown in Fig. 2-3.

The pair correlation function and the radial distribution function are important for the study of noncrystalline solids, not only because they describe the mutual relationship between the location of various atoms, but also because these can be experimentally measured after Fourier transformation of x-ray diffraction (XRD) or neutron or electron diffraction data. In diffraction experiments, the scattered intensity is recorded as a function of $Q = (4\pi \sin\theta)/\lambda$, where λ = wavelength and θ is the scattering angle. After

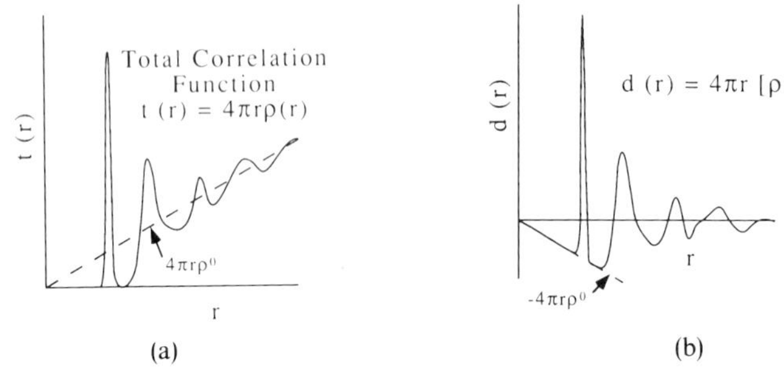

Figure 2-3. (a) Total correlation function. (b) Differential correlation function.

correction and normalization, one obtains the coherent intensity $I_{coh}(Q)$ as a function of Q, from which the total structure factor $S(Q)$ is obtained. The Fourier transform of $S(Q)$ is simply the differential correlation function $d(r)$.

The simplified pair distribution functions of the various states of matter are shown in Fig. 2-4. Ideally, the crystals exhibit a regular, periodic array of atoms represented by vertical lines, but these lines broaden somewhat to become sharp peaks because of the presence of thermal vibrations about the equilibrium positions and defects. In gases, there is no equilibrium position: gas atoms impinge upon each other and "penetrate" into each other some distance until Pauli's exclusion principle stops further penetration. The function $g(r)$, for a snapshot in time, rapidly rises and becomes a uniform value irrespective of the distance from the origin atom. (For a hard sphere model, $g(r)$ would be a step function at the hard sphere diameter.) In liquids, the probability of finding atoms is neither regularly spaced as in crystals nor uniform as in gases. For the first few neighbors, definite peaks occur at locations bearing some resemblance to the locations found in crystals. Because of variations in the interatomic distances, the peaks get broader as the distance from the origin atom increases.

Because the glassy state is an extrapolation of the liquid state to lower temperatures, it is expected that glass has a liquidlike structure. This indeed is the case, as is shown in Fig. 2-5, where the PDF of various ion pairs is shown in both liquid and glassy $45CaO{\cdot}55SiO_2$ (mol. %). In liquids, the thermal vibrations (and subsequent time-averaging) broaden the peaks relative to the glasses, where the atomic motions are relatively frozen in space. Figure 2-6 shows a comparison of the experimentally measured x-ray diffraction (XRD) spectra of vitreous silica, cristobalite, and silica gel. Note that, unlike the sharp peaks in the crystal form, peaks in vitreous silica occur in the form of a broad hump. However, the hump is located roughly at the

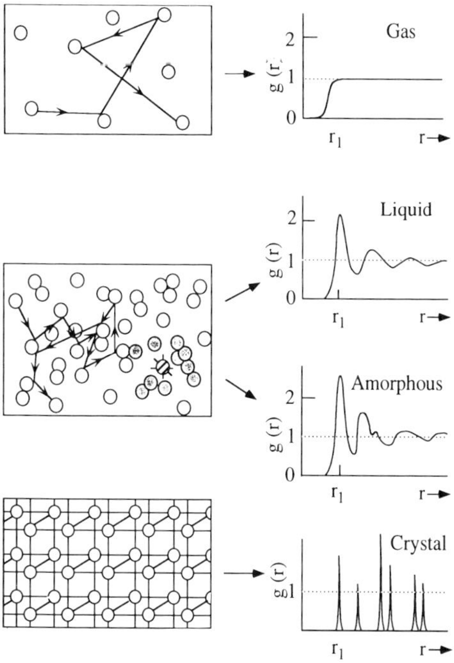

Figure 2-4. Schematic diagram of the atomic and pair distribution function for gas, liquid, amorphous, and crystal states. In gases, because of the kinetic energy, the atoms "penetrate" into each other slightly more deeper than in the condensed state. (Modified and redrawn from Y. Waseda, *Metal Phys. Seminar* **1**, 305 (1976).)

same place as the primary peak in cristobalite. In essence, *in both the liquids and the glasses, there may be some short-range order; however, the long-range order typical of crystalline arrangement is absent.* The similarity of the broad hump to that observed in silica gel led Randall *et al.* [1] to conclude that glass consisted of an aggregate of finely divided crystals (perhaps finer than in a gel). They stated that "the difference between a crystalline state, on the one hand, and the 'vitreous' and 'amorphous' on the other is one of degree, not of kind." However, the small-angle scattering typically seen in the gel spectra is nearly absent in the glass spectra. This observation led Warren [2] to suggest that the crystal size in the glass would have to be of the order of a few angstroms, hardly bigger than a single unit cell, at which point the concept of a crystal becomes meaningless. The V–T diagram illustrates the fact that there are no direct paths between a crystalline state and a glassy state using heat as a means for change. From either state, one must approach

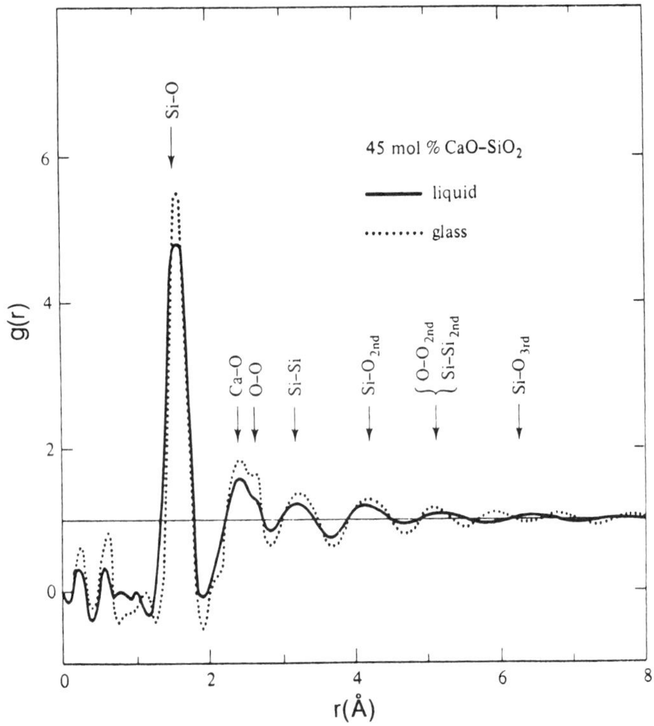

Figure 2-5. Comparison of $g(r)$ between the glassy and the liquid states for $45CaO{\cdot}55SiO_2$. (After Y. Waseda and J. M. Toguri, *MeT. Trans.* **8B**, 563 (1977).)

at least the supercooled liquid state before an excursion to the other state can be made. The relationship between the atomic arrangements in glass with respect to a crystal will be discussed in greater detail in Chapter 3.

Suggested reading: Y. Waseda, *The Structure of Non-Crystalline Materials*, McGraw-Hill Book Co., New York, 1980.

2.4. Anomalies in the *V–T* Diagram

Figure 2-1 represents the behavior typically observed for a substance which upon cooling from a liquid state either changes into a crystal or moves into the glassy state. It is interesting to raise the question whether, because of the disorder, the volume of glass will always be higher than that of the

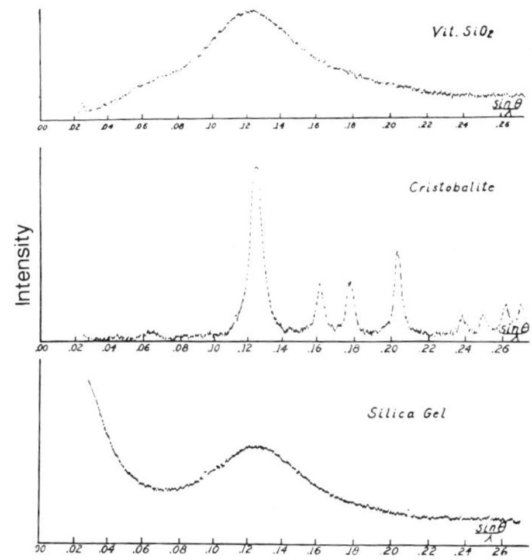

Figure 2-6. Microphotometer records of x-ray diffraction patterns of vitreous silica, cristobalite, and dried silica gel.

corresponding crystal. The answer is, "Not necessarily so!" It is possible to think that a periodic arrangement of atoms could not be accomplished in the most dense form and that the disorder allows a closer packing. An example of this is the water–ice system. It is well known that the densities of ice and water at 0°C are 0.917 g/cm^3 and 0.99987 g/cm^3, respectively. It is clear that if we could avoid crystallization of water to ice below 0°C and bring the resulting supercooled water to a glassy state (by rapid quenching), the volume of the vitreous water would still be less than that of the ice crystals. Some other examples of this type of behavior are known to exist in tetrahedral glasses such as $CdGeAs_2$, eucryptite glass-ceramic [3], and several heavy metal fluoride glasses.

Yet another and somewhat similar question is, "Will the faster-cooled glass always have a higher volume than a slower-cooled glass?" Again, the answer is, "Not necessarily so." It depends upon the shape of the V–T curve in the supercooled liquid region. It happens that the supercooled liquid curve for vitreous silica is V-shaped [4]. Slower and slower cooling causes the supercooled liquid state to move gradually from the right arm to the left arm (see Fig. 2-7). Glasses obtained by cooling from points on the right arm behave "normally," however, those cooled from points on the left arm behave "anomalously." A quickly cooled glass will actually have a lower volume than one cooled more slowly.

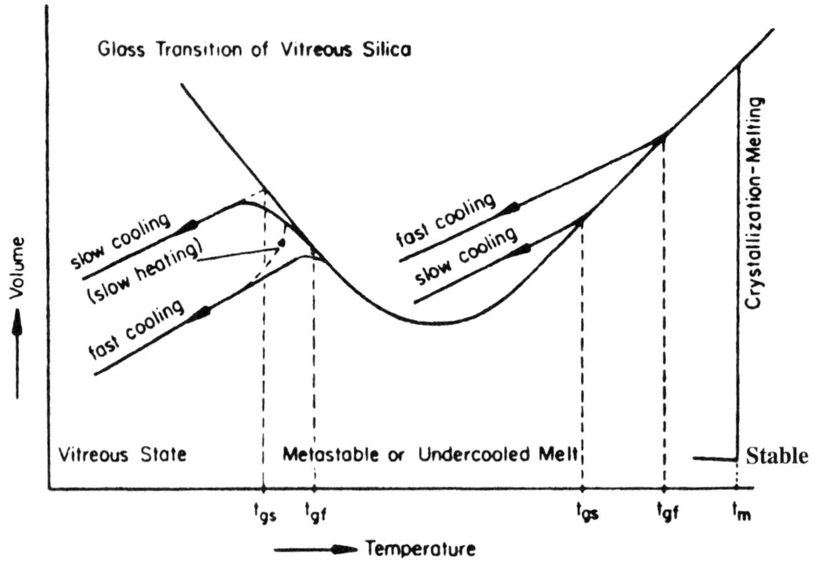

Figure 2-7. Volume versus temperature diagram for low-OH silica glass. t_{gs}, t_{gf} are glass transition temperatures for slow and fast cooling, respectively. (After R. Brückner, *Glastech. Ber.* **43**, 8 (1970).)

The ice–water and the fused silica examples, however, are rare. Fig. 2-1 represents the typical behavior. It is, perhaps, sufficient to recognize that glass maintains the structure, the volume, and hence the energy of a liquid; however, the *changes (or the derivative)* of the energy and the volume with respect to temperature and pressure are similar in magnitude to those in a corresponding crystal.

Exercises

1. Scientists normally classify matter into three groups: solids, liquids, and gases. What property or properties distinguish a solid from a liquid?

(Answer: There is no single physical property that distinguishes a solid from a liquid. A solid should have a "significantly large" shear (rigidity) modulus. A liquid should have "significantly low" viscosity. Much of this is discussed in Chapter 13.)

2. Would you classify the glassy state as the "fourth state of matter"?

(Answer: The three states of matter, viz., gas, liquid and solid, are thermodynamically stable states. By this we mean that any two of the three thermodynamic stable variables *P*, *V*, and *T* uniquely define the third variable. It will be stated in Chapter 13 that glass is not a thermodynamically stable state of matter. Some properties of glasses make them resemble solids; others make them resemble liquids. The distinction in each case is not sufficient to warrant another classification.)

3. Draw a tentative *V–T* diagram for the ice–water system, clearly showing the formation of glass and the crystal by cooling from the liquid state.

4. Draw a tentative *V–T* diagram for elemental silicon, showing the formation of the glassy state, the amorphous state, and the crystal by cooling from the liquid state. (Note: This is still a debated topic.)

References

1. J. T. Randall, H. P. Rooksby, and B. S. Cooper, *J. Soc. Glass Technol.* **15**, 54 (1931).
2. B. E. Warren, *J. Appl. Phys.* **8**, 645 (1937).
3. W. Baum, *Glastech. Ber.* **36**, 444 and 468 (1963).
4. R. Brückner, *Glastech. Ber.* **43**, 8 (1970).

Chapter 3

Glass Formation Principles

Suggested reading: H. Rawson, *Inorganic Glass-Forming Systems*, Academic Press, London, 1967.

The preceding discussion indicates that a glass may be made from a crystal by first moving the atoms from their periodic locations to nonperiodic "liquid locations" and then preventing them from falling back into periodic locations. It may be readily realized that fusion of a crystal is an easy way of moving the atoms around. However, it need not necessarily be so (see Section 20.3). The crystal can be dissolved in liquids and reacted until the atoms acquire the desired "liquid locations," and finally the excess solvent can be dried away. This latter process is the basis of glassmaking by the sol–gel route (Section 20.3.3). Likewise, in a vapor deposition technique, the atoms are deposited, nonreactively or reactively, in a disordered manner with or without the use of a substrate that is cold enough to prevent ordering. Some other methods of glassmaking that can be envisioned are bombardment of a crystal by high-energy particles, and shock-induced transformation (Section 20.3.1).

As might be expected, glassmakers' interests have focused on the melting technique since this is suitable for a wide range of compositions and can be used in the laboratory to make glasses on a scale of a few grams and in the plant on a scale of hundreds of tons per day. Table 3-1 shows a partial list

Table 3-1. A Partial List of Glasses Formed by Cooling from a Melt[a]

Elements
- S, Se
- Te(?)
- P

Oxides
- B_2O_3, SiO_2, GeO_2, P_2O_5, As_2O_3, Sb_2O_3
- In_2O_3, Tl_2O_3, SnO_2, PbO_2, SeO_2
- "Conditional" TeO_2, SeO_2, MoO_3, WO_3, Bi_2O_3, Al_2O_3, Ba_2O_3, V_2O_5, SO_3

Sulfides
- As_2S_3, Sb_2S_3
- various compounds of B, Ga, In, Te, Ge, Sn, N, P, Bi
- CS_2

Selenides
- various compounds of Tl, Sn, Pb, As, Sb, Bi, Si, P

Tellurides
- various compounds of Tl, Sn, Pb, As, Sb, Bi, Ge

Halides
- BeF_2, AlF_3, $ZNCl_3$, Ag(Cl, Br, I), Pb(Cl_2, Br_2, I_2), and multicomponent mixtures

Nitrates
- KNO_3—$Ca(NO_3)_2$ and many other binary mixtures containing alkali and alkaline earths nitrates

Sulfates
- $KHSO_4$ and other binary and ternary mixtures

Carbonates
- K_2CO_3—$MgCO_3$

Simple organic compounds
- *O*-Terphenyl, toluene, 3-methyl hexane, 2,3-dimethyl ketone, diethyl ether, isobutyl bromide, ethylene glycol, methyl alcohol, ethyl alcohol, glycerol, glucose
- As droplets only: *m*-xylene, cyclopentane, *n*-heptane, methylene chloride

Polymeric organic compounds
- Example—polyethylene (–CH_2·–), and many others

Aqueous solutions
- Acids, bases, chlorides, nitrates, and others

Metallic alloys by "splat cooling"
- Au_4Si, Pd_4Si
- Te_x–Cu_{25}–Au_5

[a] After R. H. Doremus, *Glass Science*, p. 12. Wiley-Interscience, New York, 1973. Reproduced with permission of J. Wiley & Sons.

of substances that have been brought into glassy state by cooling a melt. Some of these employed ordinary means of cooling (left in air), whereas others required high rates of cooling, for instance, by quenching in a fluid or pressing between high conductivity surfaces (splat cooling). One clearly needs to inquire what criteria lead to ready glass formation in a given substance. There are two types of **theories for glass formation: structural theories**, and the **kinetic theory**. As indicated earlier, most of these theories have been concerned mainly with glass formation in oxides.

3.1. Structural Theories of Glass Formation

There are several structural theories of glass formation: Goldschmidt's radius ratio criterion and Zachariasen's random network theory, both based on coordination number; Smekal's mixed bonding rule and Stanworth's electronegativity rule, both based on bond type; Sun's single bond strength criterion, based on bond strength; Dietzel's field strength criterion; and Phillips' topological constraint model. Some of these are now of historical interest only and are discussed by Rawson. We shall confine our attention to Zachariasen's theory and the latter three.

3.1.1. Zachariasen's Random Network Theory

In a classic paper which, perhaps, has had the most influence on the thinking in glass science, Zachariasen [1] argued that since the mechanical properties of glasses are similar to those of the corresponding crystals, the atomic forces in both must be of the same order. The diffuseness of the x-ray diffraction spectra for glasses clearly indicated that glass itself has an infinitely large unit cell. Glass would therefore consist of a three-dimensional random network. A direct consequence of the randomness would result in glass having higher internal energy than the crystal. Zachariasen believed that this difference in internal energy from that of the corresponding crystal must be small. Otherwise, there would be sufficient driving force towards crystallization. Also, this small energy difference, he suggested, required an open and flexible structure.

In ionic crystals, the cation polyhedra generally share corners. This is particularly true with cations that have a small radius but a high charge. Zachariasen believed that sharing corners is a primary requirement to achieving an open and a random structure. Thus, if a hypothetical compound A_2O_3 could crystallize in two dimensions, then the atomic arrangements in the crystal and the glass should look like those shown in Fig. 3-1. Note that both the crystalline and the glassy forms are composed of AO_3 triangles joined to each other at corners, except that the glassy form has disorder introduced by changes in the A–O–A angles (which are called the **bond angles**) and slight changes in the A–O bond length. It may also be noted that the

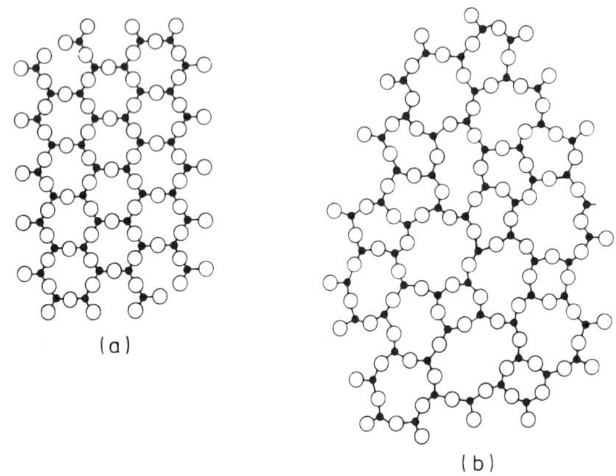

Figure 3-1. Atomic structural representation of (a) A_2O_3 crystal (b) A_2O_3 glass.

triangles or the O–A–O angles themselves need not be deformed much. It is also easy to realize why the compound AO would not form a glass (Fig. 3-2). If the cation polyhedra were triangles, then the coordination of A around O would also be 3. This clearly would imply sharing of oxygen by 3 A's, and hence an ordering would follow simply from topological restraints. Zachariasen therefore defined a glass as a "substance (that) can form extended three-dimensional networks lacking periodicity with an energy content comparable with that of the corresponding crystal network." Accordingly, he laid down **four rules for glass formation** in a compound A_mO_n:

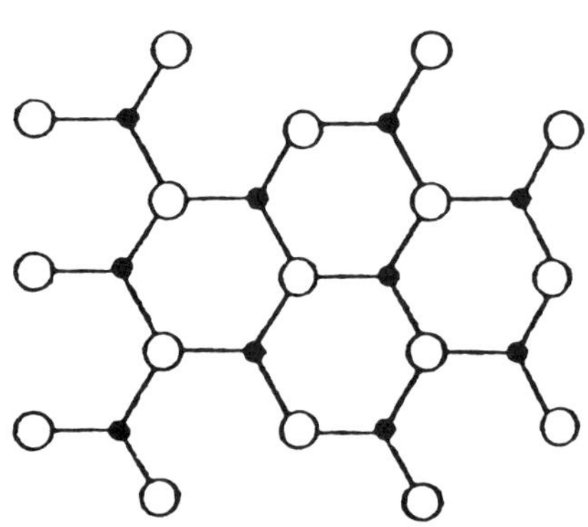

Figure 3-2. Structural representation of a hypothetical crystalline compound AO.

1. An oxygen atom is linked to no more than two atoms of A.
2. The oxygen coordination around A is small, say 3 or 4.
3. The cation polyhedra share corners, not edges, not faces.
4. At least three corners are shared.

Zachariasen stated that whether the atoms are truly ionized or not was of little consequence as long as we could use the ionic interpretation as a convenient means of description. Thus, oxides of the formula A_2O and AO (for cations of group I and II) cannot satisfy many of these rules. For instance, MgO crystallizes in the rock-salt lattice (two interlocking fcc structure). The cations form octahedra that share corners as well as edges. The oxygens are also shared by six cations. Clearly, rules 1–3 are violated, suggesting that glass formation should not occur, as is in fact the case. Rules 1, 3, and 4 are satisfied by any group III oxide A_2O_3 if the oxygens form triangles around A. B_2O_3 is an example of a glass former in this category. (Fig. 3-1b is actually a planar projection or a "pancaked version" of B_2O_3 glass). Group IV oxides (AO_2), and group V oxides (A_2O_5) would be glass formers if the oxygens formed tetrahedra around the cations. SiO_2, GeO_2, P_2O_5, and As_2O_5 are examples of these. Oxides AO_3, A_2O_7, and AO_4 can satisfy rules 1, 3, and 4 if cation octahedra are formed. However, rule 2 would be violated. Oxide MoO_3 does not usually form a glass when melted alone, but glasses containing high percentages of MoO_3 can be made by melting the oxide either with P_2O_5 or with certain other basic oxides.

Zachariasen's random network model received wide support, particularly from the x-ray diffraction work of Warren [2] and his students at MIT. They were easily able to dispel Randall *et al.*'s [3] notion of glass being an assemblage of microcrystals. As stated earlier in Chapter 2, the absence of small-angle scattering in fused silica was used to confirm the absence of any microcrystals such as those found in silica gel. The diffraction spectra from alkali silicate and borosilicate glasses also showed essentially similar features—broad humps at near-neighbor distances (instead of the sharp peaks typical of crystals) and lack of any significant small-angle scattering. As will be discussed in later chapters, Warren and co-workers even showed that the near-neighbor coordinations that could be extracted from the diffraction spectra of various glasses (single-component as well as multicomponent) were essentially in good agreement with the predictions of the random network model.

Support for the random network model in the recent years has come from the work by Cooper [4] using the concepts of topology. As Cooper indicates, Zachariasen's rules 1 and 2 are sufficient to assure geometric freedom to obtain random structures within a closure, and hence they are sufficient to give a low internal energy difference between the glass and the crystal. Rule

3, which is intended to maximize openness and randomness, may be too stringent. One could include the possibility of edge-sharing by combining rules 3 and 4 and modifying them to state that each oxygen polyhedron must be connected directly to at least three other oxygen polyhedra. Thus, tetrahedra that share an edge and two opposite corners would no longer violate the rules.

Early criticism of Zachariasen's work arose from doubts about whether particular oxides would form glasses or not. Zachariasen himself was not too sure about glass formation in P_2O_3, As_2O_3, Sb_2O_3, V_2O_5, Sb_2O_5, Nb_2O_5, and Ta_2O_5. In the first three, the oxygens were thought to make octahedra around the cations where three of the oxygens were closer to the cation than the other three. The approximation to a triangular coordination was thought to be the closest in P_2O_3, and hence the glass-forming capability of P_2O_3 was thought to be the best out of that group. As it turns out, none of these oxides are known to form glasses by ordinary means; however, all of them do form glass networks when a varying quantity of a "stabilizing" modifier oxide is added. In these glasses, the oxygen coordinations of the network forming cation oxide are generally 6–8. (For a recent review of glasses with non-classical glass network forming cations, see Kreidl [5].)

The most important challenge to the random network model came from Lebedev and his Russian co-workers [6,7] who questioned the exact meaning of the word "random." The subsequent discovery of glassy metals, semiconducting chalcogenides, and heavy metal fluoride glasses that have no oxygens at all required an overhaul in our concepts of "randomly packed" structures. Lebedev and his co-workers' criticism is discussed later in this chapter, and the discussion on the structure of glassy metals is carried out in Chapter 5.

3.1.2. Sun's Single Bond Strength Criterion

Sun [8] suggested that the process of forming a glass had to involve inability to rearrange the bonds in the liquid state during crystallization, and hence, the higher the bond strength the better the glass former. He calculated the strength of the M–O single bond in the oxide $MO_{x/y}$ using the values obtained by Sun and Huggins [9] for the energy E_d required to dissociate the molecule. Shown in Table 3-2 is a summary of his computations. Note that the dissociation energy E_d, for the compound M_yO_x is first divided by y to obtain the energy to dissociate the first coordination sphere and subsequently divided by the coordination number to obtain the strength of the single bond. Where the coordination number is known to change, the values of the single bond strength are computed for the different coordinations. In some cases, such as P, As, V, and Sb, the elements are thought to

Table 3-2. Single Bond Strengths for Oxides[a]

	M in MO_x	Valence	Dissociation energy $E_{d'}$ per MO_x (kcals)	Coordination number[b]	Single-bond strength (kcals)
Glass formers	B	3	356	3	119
	Si	4	424	4	106
	Ge	4	431	4	108
	Al	3	402–317	4	101–79
	B	3	356	4	89
	P	5	442	4	111–88
	V	5	449	4	112–90
	As	5	349	4	87–70
	Sb	5	339	4	85–68
	Zr	4	485	6	81
Intermediates	Ti	4	435	6	73
	Zn	2	144	2	72
	Pb	2	145	2	73
	Al	3	317–402	6	53–67
	Th	4	516	8	64
	Be	2	250	4	63
	Zr	4	485	8	61
	Cd	2	119	2	60
Modifiers	Sc	3	362	6	60
	La	3	406	7	58
	Y	3	399	8	50
	Sn	4	278	6	46
	Ga	3	267	6	45
	In	3	259	6	43
	Th	4	516	12	43
	Pb	4	232	6	39
	Mg	2	222	6	37
	Li	1	144	4	36
	Pb	2	145	4	36
	Zn	2	144	4	36
	Ba	2	260	8	33
	Ca	2	257	8	32
	Sr	2	256	8	32
	Cd	2	119	4	30
	Na	1	120	6	20
	Cd	2	119	6	20
	K	1	115	9	13
	Rb	1	115	10	12
	Hg	2	68	6	11
	Cs	1	114	12	10

[a] After Sun [8].

[b] Based on either known or assumed values.

be surrounded by three singly bonded oxygens and one doubly bonded oxygen. Thus, two values for the single bond strength were computed corresponding to apparent coordinations of 4 and 5. It can be seen that oxides where the single bond strength exceeds about 80 kcal/mol ($\sim$334 kJ·mol^{-1}, note 1 cal = 4.18 J) could be thought of as **glass network formers (NWF)**. Oxides where the single bond strength was less than about 60 kcal/mol were the **glass network modifies (NWM)**, and those having values between 60 and 80 kcal/mol were obviously the **intermediates**.

An important contribution made by Sun was the introduction of an "intermediate" category. Many binary glasses have now been prepared where one of the "non-classical" network-forming oxides is the "network-forming" majority constituent: the presence of the second minority oxide acts to stabilize the glass. (For a review, see Ref. 5.)

The most serious problem with Sun's classification is the exclusion of the entire range of chalcogenide glasses where the single bond strengths are generally of the order of only about 40 kcal/mol along the chain (covalent bond) and less between the chains (van der Waals forces). Likewise, a similar problem exists in explaining the glassy metals. In CO_2, the dissociation energy of the molecule is about 480 kcal/mol, which gives the strength of the single bond to be about 120 kcal/mol (CO_2 may be written as O=C=O). The absence of ready glass formation in CO_2 is attributed to the weak van der Waals forces between the CO_2 molecules, which encouraage rearrangement to yield crystallization. Despite these several obvious difficulties, the classification of substances into three divisions with regard to their glass-forming ability has been the most important contribution of Sun's single bond strength criterion.

Rawson [10] suggested that the ability for structural rearrangements to occur during crystallization should be determined by the ratio of the bond strength to the energy available at the freezing point. Since the kinetic energy itself is roughly $\frac{3}{2}RT_m$, an appropriate parameter to compare would be B_{M-O}/T_m. The higher the value, the lower the probability for bonds to break at the T_m, and hence the higher the tendency for glass formation. Rawson's most important contribution was to point out the significance of the liquidus temperature in binary and multicomponent systems that, hitherto, had been ignored almost entirely. Glass formation in such systems could be easier if the oxides formed a eutectic at temperatures lower than the melting point of any of the constituent oxides.

3.1.3. Dietzel's Field Strength Criterion

The **field strength** F of a cation is given by

$$F = Z_c/(r_c + r_{O^-})^2 = Z_c/a^2, \tag{3.1}$$

Table 3-3. Field Strengths of Various Ions[a]

Element	Valence Z	Ionic radius (for $CN = 6$) r in Å	Most frequent coordination number CN	Ionic distance for oxides a in Å	Field strength at distance of O^{2} ions Z/a^2	Function in glass structure
K	1	1.33	8	2.77	0.13	Network modifier $Z/a^2 \approx 0.1 \ldots 0.4$
Na	1	0.98	6	2.30	0.19	
Li	1	0.78	6	2.10	0.23	
Ba	2	1.43	8	2.86	0.24	
Pb	2	1.32	8	2.74	0.27	
Sr	2	1.27	8	2.69	0.28	
Ca	2	1.06	8	2.48	0.33	
Mn	2	0.91	6	2.23	0.40	
Fe	2	0.83	6	2.15	0.43	
Mn	2	0.83	4	2.03	0.49	Intermediate $Z/a^2 \approx 0.5 \ldots 1.0$
Mg	2	0.78	6	2.10	0.45	
			4	1.96	0.53	
Zr	4	0.87	8	2.28	0.77	
Be	2	0.34	4	1.53	0.86	
Fe	3	0.67	6	1.99	0.76	
			4	1.88	0.85	
Al	3	0.57	6	1.89	0.84	
			4	1.77	0.96	
Ti	4	0.64	6	1.96	1.04	
B	3	0.20	4	1.50	1.34	Network former $Z/a^2 \approx 1.5 \ldots 2.0$
Ge	4	0.44	4	1.66	1.45	
Si	4	0.39	4	1.60	1.57	
P	5	0.34	4	1.55	2.1	
B	3	0.20	3	1.36	1.63	

[a] After Dietzel [11].

where Z_c is the valency of the cation, and r_c and r_{0-} are the ionic radii of the cation and the oxygen ion in Å units. If e is the electronic charge, then Fe^2 is the force exerted by the cation on a unit point charge placed at the location of the oxygen ion. Values of the field strength for the various ions are shown in Table 3-3.

According to Dietzel [11], the network-forming cations have high field strengths (~ 1.3 to 2), the network-modifying cations have low field strengths (0.1 to 0.4), and those having F values, between 0.5 and 1.1 are the intermediates. Apparently, the high-field-strength cations repel each other strongly. At the same time, their small size requires low-coordination-number polyhedra with the oxygen. The two factors promote disordered arrangements. The low-field-strength cations, on the other hand, can come close to each other and force periodicity of their high-coordination polyhedra. In binary melts, there presumably is a "struggle for oxygen." The coordination requirements for the high-field-strength cations are met first.

Dietzel's field strength argument has been important; however, it clearly is inapplicable to non-ionically bonded solids.

3.1.4. Phillips's Topological Constraints Hypothesis

This is the only concept aimed specifically at estimating glass formability in covalently bonded solids such as the chalcogenides. The coordination of a covalently bonded atom is given by **Mott's "8-*N*" rule or the octet rule** [12], according to which the coordination number $m = 8 - N$, where N is the number of valence electrons. Thus, $m_S = m_{Se} = m_{Te} = 2$; $m_{As} = m_{Sb} = 3$; $m_{Si} = m_{Ge} = 4$. The average coordination number $\langle m \rangle$ for a covalently bonded alloy $A_a B_b C_c$ is then given by

$$\langle m \rangle = [a m_A + b m_B + c m_C]/(a + b + c). \tag{3.2}$$

For example, $\langle m \rangle_{As2Se3} = 2.4$, and if we assume a fully covalent bond in SiO_2, then $\langle m \rangle_{SiO2} = 2.67$. In essence, $\langle m \rangle$ is the coordination of a pseudo-atom that would form a structure having the same topology as the complex system.

In the valency-force-field model, a covalently bonded atom experiences bond-stretching (the α type) and bond-bending (the β type) interactions such that the total potential energy, U, is given by

$$U = (\tfrac{1}{2})\alpha \Delta r^2 + (\tfrac{1}{2})\beta r_0^2 \Delta\theta^2, \tag{3.3}$$

where Δr and $\Delta\theta$ are changes in the interatomic spacing and bond angle, respectively, from the equilibrium values r and θ. In a binary alloy $A_x B_{1-x}$, there is one bond-stretching interaction and two bond-bending interactions (angles BAB and ABA). Each of these interactions is essentially a force constant that provides a mechanical constraint.

Phillips [13] assumed these mechanical interactions to be rigid over a short-range order, i.e., the bond length r_{AB}, and the next-nearest neighbor distances r_{AA} and r_{BB} were assumed to have zero deviations from the equilibrium values. He then suggested that the *glass-forming tendency is maximized when the number of mechanical constraints N_c equals the number of degrees of freedom N_d*.

For a pseudo-atom in a covalently bonded solid with average coordination number $\langle m \rangle$, the number of α interactions is given by $\langle m \rangle/2$ (the division by 2 is performed to avoid double counting), and the number of β interactions is $\langle m \rangle (\langle m \rangle - 1)/2$ (the number of combinations out of m neighbors taken two at a time). Hence,

$$N_c = \langle m \rangle /2 + \langle m \rangle (\langle m \rangle - 1)/2 = \langle m \rangle^2/2. \tag{3.4}$$

Equating this to the allowed degrees of freedom, which is 3 (the dimensionality of the system), we get

$$\langle m \rangle_c^2/2 = 3 \quad \text{or} \quad \langle m \rangle_c = 2.45. \tag{3.5}$$

This is the "universal" critical average coordination number most favorable for glass-forming capability.

For a real system such as As_xSe_{1-x} or Ge_ySe_{1-y}, it is more instructive to compute the constraints for each of the atoms first rather than to average the coordination number. For As_xSe_{1-x}, this can be done as follows:

$$m_{As} = 3; \qquad m_{Se} = 2$$

For As: $$N_c(\text{As}) = x m_{As}^2/2 = 9x/2;$$

For Se: $$N_c(\text{Se}) = (1 - x) m_{Se}^2/2 = 2(1 - x).$$

Hence, the total number of constraints $N_c = 9x/2 + 2(1 - x)$. Setting this equal to 3 $(= N_d)$, we get $x = 0.4$, which yields

$$\langle m \rangle_c = 2.40. \tag{3.6}$$

Likewise, for the Ge_ySe_{1-y} alloy, we get

$$N_c = 4^2 y/2 + 2^2(1 - y)/2 = 3,$$

or $y = \frac{1}{6}$, which gives

$$\langle m \rangle_c = 2.33. \tag{3.7}$$

Phillips plotted the glass-forming difficulty (the magnitude of the minimum quenching rate required to obtain a glass over a reasonable time) as a function of the alloy composition in the Ge–Se system (Fig. 3-3) and showed that, indeed, the vicinity of 16 mol. % Ge had the least glass-forming difficulty. The solid curve in Fig. 3-3 was separated into a primary parabola (dashed curve) plus secondary peaks at $y = 0$ and at $y = 0.33$ (composition $GeSe_2$). The primary curve represented the effects of the short-range order on the topology and the medium-range order of the covalent network. The secondary peaks represented the effects of the long-range order, i.e., the need to avoid chemical ordering in the vicinity of the crystalline compounds.

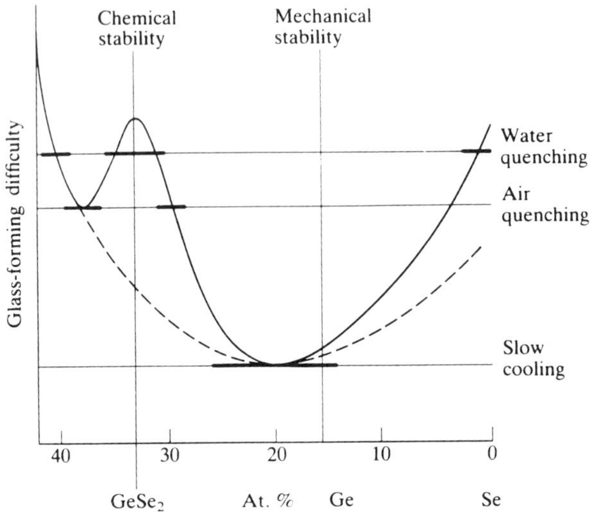

Figure 3-3. Glass-forming difficulty versus composition in the Ge–Se system. Solid horizontal bars are experimental data for the three cooling rates employed (rapid using water, medium using air, and slow cooling). Solid curve is drawn to guide the eye. Dashed curve is the prediction from Phillips' constraint theory. (After J. C. Phillips, *J. Non-cryst. Sol.* **34**, 153 (1979). Reproduced with permission of Elsevier Science Publishers.)

When $\langle m \rangle$ is less than $\langle m \rangle_c$, the glass is "underconstrained" and is appropriately called a "polymeric glass" having large "floppy" or "spongy" regions with a few rigid inclusions. On the other hand, covalent random networks having $\langle m \rangle$ greater than $\langle m \rangle_c$ are "overconstrained" and are best termed "amorphous solids" having large rigid regions with a few "floppy" inclusions (14).

According to Döhler *et al.* [15], Eq. (3.6) is correct for $\langle m \rangle \leq N_d - 1$, but needs to be modified when $\langle m \rangle > N_d - 1$. This is because the $N_d + 1$ angles in N_d dimensions are not linearly independent. The constraint equations should read:

$$N_c = \langle m \rangle/2 + \langle m \rangle(\langle m \rangle - 1)/2 \qquad \text{for } \langle m \rangle \leq N_d - 1, \quad (3.8)$$

$$N_c = \langle m \rangle/2 + (N_d - 1)(2\langle m \rangle - N_d)/2 \qquad \text{for } \langle m \rangle > N_d - 1. \quad (3.9)$$

Setting $N_c = N_d$ yields

$$\langle m \rangle_c = (2N_d)^{1/2} \qquad \text{for } \langle m \rangle \leq N_d - 1$$

and

$$\langle m \rangle_c = N_d(N_d + 1)/(2N_d - 1) \qquad \text{for } \langle m \rangle > N_d - 1.$$

For a 3-D solid and $\langle m \rangle > 2$, Eq. (3.9) gives $\langle m \rangle_c = 2.40$.

In addition to the modification of the number of constraints, Döhler *et al.* also pointed out that substantially polarizable ions would add "internal degrees of freedom" to the N_d. Likewise, the presence of van der Waals forces would tend to increase the effective coordination as well as the internal degrees of freedom due to polarizability.

Many of the topological concepts advanced by Phillips, Thorpe, Döhler *et al.*, and others are quite recent and not fully understood. As a result, they remain as yet inadequately tested. No doubt, a great deal of insight into the glass-forming tendencies of covalently bonded systems and their properties has already been gained through these topological principles. Some of these will be picked up in the discussion of the rigidity properties of the chalcogenides.

3.2. Russian Workers' Criticism of Zachariasen's Hypothesis

Suggested reading: E. A. Porai-Koshits, *J. Non-cryst. Sol.* **123**, 1–13 (1990).

As we mentioned earlier, of all the structural-characteristics-based hypotheses of glass-forming criteria, Zachariasen's random network model had the most impact on scientific thought. The challenge to this model, primarily from Lebedev, Valenkov, Porai-Koshits, and other Russian workers, was focused on the issue of the extent of randomness. Because of the chemical bonding, some ordering is clearly expected. For instance, a silicon atom has to have four oxygens as its first near neighbors, and not two or six oxygens, and definitely not other silicons. The oxygen, in turn, would be bonded to silicons. The number of second near-neighbors for a silicon would hence be somewhat similar to that in a corresponding crystal. The question is, "To what extent is this similarity to the crystalline order extended in a glass?"

Lebedev and coworkers suggested that the structure of a glass was not as random as Zachariasen had postulated, but that there was *evidence of definite short-range structures similar to those found in crystals.* (In chronological terms, Lebedev's view existed 10 years prior to Randall *et al.*'s microcrystals model, and 11 years prior to Zachariasen's random network model.) According to Lebedev:

- A silicate glass is an accumulation of submicrocrystalline formations of various silicates and silica, called "crystallites."

- These crystallites are definite chemical compounds or solid solutions: their chemical nature is determined by the phase diagram for the system that constitutes the glass.
- The crystallites are separated by amorphous layers in which the degree of disorder increases with the distance from the microcrystalline region.
- The crystallites are not mere fragments of a crystal lattice. In reality, they are considerably deformed structural formations with more or less pronounced features of a crystal lattice.

The schematics of the crystallite morphology according to Lebedev are shown in Fig. 3-4. Note the essential difference of their ideas from those of Randall *et al.* Whereas the latter had suggested that a glass was simply an assemblage of finely divided (micro) crystals, Lebedev and coworkers had advanced the idea that the structures of these seemingly crystalline regions were deformed rather than perfect lattices. Thus, a $22Na_2O{\cdot}78SiO_2$ (mol. %) glass could be thought to be composed of SiO_2, $Na_2O{\cdot}2SiO_2$, and $Na_2O{\cdot}3SiO_2$ crystallites in various percentages, depending upon the phase diagram (and the cooling conditions).

The most important experimental evidence put forward was from XRD. They showed that the radial distribution functions (RDFs) of a borosilicate glass measured using x-rays could simply be constructed by adding individual RDFs of B_2O_3 and SiO_2 in appropriately weighted proportions (see Fig. 3-5a). Likewise, if the RDF of SiO_2 were multiplied by y and subtracted

Figure 3-4. Schematic diagram of Lebedev's crystallite theory. Regions of local order (crystallites) are separated by regions of disorder.

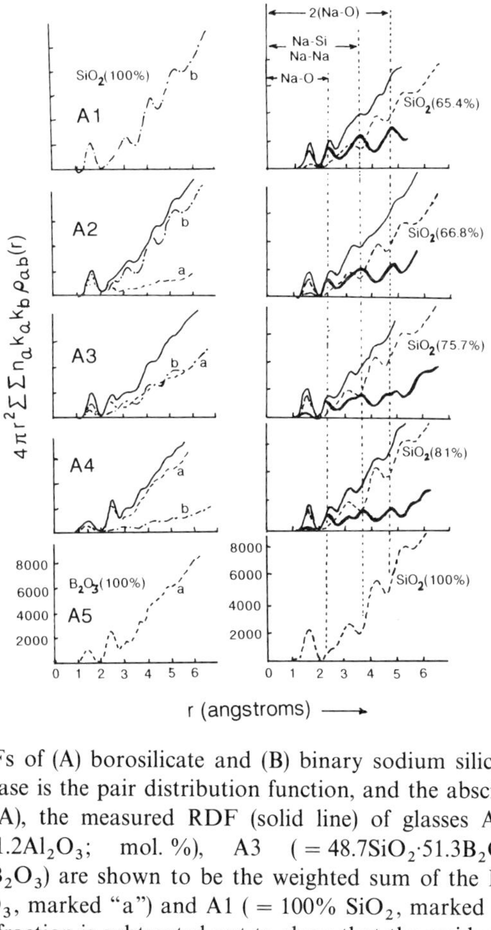

Figure 3-5. RDFs of (A) borosilicate and (B) binary sodium silicate glasses. The ordinate in each case is the pair distribution function, and the abscissa is the radial distance (Å). In (A), the measured RDF (solid line) of glasses A2 ($= 83.3SiO_2 \cdot 11.5B_2O_3 \cdot 4Na_2O \cdot 1.2Al_2O_3$; mol. %), A3 ($= 48.7SiO_2 \cdot 51.3B_2O_3$), and A4 ($= 16.9SiO_2 \cdot 83.1B_2O_3$) are shown to be the weighted sum of the RDFs of glasses A5 (= 100% B_2O_3, marked "a") and A1 (= 100% SiO_2, marked "b"). In (B), the RDF of the silica fraction is subtracted out to show that the residual (heavy line) in each case represents Na_2O, which is roughly the same in each case. [Redrawn in portion for clarity from E. A. Porai-Koshits in *The Structure of Glass* **1**, pp. 29 and 31, Consultants Bureau, New York, 1958. Reproduced with permission of Plenum Press.)

out of the RDFs of various $xNa_2O \cdot ySiO_2$ glasses, then the residuals (indicative of the distribution of Na) had essentially similar shape, suggesting that compounds of sodium present were of similar type in all the different glasses (see Fig. 3-5b).

By 1958, replica electron microscopy of freshly fractured glass surfaces, pioneered by Vogel [16], showed clear evidence of surface "roughness" uncharacteristic of a random, homogeneous phase. An example of this

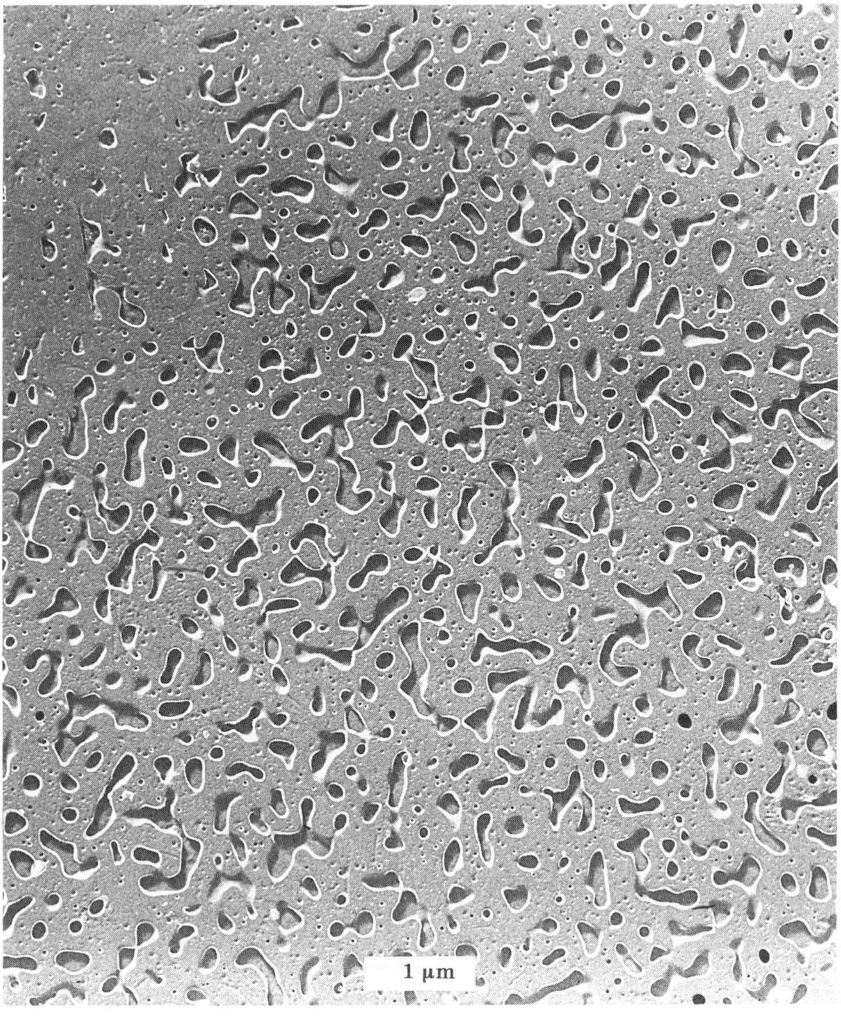

Figure 3-6. Replica transmission electron micrograph of phase separation in $77B_2O_3 \cdot 18PbO \cdot 5Al_2O_3$ glass. (Photograph Courtesy of W. Dumbaugh, Corning Inc., Corning, New York.)

so-called **microstructure** (more appropriately, the **nanostructure**) is shown in Fig. 3-6. (To prepare a replica, see Section 4.3.)

No evidence of a microstructure could ever be found in replicas of fused silica. It became clear that *the degree of randomness varied from glass to glass*: some were closer to Zachariasen's picture, while others had more microcrystalline-type, but not necessarily "crystalline," inhomogeneities. The understanding of the occurrence of such large degree of variations was achieved once the kinetic theory of glass formation and the concepts of phase separation and liquid immiscibility were understood. The kinetic theory is discussed next: phase separation and liquid immiscibility are discussed in the next chapter.

3.3. The Kinetic Theory of Glass Formation

It is generally agreed that many of the liquids that would otherwise crystallize during normal cooling can be brought to a glassy state by a more rapid cooling. We should recognize that the question, "Which compounds can form glass?" is only academic; the real question ought to be, "At what rate should a given liquid be cooled to bring it into a glassy state?" It is now well established that *all liquids*, including water and molten metals, *can be vitrified provided that the rate of cooling is rapid enough to avoid crystallization (also called devitrification) below the freezing point.* Crystallization requires *first* the formation of a measurable number of nuclei, and *then* the occurrence of a measurable crystal growth rate. Note the emphasis on the sequence of events. In some situations, there may be sufficient nuclei present at the surface. In such cases, a small crystal growth rate could lead to detectable crystallization. To avoid crystallization, one needs to avoid either or both of the two. The following steps are needed to determine whether a system will form a glass (i.e., will not have a certain minimum volume fraction crystals):

1. Calculate the rate of nucleation, I, as a function of temperature.
2. Calculate the rate of crystal growth, u, as a function of temperature.
3. Combine (1) and (2) to determine the volume fraction of crystallization when the mass is held at a given temperature for a given period of time. This is normally accomplished by using the Johnson–Mehl–Avrami equation and developing T–T–T (time–temperature–transformation) diagrams. Let us consider each of these topics.

3.3.1. The Nucleation Rate

A nucleus is a precursor to a crystal. (A crystal is an assemblage of atoms in a regular, periodic array. In addition, we will now add here that a crystal has recognizable growth habit planes.) A nucleus is also a periodic assemblage of atoms, but it does not have recognizable growth habit planes. Nucleation occurs because the atoms are constantly vibrating and moving around as a result of the thermal energy in the system. If every atomic vibration allowed the atom to join to a nucleus, the nucleation rate, I, would simply be $n\nu$ per unit volume per second, where n is the number of atoms (per unit volume) and ν is the atomic vibration frequency (per second). There are **two barriers** to accomplishing this. The first is the "**kinetic barrier**," ΔE_D, which is the "activation" energy required for an atom to cross the liquid–nucleus interface. This involves, perhaps, the breaking of bonds with its nearest neighbors and certainly some realignment into the more ordered structure within the surface of the nucleus (barrier to diffusive motion over small distances including a barrier to realign). The second is the "**thermodynamic barrier**," W^*, which is the net free energy change in the system after a nucleus has formed. Each of these barriers introduces a Boltzmann probability factor to the total maximum rate, $n\nu$. In other words, I, is given by

$$I = n\nu \exp(-NW^*/RT)\cdot\exp(-\Delta E_D/RT), \tag{3.10}$$

where N is Avogadro's number, R the gas constant, and T the absolute temperature. Here I is written as the number of nuclei formed per unit volume per second.

We can define two types of nucleation:

1. **Homogeneous nucleation:** The chance assembly of a number of atoms to form a group that leads to the lowering of the net free energy of the system. The probability of such an assembly is equal everywhere throughout the material.
2. **Heterogeneous nucleation:** The chance assembly on preexisting surfaces of "foreign" materials.

When atoms get together to form a cluster, two effects on the energy of the system should be considered: first, the lowering of the volume free energy due to ordering in a crystalline arrangement, and second, the increase in surface energy because the system now contains separate entities where the structure on the left side of an interface is not the same as the structure of the right side of that interface. If the assembly is approximated as a sphere of radius r, the first energy term is $\frac{4}{3}\pi r^3 \Delta G_x$, where ΔG_x is the change in the free energy in transforming a unit volume of the liquid to a unit volume of the crystal (the "free energy of crystallization") and is a negative number below T_m. The second term is given by $4\pi r^2 \sigma$, where σ is the surface energy of the interface (per unit area). The net change in energy, W, for the cluster

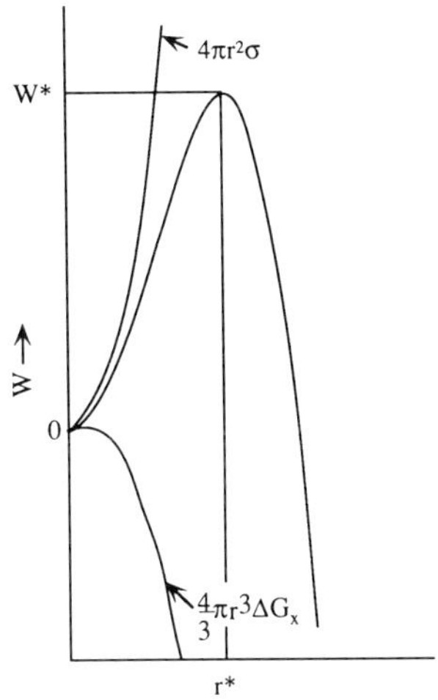

Figure 3-7. Change in free energy of a spherical nucleus as a function of radius.

is given by

$$W = \tfrac{4}{3}\pi r^3 \Delta G_x + 4\pi r^2 \sigma \tag{3.11}$$

and is plotted as a function of r in Fig. 3-7. Initially, when r is small, the square term dominates over the cube term, and hence the net energy tends to increase. This means that small-sized assemblies of atoms (embryo) are not stable. At large values of r, the volume free energy of ordering dominates and leads to stable nucleation. Somewhere in between, at r^*, the energy, W, reaches a maximum value W^*. At this point, the probability of the survival of an embryo is about 50%. An embryo from this point onwards may be called a nucleus. The radius, r^*, of the **critical-sized nucleus** is determined by differentiating W with respect to r and setting equal to zero:

$$dW/dr = 4\pi r^2 \cdot \Delta G_x + 8\pi r\sigma = 0 \qquad \text{when} \qquad r = r^*;$$

hence

$$r^* = -2\sigma/\Delta G_x. \tag{3.12}$$

(Note that by our definition, ΔG_x is a negative value below T_m.) Substituting (3.12) in (3.11) yields the maximum value W^* (which may now be called "the work of nucleation") as

$$W^* = 16\pi\sigma^3/3(\Delta G_x)^2. \tag{3.13}$$

When "foreign" surfaces are present as prospective nucleation sites (see Fig. 3-8), the supercooled liquid attempts to nucleate (labeled "crystal cluster") in contact with the substrate. If we assume that the cluster is in the form of

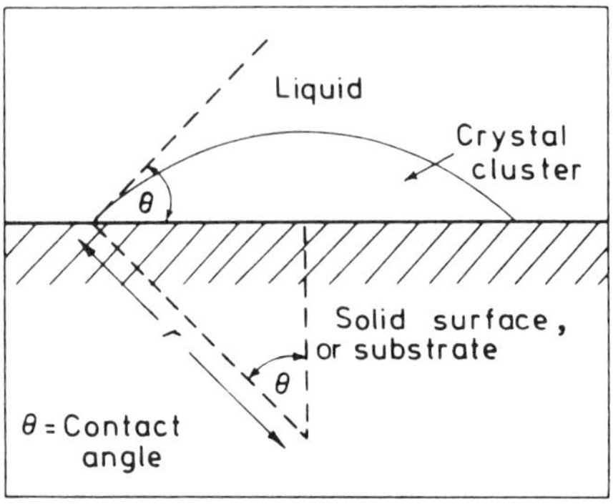

Figure 3-8. Formation of a crystalline cluster on a solid substrate.

a segment of a sphere of radius r, then the volume energy term for this nucleated droplet is $V_x(\Delta G_x)$, where V_x is the volume of the segment. The surface energy term for the segment should be

$$A_{LX} \cdot \sigma_{LX} + [\pi r^2 \sin^2 \theta](\sigma_{XS} - \sigma_{LS}), \tag{3.14}$$

where θ is the contact angle between the substrate, the crystal, and the liquid; σ_{LX} is the surface energy of the liquid–crystal interface; σ_{XS} is the surface energy of the crystal–substrate interface; and σ_{LS} is the surface energy of the liquid–substrate interface.

Hence, if $\sigma_{LS} < \sigma_{XS}$, then the net surface energy of the liquid–crystal–substrate is always smaller than $A_{LX} \cdot \sigma_{LX}$ (which represents the surface energy of a homogeneously nucleated crystal). The net change in energy, W, is therefore given by

$$W = V_x \Delta G_x + A_{LX} \cdot \sigma_{LX} + [\pi r^2 \sin^2 \theta](\sigma_{XS} - \sigma_{LS}). \tag{3.15}$$

Now,

$$A_{LX} = 2\pi r^2(1 - \cos\theta) \quad \text{and} \quad V_x = [\pi r^3/3](1 - \cos\theta)^2(2 + \cos\theta).$$

Also, by balance of forces at the edge of the crystalline cluster,

$$\sigma_{LS} = \sigma_{XS} + \sigma_{LX} \cos\theta. \tag{3.16}$$

Upon substitution in (3.15), and setting the differential with respect to r equal 0, we get for the critical radius

$$r^* = -2\sigma_{LX}/\Delta G_x, \tag{3.17}$$

which is the same as that for homogeneous nucleation, but

$$W^*_{\text{hetero}} = [16\pi\sigma^3_{\text{LX}}/3(\Delta G_x)^2]\cdot[(1 - \cos\theta)^2(2 + \cos\theta)/4] \tag{3.18}$$

$$= W^*_{\text{homo}}[(1 - \cos\theta)^2(2 + \cos\theta)/4]. \tag{3.19}$$

Hence, when

$$\begin{aligned} &\theta = 0^\circ \text{ (fully wetting condition)} && \Rightarrow W^*_{\text{hetero}} = 0, \\ &\theta = 90^\circ && \Rightarrow W^*_{\text{hetero}} = W^*_{\text{homo}}/2, \\ &\theta = 180^\circ \text{ (fully nonwetting condition)} && \Rightarrow W^*_{\text{hetero}} = W^*_{\text{homo}}. \end{aligned} \tag{3.20}$$

As one would expect, the last situation is equivalent to the nonexistence of the "foreign" substrate. Note also that, *in the presence of a perfectly wetting substrate, there is no thermodynamic barrier to nucleation*, i.e., any size of the embryo will grow. Only the kinetic barrier exists, i.e., the atoms need only to be transported from a liquid-type location to a crystal-type location.

For a perfectly wetting foreign substrate, one of the requirements for heterogeneous nucleation is the matching of the lattice spacing of the exposed planes of the substrate to the growth habit planes of the crystal (in other words, "the registry of the two lattice planes is perfect"). The growth of the crystal would clearly end up being oriented, hence the term **oriented overgrowth**, or more generally, **epitaxial growth**. In most real situations, the registry between the two is rarely perfect. Epitaxial growth may still take place as long as the disregistry is not very large (perhaps about 10%). The growing crystal may start out coherently, which will necessarily lead to a development of strains at the interface. With continuing growth, the lattice spacing gradually becomes close to the strain-free value. If the disregistry is large, then the strains developed at the interface may cause its "fracture." In other words, epitaxial growth on the foreign substrate will not occur by the heterogeneous nucleation mechanism.

Now, consider the thermodynamic relation

$$\Delta G = \Delta H - T\Delta S, \tag{3.21}$$

where ΔG, ΔH, and ΔS are the free energy, the enthalpy, and the entropy of a change, respectively. If we assume that ΔH and ΔS are independent of T in the vicinity of the melting point, then

$$d(\Delta G)/dT = -\Delta S. \tag{3.22}$$

If we integrate both sides from T_m to T, then

$$\Delta G_T - \Delta G_{Tm} = -\Delta S(T - T_m).$$

ΔG_{Tm} is the change in free energy at the melting point and is simply zero. ΔG_T can be identified as ΔG_x. The entropy change ΔS (from liquid to crystal) is $[-\Delta S_f]$, where ΔS_f is called the entropy of fusion. Hence, ΔG_x (the free energy of crystallization) for a temperature interval $\Delta T(= T - T_m)$ away from the melting point T_m can be written as

$$\Delta G_x = \Delta T \cdot \Delta S_f. \tag{3.23}$$

Note that ΔS_f is a positive number. ΔT is a negative number for supercooled liquids, and hence ΔG_x is negative below T_m, as should be the case. (Many authors often use the term "undercooling" to imply $T_m - T$.) Again, since $\Delta G = 0$ at T_m, we get from (3.21) $\Delta S_f = \Delta H_f/T_m$, where ΔH_f is the heat of fusion. It follows that

$$\Delta G_x = \Delta T \cdot \Delta H_f/T_m. \tag{3.24}$$

Combining (3.24) with (3.10), the expression for the nucleation rate, I, can be rewritten as

$$I = nv \exp[-(N/RT)(16\pi\sigma^3/3\Delta H_f^2)(T_m/\Delta T)^2] \exp[-\Delta E_D/RT]. \tag{3.25}$$

(Caution: The ΔH_f used in (3.25) is per unit volume basis. To convert from the usually tabulated value in per gram-mole units, divide by the molar volume.)

Initially, when $\Delta T = 0$, the first term is equal to zero (thermodynamically controlled). Eventually, as T decreases, $\exp(-\Delta E_D/RT) \to 0$ (kinetically controlled). As a result, the I vs. T curve is a skewed bell-shape reaching a maximum in I somewhere below T_m and reaching zero asymptotically at both the high- and low-temperature ends (Fig. 3-9). It is customary to show a "metastable zone of supercooling" below T_m to correspond to a region where the nuclei present are so few that they cannot be detected by a measuring instrument.

3.3.2. Crystal Growth Rate

Suggested reading: I. Gutzow, in *Crystal Growth and Materials* (E. Kaldis and H. J. Scheel, eds.), pp. 380–520. North-Holland, Amsterdam, 1976.
D. R. Uhlmann, in *Advances in Nucleation and Crystallization in Glasses* (L. L. Hench and S. W. Freiman, eds.), pp. 91–115. American Ceramic Society, Columbus OH, 1971.

Once a critical-sized nucleus has formed, crystal growth may occur by advancement of deposited atomic layers. The growth would clearly depend upon how rapidly atoms can diffuse from liquid regions to the interface. A kinetic treatment according to Turnbull [17] can be discussed as follows.

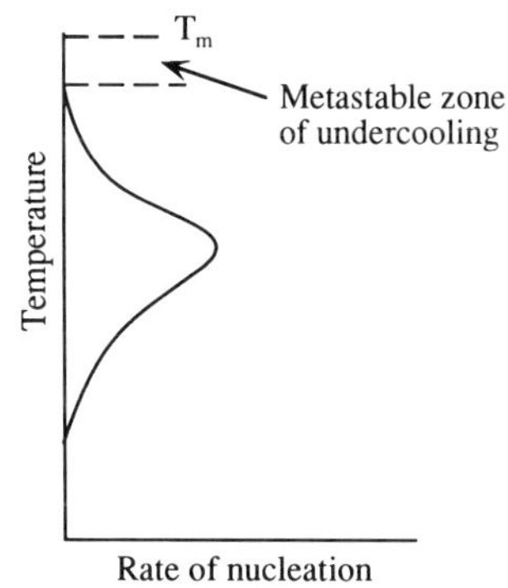

Figure 3-9. Variation of the nucleation rate, I, with temperature, T.

The atoms near the interface can be viewed as located in potential wells (Fig. 3-10) where the depth of the wells is representative of the free energy of the atom in a particular state. Thus, the well is deeper for the atom in the crystalline order than for the liquid location. The difference between the two depths is, essentially, the free energy of crystallization ΔG_x. The barrier height between the two is labeled $\Delta E'$ and is an activation energy for motion. Note that two kinetic barriers, $\Delta E'$ and ΔE_D, used for nucleation need not be the same. Atomic movements during nucleation are far more local than during crystal growth. If v is the atomic jump frequency, then the frequency v_{lx} of a successful jump from the liquid to the crystal is given by

$$v_{lx} = v \exp(-\Delta E'/RT), \tag{3.26}$$

and that from the crystal to the liquid by

$$v_{xl} = v \exp[-(\Delta E' - \Delta G_x)/RT]. \tag{3.27}$$

(Note that by our sign convention, the activation energy $\Delta E'$ is a positive number; however, the free energy of crystallization is negative when $\Delta T = T - T_m$ is negative.) The net crystal growth u is equal to $a(v_{lx} - v_{xl})$, where

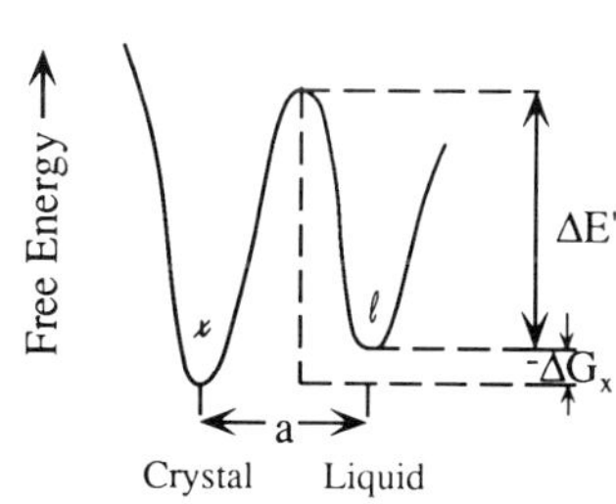

Figure 3-10. Free energy versus position diagram for atoms or molecules crossing a crystal–liquid interface.

a is the distance between the two sites. Hence,

$$u = a\nu \exp(-\Delta E'/RT)[1 - \exp(\Delta G_x/RT)]. \tag{3.28}$$

At $T > T_m$, ΔG_x is positive, hence $\{1 - \exp(\Delta G_x/RT\}$ is a negative number leading to a negative value for u. This is understandable, since at $T > T_m$, crystals start dissolving. At T_m, $\Delta G_x = 0$, which yields $u = 0$ and is essentially the definition of T_m. As the temperature T is lowered, ΔG_x becomes increasingly negative, with the result that u increases. Ultimately, at very low values of T, the kinetic terms, $\exp(-\Delta E'/RT)$, approaches zero faster, and the crystal growth rate once again vanishes asymptotically.

Turnbull and Cohen [18] have suggested that mechanisms controlling the activation energy for the rate of crystal growth should be the same as those that control diffusive flow. In other words, $\Delta E'$ is also the activation energy for diffusion. Since the diffusion coefficient, D, may be written as (see Eq. (14.30))

$$D = a^2\nu \exp(-\Delta E'/RT),$$

Eq. (3.28) may be rewritten, incorporating Eq. (3.24), as

$$u = (D/a)(1 - \exp[\Delta H_f \cdot \Delta T/RTT_m]). \tag{3.29}$$

Equation (3.29) is generally referred to as the **Wilson–Frenkel theory**. Utilizing the Stokes–Einstein relation between the diffusion coefficient and the viscosity η, $D = RT/3N\pi a\eta$ (where we have assumed the average jump

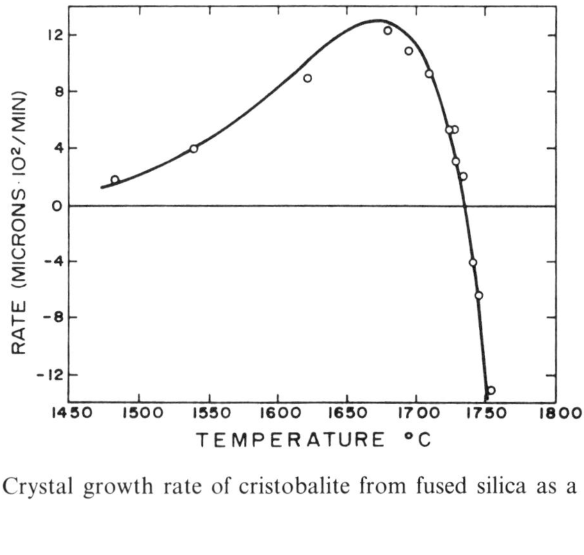

Figure 3-11. Crystal growth rate of cristobalite from fused silica as a function of temperature.

Table 3-4. Maximum Crystal Growth Velocities and Viscosities of Glass-Forming Liquids[a]

Materials	Melting point (°C)	Maximum crystallization velocity (cm/s)	Temperature of maximum *u* (°C)	Log viscosity at m.p. (P)
Vitreous silica, SiO_2	1,734	$2.2(10)^{-7}$	1,674	7.36
Vitreous germania, GeO_2	1,116	$4.2(10)^{-6}$	1,020	5.5
Phosphorous pentoxide P_2O_5	580	$1.5(10)^{-7}$	561	6.7
Sodium disilicate $Na_2O{\cdot}2SiO_2$	878	$1.5(10)^{-4}$	762	3.8
Potassium disilicate $K_2O{\cdot}2SiO_2$	1,040	$3.6(10)^{-5}$	930	
Barium diborate $BaO{\cdot}2B_2O_3$	910	$4.3(10)^{-3}$	849	1.7
Lead diborate $PbO{\cdot}2B_2O_3$	774	$1.9(10)^{-4}$	705	1.0
1,2-Diphenylbenzene	55.5	$2.5(10)^{-3}$	38	−0.46
1,3,5-Tri-α-naphthylbenzene	197	$9.3(10)^{-5}$	175	−0.34
Glycerol	18.3	$1.8(10)^{-4}$	−6.7	1.0
Salol	43	$6.4(10)^{-3}$	20	
Polyethylene adipate	55	$9.5(10)^{-6}$	26	
Toluene	−95	>0.5		
Ethyl ether	−166	>0.13		
Methyl alcohol	−98	>0.09		

[a] After R. H. Doremus, *Glass Science*, p. 15, Wiley-Interscience, New York, 1973. Reproduced with permission of J. Wiley & Sons.

distance a to be roughly a molecular diameter; see Eq. (14.16)), we get

$$u = (fRT/3N\pi a^2\eta)(1 - \exp[\Delta H_f \cdot \Delta T/RTT_m]), \tag{3.30}$$

where f is the fraction of the surface area on which growth sites may be available. Note that the crystal growth is clearly low if the viscosity at the liquidus is quite high.

The experimentally measured variation of u with T for fused silica [19] as an example is plotted in Fig. 3-11: its form at first glance clearly appears to agree with that predicted by Eq. (3.30). Table 3-4 (after Ref. 20) lists the maximum crystal growth velocities for a few substances along with their viscosities at the melting point. One can readily understand on a qualitative basis why silica is a good glass former. (The maximum crystal growth velocity in B_2O_3 is presumably even lower.) Upon the addition of small amounts of sodium impurity, the crystal growth rate in silica increases considerably (Fig. 3-12), in addition to a decrease in the temperature at which the maximum occurs (from ~1,600°C to ~1,400°C).

In Fig. 3-13, both I and u are plotted together against T. Note that the $I - T$ peak occurs at temperatures lower than the $u - T$ peak. Since nucleation must precede crystal growth for crystallization to occur, it is clear that *the extent of the overlap between the two curves, shown as the hatched area, gives, in general, the capability of a liquid to form glass upon cooling: the smaller the overlap, the easier the glass formation.* One can also recognize the possibility of having little or no overlap between the two curves in

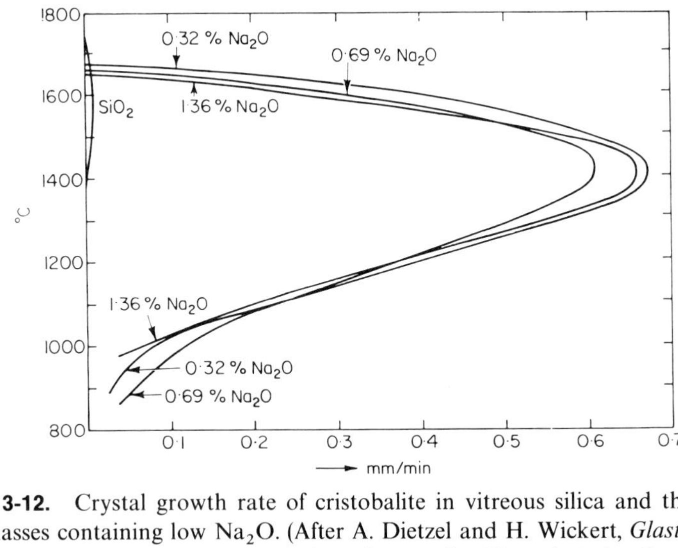

Figure 3-12. Crystal growth rate of cristobalite in vitreous silica and three soda silica glasses containing low Na_2O. (After A. Dietzel and H. Wickert, *Glastech. Ber.* **29**, 1 (1956). Reproduced with permission of Deutsche Glastechnische Gesellschaft.)

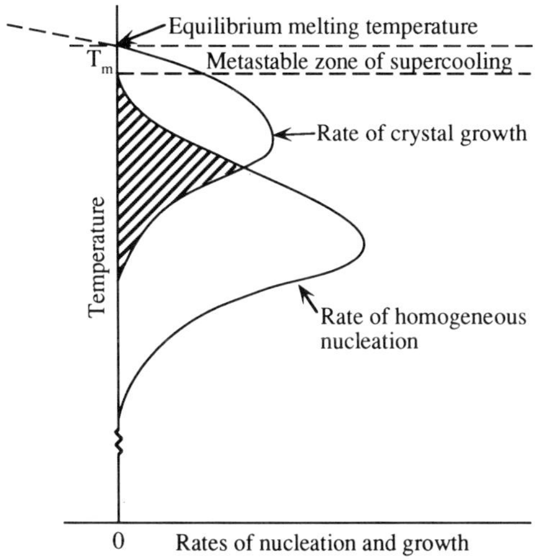

Figure 3-13. Variation of the crystal growth rate, u, and nucleation rate, I, with temperature, T.

carefully tailored compositions, forming a glass during first cooling past the two curves, and then reheating to the crystal growth maximum at which point crystallization occurs at pre-existing nucleation sites. For instance, copper-containing glasses, when melted under reducing conditions, are colorless during first cooling. They develop brilliant ruby-red colors upon a subsequent suitable heat treatment. Glass artisans know this technology as "striking of a ruby." Actually, an entire class of colored glasses is made by striking. The colors are produced by light scattering from the precipitated crystals (hence termed **precipitation colors**). Besides copper, Au, Ag and ZnS–Se colors in glass develop by this process. Additionally, the possibility of making **glass-ceramics** exists. The glass is given a specific nucleation treatment, and then controlled crystallization is carried out subsequently until a desired effect in terms of a glass–ceramic composite is produced. The nucleation itself can be bulk homogeneous nucleation or a bulk heterogeneous process by intentionally adding nucleating agents (which reduce the wetting angle and hence aid nucleation—see Eq. (3.18). We shall discusss glass ceramics in further detail in Chapter 4.

The foregoing model for **normal crystal growth** assumed that arriving atoms have a uniform probability of attaching or leaving any site on the crystal–liquid interface. The model would apply when there is a large number of approximately equivalent sites available, particularly when the surface is atomically very rough. Such is believed to be the case for simple glass-forming melts such as SiO_2 and GeO_2. It may be shown that du/dT is finite as

$T \to T_m$, and the slope continues during melting ($T > T_m$), as shown in Figs. 3-12 and 3-13. On the other hand, if the surface is atomically smooth, the atoms that arrive later may attach preferentially to those that arrived earlier. Hence, growth in such as case would occur first by spreading of a monolayer on the surface (adsorbing), and subsequently by nucleation of a new layer above the first one. This mechanism is referred to as **surface nucleated growth**.

For a one-component liquid, the fraction f of such sites is given by [21]

$$f = -\Delta T/2\pi T_m. \tag{3.31}$$

The resultant crystal growth rate is then given by Eq. (3.30). It is left as an exercise for the reader to show that, in such a case, $du/dT \to 0$ as $T \to T_m$. During melting, one would normally observe a finite slope. Real surfaces are generally between the normal and the surface nucleated growth types. The density of imperfections such as kinks and steps is limited. Newly arriving atoms attach themselves to these kinks and steps. An example of growth by covering of the steps is the screw dislocation mechanism, shown in Fig. 3-14, which essentially leads to a **spiral growth**.

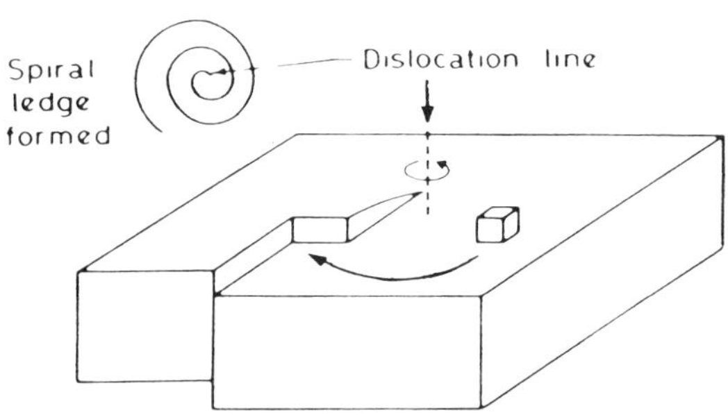

Figure 3-14. Spiral growth of crystals at the termination of a screw dislocation.

3.3.3. *T–T–T* Diagram

To develop a quantitative model for the kinetics of crystallization, one needs to combine the I–T and the u–T curves to give the fraction, X, of material crystallized as a function of heat treatment time, t. This is accomplished by using the Johnson–Mehl–Avrami equation [22], which is

$$\begin{aligned} X = V_x/V_0 &= 1 - \exp[-\pi I u^3 t^4/3] \\ &\cong \pi I u^3 t^4/3, \end{aligned} \tag{3.32}$$

for small values of the argument, where V_x is the volume of crystals, and V_0 is the entire volume. If we assume the functional dependence of I and u upon T as in Fig. 3-13, then we can construct curves that yield the values of times,

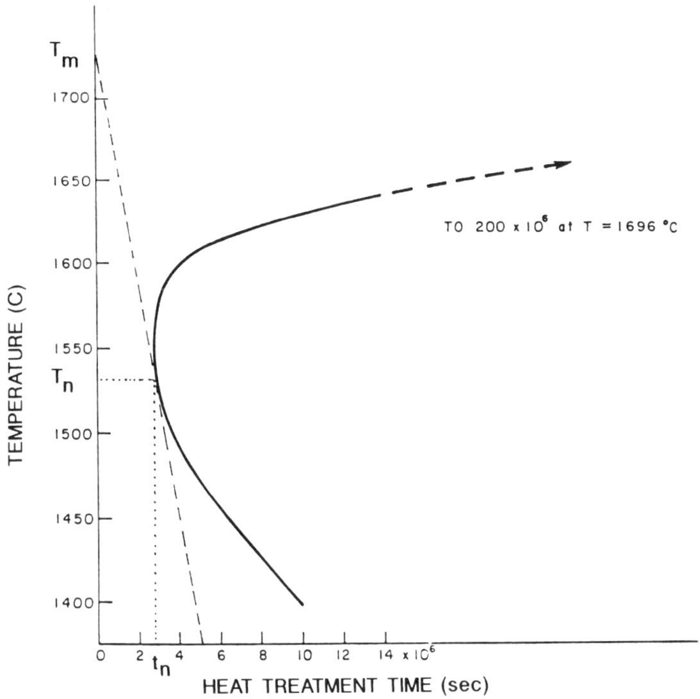

Figure 3-15. Time–temperature–transformation diagram for silica corresponding to a volume fraction crystallized of 10^{-6}. (After D. Uhlmann, *J. Non-cryst. Sol.* 7, 337 (1972). Reproduced with permission of Elsevier Science Publishers.)

t, necessary to cause a given volume fraction crystallization, X, in an isothermal heat treatment at temperature T. Such curves are called ***T–T–T* (*time–temperature–transformation*) *diagrams*.** When we examine the criteria for glass formation, X is usually taken to be 10^{-6}, which is generally the instrumental limit for detecting the presence of crystals. A typical *T–T–T* diagram, generally in the shape of an (ugly) nose, is shown in Fig. 3-15. Near T_m, the time required is essentially infinite (both I and u approach zero because of the lack of a thermodynamic driving force). Likewise, the time required at low temperatures is again very large because of the low atomic mobility. At intermediate temperatures, the times required for detectable crystallization are finite the lowest being t_n for a heat treatment at the nose temperature, T_n. As long as the heat treatments are such that one is to the left of the nose, detectable crystallization is avoided. It is clear, then, that if we start cooling from T_m, *glass formation will occur as long as the cooling rate is faster than the rate given by the tangent at the nose.* In other words,

Table 3-5. Examples of Critical Cooling Rates (°C/s) for Glass Formation

Material	Homogeneous nucleation	Heterogeneous nucleation contact angle (deg)		
		100	60	40
SiO_2 glass[a]	9×10^{-6}	10^{-5}	8×10^{-3}	2×10^{-1}
GeO_2 glass[a]	3×10^{-3}	3×10^{3}	1	20
$Na_2O{\cdot}2SiO_2$ glass[a]	6×10^{-3}	8×10^{-3}	10	$3 \times 10^{+2}$
Salol	10			
Water	10^{7}			
Ag	10^{10}			
Typical metal[a]	9×10^{8}	9×10^{9}	10^{10}	5×10^{10}

[a] After P. I. K. Onorato and D. R. Uhlmann, J. Non-Cryst. Sol., 22(2), 367–378 (1976).

the essential argument of the kinetic theory for glass formation is that the cooling rate dT/dt be greater than or equal to the critical cooling rate, $(dT/dt)_c$, where the latter is given by

$$(dT/dt)_c = \Delta T_n/t_n, \qquad \text{where} \qquad \Delta T_n = T_m - T_n. \tag{3.33}$$

The maximum thickness, Y_c, of material transformed into a glass can be estimated using the relation

$$Y_c \cong [D_{th} t_n]^{1/2}, \tag{3.34}$$

where D_{th} is the thermal diffusivity. Values of the critical cooling rates for several substances are given in Table 3-5. We can readily realize why it is difficult to form glass out of metals by normal cooling techniques (quenching in air cools most materials at 1–10°C/s; the rates can be raised to about 10^3°C/s if a liquid-medium quenching is used). Indeed, one has to employ quenching techniques such as splat-cooling or melt-spinning, where cooling rates as high as 10^5–10^8°C/s can be achieved. Several metals and alloys have now been brought into glassy form using such techniques and are being employed in commercially useful products. A good source of reading on glassy metals is given by Elliott [23].

The T–T–T diagram shown in Fig. 3-15 represents a simple situation involving isothermal heat treatments. More often, a molten mass of liquid is cooled continuously. Modifications of the T–T–T diagram, when the cooling rate is either a constant or proportional to T itself (logarithmic cooling), have been worked out [24] and are shown, for example, for anorthite in Fig. 3-16. In general, the nose is shifted to the right, allowing greater opportunity for the glass to form.

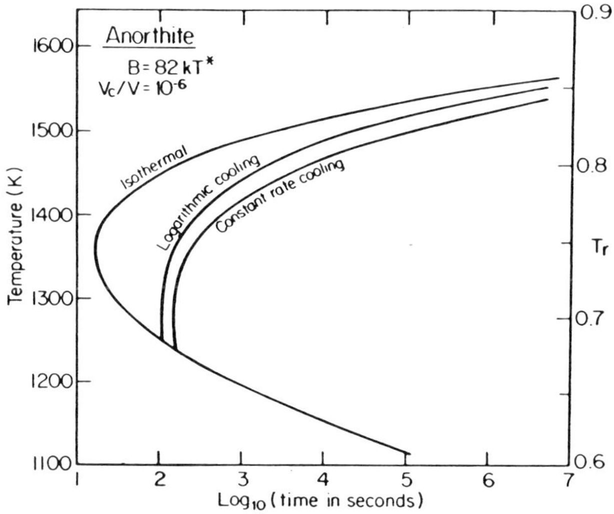

Figure 3-16. Continuous cooling curves for constant-rate and logarithmic cooling for anorthite. Volume fraction crystallized = 10^{-6}. (After D. Uhlmann, *J. Amer. Ceram. Soc.* **66**(2), 95 (1983). Reproduced with permission of The American Ceramic Society.)

Often the Johnson–Mehl–Avrami equation is written as

$$X = 1 - \exp(-kt^n), \tag{3.35}$$

where the value of n can vary between 1 and 4 to account for preferential growth in one, two, or three directions (for needlelike, platelike, or polyhedral-shaped crystals). Presumably, the value of n remains unchanged with temperature so long as the mechanism for nucleation and growth does not change.

To summarize, *the kinetic theory of glass formation does not address the question as to what structural characteristics of substances encourage ready glass formation. It assumes that all substances can be brought into an amorphous state. The only question it addresses is what minimum cooling rate is required to avoid a perceptible degree of crystallization* (*or devitrification*). Although there is general agreement on this concept, a quantitative agreement between theoretical predictions of the rate of crystallization and experimental measurements has only been achieved in a very few cases. In fact, in several cases, the disagreement is of several orders of magnitude. The clear difficulty is in the detailed understanding and estimation of the parameters required for calculating the work of nucleation, W^*, and the nucleation rate, I. A nucleus is so small that, firstly, its own composition can hardly be expected to be homogeneous, and secondly, because the atomic potentials can be "felt" over distances larger than the size of the nucleus, the concept of a surface energy for the nucleus is somewhat difficult to understand. Our use of ΔG_x in Eq. (3.11) was quite loose. The free energy of crystallization is measured when pressure remains constant. However, it is clear that when nucleation involving curved surfaces occurs, the nucleus has an excess pressure over the surroundings. (For this reason, our simple derivation of the "work of nucleation," W^*, is not very precise. A more rigorous derivation is given by Landau and Lifshitz [25].) The question of homogeneous versus heterogeneous nucleation has not been resolved; the container walls may provide sufficient nuclei to throw all estimations off. We expect that containerless melting experiments carried aboard the space shuttle will help greatly to reduce some of the discrepancies between theory and experiment on the kinetics of crystallization in supercooled liquids.

Exercises

1. Apply Zachariasen's rules to glass formation in BaO.
2. Apply Zachariasen's rules to glass formation in Cu_2O, assuming tetrahedra could form as OCu_4.
3. How would Fig. 3-15 be modified if the lowest detectability limit of the percent crystallinity measurement were (a) 10^{-8}, (b) 10^{-2}, instead of 10^{-6}?
4. The I and U vs. T curves for two glasses (a) and (b) are as shown below (Fig. Q3-4). Draw a tentative T–T–T diagram for (i) isothermal heat treatment, (ii) constant rate cooling from the melting point.
5. Maximum crystal growth rates of cristobalite in vitreous silica shown in Figs. 3-11, 3-12 and Table 3-4 differ by a factor of 100 or more. Why?

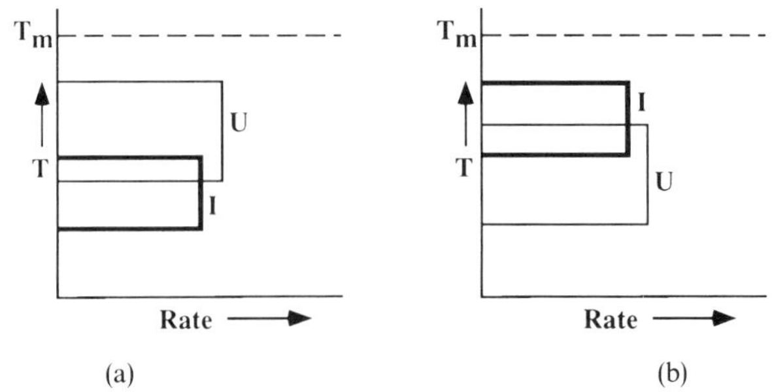

Figure Q3-4.

References

1. W. H. Zachariasen, *J. Chem. Soc.* **54**, 3841–3851 (1932).
2. B. E. Warren, *J. Am. Ceram. Soc.* **17**, 249 (1934).
3. J. T. Randall, H. P. Rooksby, and B. S. Cooper, *J. Soc. Glass Technol.* **15**, 54 (1931).
4. A. R. Cooper, *J. Non-cryst. Sol.* **49**, 1–17 (1982).
5. N. J. Kreidl, *Adv. in Fusion & Proc. of Glass III* (A. K. Varshneya, D. F. Bickford, and P. P. Bihuniak, eds.), *Ceramic Transactions*, **29**, 175–189 (1993).
6. A. A. Lebedev, *Trans. Opt. Inst. Petrograd*, No. 10, p. 2 (1910).
7. N. N. Valenkov and E. A. Porai-Koshits, *Z. Krist.* **95**, 195 (1936).
8. K.-H. Sun, *J. Am. Ceram. Soc.* **30**, 277 (1947).
9. K.-H. Sun and M. L. Huggins, *J. Phys. Colloid Chem.* **51**, 438–443 (1947).
10. H. Rawson, *Proc. IV Intl. Cong. Glass, Paris*, pp. 62–69. Imprimerie Chaix, Paris, 1956.
11. A. Dietzel, *Z. Electrochem.* **48**, 9–23 (1942).
12. N. F. Mott, *Phil. Mag.* **19**, 835 (1969).
13. J. C. Phillips, *J. Non-cryst. Sol.* **34**, 153–181 (1979).
14. M. F. Thorpe, *J. Non-cryst. Sol.* **57**, 355 (1983).
15. G. H. Döhler, R. Dandoloff, and H. Blitz, *J. Non-cryst. Sol.* **42**, 87 (1980).
16. W. Vogel, *Silikattechnik* **9**, 323 (1958).
17. D. Turbull, in *Solid State Physics*, Vol. 3 (F. Seitz and D. Turnbull, eds.), pp. 226–306. Academic Press, New York, 1956.
18. D. Turnbull and M. H. Cohen, in *Modern Aspects of the Vitreous State*, Vol. 1 (J. D. Mackenzie, ed.), pp. 38–62. Butterworths, London, 1960.
19. F. E. Wagstaff, *J. Am. Ceram. Soc.* **52**, 650 (1969).
20. R. H. Doremus, *Glass Science*, p. 15. Wiley-Interscience, New York, 1973.
21. W. B. Hillig and D. Turnbull, *J. Chem. Phys.* **24**, 914 (1956).
22. W. A. Johnson and R. F. Mehl, *Trans. AIMME* **135**, 416 (1939); M. Avrami, *J. Chem. Phys.* **7**, 1103 (1939); *Ibid.*, **8**, 212 (1940); *Ibid.*, **9**, 177 (1941).
23. S. R. Elliott, *Physics of Amorphous Materials*, 1st Ed. Longman Scientific & Technical, Essex, 1990.
24. D. R. Uhlmann, *J. Am. Ceram. Soc.* **66**(2), 95–100 (1983).
25. L. D. Landau and E. M. Lifshitz, *Statistical Physics*, 3rd Ed., Part 1, p. 533. Pergamon Press, Oxford, 1980.

Chapter 4

Glass Microstructure: Phase Separation and Liquid Immiscibility

Glass often appears homogeneous. There is considerable evidence, however, that, the microstructure of glass, on the scale of a few hundred atoms, is not as homogeneous as that of a perfect single crystal or a liquid solution. In the preceding chapter, we noted two processes leading to the development of inhomogeneous microstructure in glass. One of these was the crystallization or devitrification where definite crystalline forms nucleate and grow from a supercooled liquid mass. To cause a molten liquid mass to devitrify, the mass needs to be heat-treated below the liquidus and above the glass transition. By cooling sufficiently rapidly through this range of temperatures to below T_g, the liquid turns into an amorphous solid, which is called glass. The second type of microstructural inhomogeneity was discussed in context with the Lebedev crystallites theory. The crystallites were not microcrystals; they possessed distorted lattices, but did have definite chemical compositions determined by the phase equilibrium diagram applicable to the system. (Recall that a phase is a homogeneous region across which the properties are uniformly constant, and which is bounded by a surface having a distinct surface energy. As long as the distorted crystalline lattices are "homogeneous" and are connected to the rest of the amorphous structure through an identifiable boundary, they qualify for being called a separate phase.) We also noted that not all glasses are multiphased. Some, notably fused silica, clearly lack microstructure even on a scale of a few angstroms. In essence, real glasses appear to span a continuous range of microstructures; Zachariasen's continuous random network being one extreme, and inhomogeneous microstructure such as Lebedev's crystallites picture as well as "spinodal" microstructure (to be discussed later in this chapter) being the other extreme. This picture of continuous morphology as an inhomogeneous structure, falling under the topic called **phase separation**, is the subject of this chapter. It is best understood by beginning with a study of the thermodynamics of mixing.

4.1. Thermodynamics of Mixing

Assume that components A and B are mixed at a temperature, T, and pressure, P (which is generally atmospheric pressure), to yield a solution of a certain composition. The system containing components A and B clearly has three options: to stay unmixed, i.e., remain no more than a physical mixture; to mix on the atomic scale and form a complete solution of A–B; or to mix in preferential ratios to form various compounds of A–B. The choice, *in equilibrium under isothermal and isobaric conditions*, depends on the sign of the term ΔG_m, which is the change in the Gibbs free energy of the system upon mixing. The change is defined as free energy of the mixed system, G_{AB}, minus the free energy of the unmixed system. In other words,

$$\Delta G_m = G_{AB} - [(1 - c)G_A^0 + cG_B^0], \tag{4.1}$$

where G^0 are the free energies of the pure components and c is the mole fraction of component B (Fig. 4-1). If ΔG_m is positive, then the system will remain separated. If, on the other hand, ΔG_m is negative for a certain range of compositions, then mixing will take place in that range. In the unlikely event that $\Delta G_m = 0$, none of the options will be a preferred choice over the other.

The free energy change upon mixing, or simply the free energy of mixing, is given by

$$\Delta G_m = \Delta H_m - T\Delta S_m, \tag{4.2}$$

where ΔH_m and ΔS_m are the enthalpy (or simply the heat) and the entropy of mixing, respectively. (Again, the reader should bear in mind that changes in thermodynamic variables are defined as the value of the variable in the final state minus the value of the variable in the initial state. It will become readily apparent that, in most cases, it is more convenient to plot the changes rather than the absolute values of the thermodynamic variables. For instance, since $(\partial G^0/\partial T) = -S^0$, and S^0 is positive, G^0 rises with decreasing tempera-

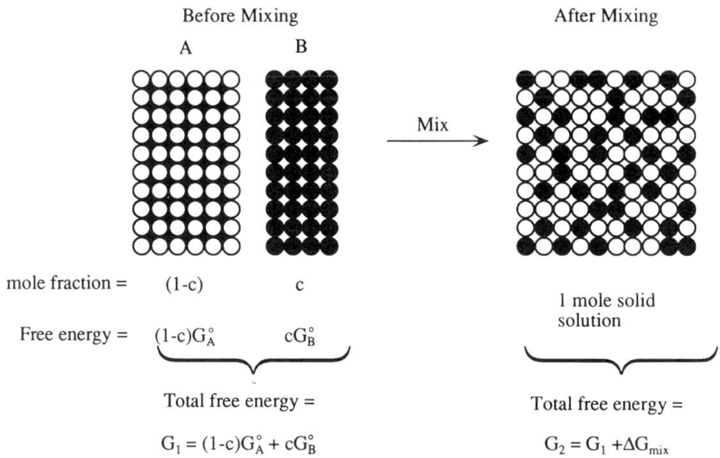

Figure 4-1. Free energy of mixing.

ture. However, ΔG_m only need to be plotted relative to zero at all temperatures.) Now, the entropy of mixing ΔS_m in a two-component system is given by

$$\Delta S_m = -R[(1-c)\ln(1-c) + c\ln c], \tag{4.3}$$

where R is the gas content. Since $c < 1$, the expression in the square brackets in Eq. (4.3) is a negative number, and hence ΔS_m is always a positive number. In other words, the entropy always increases upon mixing.

For ideal solutions, $\Delta H_m = 0$. Hence, ΔG_m is negative at all temperatures, and the resultant material will be a mixed solution of the molar formula $A_{(1-c)}B_c$.

For regular solutions, ΔS_m is given by Eq. (4.3) and $\Delta H_m \neq 0$. For nonregular solutions, $\Delta H_m \neq 0$, and ΔS_m is not symmetric about $c = 0.5$ as implied by Eq. (4.3). For simplicity, let us focus our attention on regular solutions. The heat of mixing for regular solutions, according to Hildebrand [1], is given by

$$\Delta H_m = \alpha c(1-c), \tag{4.4}$$

where α = the excess interaction energy $= NZ[E_{AB} - (E_{AA} - E_{BB})/2]$; N = Avogadro's number; Z = coordination number, and E's are the energies of the various bonds between the atoms. According to this model then, ΔH_m is also symmetric about $c = 0.5$. There are two possibilities depending on whether ΔH_m is negative or positive (which depends upon the sign of α).

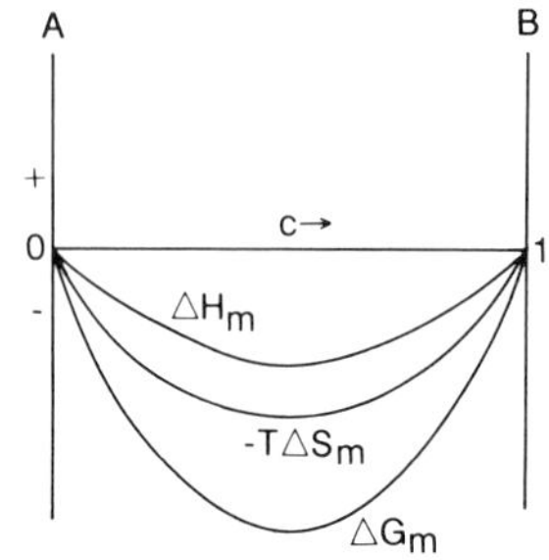

Figure 4-2. Free energy of mixing when the heat of mixing is negative.

In case I, when ΔH_m is negative, i.e., when the system gives off heat upon mixing (exothermic), ΔG_m is negative at all temperatures (Fig. 4-2). This implies that a mixed AB system will again be encouraged.

In Case II, when ΔH_m is positive, i.e., when mixing is endothermic, the shape of the ΔG_m variation with c depends upon the temperature, T. At very high temperatures, the term $-T\Delta S_m$ is greater than ΔH_m, and so ΔG_m is negative everywhere (Fig. 4-3). In other words, a system is always single-phased at sufficiently high temperatures. As the temperature is decreased to T_1, ΔG_m begins to look like a saddle (Fig. 4-4a). Although the mixed system has a lower ΔG_m than the unmixed end compositions A and B, it is clear that bet'''een points "a" and "b" (at which the common tangent representing lowest free energy touches the two free energy lobes), ΔG_m of the fully mixed system is higher than those of "a" and "b." In other words, *the free energy G_M of composition M lying between* "a" *and* "b" *can be lowered to $G_{M'}$ by separating into compositions* "a" *and* "b." The line joining "a" and "b" is

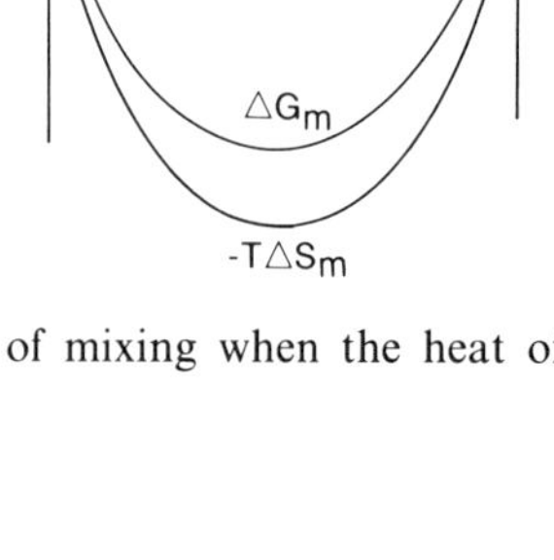

Figure 4-3. Free energy of mixing when the heat of mixing is positive, but the temperature is high.

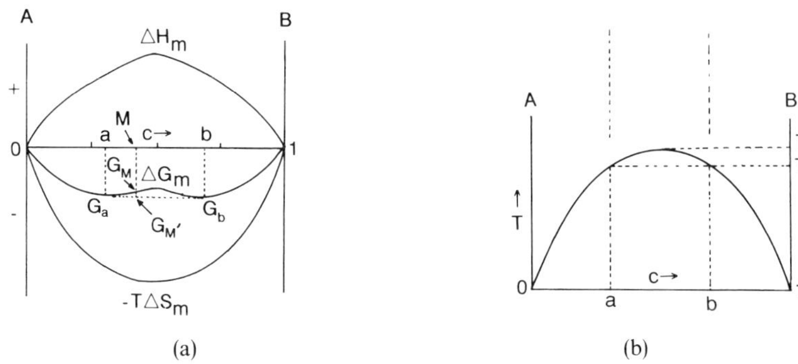

Figure 4-4. (a) Free energy of mixing when the heat of mixing is positive, and $T_1 < T_c$. (b) Phase diagram corresponding to Figs. 4-3 and 4-4a.

called a **tie-line**. In multicomponent systems, all compositions lying on the tie-line ab will tend to separate into the termini compositions "a" and "b." The amounts of "a" and "b" are determined by the lever rule: the amount of "a" times the distance a–M is equal to the amount of "b" times the distance b–M.

The corresponding phase equilibrium diagram is shown in Fig. 4-4b. Above the **critical temperature** T_c (also called the **upper consolute temperature**), the system is single-phased for all compositions. At T_1 ($<T_c$), the composition M splits into two phases of compositions "a" and "b."

Note that the discussion so far has made no assumptions regarding the kinetics of the separation process or the mechanisms by which it takes place. Atoms only have to possess sufficient mobility to separate over the experimental times. In a simple situation, M could simply be a liquid that undergoes demixing as it cools. Neither "a" nor "b" has to be crystalline; they merely have to be some compositions at which the free energy displays minima. *When both* "a" *and* "b" *are liquids, the phase separation process is termed "liquid-liquid immiscibility."* The immiscibility is termed **stable** or **metastable** depending upon whether it occurred above the liquidus or below the liquidus. Upon rapid cooling after phase separation, the material will be a composite of two glasses, "a" and "b." If M is nearer to "a," one would have to nucleate "b" to make a radical change in composition from M to "b." Since the amount of "a" will be larger than that of "b"—the microstructure would look like droplets of "b" in a matrix of "a." This is the case of the nucleated growth of phase separation. When there is sufficient fluidity in the system, the lighter phase may float upon the denser phase.

Lebedev and his Russian co-workers referred to compound "b" as a "crystallite" having a "distorted lattice," and not the microcrystal thought of by Randall, Rooksby, and Cooper. It may be understood that both "a" and "b" may have liquidlike structure of the type described by Zachariasen, yet the overall glass is compositionally inhomogeneous.

The locus of "a" and "b" on the phase diagram is called the **immiscibility dome phase boundary** or the **binodal** (the region inside is sometimes referred to as the **miscibility gap**). Within the immiscibility dome, there exists a spinodal dome that is the locus of points "d" and "e," which are inflection points on the G (or the ΔG_m) curve (Fig. 4-5). The inflection points have an interesting property. This can be seen as follows.

Let us define **G** as the total Gibbs free energy of a binary A–B system in which the free energy density (free energy per molecule) of a uniform solution of composition, c, is $G(c)$. We assume that $G(c)$ is a continuous function of the composition c (which is the mole fraction of B). For an infinitesimal fluctuation in composition, h, about the mean composition, $\bar{c}$, the free energy $G(c)$ can be expanded using a Taylor series as

$$G(c) = G(\bar{c} + h) = G(\bar{c}) + h[dG/dc] + [\tfrac{1}{2}!]h^2[(d^2G/dc^2)] + [\tfrac{1}{3}!]h^3[(d^3G/dc^3)] + \cdots \tag{4.5}$$

The total energy of the system may be written as

$$\mathbf{G} = N_v \int_V G(c)dv \tag{4.6}$$

where N_v is the number of molecules per unit volume. Since fluctuations must average to zero over the entire system, i.e.,

$$\int_V h\, dv = 0,$$

the change in the free energy of the system with fluctuations over a perfectly homogeneous system is given by (neglecting higher-order terms)

$$\Delta\mathbf{G} = \mathbf{G}(c) - \mathbf{G}(\bar{c}) = \tfrac{1}{2}(d^2G/dc^2) \int_V h^2\, dv. \tag{4.7}$$

Now, since h^2 is always positive, it can be seen that when $d^2\mathbf{G}/dc^2$ (= the curvature of the **G** vs. c curve) is positive, minor fluctuations in the composition M lead to an increase of the free energy, and hence such fluctuations die out. This is the situation when the composition M lies between a–d and e–b (see Fig. 4-5b). It can be seen that minor composition changes about the composition M lead to the increase of the free energy from G_M to $G_{M'}$. One clearly needs to nucleate a composition, p, far away such that the average energy would be lowered to $G_{M''}$. Regions a–d and e–b

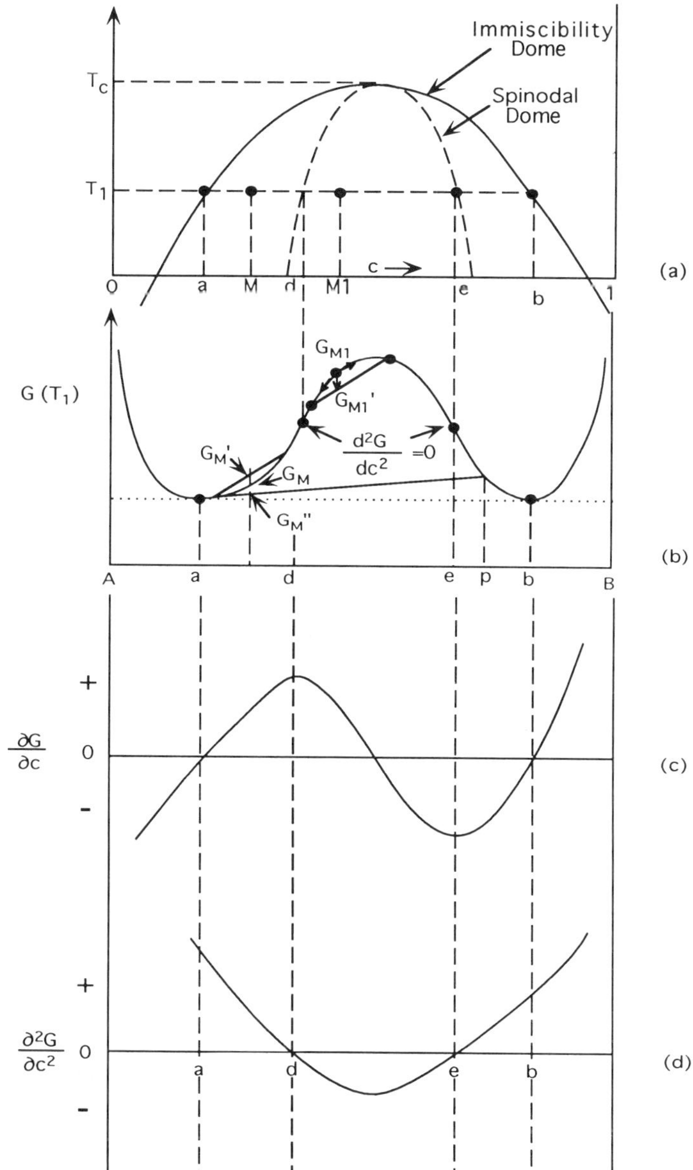

Figure 4-5. Details of the (a) phase diagram, (b) free energy of mixing at T_1, (c) the first derivative of the free energy at T_1, and (d) the second derivative of the free energy at T_1 when immiscibility occurs.

are thus regions of metastability with respect to compositional changes. Thermodynamics require M to separate to attain the lowest-energy state that is a mixture of a and b. However, because of the energy barrier, the kinetics may be sluggish. On the other hand, if $d^2\mathbf{G}/dc^2$ is negative, then $\Delta\mathbf{G} < 0$. This happens if the starting composition, M_1, lies within d–e (see Fig. 4-5a, d). It also means that any small fluctuation in composition M_1 leads to lowering of the free energy from G_{M1} to $G_{M1'}$, and hence such fluctuations grow with time. The system decomposes spinodally. Since composition fluctuations are always present because of thermal fluctuations, there is no need for a nucleated growth of the phase separation. Thus, *the inflection points* "d" *and* "e," *at which* $d^2\mathbf{G}/dc^2 = 0$, *mark the boundary of instability with respect to small fluctuations*, and hence the locus of "d" and "e" is termed the **chemical spinodal** (or simply the **spinodal**), and the points "d" and "e" are called **spinodes**. Within the spinodal, the fluctuation that grows may itself decompose further depending upon whether its composition is within the inflection points or outside. Given sufficient time at temperature T_1, the compositions of the separated phases will gradually change until the

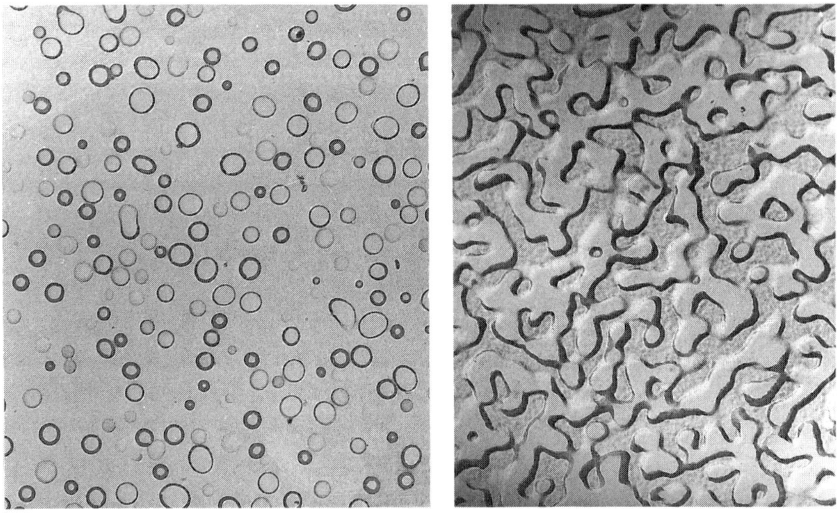

(a) (b)

Figure 4-6. Microstructure of (a) soda lime silica glass heated at 740°C for 7.25 h (14,000 ×) showing nucleated droplet phase separation; (b) Vycor glass heated at 700°C for 5.5 h (24,000 ×) showing spinodal decomposition. (After H. Rawson, *Properties and Applications of Glass*, Fig. 10. Elsevier Science Publishers, Amsterdam, Netherlands, 1980.) Reproduced with permission of the Publisher. Photographs Courtesy of H. Rawson, University of Sheffield, Sheffield.

Table 4-1. Distinction between Phase Separation Mechanisms[a]

Nucleation and growth	Spinodal decomposition
Invariance of second-phase composition with time at constant temperature	Continuous variation of both extremes in composition with time until equilibrium compositions are reached
Interface between phases is always same degree of sharpness during growth	Interface between phases initially is very diffuse, eventually sharpens
Tendency for random distributions of particle sizes and positions in matrix	Regularity of second-phase distribution in size and position characterized by a geometric spacing
Tendency for separation of second-phase spherical particles with low connectivity	Tendency for separation of second-phase, nonspherical particles with high connectivity

[a] After J. W. Cahn and R. J. Charles, Phys. Chem. Glasses, 6(5), 181–191 (1965).

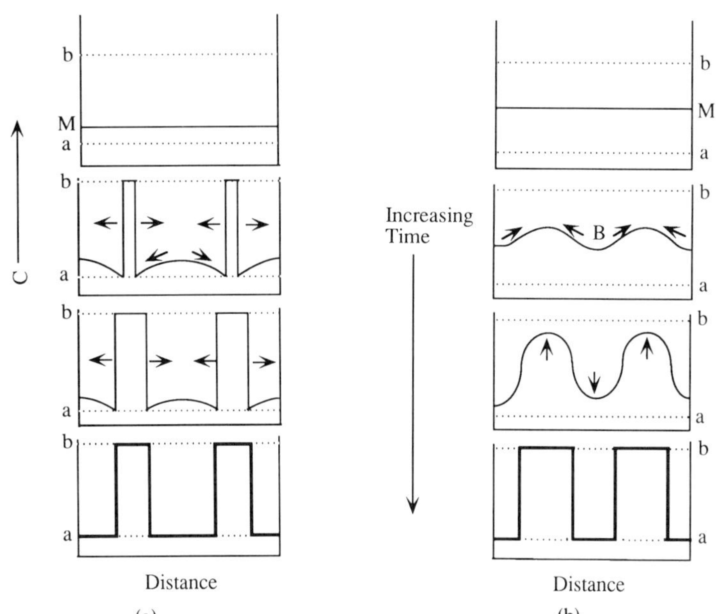

Figure 4-7. Growth of the phase separation for (a) nucleated-type and (b) spinodal-type mechanisms with time at a given temperature. Note that in the nucleated-type mechanism, the droplets have a fixed composition "b" with an interfacial composition "a." The growth is in the physical size of the droplet. In the spinodal-type mechanism, a small fluctuation in composition gradually grows over a period of time via "uphill" diffusion. The terminal compositions in each case are "a" and "b." The ratio of the two phases depends upon the starting composition.

equilibrium compositions "a" and "b" have been reached. Again, the separated phases do not possess crystalline order. Figure 4-6 contrasts the microstructure of a spinodally decomposed glass from that of a nucleated phase-separated glass. *The primary difference from the nucleated phase separation is the presence of interconnected, wormlike morphology.* This distinction between the microstructures, however, is likely to decrease as the time allowed for phase separation increases.

Table 4-1 lists the differences between the characteristics of nucleated separation and those of spinodal decomposition. Figure 4-7 schematically shows the development of nucleated separation and spinodal decomposition. Note how, in the case of the spinodal decomposition, growth occurs by some material moving against its own concentration gradient—a phenomenon called uphill diffusion. The diffusion coefficient, D, is proportional to $d^2\mathbf{G}/dc^2$, and so wherever the curvature is negative, D is negative (i.e., the material moves up the concentration gradient). The theory of spinodal decomposition was developed by Cahn [2], Cahn and Hilliard [3], and Hillert [4] for metals, and was later extended to ceramic systems by Cahn and Charles [5].

4.2 More Formal Analysis of Phase Separation

We should not have used the Gibbs free energy G in our analysis. This is because minimization of G is not an appropriate criterion when considering a system where the pressure may change. We assumed isobaric conditions. However, when considering nucleation, the surface energy can change the internal pressure of the nucleus. Hence, the work of nucleation is not given by considering changes in the Gibbs potential. The proper energy potential to minimize is Helmholtz free energy, $F = U - TS$ (U = the internal energy), which assumes isothermal and isochoric (isovolumic) conditions and allows the pressure to vary. (Changes in Gibbs potential at constant P and T give a system capability to do work *other than* volume work, whereas the changes in Helmholtz potential at constant T give the system capability to do any form of work *including* volume work.) In condensed systems, the correction due to the pressure terms may be small, and hence the Gibbs and Helmholtz free energy curves do not differ significantly. Calculations involving G are preferred over those involving F primarily because enthalpy data are more readily available than the internal energy data. Nonetheless, if we pursue Cahn's analysis and write the expression for $\mathbf{F}(c)$ in an isotropic, incompressible system under isothermal conditions (see Ref. 3, and the footnote on

p. 500 in Ref. 6), we obtain

$$\mathbf{F}(c) = \int_V [F(c) + \omega(\nabla c)^2]dv, \tag{4.8}$$

where $F(c)$ is the Helmholtz free energy per unit volume (no need to include N_v), and $\omega(\nabla c)^2$ represents the contribution to the energy due to the spatial composition gradient, which in turn gives rise to surface tension. ω is positive for phase-separating systems; otherwise, we will have negative surface tension. As before,

$$\Delta\mathbf{F}(c) = \mathbf{F}(c) - \mathbf{F}(\bar{c}) = \int_V \{\tfrac{1}{2}(d^2F/dc^2)h^2 + \omega(\nabla c)^2\}dv. \tag{4.9}$$

One now writes the composition fluctuation, h, as a sum of Fourier components of wavelength λ. Because of the orthogonality of the Fourier terms, their contribution to $\Delta\mathbf{F}(c)$ is the sum of the contributions from individual components. For one such component in one dimension, let us write $h = c - \bar{c} = A\cos(2\pi x/\lambda)$, where x is the length scale. Then $dc/dx = -[2\pi A/\lambda]\sin(2\pi x/\lambda)$. If volume $V = K\lambda$, where K = cross-sectional area, then substituting in (4.9) and integrating with respect to x from 0 to λ, we get

$$\Delta\mathbf{F}(c) = [VA^2/4][(d^2F/dc^2) + 8\omega\pi^2/\lambda^2]. \tag{4.10}$$

This shows that when $d^2\mathbf{F}/dc^2 > 0$, the solution is stable with respect to minor fluctuations in composition. However, when $d^2\mathbf{F}/dc^2$ is negative, then the system is unstable with respect to composition fluctuations that have wavelengths greater than the critical wavelength λ_c, given by

$$\lambda_c = [-8\pi^2\omega/(d^2F/dc^2)]^{1/2}. \tag{4.11}$$

The effect of the surface energy contribution is thus to prevent the system from decomposing into too fine a scale. Although there is no thermodynamic barrier to phase separation for fluctuations whose size is above a certain minimum wavelength, the decomposition remains diffusion-controlled.

To consider an added complexity, we may now allow the system to develop elastic strain energy as well (because the molar volume of the growing nucleus may be different from the space it grew in). Strictly speaking, because Helmholtz free energy minimization assumes isochoric (isovolumic) conditions, it is not an appropriate criterion for handling this situation. Cahn assumed isochoric conditions by imagining spaces in one phase to be occupied by the second phase, and calculating the elastic energy necessary to do so. For such an assumption, the Helmholtz free energy, Eq. (4.8), must be modified. This is done for the case of a coherent strain by introducing

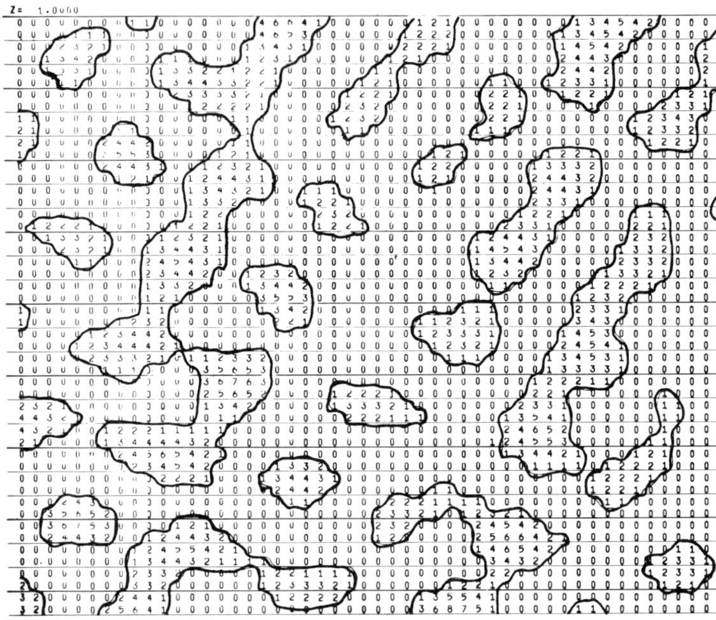

Figure 4-8. Computer simulation of a spinodal structure in an isotropic material obtained by adding 100 random sine waves of wavelength λ. (Photograph courtesy of R. J. Charles, General Electric Co. Schenectady, New York.)

the elastic energy per unit volume, $h^2\beta^2E/(1 - \nu)$, where β is the linear dilatational coefficient per unit compositional change, as a third term on the right-hand side in (4.8); E and ν are the Young's modulus and the Poisson ratio, respectively. The changes in the energy diagrams are significant when β is significantly large and when volume relaxational times are larger than short-distance diffusional times.

A computer simulation by Cahn for the case of negligible elastic strain, but including the gradient energy term (i.e., using Eq. (4.8)), for $c = 0.5$ with 100 random sine waves having wavelength λ is shown in Fig. 4-8. The points in the plot show regions where $c > 0.5$. Cahn pointed out that a high degree of interconnectivity of phases could exist for a wide range of composition ($0.15 < c < 0.85$) when decomposition occurs spinodally. Although there is a possibility that some interconnectivity between nucleated particles could occur during late stages of growth, in general the interconnectivity factor in the morphology distinguishes the two mechanisms. There have been suggestions that wormlike structures do not represent true spinodal decomposition, and that the spinodal microstructure must consist of quite regularly spaced phases, as shown in Fig. 4-9. Such suggestions are incorrect.

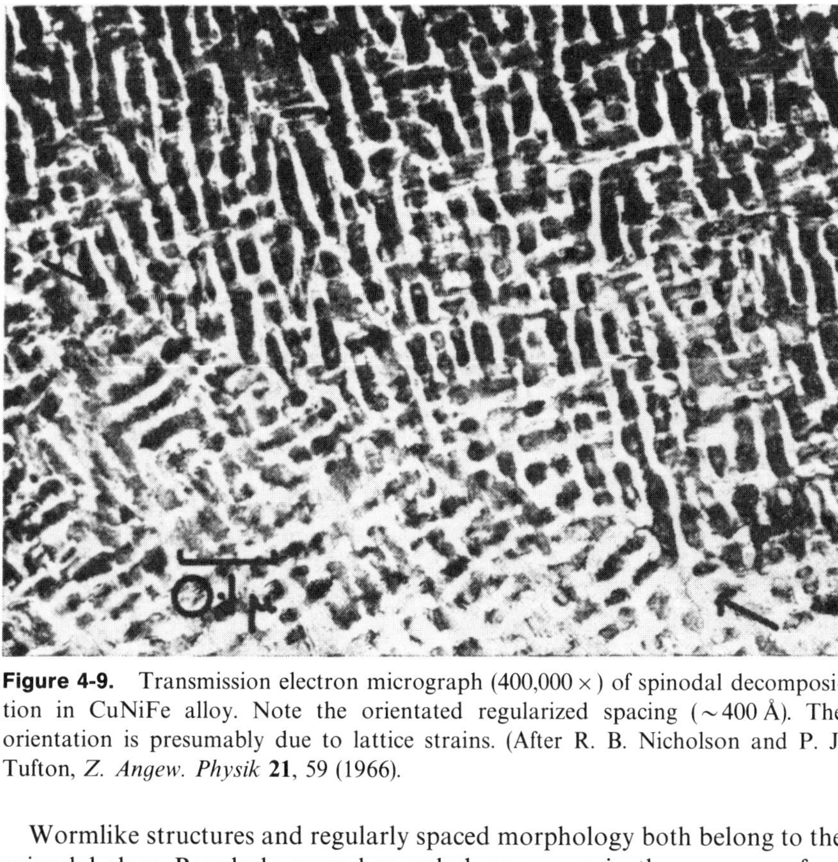

Figure 4-9. Transmission electron micrograph (400,000 ×) of spinodal decomposition in CuNiFe alloy. Note the orientated regularized spacing (~400 Å). The orientation is presumably due to lattice strains. (After R. B. Nicholson and P. J. Tufton, *Z. Angew. Physik* **21**, 59 (1966).

Wormlike structures and regularly spaced morphology both belong to the spinodal class. Regularly spaced morphology occurs in the presence of an elastic strain energy. In most glasses, volume relaxation of the elastic strain energy occurs faster than the diffusion kinetics involved in the spinodal decomposition process, and hence cases of regularly spaced morphology in glasses may be rare. (Are there any?) On the other hand, because of a lack of volume fluidity in most metals during spinodal decomposition, the unrelaxed elastic strain energy frequently causes the appearance of a regularly spaced morphology (Fig. 4-9).

4.3. Electron Microscopy to Observe Phase Separation in Glass

Because glass is an insulator of charge, examination of glass using direct transmission of high-energy electrons is not recommended generally. Overheating of the sample and mobility of alkali ions under the influence of

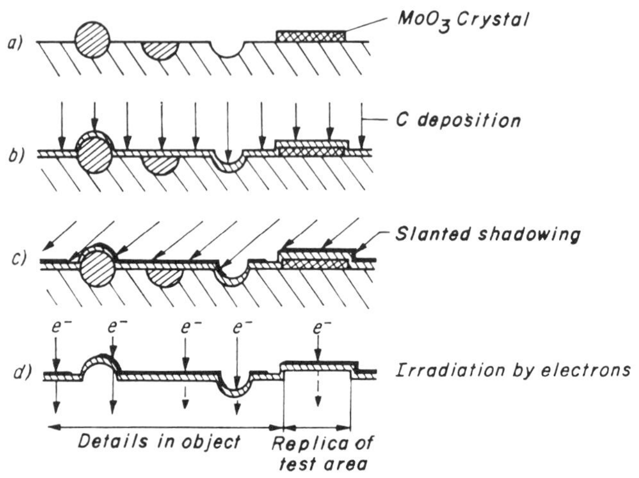

Figure 4-10. Modified carbon replica method using MoO_3 crystal test plane. (After W. Skatulla and L. Horn, *Exp. Techn. Phys.* **8**, 1 (1960).

electric field [7], which the penetrating electrons set up, make it quite difficult to isolate fact from artifact. Also, occasionally the electrons can induce reactions within the glass (for instance, PbO in glass can be reduced to metallic Pb).

The best way to carry out electron microscopy of glass is to perform carbon replication of a freshly fractured surface. Figure 4-10 shows the schematics of a modified technique according to Skatulla and Horn [8]. After obtaining a fractured surface of glass in vacuum, a tiny plate of MoO_3 is placed on this surface, and about 300 Å of carbon film is vacuum-deposited vertically. The purpose of the MoO_3 is to provide a smooth surface for comparison when extremely fine granularity of the microstructure is to be distinguished from, for instance, photographic film granularity. Following the carbon deposition, a Pt–Ir film is deposited at a slant. Subsequent immersion in water or very dilute HF allows the film to float off and be picked up on a copper grid. Occasionally, a 15s etching in 2% HF of the freshly fractured glass surface is carried out to develop the texture. Again, because of the corrosiveness of water on glass, sometimes the developed texture cannot be relied upon as being the true microstructure of glass. Vogel suggests that the most reliable procedure is to eliminate any etching and to perform all the steps starting from fracturing up to replication directly in high vacuum in one sequence without ever exposing the fractured surface of glass to laboratory air. A typical example of such a practice is shown in

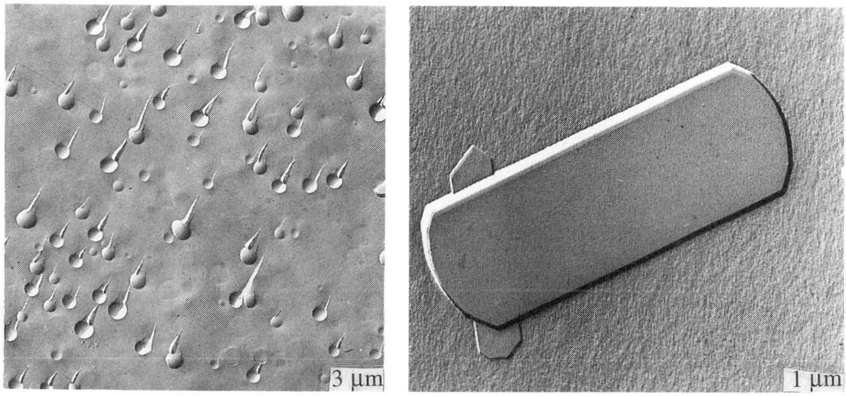

(a) (b)

Figure 4-11. (a) Replica showing droplet- shaped immiscibility regions in a fracture plane. Some droplets are pits with dark shadow on one side, others are hills with the shadow on the other side, and some are fractured through the middle. (After W. Vogel, *Chemistry of Glass*, Fig. 5.25, p. 65. Amer. Ceram. Soc., Columbus, Ohio, 1985. (b) Replica of a weakly immiscible region. (Compare with the texture of the two MoO_3 test planes.) (After W. Vogel, *Chemistry of Glass*, Fig. 5.26, p. 65. Amer. Ceram. Soc., Columbus, Ohio, 1985. Photographs Courtesy of W. Vogel, Otto Schott Institute, Jena.)

Fig. 4-11, where one notices a near-random distribution of droplets and holes along with fracture tails.

4.4. Observations of Phase Separation in Glass

It should be apparent that phase separation phenomena in glass are a fundamental consequence of the thermodynamics of two-component and multicomponent melts. Hence, observations of phase separation phenomena should be common. Indeed, the entire book by Mazurin and Porai-Koshits [9], and a large portion of Vogel's book [10], describe phase separation observations in glass. One really ought to wonder under what circumstances a glass *does not* phase-separate! The basic rules to obtain a less separated glass is to cool rapidly from a single-phased liquid region, or to add a large number of components to increase the entropy term so that mixing is encouraged. Most binaries and ternaries involving SiO_2, B_2O_3, and GeO_2 show a strong tendency to phase-separate. Apparently, some silicates containing larger cations such as K^+, Rb^+, and Cs^+ show a decreased tendency.

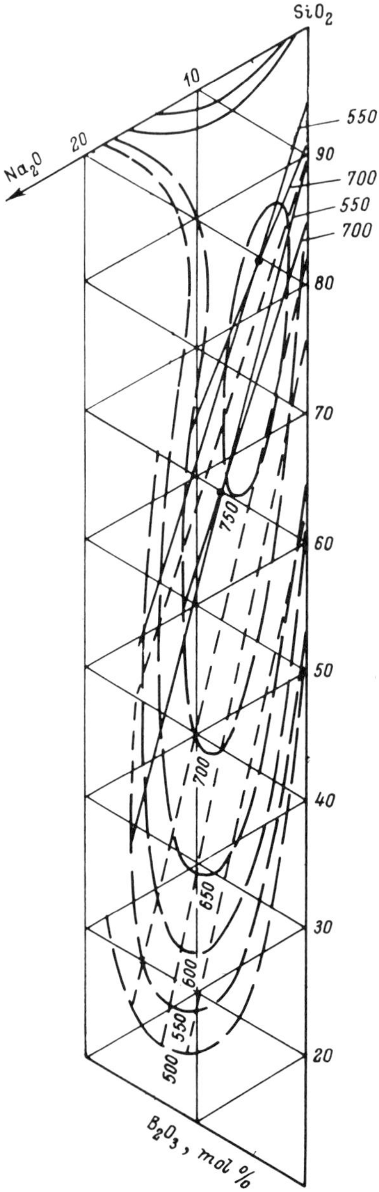

(a)

Figure 4-12. (a) Tie-line directions of immiscibility in the Na_2O–B_2O_3–SiO_2 system. (After O. V. Mazurin, *Phase Separation in Glass*, Fig. 59. 155. Elsevier Science Publishing Co., Amsterdam, Netherlands, 1984.)

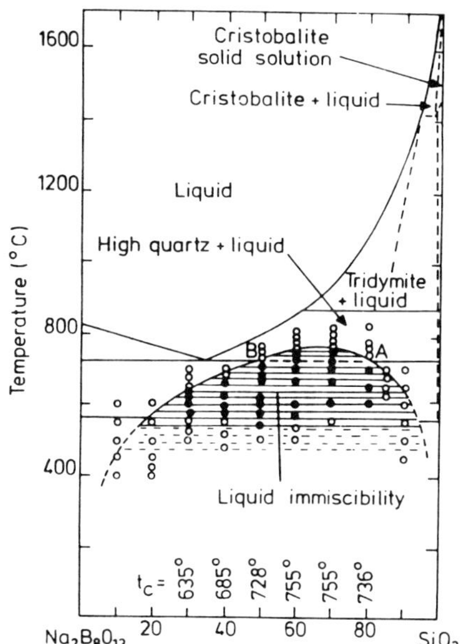

(b)

Figure 4-12. (b) Phase diagram of a part of the $Na_2O{\cdot}4B_2O_3$–SiO_2 system. (After T. J. Rockett, W. R. Foster, and R. G. Ferguson, *J. Amer. Ceram. Soc.* **48**, 329 (1965). Reproduced with permission of the American Ceramic Society.)

The observations of phase separation in the Na_2O–B_2O_3–SiO_2 glass system have been one of the most important and the most thoroughly investigated [11]. Shown in Fig. 4-12 is the ternary phase diagram where temperature cuts through the immiscibility dome are shown as elliptical lines. (Shown in Fig. 4-12b is the observation that the dome itself is below the liquidus. In other words, the liquid immiscibility in this system is metastable.) Note that the major axis is lined up roughly along the 16%$Na_2O{\cdot}$84%-B_2O_3–100%SiO_2 tie line. (One may be cautioned that the tie line compositions on the silica-rich side are approximate only.) A specimen of glass on the B_2O_3-rich side of the dome, when heat-treated, develops a droplet structure (Fig. 4-13a). On the other hand, a composition near the center of the dome displays spinodal decomposition even when initially cooled (Fig. 4-13b). As shown in Fig. 4-14, the two phases turn out to be a Na_2O–B_2O_3 rich phase and a SiO_2-rich phase containing roughly 96% SiO_2. This decomposition was exploited by Corning Glass Works as an interesting technology known as the **Vycor process**. Compositions within the field 55–75%SiO_2, 20–35%B_2O_3, 5–10%Na_2O are melted around 1,500°C and formed into glass products by conventional pressing, blowing, or drawing

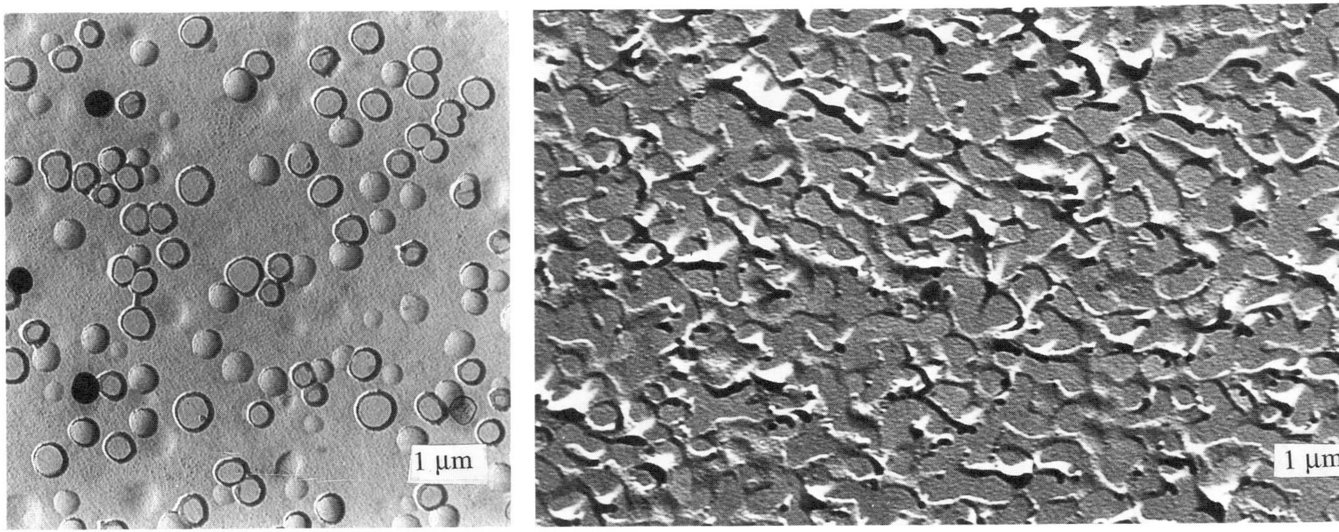

(a) (b)

Figure 4-13. (a) Sodium borosilicate glass from the left portion of the immiscibility region in Fig. 4-12b after reheating. Note the nucleated type of phase separation. (After W. Vogel, *Chemistry of Glass*, Fig. 7.21, p. 111. Amer. Ceram. Soc., Columbus, Ohio, 1985.) (b) Sodium borosilicate glass from the center of the immiscibility region in Fig. 4-12b. Note the separation by spinodal decomposition. (After W. Vogel, *Chemistry of Glass*, Fig. 7.22, p. 111. Amer. Ceram. Soc., Columbus, Ohio, 1985. Photograph Courtesy of W. Vogel, Otto Schott Institute, Jena.)

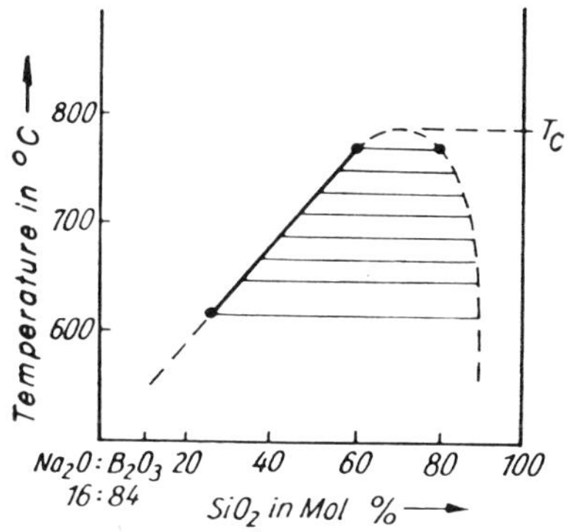

Figure 4-14. Asymmetric immiscibility dome in the Na_2O–B_2O_3–SiO_2 system. (After W. Vogel, *Chemistry of Glass*, Fig. 7.24, p. 112. Amer. Ceram. Soc., Columbus, Ohio, 1985. Photograph Courtesy of W. Vogel, Otto Schott Institute, Jena.)

techniques. They are then heat-treated at around 500–600°C to develop compositions approaching immiscibility boundaries and having an interconnecting microstructure. The glass at this point becomes opalescent. It is then immersed in 3N H_2SO_4 at about 90°C, which dissolves out the continuous sodium borate phase at the rate of 0.8–1.5 mm/h and leaves a 96% silica skeleton behind. At this point, the pore diameter is about 20–50 Å, and the glass has roughly 25–40% total porosity. This porous "thirsty" glass can be used as a support for catalysts, for encapsulating nuclear waste products, or for time-release glass capsules (for instance, in biological and agricultural engineering, where drugs and soil nutrients would be released through the micropores over a designed life of the capsule). Upon heat-treating the porous object at around 1,100°C, a clear compacted 96% silica glass (with some residual porosity), known as Vycor, results. Vycor glasses are used as substitutes for the more expensive pure silica products where optical transmission is not a concern.

Laboratory chemical glassware of the Pyrex type is made using the sodium borosilicate compositions near the right edge of the immiscibility dome shown in Fig. 4-14. The products are cooled rapidly enough to keep the droplet phase to no greater than 20–50 Å size. The fine droplet structure allows a high chemical resistance of the final product.

The observations [12] of phase separation in BaO–B_2O_3–SiO_2 glass are also interesting, Figure 4-15 shows the ternary phase diagram (5) where temperature has been plotted on the vertical axis. The immiscibility dome stretches along the B_2O_3–SiO_2 binary with the locus of the upper consolute at about 18% BaO. Figure 4-16 shows the development of as many as six different microphases because of a stepwise unmixing. The initial

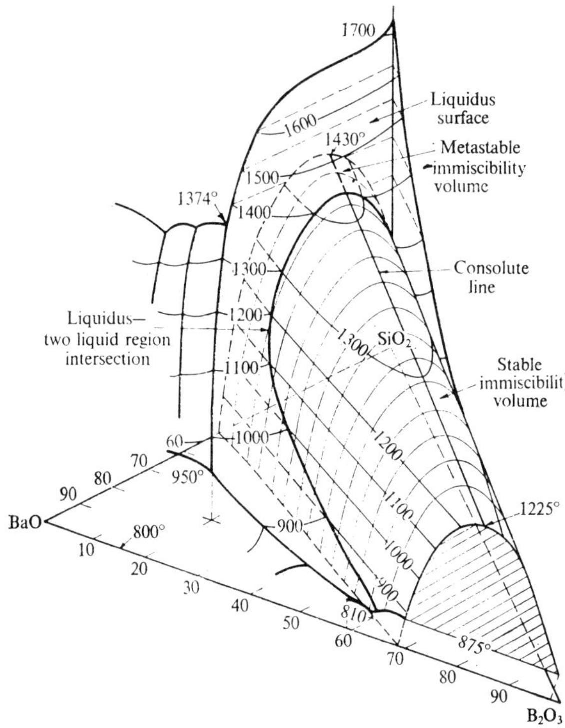

Figure 4.15. Space model showing the liquidus/two-liquids region intersection for the system $BaO–B_2O_3–SiO_2$. (After J. W. Cahn and R. J. Charles, *Phys. Chem. Glasses* **6**(5), 181 (1965). Reproduced with permission of The Society of Glass Technology.)

composition of the melt is slightly to one side of the upper consolute. The first heat treatment develops droplets, Tr_1, in matrix, M_1; the second treatment at a slightly lower temperature causes the matrix to further split into droplets, Tr_2, and matrix, M_2. In the next treatment at a lower holding temperature, Tr_1 droplets show separation to Tr_4 and balance matrix droplets, Tr_3. In this way, sequential separation could be identified over five heat treatments.

4.5. Observations of Controlled Crystallization in Glass (Glass–Ceramics)

In Chapter 3 our discussion on crystallization focussed on the normal sequence of crystallization from a supercooled melt. Bulk nucleation could

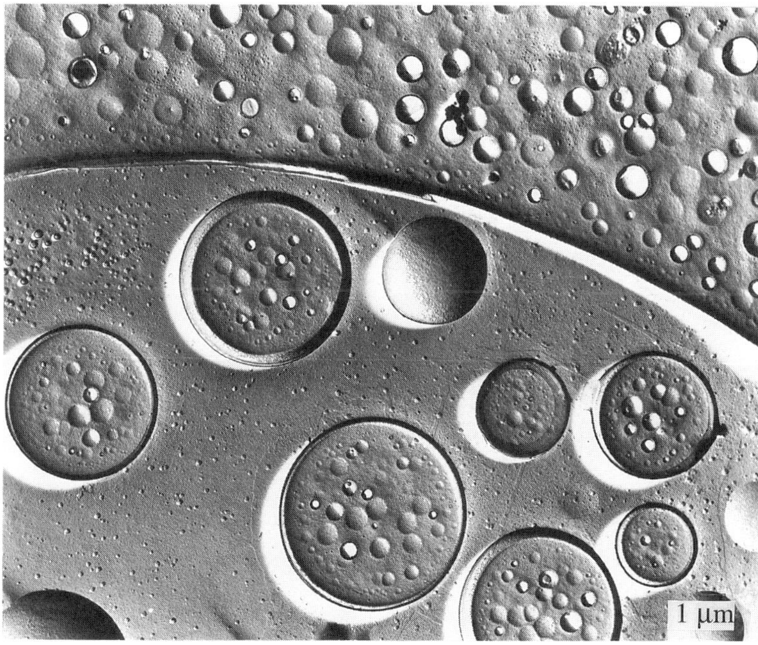

(a)

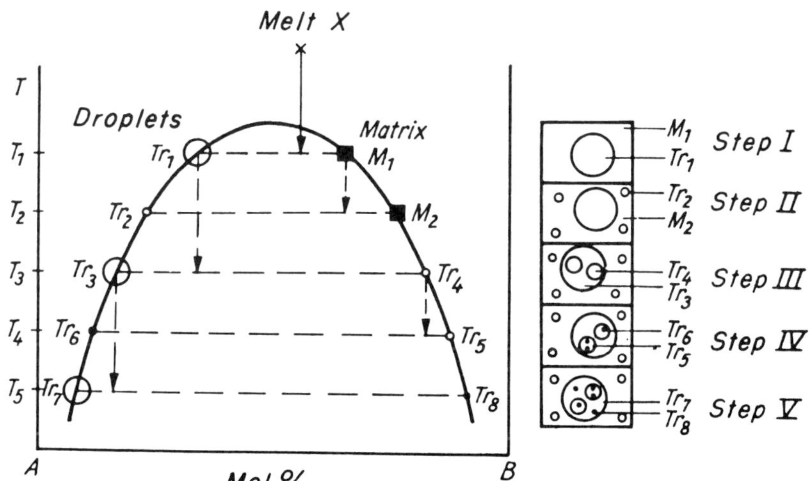

(b)

Figure 4.16. Stepwise immiscibility in a barium borosilicate glass. (a) Replica electron micrograph. (b) Suggested explanation. (After Vogel, Ref. 10. Photograph Courtesy of W. Vogel, Otto Schott Institute, Jena.)

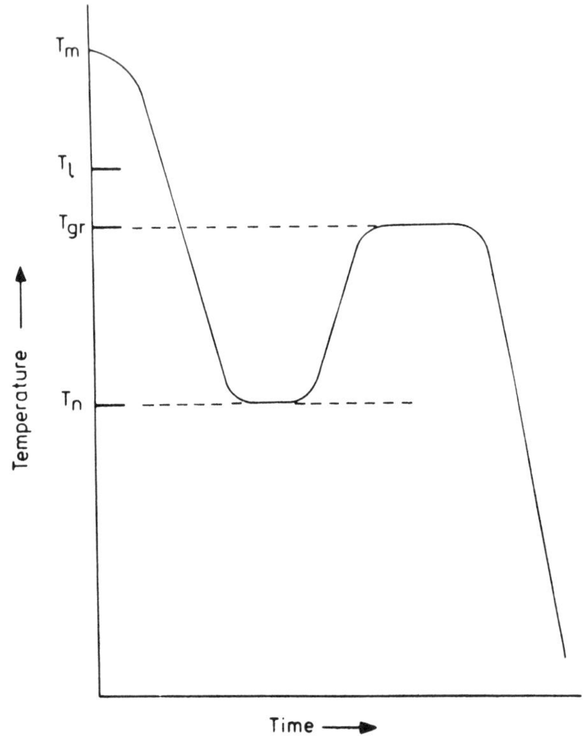

Figure 4-17. A typical heat treatment schedule for the glass–ceramic process. T_m = the melting temperature; T_ℓ = the liquidus temperature; T_n = temperature corresponding to maximum nucleation rate; T_{gr} = maximum crystal growth rate temperature.

occur at random times and at random locations (including the surface). As a consequence, crystal growth would be uncontrolled. The mechanical properties of such glass–crystal composites are generally poor. On the other hand, when a glass is given a particular nucleation treatment to confine the nucleation to within a certain time interval, and such nuclei are allowed to grow till they reach a desired size (for instance, by following a schedule as shown in Fig. 4-17), then the resulting composite can have valuable thermal and mechanical properties. Products of this type are termed glass–ceramics, and instead of "devitrification" or "crystallization", one uses the term **ceramming** of glass. Often the ceramming process might be carried out to virtually 99%+ ceramic with very little residual glass layer left. Common oven-to-table Corningware and Macor (a machinable glass–ceramic) and Schott's Zerodur are good examples of commercial glass–ceramics.

The key to making a successful glass–ceramic is controlled bulk nucleation. For good mechanical properties, the crystal size ought to be around a micron, and hence the density of nucleation sites must be very high (about $10^{12}/cm^3$). The system may have naturally occurring immiscibility where the separating phases are finely sized (microphases), or this could be enhanced by the addition of low-solubility nucleating agents in the glass batch. Among the common nucleating agents are TiO_2, ZrO_2, fluorides, phosphates, Ag, Au, etc. In effect, the microphase separation starts out as homogeneous nucleation sites, which could be induced to convert to crystals. Often the microphases soon become sites for heterogeneous nucleation for other crystalline phases to grow (epitaxial growth, see Section 3.3.1).

Some typical glass–ceramic compositions are shown in Table 4-2, along with their properties before and after ceramming. Glass–ceramics are generally much stronger than ceramics made using conventional fabrication techniques, primarily because of a very uniform particle size distribution and near-zero porosity. They also have a higher fracture toughness relative to the parent primarily because of deflection of fracture fronts around the crystalline particles. Because the product is essentially a composite of glass and ceramic, the thermal expansion and softening behavior are different from the parent. (They are essentially the volume average of the residual glass and crystal properties.) In some systems, the thermal expansion can be tailored by stopping the extent of crystallization at some stage. For instance, in Li_2O–Al_2O_3–SiO_2-based glass–ceramics, the relative ratio of β-eucryptite (strongly negative expansion coefficient, α) or β-spodumene (slightly positive α) or α-quartz (highly positive) determines the final expansion coefficient of the product. Thus, some of the commercial applications of glass–ceramics such as cookware and missile cones utilize the high strength and high toughness relative to the parent glass, and a near-zero expansion coefficient of the final product allowing it to be very resistant to thermal shock. If the crystal size is controlled to a few hundred angstroms, and if the refractive indices of the crystals and the matrix were well matched, then a transparent glass–ceramic, such as Corning's Visionware, can be obtained.

The expansion tailoring also provides an interesting opportunity in glass sealing technology. Many of the "devitrifiable glass frits" used in electronic applications utilize the flow properties of the glass to achieve bonding. The subsequent change to glass–ceramic form achieves a proper thermal expansion match to the substrate, a high strength, and a greatly increased upper service temperature. Nearly all television tubes incorporate sealing of the face panel to the back funnel using a devitrifiable frit. The sealing is carried out at temperatures as low as about 400°C, which allows the glass parts to retain the optics. After sealing and "ceramming," the tubes may be vacuum-baked at temperatures as high as 500°C to clean the inside surface.

Table 4-2. Some Glass-Ceramic Compositions and Their Properties before and after Ceramming[a]

Weight percentage composition[b]					Thermal expansion coefficients × 10^7/°C (20–400°C)		Dilatomeric softening temperatures (°C)	
SiO_2	Li_2O	MgO	Al_2O_3	K_2O	Glass	Glass-ceramic	Glass	Glass-ceramic
81.0	12.5	—	4.0	2.5	84.3	113.0	540	910
77.5	12.5	—	10.0	—	78.0	50.0	530	930
62.1	1.9	17.6	18.4	—	42.0	50.3	790	1,000
66.4	10.0	3.0	20.6	—	63.4	0.7	640	1,000
60.2	8.5	2.8	28.5	—	60.5	−42.4	720	1,000
76.7	14.7	8.6	—	—	92.8	92.5	525	890
62.4	12.4	25.2	—	—	78.8	74.8	600	930

[a] After P. W. McMillan, *Glass-Ceramics*, p. 99. Academic Press, London 1964.
[b] The glasses also contain a small proportion of P_2O_5 to act as a nucleation catalyst.

Photogrey glasses utilize light-induced, reversible bulk crystallization of very fine Ag particles. In some compositions, exposure to UV light causes irreversible crystallization of silver particles. These photosensitive glasses have found use in optical circuits and art impression.

Glass–ceramics containing apatite resemble bone substance in composition, and hence have found use for bone prostheses and dental implants. Body tissue can grow within the glass–ceramic and form a natural bond with a prosthesis. This is not possible with a traditional metal prosthesis.

The glass–ceramic process has also been used [13] to develop high-temperature superconductors in the $YBa_2Cu_3O_{6+\delta}$ and the $Bi_2Sr_2CaCu_2O_8$ systems. The process would allow the use of a host of continuous glass-forming processes to form bulk superconductor products that have a high mechanical strength (the lack of which has been one of the primary reasons for the absence of commercially useful bulk superconductors) and a high resistance to atmospheric degradation because of the residual glass layer acting as a diffusion barrier (the other reason for the absence of useful bulk products).

Exercises

1. Draw a tentative V–T diagram of a spinodally phase separated mass that is cooled into a glassy state. (Note that liquid immiscibility may occur at temperatures beyond T_m.)

2. The phase diagram for A–B is as shown (Fig. Q4-2). Consider the composition A_3B_2, which is being cooled from a high temperature. Sketch its microstructure qualitatively, and identify the compositions when an initially single-phased specimen is heated rapidly from room temperature to

(i) 1,000°C for a short time and quenched,
(ii) 800°C for a long time and quenched,
(iii) 700°C for a long time and quenched,
(iv) 700°C for a short time and quenched,
(v) 550°C for a long time and quenched,
(vi) 550°C for a short time and quenched.

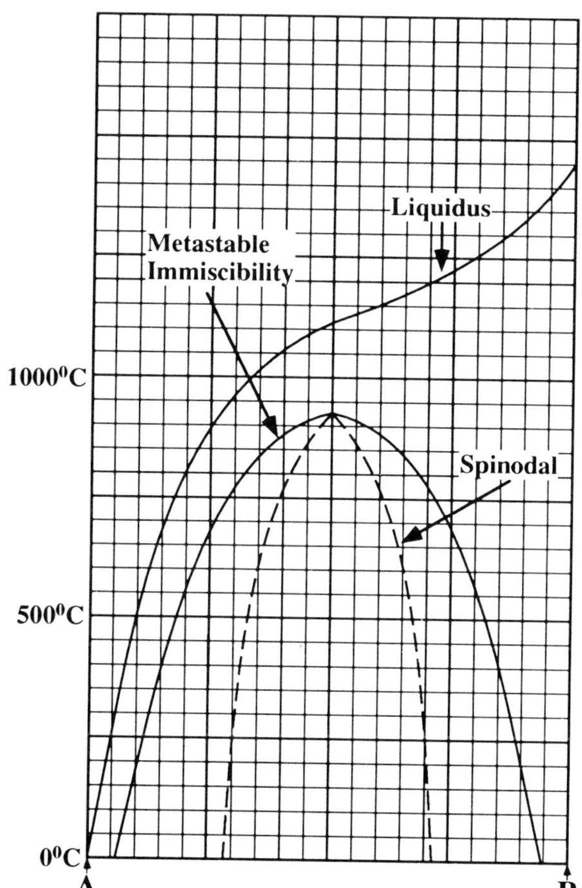

Figure Q4-2.

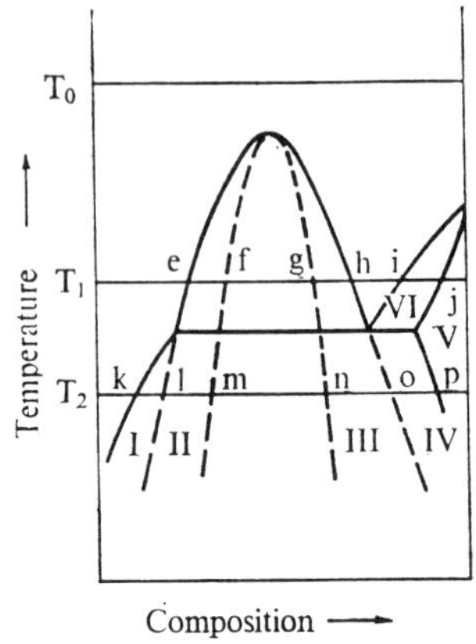

Figure Q4-3.

3. For the phase diagram shown (Fig. Q4-3), sketch the free energy vs. composition diagrams at temperatures T_0, T_1, and T_2. (Answer: See Cahn and Charles. Ref. 5)

References

1. J. H. Hildebrand, *J. Am. Chem. Soc.* **51**, 66 (1929).
2. J. W. Cahn, *Trans. Met. Soc. AIME* **242**, 166–180 (1960).
3. J. W. Cahn and J. E. Hilliard, *J. Chem. Phys.* **31**(3), 688–699 (1959).
4. M. Hillert, *Acta. Met.* **9**, 525 (1961).
5. J. W. Cahn and R. J. Charles, *Phys. Chem. Glasses* **6**(5), 181–191 (1965).
6. J. E. Hilliard, in *Phase Transformations*. Ed. H. I. Aaronson, *American Society for Metals*, Metals Park, OH, (1970), ASM, 1970.
7. A. K. Varshneya, A. R. Cooper, and M. Cable, *J. Appl. Phys.* **37**(5), 2199 (1966).
8. W. Skatulla and L. Horn, *Exp. Techn. Physik* **8**, 1–9 (1960).
9. O. V. Mazurin and E. A. Porai-Koshits, Eds., *Phase Separation in Glass*. North-Holland, Amsterdam, 1984.
10. W. Vogel, *Chemistry of Glass*. Amer. Ceram. Soc., 1985.
11. See, for example, Ref. 10, p. 113; T. H. Elmer and M. E. Nordberg, *J. Am. Ceram. Soc.* **41**, 517–520 (1958); and W. Vogel, *Bull. of C.G.C.R.I. India* **2–3**, 95–107 (1975).
12. W. Vogel, W. Schmidt, and L. Horn, *Z. Chem.* **9**, 401–440 (1969).
13. A. Bhargava, A. K. Varshneya, and R. L. Snyder, *Mat. Lett.* **8**(1–2), 41–45 (1989); *Ibid.*, **11**(10–12), 313–316 (1991).

Chapter 5

Glass Compositions and Structures

Having looked at the glass-forming criteria and the morphological features in glasses, we will now look into the atomic configurations that are thought to be present in various glass compositions of scientific and commercial interest. It should be clearly understood that the proposed configurations are those believed generally. Occasional challenges to such beliefs do exist.

The chapter is introduced by first discussing the various ways in which glass formulae are presented. In drawing the atomic configurations, two schemes are used: (i) bonded open and solid circles with missing third dimension, and (ii) chemical symbols with the appropriate number of bonds to sketch a 2-D projection (a 'pancaked' version). The second scheme is preferred. The reader is encouraged to practice the drawing of glass network configurations: one picture is worth a thousand words!

5.1. Presentation of Glass Formulae

Glass formulations may be presented on a mol %, wt. %, mole fraction, mole formula, or weight fraction basis. In oxide glasses, the convention is to list the glass network modifiers in increasing valency order ending with glass network formers. For instance, a glass formed by melting one mole of CaO, one mole of K_2O, and five moles of SiO_2 together is written in the mole formula presentation as $K_2O{\cdot}CaO{\cdot}5SiO_2$.

To convert this to a mole fraction basis, note that there are seven molecules in all, and hence the formula will be $0.143K_2O{\cdot}0.143CaO{\cdot}0.715SiO_2$ (neglecting the rounding-off error). The conversion to mol % is obtained simply by multiplying this formula by 100, i.e., the formula will be $14.3K_2O{\cdot}14.3CaO{\cdot}71.5SiO_2$ (mol %).

To convert this to a weight percent (abbreviated wt. %) basis, one notes the atomic weights: K = 39, O = 16, Ca = 40, and Si = 28. Thus, one formula weight of $K_2O{\cdot}CaO{\cdot}5SiO_2$ has 94 grams K_2O, 56 grams CaO, and 300 grams SiO_2, totaling 450 grams. This yields $20.9K_2O{\cdot}12.4\,CaO{\cdot}66.7SiO_2$ (wt. %). The reverse process, i.e., conversion from wt. % to mol %, should be obvious.

The reader may readily show that a soda lime glass composition does not change much in converting from mol % to wt. %.

Sometimes, glasses have minor constituents, added deliberately or by way of contamination. Colorants and glass fining agents are good examples of the deliberate additions. On the other hand, except for the production of amber glass, Fe is invariably a contaminant (coming primarily from the tools used for the mining of the raw materials). The minor constituents are often listed at the end of the glass formula, presented as oxides (e.g., FeO, As_2O_3), elements (e.g., Co, F), and groups (e.g., SO_4^{2-}). Where an ion may be present in the multivalent form (e.g., Fe^{2+} and Fe^{3+}), and the exact ratio of the two is not known, it is conventional to lump the two into one form. (The student may therefore show that the glass composition may occasionally not add up to 100% despite all the accuracy of a chemical analysis.)

Unfortunately, the convention of presenting the oxide glasses is not maintained for the case of other glass systems. In fluoride glasses, the network former is listed first, with no particular order to the rest. In chalcogenides, the listing is done in order of the chemical group number. Fortunately, this leads to the glass chain former being listed last. In metallic glasses, the listing is done in increasing order of content.

The reader may also recognize that the wt. % formulation is convenient for glass-melting, etc., whereas the mole formula and the mol % presentations give better intuitive feeling for the glass structure and properties.

5.2. Silica Glass

Suggested reading: A. C. Wright, A. G. Clare, D. I. Grimley, and R. N. Sinclair, *J. Non-cryst. Sol.* **112**, 33–47 (1989).

The structure of silica glass is the simplest of all the glass structures, yet many of the details are not yet fully understood.

As shown in Fig. 3-1, glass consists of slightly distorted SiO_4 tetrahedra joined to each other at corners. The basic unit is shown in Fig. 5-1a. Each oxygen acts as a bridge between neighboring tetrahedra (Fig. 5-1b), and hence is called a **bridging oxygen**, or simply a **BO**. Nearly 100% of the oxygens are bridging except for some defect sites and for those associated with impurities in commercial specimens. Angle β between two neighboring tetrahedra ($= Si_1$–O_4–Si_2 angle) is called the **bond angle**. Torsion angles α_1 and α_2 are, respectively, the azimuth angle between Si_1–O_1 and O_4–Si_2 when projected on the O_1–O_2–O_3 plane, and the rotation angle about the O_4–Si_2 axis. (The subscripts are used to identify the various atoms in Fig. 5.1b.) The glass is then made up of the SiO_4 tetrahedra joined to each other at the bridging oxygen corners (Fig. 5-1c). *The disorder of the glass structure is ascribed mostly to the variations in the bond angle β and the torsion angles α_1 and α_2, and to some extent to the variation in bond lengths.*

Figure 5-1a describes the short-range (range I) structure extending out to about 2.5 Å. It includes the types of the various atoms, their local coordination sphere, and the two nearest-neighbor distances, Si_1–O_1 and O_1–O_2. The range II structure ($\sim$2.5–5 Å) is described by the bond angle and the torsion angles (and hence, the distances to the atoms in a neighboring tetrahedron) (Fig. 5-1b). The intermediate range (range III) primarily constitutes the shortest-path ring size in Fig. 5-1c (ring statistics) and topological information. The long-range structure (beyond $\sim$10–20 Å), depicted as complete Fig. 5-1c, describes density and phase composition fluctuations.

The pair distribution function, curve A, extracted from XRD data by Mozzi and Warren (1), is shown in Fig. 5-2a. The peaks become broader as the distance between a pair of atoms increases because of the effect of randomness. The locations of the peaks give the various ion pair distances, which are listed in Table 5-1. The first peak corresponds to the Si–O first coordination shell. The area under the peak yields an oxygen coordination of $\sim$4.0 about each silicon, indicating that the Si–O polyhedra are rarely anything but tetrahedra. The average tetrahedral angle O_1–Si_1–O_2 is $\sim 109^\circ$. One may, however, note that the O_1–O_2 peak describing the short-range structure in Fig. 5-2a is already not too well defined. As a result, after achieving an empirical fit (curve B) to the short-range structure, a large unexplained residual (curve C) corresponding to the intermediate-range structure remains in Fig. 5-2a. Thus, the accuracy of the medium- and long-range structural information extracted from the XRD data could be questioned. Fortunately, the O_1–O_n ($n = 2, 3, 4$) component correlation can be resolved quite well using neutron diffraction (Fig. 5-2b). The nearest-neighbor distance is the Si_1–O_1 bond at 1.62 Å. The tetrahedral edge O_1–O_2

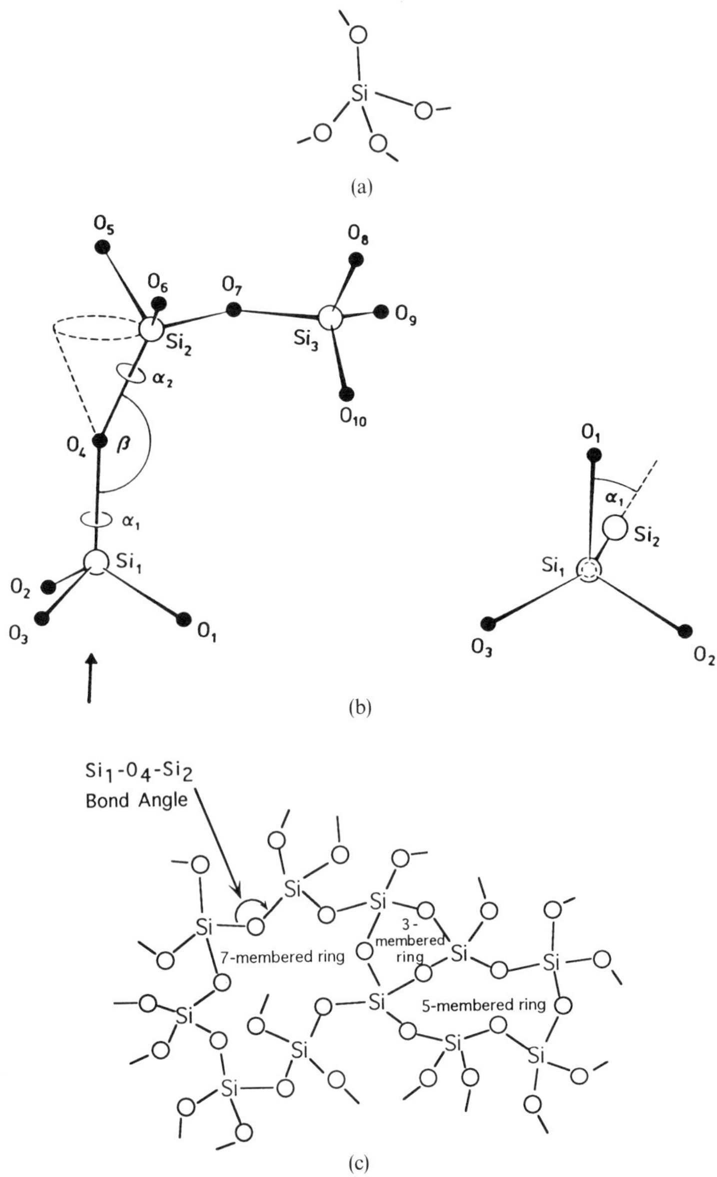

Figure 5-1. (a) Basic building block for silica glass. SiO_4 tetrahedron with all four oxygens bridging between neighboring tetrahedra. (b) Definition of the bond angle β and the torsion angles α_1 and α_2. (After A. C. Wright, G. A. N. Connell, and J. W. Allen, *J. Non-cryst. Sol.* **42**, 69 (1980). Reproduced with permission of Elsevier Science Publishers.) (c) Schematics of the silica glass network, defining the "bond angle" and the ring structures.

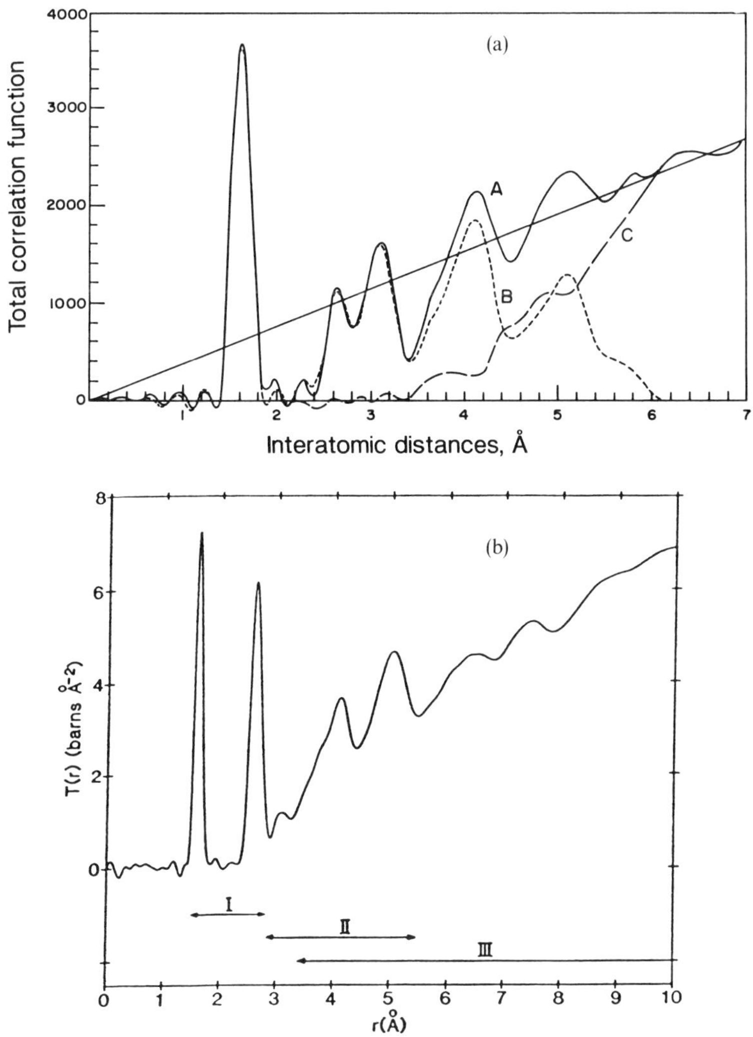

Figure 5-2. (a) The total correlation function for fused silica, measured by Mozzi and Warren (1) using x-ray. A is the measured curve; B is the sum of the calculated curve for the six contributions: Si_1–O_1, O_1–O_2, Si_1–Si_2, Si_1–O_6, O_1–O_6, and Si_1–Si_3. Curve C is the difference between A and B. (Reproduced with permission of Munksgaard International Publishers Ltd., Copenhagen.) (b) The total correlation function of fused silica measured using neutron diffraction. The roman numerals define the various ranges of structures. Note a better definition of the O_1–O_2, and not so good a definition of the Si_1–Si_2 peak. (After A. C. Wright, R. A. Hulme, D. I. Grimley, R. N. Sinclair, S. W. Martin, D. L. Price, and F. L. Galeener, *J. Non-cryst. Sol.* **129**, 213 (1991). Reproduced with permission of Elsevier Science Publishers.

Table 5-1. Ionic Pair Distances in Vitreous Silica

Si_1–O_1	1.62 Å
O_1–O_2	2.65 Å
Si_1–Si_2	3.12 Å
Si_1–O_6	4.15 Å
O_1–O_6	~5.0 Å
Si_1–Si_3	~5.0 Å

is about 2.65 Å. The peak at about 5.8–6 Å in Fig. 5-2b is, perhaps, due mostly to silicon–third oxygen (e.g., Si_1–O_8) distance. Beyond about 6 Å, the pair distribution function tends to approach the random probability and it becomes hard to assign the various peaks. *The bond angle distribution from the XRD data ranges between* 120° and 180° (Fig. 5-3) *with a maximum at* ~144°. The number of atoms within a ring may vary considerably—generally from 5 to 10 or more. The ring statistics, however, are quite sensitive to the structural model employed.

Modeling of silica glass structures has been attempted by Bell and Dean [2] using polystyrene spheres to represent silicon and oxygen atoms and 1/16 inch steel wires to represent the bonds between them (Fig. 5-4). Evans and

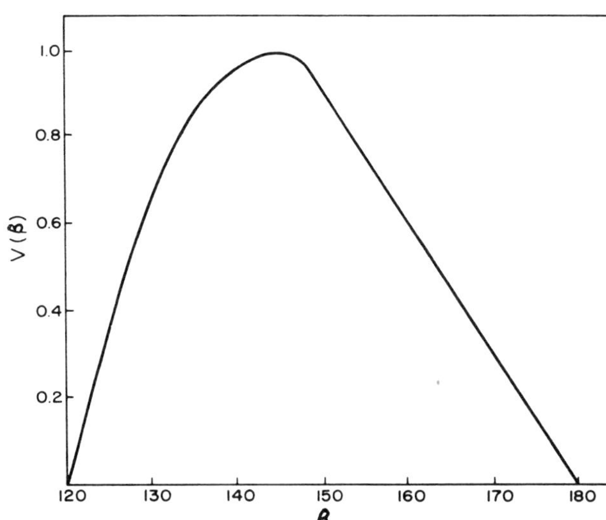

Figure 5-3. The distribution of the Si–O–Si bond angles in fused silica (after Mozzi and Warren [1]. The ordinate $V(\beta)$ is the fraction of bonds with bond angles β normalized to the most probable angle. (Reproduced with permission of Munksgaard International Publishers Ltd., Copenhagen.)

Figure 5-4. Continuous random network model of vitreous silica structure. (After R. J. Bell and P. Dean, *Nature* (London), **212**, 1354 (1966).

King [3] used tetrapods and springs to construct the tetrahedra. The basic rules for the construction of the models were adopted from Zachariasen's random network theory. After the construction, the distances between the various ions were measured to give the radial distribution function for various ion pairs in the model. Fig. 5-5 shows the excellent agreement obtained by Bell and Dean between the silica RDF obtained by XRD and that computed for the model.

Computer simulations of silica glass structure have also been attempted using the method of **molecular dynamics** (MD) by Soules [4], the **Monte Carlo technique** by Wooten and Weaire [5], and **normal coordinate analysis** by Cherukuri [6]. In the method of molecular dynamics, one starts with a certain number of constituent ions with given mass numbers that occupy random positions in a defined box such that the density at room temperature matches the experimentally observed value. The potentials at each ion are then described. These may be two-body, three-body, or multibody potentials. The derivative of the potential with respect to the interionic distance gives an interionic force. Forces also arise randomly because of the kinetic energy of

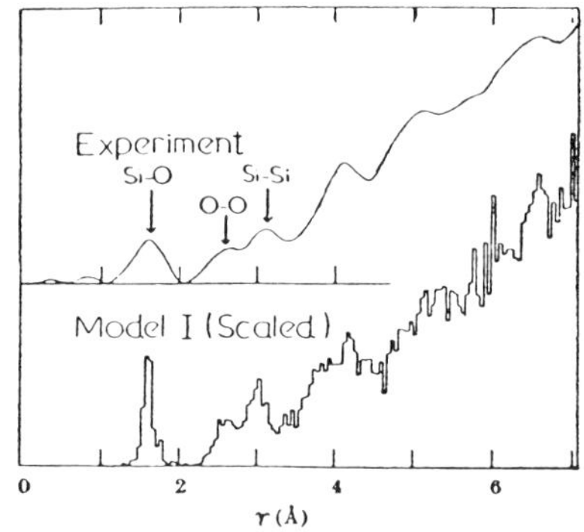

Figure 5-5. Comparison of the experimental x-ray RDF for fused silica and that measured for the Bell and Dean model. (After R. J. Bell and P. Dean, *Nature (London)* **212**, 1354 (1966). Reproduced with permission of Nature Publishing Company, Washington D.C.)

the ensemble (determined by the prescribed temperature). The vectorial sum of all the forces, and the application of Newton's laws of motion, then gives for each ion a vectorized acceleration that, upon integrating twice, gives the change in position of each ion over a given time step. The box is first thermalized for a predetermined time at a very high temperature (generally as much as 5,000 °C). The temperature is lowered, and thermalization again carried out. In this way the box is brought essentially to room temperature. Finally, all the relevant information regarding structural details, such as pair distribution function, bond angles, force constants, vibrational frequencies, etc., and dynamic parameters such as the diffusion coefficient of the various ions, are computed. To make representative calculations using MD techniques, one often needs as many as 1,000 ions and 5–10,000 time steps of 10^{-15} seconds each. In order to make the relatively small box containing 1,000 atoms more representative of a much larger volume of material, the device of 'periodic boundary conditions' on the walls is imposed. Thus, if calculations predict an atom leaving one face of the box with a particular velocity, the program automatically introduces an identical atom entering the opposite face with the same velocity. The small value of the time step (very much less than a jump frequency interval) is necessary to avoid errors in numerical integration. It is clear then that even with the fastest computers it is difficult to allow any more than a few picoseconds for the glass to cool down from a high temperature to room temperature—which does not represent reality. One further difficulty involved with MD calculations is the simulation of other than ionic potentials that are spherically symmetric. (Recent advances in this direction include the use of Stillinger and Weber three-body potentials [7].) In spite of the inability to exactly

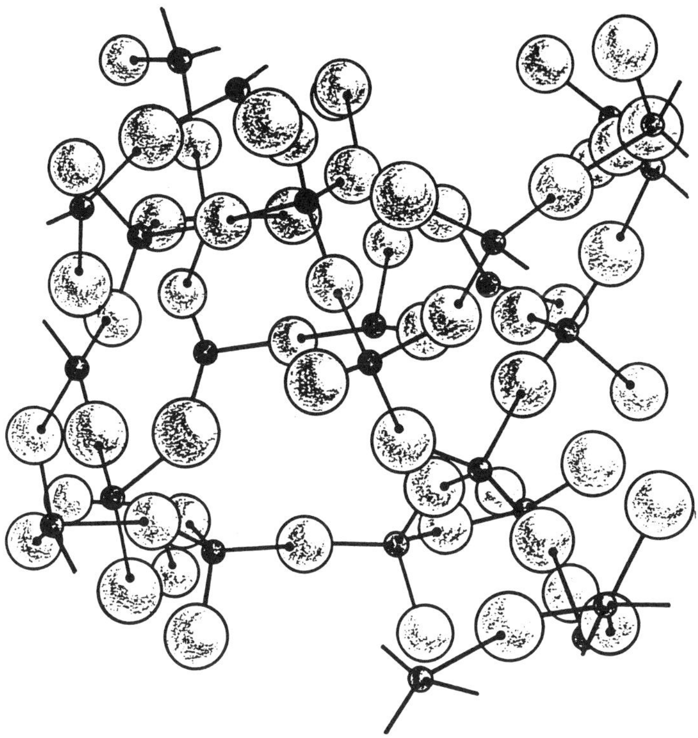

(a)

Figure 5-6. (a) Molecular dynamic simulation of the vitreous silica structure. (After T. F. Soules, *Glass Sci. & Tech.* **4A**, Fig. 31, p. 318 (M. Tomozawa and R. H. Doremus, eds.). Academic Press, New York, 1990.)

simulate the interionic potentials and the limitations to computer power currently available, the molecular dynamics method has proven remarkably effective in simulating glass structures and in gaining some insight into factors that influence glass properties. Fig. 5-6a shows a computer simulation of the silica glass structure. The individual pair distribution functions are shown in Fig. 5-6b. Note that the ionic distances match those in the experimentally measured RDF. Detailed analysis shows that the ion co-ordinations are also duplicated quite well. However, the mismatch observed in the sharpness of the peaks suggests that the variations in bond length and bond angles may not be correct. This would clearly affect the agreement over the medium-range order; in particular, the ring statistics may not be predicted accurately.

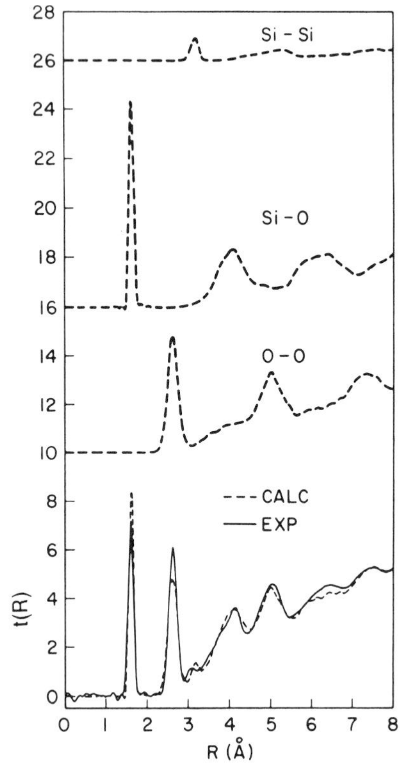

(b)

Figure 5-6. (b) Individual pair correlation functions, $t(r)$, for Si–O, Si–Si, and O–O obtained from MD calculations. A calculated composite RDF is compared with that measured using neutron diffraction. (After T. F. Soules, *Glass Sci. & Tech.* **4A**, Fig. 33, p. 320 (M. Tomozawa and R. H. Doremus, eds.). Academic Press, New York, 1990.)

Because of the high degree of connectivity, the values of mass transport properties, e.g., fluidity, in silica glass are very low (i.e., the viscosity is high).

5.3. Boric Oxide Glass

In B_2O_3 glass, the oxygen coordination around each B is only 3, and hence the basic structural unit is a BO_3 triangle. It is believed that B is slightly above the plane of the three oxygens. Again, all the oxygens are bridging between neighboring triangles. Several authors believe that the basic building

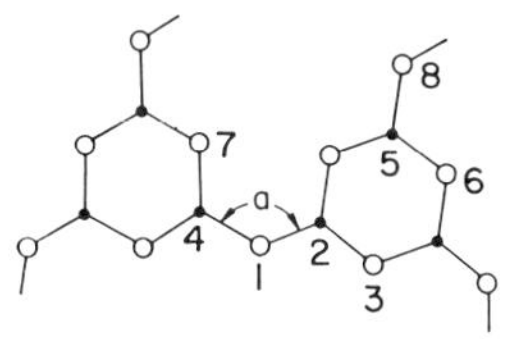

Figure 5-7. Configuration of the boroxol ring.

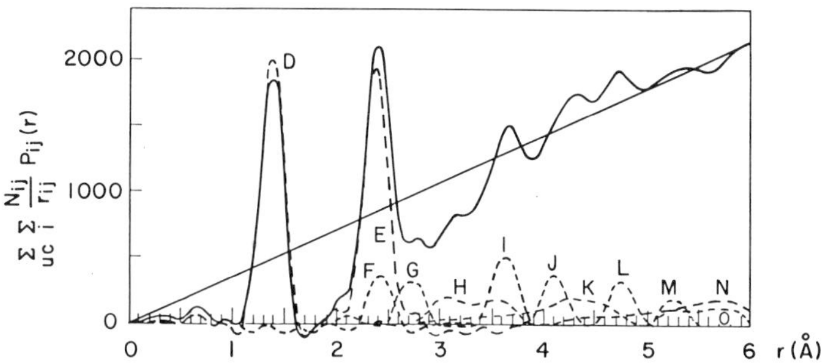

Figure 5-8. Total correlation function of vitreous B_2O_3. Dashed curves correspond to various ion pairs shown in Fig. 5-7 with distances (Table 5-2) suggested by Mozzi and Warren [8]. Note that MD calculations reveal the B–B peak to be at 0.273 nm (where peak G is located). (Reproduced with permission of Munksgaard International Publishers Ltd. Copenhagen.)

Table 5-2. Ion Pair Distances in Vitreous B_2O_3

Peak	Kind of atoms	Designation on figure	Interatomic distance	$\sum_{uc}\sum_{i} N_{ij}$
D	B–O	1-2	0.137 nm	12
E	O–O	1-3	0.237	12
F	B–B	2-4, 2-5	0.243	6
G	B–O	2-6	0.274	4
H	B–O	2-7	variable	8
I	B–O	2-8	0.363	8
J	O–O	3-8	0.410	4
K	O–O	3-7	variable	8
L	O–O	1-8	0.475	4
M	B–O	4-6	0.525	4
N	B–O	5-7	variable	16
O	B–O	4-8	variable	8

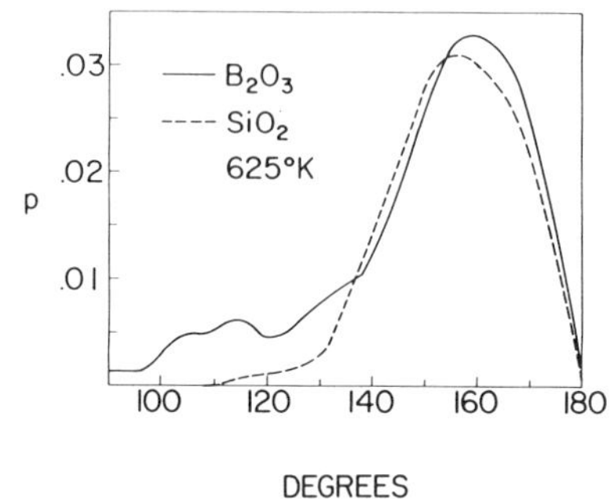

Figure 5-9. The distribution of the B–O–B bond angles in vitreous B_2O_3 compared to the Si–O–Si bond angles in fused silica (using MD calculations). The ordinate is the absolute fraction of bonds per degree angle.

block in the boric oxide glass is the **boroxol group** shown in Fig. 5-7. A pair distribution function plot [8] from XRD is shown in Fig. 5-8, and the various interionic distances are tabulated in Table 5-2. Note that the G peak at 0.274 nm (=2.74 Å) distance was assigned to the B-cis O in the boroxol group, and the B–B peak (F peak) corresponding to a distance of 0.243 nm was assumed to be concealed under the E peak. Molecular dynamic calculations (9), on the other hand, show that the B–B distance is 0.272 nm (and

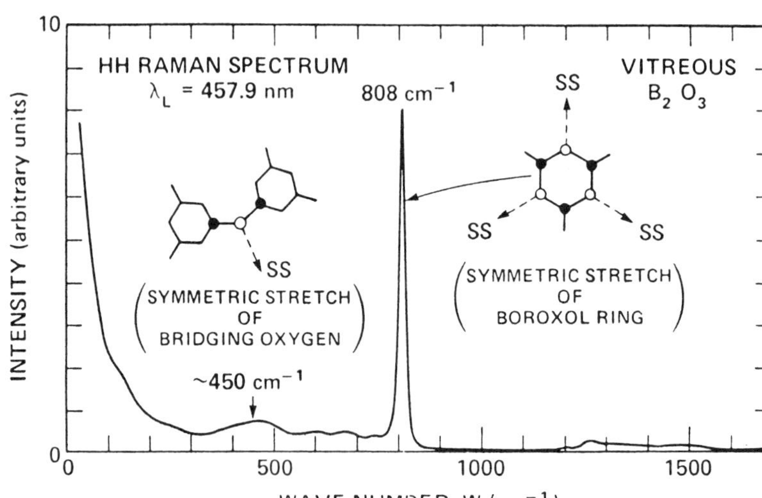

Figure 5-10. Assignment of the two highly polarized modes in the Raman spectrum of vitreous B_2O_3. (After F. L. Galeener, in *The Structure of Non-crystalline Materials 1982* (P. H. Gaskell, J. M. Parker, and E. A. Davis, eds.), Fig. 5, p. 343. Taylor & Francis, Inc., New York, 1983.) Reproduced with permission of the publisher.)

not 0.243 nm), and hence is the G peak. The absence of a significantly large population of 120° angles in the B–O–B bond angle distribution (Fig. 5-9) also suggests the presence of only a low number of the boroxol groups.

Perhaps the strongest support to the concept of boroxol rings has been derived from the Raman spectrum of B_2O_3 glass shown in Fig. 5-10. The peak at 808 cm^{-1} is ascribed to the symmetrical stretching of the boroxol ring.

Some research has shown that the equilibrium (boroxol)$\Leftrightarrow BO_{3/2}$ triangle shifts to the right with increasing temperature. This would explain why the high-temperature structures obtained by MD techniques do not show much concentration of the boroxol groups. Whatever the nature of the ring configurations is, the triangular connections in the B_2O_3 glass do not make a very rigid structure. Hence, many transport properties of B_2O_3 glass have high values typical of fluid structures.

5.4. Alkali Silicate Glasses

The range of glass formation in the various alkali silicates determined by means of conventional melting techniques is continuous from SiO_2 up to about 36 mol % Li_2O, about 58 mol % Na_2O, and about 55 mol % $K_2O/Rb_2O/Cs_2O$.

As indicated in Section 3.1.2, *the alkalis are network modifiers; they enter the glass as singly charged cations and occupy interstitial sites.* The unit positive charge is satisfied by an ionic bond to an oxygen. This is accomplished by breaking a bridge and attaching an oxygen (provided by the alkali oxide M_2O) to the broken bridge, as shown by the reaction

$$-\overset{|}{\underset{|}{Si}}-O-\overset{|}{\underset{|}{Si}}- + M_2O = -\overset{|}{\underset{|}{Si}}-O^{-}M^{+} \quad M^{+}{}^{-}O-\overset{|}{\underset{|}{Si}}- \tag{5.1}$$

Thus, *each alkali ion is expected to create one "nonbridging oxygen (NBO)."* A schematic representation of an alkali silicate network is shown in Fig. 5-11. *The creation of NBO in the network reduces the connectivity*, and hence, mass-transport-related properties such as fluidity (inverse of viscosity), diffusion, electrical conduction, and chemical corrosion increase. For similar reasons, the structure acquires the freedom to expand, for instance, its thermal expansion coefficient increases.

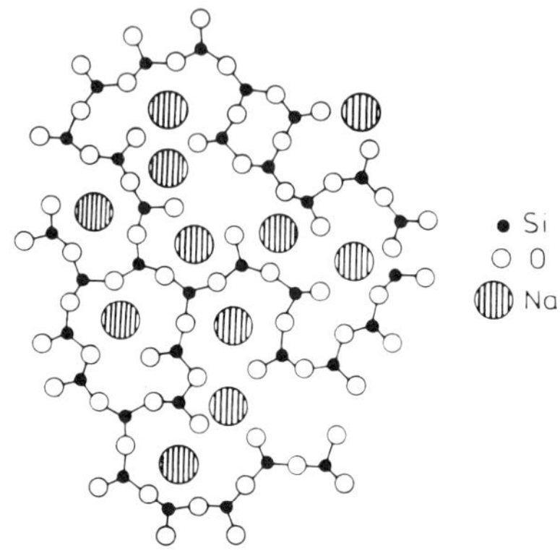

Figure 5-11. Two-dimensional representation of the structure of sodium silicate glass. (The fourth oxygen with every Si is presumed to be out of the plane of the paper.) Note that some oxygens (bridging oxygens) are coordinated to two silicons, others to only one. In the latter case (nonbridging oxygen), a sodium ion will be found in the vicinity for charge compensation.

To determine the fraction of NBO, one first converts the glass formula to a mole basis, and proceeds as follows:

Example: For $20Na_2O{\cdot}80SiO_2$ (mol %) glass

Since # Na^+ = 40, hence # NBO = 40.
Since total oxygens = 180, # tetrahedra (□) = # silicon atoms = 80,
hence f_{NBO} = fraction of NBO of the total oxygens = 40/180 = 0.22
f_{BO} = 140/180 = 0.78, and # BO/□ = 140/80 = 1.75

Y = # of corners shared/□ = 2 × # BO/□ = 3.5 (because each BO is shared by two □)

Y can also be calculated using the formula

$$Y = 6 - 200/p, \quad \text{where } p = \text{mol \% SiO}_2 \tag{5.2}$$

Using Eq. (5.2), it can be readily seen that $Y = 2$ at 50 mol % SiO_2. The tetrahedra then form linear chains or isolated rings—there is a clear breakdown of the 3-D network at 50 mol % SiO_2. At 40% SiO_2, $Y = 1$. This implies that only one corner/□ is shared. The structure of an alkali silicate then contains isolated pairs of tetrahedra (or a mixture of isolated single tetrahedra and longer chains), as shown in Fig. 5-12. Apparently, as stated earlier, glass formation continues to occur even in these less than 3-D connected structures. It has been suggested that bonding in glasses with

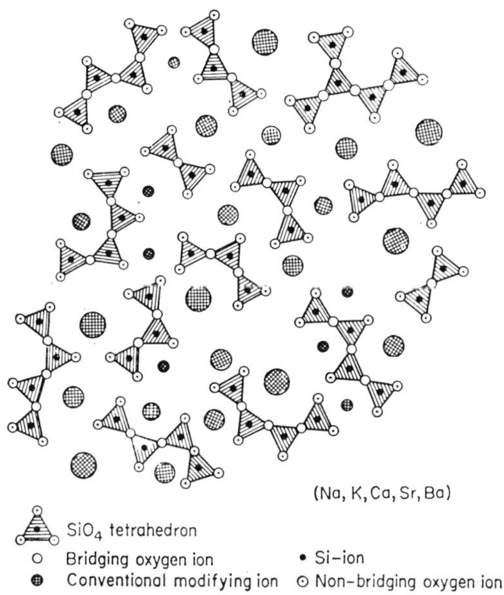

Figure 5-12. Two-dimensional representation of an "invert glass."

$Y = 2$ or less is accomplished by the van der Waals forces between the alkali ions, thus "inverting" their role; hence the name **invert glass** for this class of glasses. A plot of log η versus Y, where η is the viscosity, is shown in Fig. 5-13 and suggests the increase in the tightness of bonding with alkali increasing below $Y = 2$.

In recent years, one other powerful tool developed to elucidate the environment of the various nuclei in glass is the **magic-angle spinning nuclear magnetic resonance** [10] **(MAS-NMR)** technique. In an NMR experiment, a nucleus possessing a magnetic moment is spun at an angle θ to an externally applied magnetic field H_0, causing the nucleus to precess at Larmor frequency about the field axis. An oscillating electromagnetic field of the same frequency is then allowed to interact. The resulting nuclear resonance then causes absorption of energy from the oscillating field proportional to the Boltzmann distribution of the nuclear spin populations and the effective nuclear relaxation times. The NMR spectra are affected by the chemical-shift anisotropy interaction resulting from the shielding of H_0 by the electrons surrounding the resonating nucleus. Any change in the local electron density around the nucleus, such as those resulting from a coordinate change, can be detected. It is found that the Hamiltonians describing the dipolar, chemical shift, and first-order quadrupolar interactions are all multiplied by a factor $\beta = (3 \cos^2 \theta - 1)$. When $\theta = 54.74°$, $\beta = 0$, which removes, in effect,

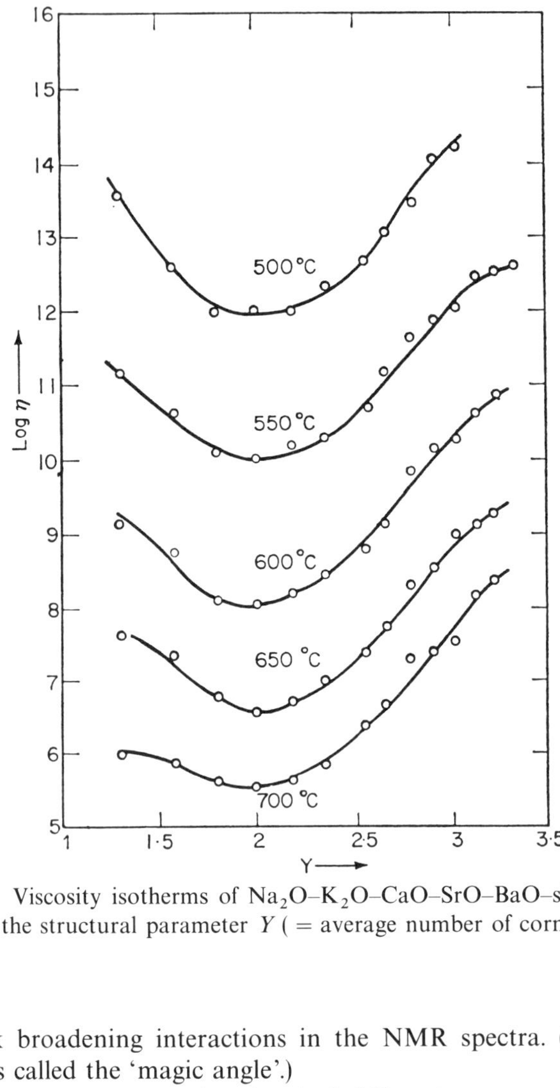

Figure 5-13. Viscosity isotherms of Na_2O–K_2O–CaO–SrO–BaO–silicate glass as a function of the structural parameter Y (= average number of corners shared per tetrahedron).

all the peak broadening interactions in the NMR spectra. (That is why $\theta = 54.74^\circ$ is called the 'magic angle'.)

In alkali silicate glasses, the chemical-shift anisotropy on ^{29}Si yields information regarding the second neighbors, particularly the identification of bridging and nonbridging oxygens. The local configuration around the ^{29}Si atom is expressed as Q_n (n = 0 to 4) units, where Q_0 represents all NBO's (isolated SiO_4 tetrahedron), and Q_4 represents all BOs. At any value of x $(0 < x < 1)$ in $(M_2O)_x{\cdot}(SiO_2)_{1-x}$ glasses, statistical arguments for a random model would predict the presence of all Q_n the relative numbers of which are determined by x. The ordered model, on the other hand, predicts for certain stoichiometric compositions only specific Q_n units, for instance, Q_3 at $x = 0.33$, and Q_2 at $x = 0.5$ (see example above). The results of MAS-NMR for $0.33 < x < 0.5$ support the ordered model: at $x = 0.33$, only Q_3 are present which transform completely to Q_2 at $x = 0.5$. In the intermediate range of compositions, both Q_3 and Q_2 are present. The amount of Q_4 present, if any, is much less than that expected from the random model. Some recent results of Maekawa *et al.* [10b] for the Li, Na, and K silicates suggest an equilibrium reaction:

$$2Q_n \Leftrightarrow Q_{n-1} + Q_{n+1} \ (n = 3, 2, 1 \text{ for Na and K, and } n = 3, 2 \text{ for Li}).$$

A plot of $\%Q_n$ against mol % alkali for the three systems is shown in Fig. 5-14.

In summary, it may be confidently said that the addition of one alkali ion creates one NBO. However, *the occurrence of the NBOs in the network is not entirely random.* Measurements of ionic diffusivity and electrical conductivity strongly suggest that alkalis, and hence the NBOs, occur in pairs. (This is the basis of writing the reaction [5.1] shown earlier.) Calculations of

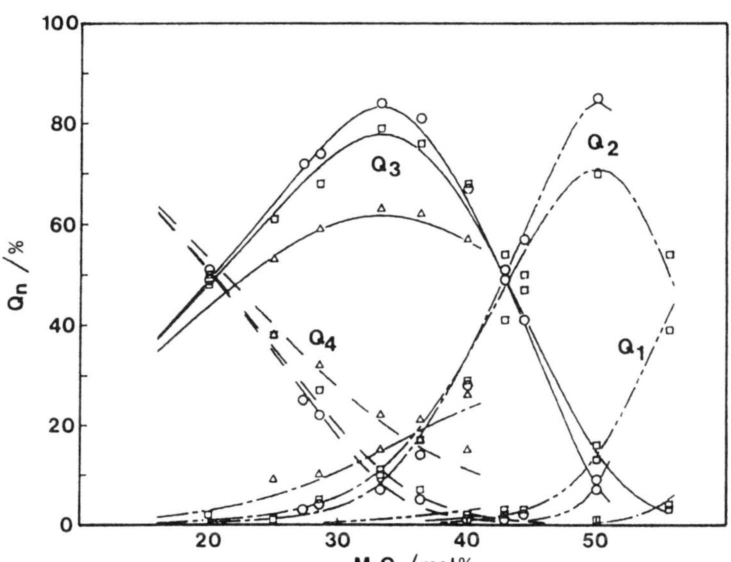

Figure 5-14. Experimentally determined Q_n distribution using MAS-NMR in lithium (△), sodium (□), and potassium (○) silicate glasses as a function of mol % alkali oxide. Curves represent theoretical fitting. (After H. Maekawa, T. Maekawa, K. Kawamura, and T. Yokokawa, *J. Non-cryst. Sol.* **127**, 53–64 (1991). Reproduced with permission of Elsevier Science Publishers.)

the pair distribution function using the method of MD, performed by Soules, also suggest larger than random probability for the occurrence of paired alkalis.

5.5. Alkali–Alkaline Earth–Silicate Glasses

The range of glass formation in alkali–alkaline earth–silicate glasses is shown on the ternary composition space in Fig. 5-15. In general, the various alkalis and alkaline earths can be introduced to a little over 50 mol % total.

The alkaline earth cations are bivalent, and like the alkalis, *they occupy interstitial spaces as glass network modifiers*. The addition of one alkaline earth creates two nonbridging oxygens. The connectivity in the process is still maintained to a large extent, and hence alkaline earth additions to alkali silicates greatly stabilize the glass network. (For instance, the glass no longer readily dissolves in water!) The phase equilibrium diagram in the soda-lime system is shown in Fig. 5-16 (%Na_2O is calculated by difference from 100). The lowest liquidus temperature (725°C) occurs at the ternary eutectic joining the quartz, devitrite ($Na_2O{\cdot}3CaO{\cdot}6SiO_2$), and sodium disilicate ($Na_2O{\cdot}2SiO_2$) phase fields. Superimposed upon the figure are the lines of equal crystal growth rate [11] (the numbers are seconds for a 10-micron

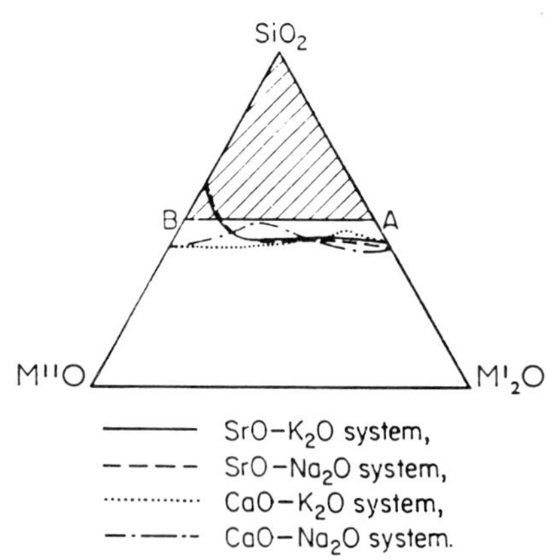

Figure 5-15. Glass-formation regions in various ternary silicate glasses. M′ is a monovalent atom, and M″ is a divalent atom. (After M. Imaoka, *Adv. in Glass Tech.*, Part I, p. 149. Plenum Press, New York, 1962. Reproduced with permission of the publisher.)

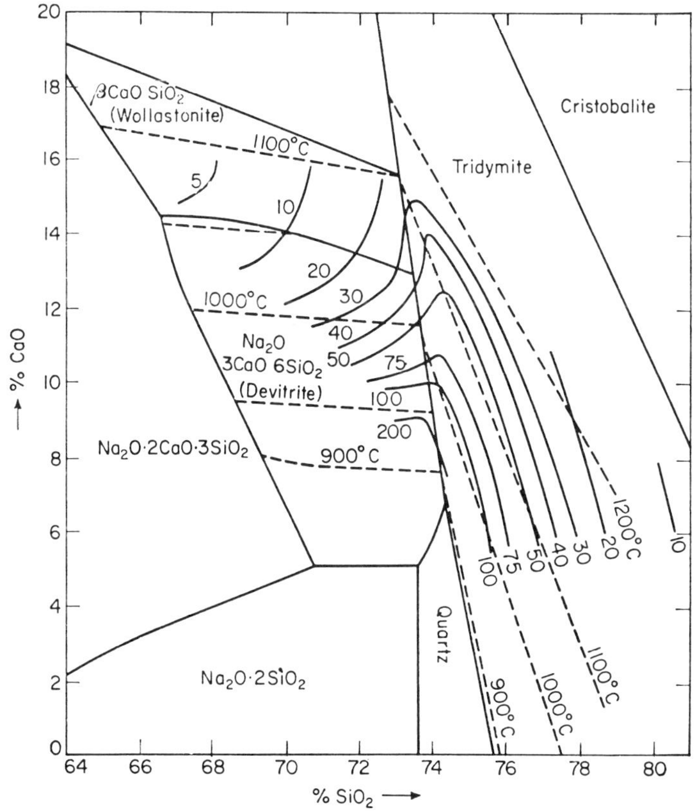

Figure 5-16. Phase diagram in the Na_2O–CaO–SiO_2 system (Na_2O is given by the difference from 100 wt. %). Lines of equal crystal growth rate (seconds for a 10 micron growth) are superimposed. (After H. Rawson, *Inorganic Glass-Forming Systems*, Fig. 35, p. 88. Academic Press, London, 1967.)

crystal to grow). The glass-forming region of commercial interest is about half of the devitrite phase field near the ternary eutectic. The calculations of f and Y in the alkali–alkaline earth–silicate glasses can be made in the same manner as for the alkali silicates.

5.6. Alkali Borate Glasses

The boron ion is trivalent positive and is a glass former. The introduction of an oxygen from a modifier oxide to boric oxide glass brings about one of the two possibilities:

(a) *Create a nonbridging oxygen*, as in the silicate glasses, by forming $BO_{2/2}O^-M^+$. (The "/2" subscript is used to identify the connection to two cations.)

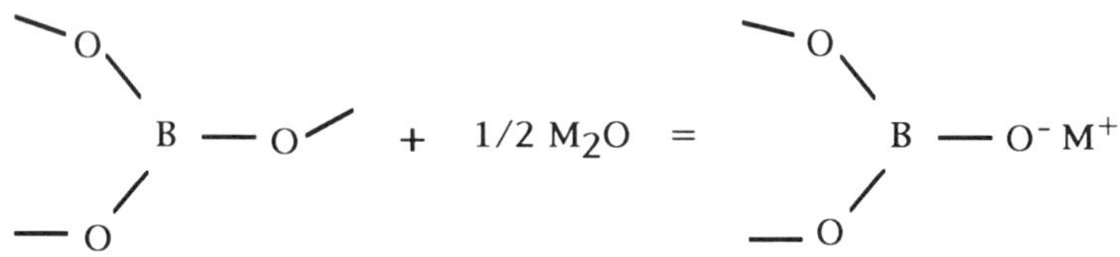

The oxygen coordination around the boron remains three.

(b) *Convert boron from a 3-coordination state* ("B_3 *state*") *to a 4-coordination state* ("B_4 *state*"). Boron in a tetrahedrally coordinated state is known to exist in some crystalline borates.

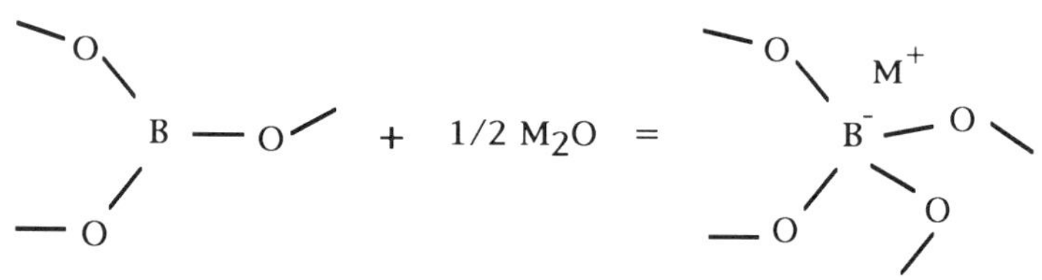

In the BO_3 group, the oxygens are fully bridging, and hence one negative charge from each oxygen satisfies the three positive charges on the boron ion. After the conversion from B_3 to B_4, all the oxygens remain bridging; the extra negative charge on the $[BO_{4/2}]^-$ group is satisfied by an alkali M^+ ion in the vicinity. The electron transfer from the M atom occurs as a distributed charge density over a large-effective-diameter BO_4 group, and not localized between the M atom and any specific oxygen. Since this association is somewhat loose, the alkali ion is expected to become more mobile. At the same time, the connectivity of the network increases, and hence flow-related properties decrease (i.e., viscosity increases) and thermal expansion decreases.

It is possible that, like the boroxol $\Leftrightarrow 3BO_{3/2}$ equilibrium, the $[BO_{4/2}]^-M^+ \Leftrightarrow BO_{2/2}O^-M^+$ equilibrium also shifts to the right as the temperature increases.

5.7. The Boron Anomaly

Gooding and Turner's data [12] on the thermal expansion coefficient of sodium borate glasses as a function of the added alkali is shown in Fig. 5-17a. Note that the expansion coefficient shows a marked minimum at about

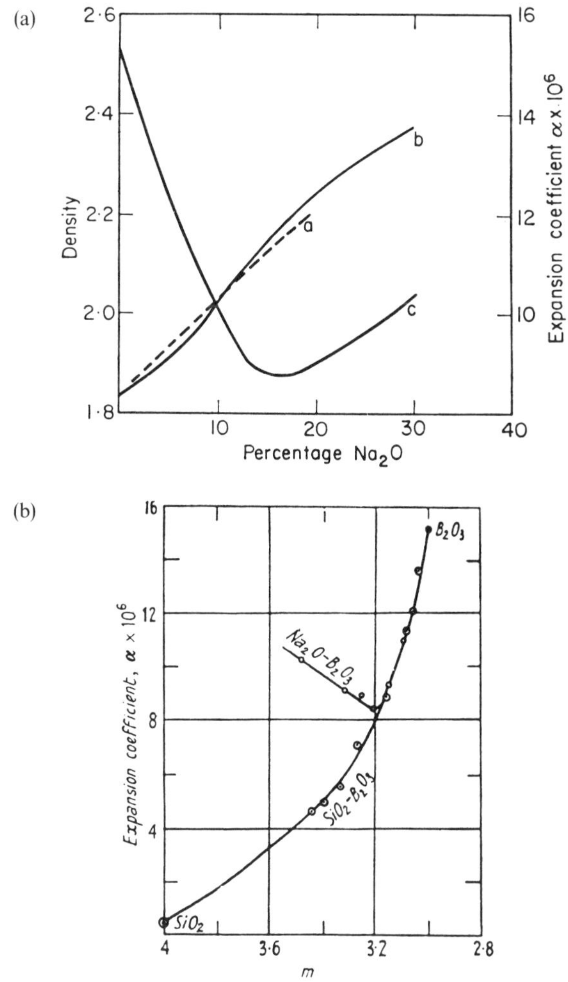

Figure 5-17. (a) Thermal expansion coefficient and density of Na_2O–B_2O_3 glasses: curve "a," calculated density; curve "b," measured density; curve "c," measured thermal expansion coefficient. (b) Thermal expansion coefficients plotted against average coordination number m for the systems SiO_2–B_2O_3 and Na_2O–B_2O_3.

16 wt. % Na_2O. (More recent data also show that, in addition to the minimum in the expansion coefficient, the glass transformation temperature T_g at which viscosity = 10^{12} Pa·s shows a maximum.) This extremum behavior (instead of the almost linear change with composition found in silicate glasses) was termed "**the boron anomaly**." Biscoe and Warren [13] explained the anomaly by suggesting that, initially, *the addition of each alkali ion to*

boron oxide causes one boron to change from the B_3 state to the B_4 state. No NBOs are created. The increased connectivity, hence, causes the thermal expansion to decrease and viscosity to increase. The conversion stops at around 16 mol % added alkali. Thereafter, further additions cause the production of NBOs, and a resultant increase in expansion and decrease in viscosity. To support this hypothesis, Warren [14] plotted the expansion coefficient of borosilicate glasses and sodium borates together against an average coordination number, m, for the network-forming cations in the glass (see Fig. 5-17b). The borosilicate data plot as a smooth curve between $m = 4$ for SiO_2 at one end and $m = 3$ for B_2O_3 at the other end. The m value for the sodium borates was calculated using the preceding premise that one added alkali raises one boron from the B_3 state to the B_4 state. The sodium borates superimpose well on the curve from $m = 3$ to about $m = 3.17$. At higher values of m, the data depart sharply upwards. The value $m = 3.17$ corresponds to about 13 wt. % Na_2O. Warren's notion of 3 to 4 coordination change for boron and its subsequent termination as the explanation of the origin of boron anomaly met wide acceptance.

Later, Bray and O'Keefe [15] experimentally determined the fraction, N_4, of the 4-coordinated boron ($= \#B_4/\#$total borons) in the alkali borates using NMR. Their results are shown in Fig. 5-18. For $xM_2O{\cdot}(1 - x)B_2O_3$ glass, the total number of borons are $2(1 - x)$, and if every M^+ converts one boron from B_3 to B_4, then $\#B_4 = \#M = 2x$. This should yield $N_4 = x/(1 - x)$. The solid curve on the figure corresponds to $N_4 = x/(1 - x)$.

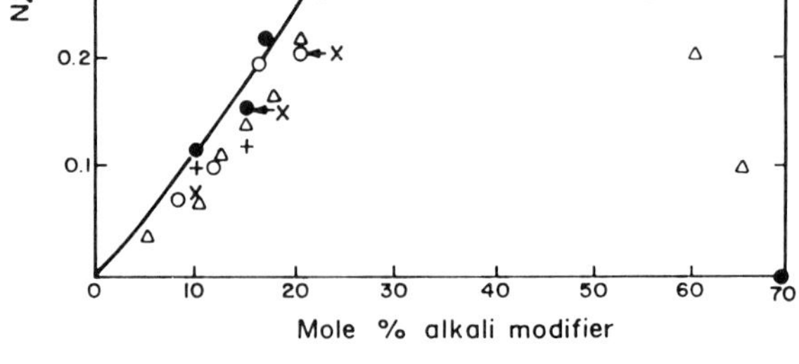

Figure 5-18. The fraction N_4 of 4-coordinated boron atoms in alkali borate glasses as a function of mol % alkali. Na_2O (•); K_2O (○); Li_2O (△); Rb_2O (+); Cs_2O (×). (After Bray and O'Keefe [15]. Reproduced with permission of the Society of Glass Technology.)

Note that N_4 does increase with added alkali as suggested by Warren. However, instead of stopping at about 16 mol % alkali, the $B_3 \rightarrow B_4$ conversion appears to continue right up to about $N_4 = 0.45$ and only decreases thereafter, approaching zero at around x = 70 mol %. Although this new experimental evidence confirmed the general idea of coordination changes of the network-forming cations in glass, it did suggest that the anomalous behavior of properties around 16 mol % added alkali had little to do with this conversion. One needed to look into other causes for the origin of the boron anomaly. More recent (perhaps more precise) measurements (16) of the expansion coefficient and the viscosity indicate that the property anomalies occur over a much wider range of composition, *viz.*, from about 13% to as high as 30% (instead of the previously thought narrow range around 16%). The proposed explanation currently for the boron anomaly is to regard the anomaly as a manifestation of the various atomic groups (Fig. 5-19) in which boron may be found as a function of the various alkali additions (Fig. 5-20).

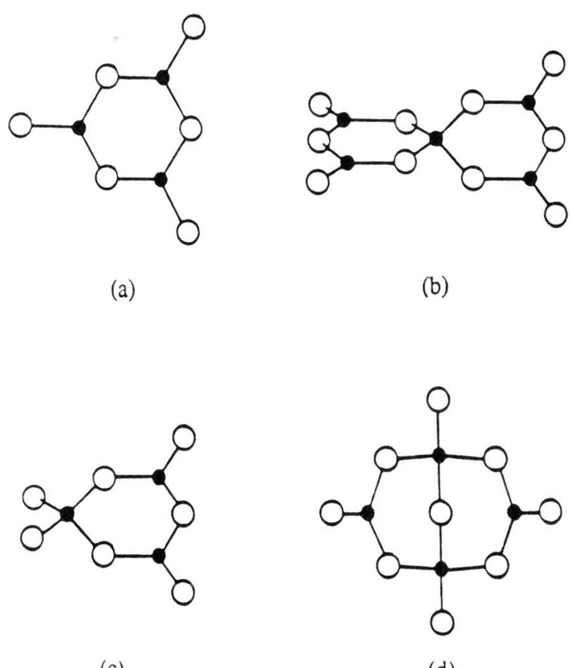

Figure 5-19. The various borate groups postulated to exist in alkali borate glasses having less than 34 mol % alkali. (a) The boroxol group. (b) The pentaborate group. (c) The triborate group. (d) The diborate group.

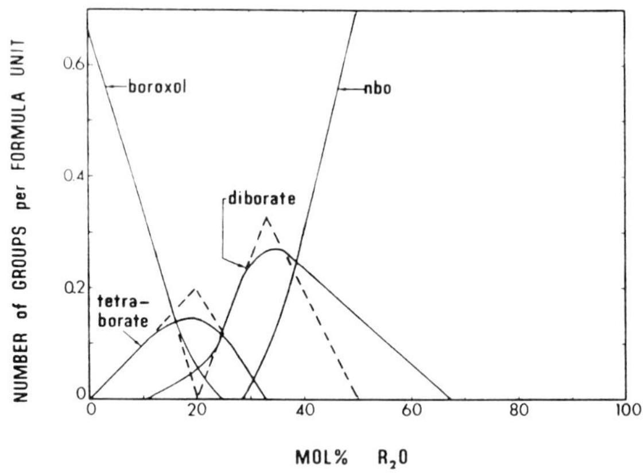

Figure 5-20. Structural models for alkali borate glasses. Note that the boroxol units disappear rapidly with alkali addition, giving rise to diborate units around 20 mol % alkali. (After Ref. 16. Reproduced with permission of the American Ceramic Society.)

If we express the molecular composition of the alkali borate glass as $RM_2O{\cdot}B_2O_3$, where $R = x/(1-x)$, then $N_4 = R$, and $f_{NBO} = 0$ up to $R = 0.5$. When $R > 0.5$, N_4 can be estimated by using Gupta's approximate relation (17)

$$N_4 = (3 - R)/5. \tag{5.3}$$

To calculate structural configurations in the alkali borates, one first needs to compute R. The computation of N_4 is then carried out. Finally, the f_{NBO} is calculated by any leftover alkalis.

Example 1: Calculate the structural configurations in $20\,Na_2O{\cdot}80B_2O_3$ (mol %) glass.

We note that $R = 20/80 = 0.25$, which is less than 0.5.
Hence, $N_4 = R = 0.25$ and $f_{NBO} = 0$. Also, $N_3 = 1 - N_4 = 0.75$.
Average coordination number of boron $= 3 \times 0.75 + 4 \times 0.25 = 3.25$.

Example 2: Calculate the structural configurations in $40K_2O{\cdot}60B_2O_3$ (mol %) glass.

In this case, $R = 40/60 = 0.666$, which exceeds 0.5.
Hence, from Eq. (5.3), $N_4 = (3 - 0.666)/5 = 0.467$ (and $N_3 = 0.533$).
This yields $\#B_4 = 120 \times 0.467 \cong 56$.

Since one B_4 requires one K^+, this number of B_4 requires 56 K^+, leaving 24 K^+ to create 24 NBOs. Total number of oxygens are $40 + 180 = 220$. Hence, $f_{NBO} = 24/220 = 0.11$.
Average coordination number of boron $= 3 \times 0.533 + 4 \times 0.467 \cong 3.47$.

5.8. Alkali Borosilicate Glasses

In alkali borosilicate glasses, $RM_2O{\cdot}KSiO_2{\cdot}B_2O_3$, there are two network formers: silicon and boron. The added alkali may associate either with silicon, creating an NBO as SiO^-M^+, or with boron, presumably converting a B_3 to a B_4 and creating no NBOs in the process. Again, using NMR techniques, it has been shown [18] that alkali prefers to associate with the boron as long as $R < 0.5$. Thereafter, the alkali distribution is partitioned between boron and silicon, creating NBOs depending upon the K value. The behavior was confirmed using molecular dynamics calculations [9]. The variation of N_4 with R and K is shown in Fig. 5-21. Shown in Fig. 5-22 is a 3-D projection image for a borosilicate glass using MD techniques. The reader may be able to see the linked BO_3, BO_4, and SiO_4 groups by viewing a photocopy after adjusting the separation between the left and the right images under a portable stereographic viewer.

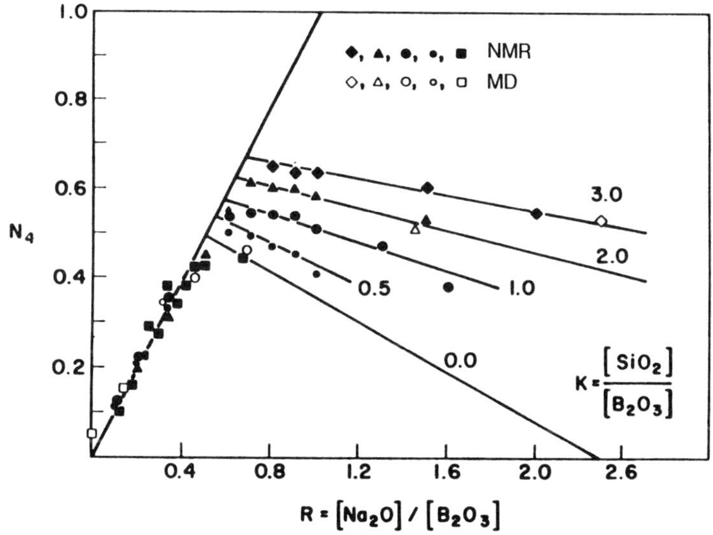

Figure 5-21. Fraction of tetrahedrally coordinated borons (N_4) as a function of mole Na_2O/B_2O_3 and SiO_2/B_2O_3 ratios. (After Ref. 9. Reproduced with permission of the American Ceramic Society.)

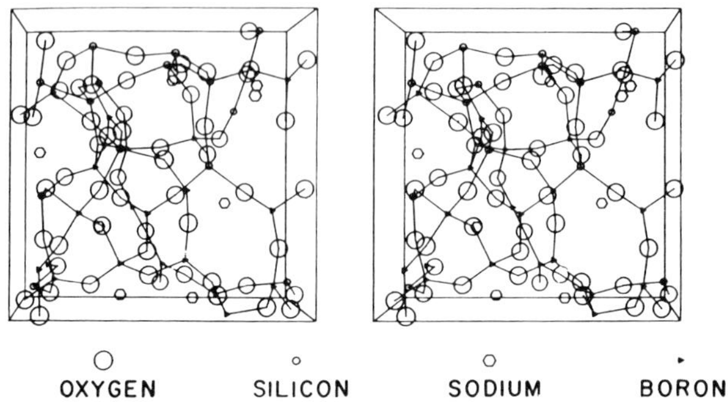

Figure 5-22. Stereographic image of a group of atoms in a $Na_2O–B_2O_3–SiO_2$ glass at 625 K. (After Ref. 9.)

5.9. Alkali Aluminosilicate Glasses

In alkali aluminosilicate glasses, the trivalent aluminum ion does not always act as a network former. The structural configurations depend upon the $[Al_2O_3/M_2O]$ ratio. Apparently, when $[Al_2O_3/M_2O] < 1$, the Al^{3+} ion goes in as a network former having tetrahedral coordination. Like the $[BO_4]$ group, the excess unit negative charge on the $[AlO_4]$ group is satisfied by an alkali ion in the neighborhood. Thus, the addition of an aluminum ion to an alkali silicate glass removes one NBO. At $[Al_2O_3/M_2O] = 1$, there are no nonbridging oxygens in the structure. For further additions of Al^{3+}, i.e. when $[Al_2O_3/M_2O] > 1$, the Al^{3+} ion enters the network as a modifier ion in octahedral coordination. Presumably three of the oxygens are non-bridging and three (already) bridging. This arrangement is shown in Fig. 5-23.

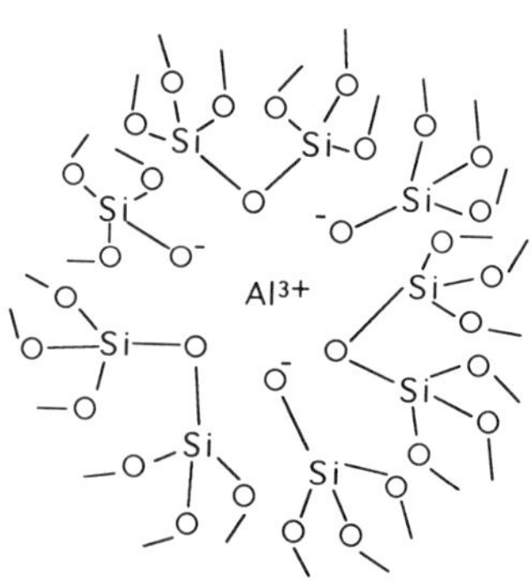

Figure 5-23. Suggested structure of an alkali aluminosilicate glass with Al^{3+} as a network modifier showing octahedrally coordinated Al^{3+}.

According to Lacy [19], when $[Al_2O_3/M_2O] > 1$, the formation of AlO_6 groups is highly unlikely because of packing difficulties. He favors the formation of triclusters where an oxygen is shared between three tetrahedra: one AlO_4 and two SiO_4 or two AlO_4 and one SiO_4. (Note that oxygen bridging to three network formers violates Zachariasen's rules.) The first type of tricluster is electrically neutral; the second would require an associated alkali. These triclusters are shown in Fig. 5-24. In principle, the triclusters appear equivalent to an Al^{3+} ion acting as a modifier ion and bonded to three nonbridging oxygens (shown in Fig. 5-24c). Thus, the basic question is whether an Al^{3+} ion is surrounded by one (already) bridging oxygen or two or three. Experiments to understand the Al coordination sphere need to be carried out.

Because of the blocking effect of the Al ion in the interstices, electrical conductivity and the diffusion coefficient of the alkali ion decrease. In Fig. 5-25 is shown the activation energy for electrical conduction in Na_2O–Al_2O_3–SiO_2 glasses against the Al_2O_3/M_2O ratio [20]. The increase in the activation energy at initial additions of alumina is attributed to the tightening of the structure due to the gradual disappearance of NBO. With further additions of alumina, the activation energy starts decreasing because of the looseness of bonding between the alkali ion and the $[AlO_4]^-$ group. When $[Al_2O_3/M_2O] > 1$, the activation energy starts increasing again because of the blocking effect of the octahedrally coordinated Al^+ ion in the interstices.

To calculate the structural configurations in the $M_2O{\cdot}Al_2O_3{\cdot}SiO_2$ glasses, one first needs to associate the alkali with available Al ions. Any residual alkalis then create nonbridging oxygens on the silicons. If the number of Al ions is greater than the number of alkali ions, then after all the available alkalis are associated with aluminum ions, each residual Al ion creates three NBO.

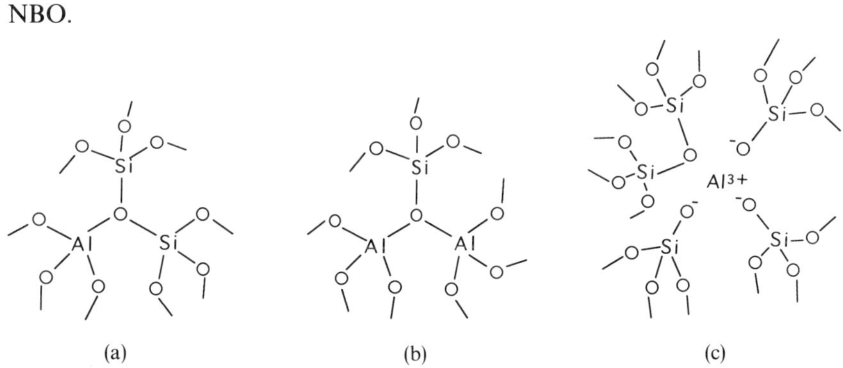

Figure 5-24. Tetrahedrally coordinated Al^{3+} in tricluster arrangements (a) and (b) according to Lacy. Note that (c) is equivalent to (a).

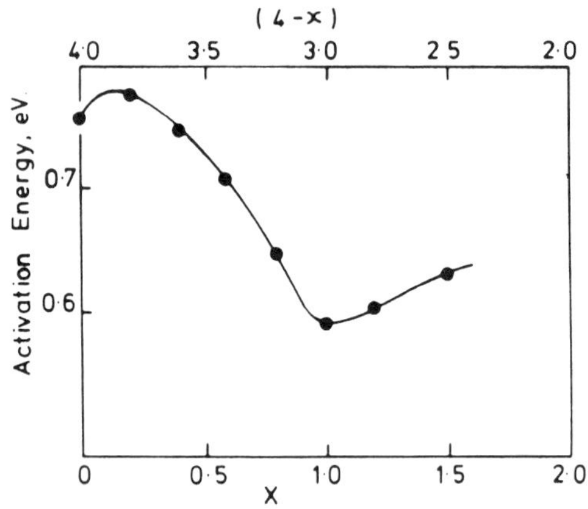

Figure 5-25. Effect of composition on the activation energy for dc conduction in the $Na_2O{\cdot}xAl_2O_3{\cdot}2(4\text{-}x)SiO_2$ glass system. (After Isard (20). Reproduced with permission of the Society of Glass Technology.)

Example 1: Calculate the structural configurations in a $15Na_2O{\cdot}10Al_2O_3{\cdot}75SiO_2$ glass.

In this case $[Al_2O_3/M_2O] < 1$; hence, $\#Al\square = 20$, which absorb 20 Na ions.
Remainder #Na ions = 10, which create 10 NBO. Hence, $f_{NBO} = 10/195$.
Total number of tetrahedra = 20 (for Al) + 75 (for Si) = 95.

Example 2: Calculate the structural configurations in a $20Na_2O{\cdot}20Al_2O_3{\cdot}60SiO_2$ glass.

In this case $[Al_2O_3/M_2O] = 1$; hence, $\#Al\square = 40$, which absorb 40 Na ions.
Remainder #Na ions = 0. Hence, $f_{NBO} = 0$.
Total number of tetrahedra = 40 (for Al) + 60 (for Si) = 100.

Example 3: Calculate the structural configurations in a $10Na_2O{\cdot}15Al_2O_3{\cdot}75SiO_2$ glass.

In this case $[Al_2O_3/M_2O] > 1$; hence, $\#Al\square = \#Na = 20$.
Remainder #Al ions (octahedral) = 10, which create 30 NBO.
Hence, $f_{NBO} = 30/205$.
Total number of tetrahedra = 20 (for Al) + 75 (for Si) = 95.

5.10. Lead, Bismuth, and Thallium Silicate or Borate Glasses

PbO can be present up to as much as 92 wt. % in lead silicate glasses, and perhaps up to 94% in borates. If Pb were present as network-modifier ions only, then the 92% PbO would be beyond the limit where all the SiO_4 tetrahedra are connected. Based upon XRD of $PbO–SiO_2$ glasses of up to 60 mol % (84.6 wt. %) PbO, Bair [21] suggested (by extrapolation) that for the $2PbO{\cdot}SiO_2$ composition all the SiO_4 tetrahedra must be connected to each other via a Pb atom, thus allowing a glass network-forming role for Pb where all the oxygens are bridging and connected to one Si and one Pb atom. Refined XRD work [22] later on showed that the first Pb–Pb distance was 3.9 Å, and the coordination of Pb around Pb varied between 6 and 9. A possible structure for $2PbO{\cdot}SiO_2$ is shown in Fig. 5-26a. Note that, because of the 2-coordination of Pb, the structure contains substantially large holes in which presumably more PbO could be accommodated, if needed. NMR work, however, has revealed the presence of PbO_4 pyramids where Pb forms the vertex. Current belief is that, as in crystalline PbO, high-lead glasses may

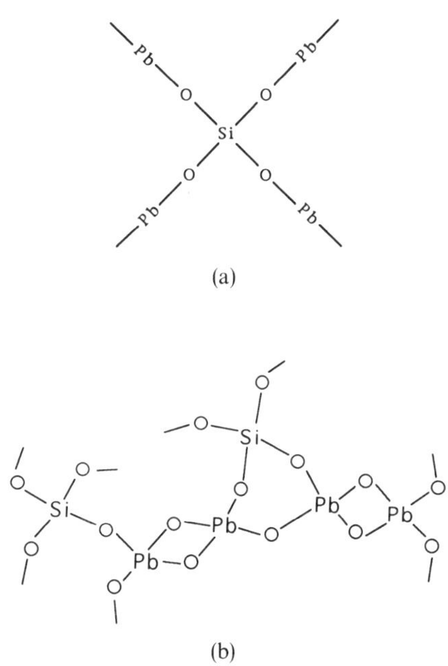

Figure 5-26. Structure of lead silicate glasses showing (a) two-coordinated Pb, (b) tetrahedrally coordinated Pb. (After Mydler *et al.* [22]. Redrawn with permission of the Society of Glass Technology.)

have Pb–O–Pb bonds in the form of twisted PbO_4 pyramids (Fig. 5-26b). Apparently, four Pb–O distances are 2.3 Å and four other Pb–O distances are 4.3 Å: the asymmetry of the pyramids contributes towards randomness.

As in the PbO–SiO_2 glasses, glass formation in Bi_2O_3–SiO_2 and Tl_2O–SiO_2 glasses can occur up to very high "modifier" content. Because the Bi^{3+} and the Tl^+ ions are very large, they are highly polarizable. The high polarizability apparently leads to a high degree of asymmetry in the cation–oxygen polyhedra, and perhaps helps glass formation.

5.11. Phosphate Glasses

Phosphorus is a network-forming element. However, only P_2O_5 is known to form a glass. All three allotropic forms of P_2O_5 (hexagonal H, orthorhombic O, and tetragonal T) can form glasses after melting and cooling. The H form slowly converts to the O form around 380°C, which itself is metastable with respect to the T form from the melting point (580°C) down to 70°C. Because of the slow conversion to the T form, the properties of glass such as the viscosity and refractive index depend upon the parent crystalline form, the melting time, and the cooling rate. In both the crystals and the glasses, the basic unit of structure is the PO_4 tetrahedron. Because P is a pentavalent ion, one oxygen from each tetrahedron remains nonbridging to satisfy charge neutrality of the tetrahedron (Fig. 5-27). In this respect, P_2O_5 differs from SiO_2.

Because of the presence of the nonbridging oxygen, which can act as a terminator, P_2O_5-containing glasses can form isolated ring (cyclic) molecules. It is believed that the melt obtained from the hexagonal form initially contains isolated P_4O_{10} molecules (Fig. 5-27b), which gradually link up to form larger-sized (cyclic) molecules, then linear-chain and finally layered-sheet structures. In P_2O_5, the connections are made only at three corners (unlike SiO_2, where all four corners are connected) (Fig. 5-27a); hence, the P_2O_5 glass is significantly less rigid (or is more fluid) compared to SiO_2 glass. Presumably, like the B_2O_3 glass, the P_2O_5 glass in the sheet form consists of layers of oxygen polyhedra with weak van der Waals attraction between the layers. The rigidity of the structure is increased by the addition of alkalis or alkaline earths. The extent of glass formation in alkali and alkaline earth binary phosphates is shown in Table 5-3 and is generally larger than in the silicates or the borates. The range is larger when, for instance, mixed alkalis are present. It is believed that the ring molecules are broken by the modifier ions to form linear phosphate chains with varying chain lengths depending upon the modifier ion/phosphorus ratio. For instance, in

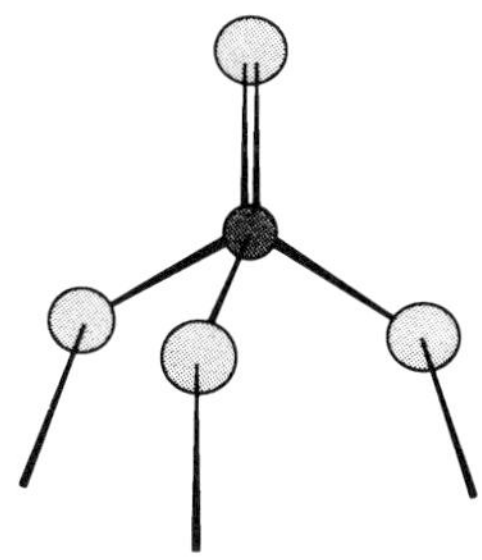

a

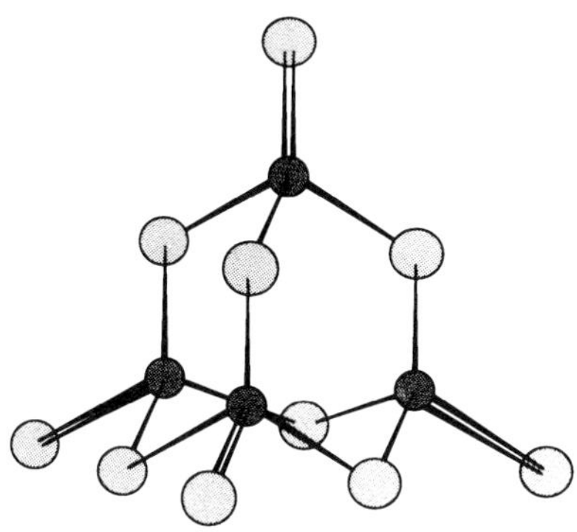

b

Figure 5-27. Structures of (a) PO_4 and (b) P_4O_{10} molecules. Darker-shaded circles represent P.

$Na_2O–P_2O_5$ glasses with a Na/P ratio exceeding 1, linear chains of the general formula $Na_{n+2}P_nO_{3n+1}$ and cyclic chains of the general formula $(NaPO_3)_n$ are found. Some examples of linear chain configurations in alkali and alkaline earth phosphates are shown in Fig. 5-28.

The measurement of chain lengths has been accomplished by dissolving the glass in water and using paper chromatography techniques, notably by Westman [23]. (It has been shown that the dissolution process in water does not create or destroy the polymeric structure already present in the glass.) The distribution of chain lengths in various sodium phosphate glasses with $Na/P \geq 1$ is shown in Fig. 5-29. The percentage of total phosphorus present in linear chains with various values of n are plotted against the parameter n_{av} (=number average chain length), and T ($=100/n_{av}$, the number of molecules per 100 P atoms). The numbers on the curves give the values of

Table 5-3. Regions of Glass Formation in Binary Phosphate Systems[a]

Modifier	Maximum percentage of modifier (mole per cent)	
	Elyard *et al.*[b]	Imaoka[c]
K_2O	—	47
Na_2O	—	60
Li_2O	—	60
BaO	57	58
SrO	56	56
CaO	58	56
MgO	65	60
BeO	—	66
Ag_2O	—	66
Tl_2O	—	50
ZnO	71	64
CdO	66	57
PbO	66	62

[a] After Rawson.

[b] Melted on scale of approximately 20 mg on electrically heated platinum alloy wire. Melt quenched by switching off the heating current. C. A. Elyard, P. L. Baynton, and H. Rawson, *Glastech. Ber.*, 32K, V, 36–43 (1960).

[c] Melted on scale of 1–3 g in platinum crucible. Melt cooled in the crucible. M. Imaoka, in "Advances in Glass Technology, Part I", pp. 149–164, Plenum Press, New York (1962).

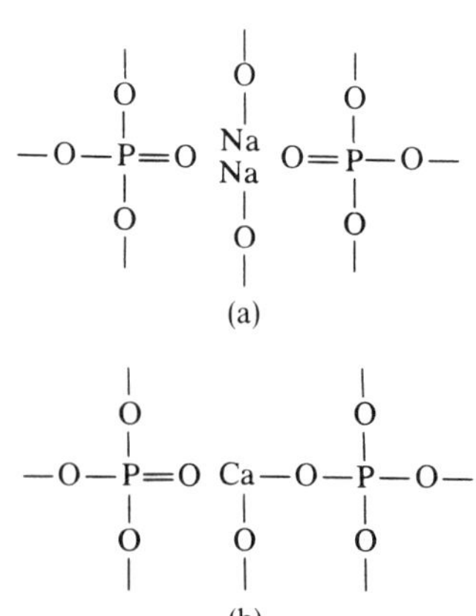

Figure 5-28. Crosslinking of P_2O_5 by (a) alkali and (b) alkaline earth oxides.

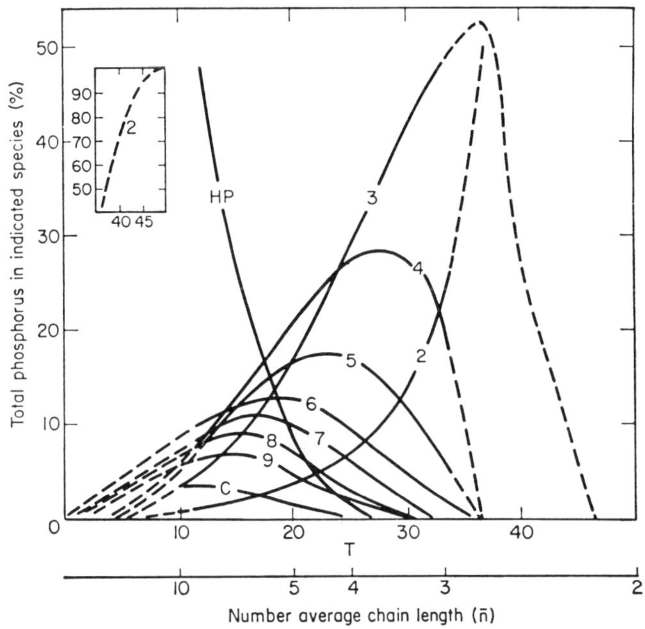

Figure 5-29. Constitution of sodium phosphate glasses. HP = hypolyphosphite; C = total cyclic phosphates. Inset shows the extension of the pyrophosphate curve ($n = 2$). Dashed lines on the right are based on results for rapidly flame-fused beads. Dashed lines on the left are extrapolations. (After Westman (23). Reproduced with permission of Butterworth-Heinemann Ltd., Oxford.)

n from 2 to 9. Curve HP ("hypolyphosphate") gives the percentage in linear molecules with $n > 9$, and the curve C is the percentage in cyclic molecules. If the number of cyclic molecules is ignored, then n_{av} may be calculated using the formula $n_{av} = 2/[(Na/P) - 1]$, which shows that the composition $Na_2O \cdot P_2O_5$ with Na:P = 1:1 has essentially a single infinite-chain-length molecule ($n_{av} = \infty$).

The charge compensation for the PO_4 tetrahedron can be accomplished exactly by replacing one pentavalent P by one trivalent Al or B. Thus, the structure of $AlPO_4$ is isomorphic to SiO_2 (written as $SiSiO_4$). In fact, in the crystalline form, analogous allotropic forms of $AlPO_4$ exist for all allotropic forms of silica (including high and low forms). Yet, $AlPO_4$ can only be brought to glassy state by vapor phase condensation. The reasons for this anomaly (relative to silica) are not clearly understood. It is suggested [24] that with the addition of Al_2O_3 to P_2O_5, the size of the tetrahedral AlO_4–PO_4 clusters continue to grow; however, full connectivity is still not obtained even at the stoichiometric $AlPO_4$ composition, allowing structural rearrangement, and thus crystallization, to occur. Experimental problems in the melting of $AlPO_4$ are augmented because of the rapid volatilization of P_2O_5.

5.12. Other Oxide Glasses

Among other oxide glasses, only GeO_2 is known to form a glass by normal cooling of a melt. The glass has a tetrahedral quartzlike structure, rather than the more open cristobalite-like structure of silica glass. This presumably results from the fact that Ge/O radius ratio is 0.414, which is on the border between tetrahedral and octahedral structures. Neutron diffraction studies show that the primary Ge–O bond is 1.73 Å (about 8% larger than the Si–O bond). Other bond distances are O–O = 2.83 Å, Ge–Ge = 3.45 Å; the average Ge–O–Ge bond angle is 133° (compared to about 144° for the Si–O–Si bond angle in silica glass).

Upon the addition of modifier alkalis, the coordination number of Ge changes from 4 to 6. As shown by Kamiya and Sakka (25), the conversion continues to occur up to about 35% GeO_6 at around 30% added alkali. Thereafter, a sharp decrease is observed (Fig. 5-30). Not all of the oxygens in the GeO_6 group are bridging.

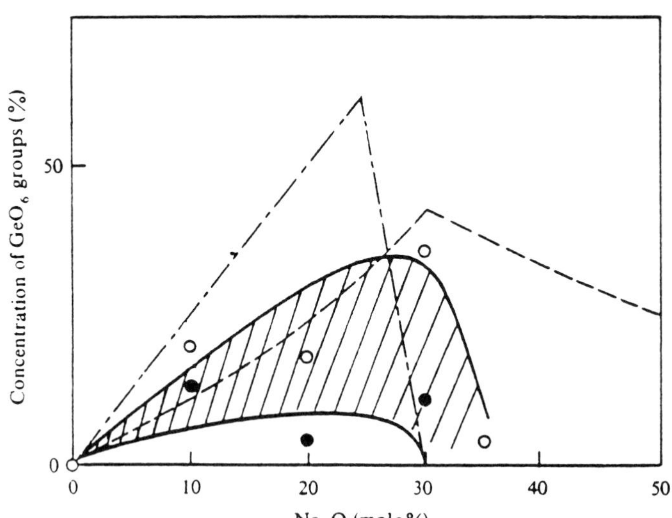

Figure 5-30. Concentration of GeO_6 groups as a function of Na_2O. Open circles and filled circles are calculated upper and lower limits. (After Kamiya and Sakka [25]. Reproduced with permission of the Society of Glass Technology.)

Al_2O_3, As_2O_3, Sb_2O_3, Bi_2O_3, V_2O_5, TiO_2, TeO_2, etc., are not known to form glasses readily. However, some such as As_2O_3, Sb_2O_3, and V_2O_5 have been claimed to form glasses by vapor-phase condensation, or by splat-cooling techniques. Many binary glasses that include a large amount of these oxides can be formed, particularly with alkalis, alkaline earths, PbO, and other glass network formers. For example, Table 5-4 shows the glass formation range in binary tellurite glasses. The coordination number of Al in aluminate glasses is thought to be 4. However, such studies are not conclusive. In arsenates, antimoniates, and bismuthates, it has been found that the primary polyhedra are pyramids of MO_3 with M slightly out of the oxygen triangular plane (much as in B_2O_3). In vanadates, the vanadium is present as VO_4 tetrahedra having three bridging oxygens and one doubly bonded nonbridging oxygen (as in PO_4). In tellurites, Te^{4+} ion may occur in three or four coordinated structures.

Table 5-4. Range of Glass Formation in Tellurite Systems[a,b]

Metal oxide	Range (mol %)
Li_2O	12.2–34.9
Na_2O	5.5–37.8
K_2O	6.5–19.5
Rb_2O	5.6–21.0
BeO	15–27
MgO	11–35
CaO	—
SrO	9.2–13.1
BaO	8.0–35.7
ZnO	17.3–37.2
Al_2O_3	7.6–16.8
Tl_2O	13.0–38.4
PbO	12.8–22.6
Nb_2O_5	2.2–24.0
Ta_2O_5	1.4–15.3
WO_3	8.5–44
La_2O_3	4.5–9.5
TiO_2	6.2–18.9
ThO_2	5.2–11.0

[a] From W. Vogel, H. Bürger, B. Müller, G. Zerge, W. Müller, and K. Fortel, *Silikattechnik*, **25**(6), 205–209 (1974).

[b] 20–100 g, cooled at 8–10°C/s through transformation.

5.13. Amorphous Silicon and Germanium

In amorphous silicon (abbreviated *a*-Si) and *a*-Ge, the local atomic coordination number = 4 (from the "8-*N*" rule, Section 3.1.4). The structure of these amorphous solids is hence believed to be a continuous random network (CRN) of tetrahedrally bonded atoms. The structure is similar to the diamond lattice in that each atom is four-coordinated, bond lengths are constant, and there are no dangling bonds. However, in contrast to the diamond structure, there presumably is a spread of bond angles, and of course, there is no long-range periodicity.

Polk (26) was the first to attempt to construct a model of *a*-Si that observed the preceding criteria. Metal rods of equal length representing the bonds snapped into a plastic section representing the atom. The plastic section

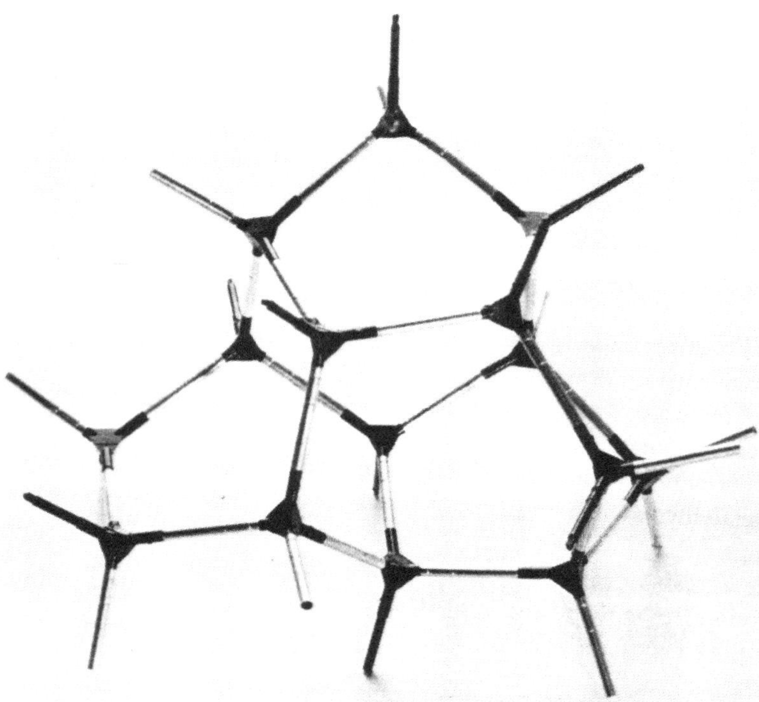

Figure 5-31. Polk model for the structure of *a*-Si. Five-membered rings may be seen at the top (atoms 10, 11, 13, 14, 15) and lower right (atoms 3, 4, 5, 12, 7). A six-membered ring in "chair" configuration is on edge at the right (atoms 4, 5, 6, 11, 12, 13). "Boat"-configured six-membered rings are at the left (atoms 1, 2, 10, 15, 8, 9 and atoms 7, 3, 2, 8, 15, 10).

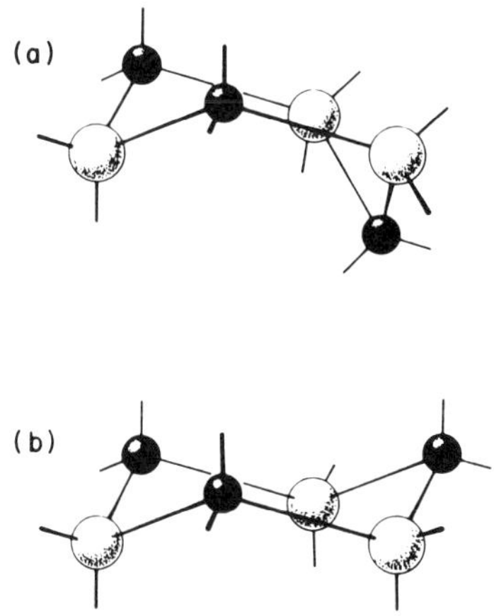

Figure 5-32. (a) "Boat" and "chair" configurations of six-membered rings.

could be bent somewhat, allowing slight variation in the tetrahedral angles. Polk showed that for the fixed equal bond lengths, the variation in $\Delta\theta$ in the tetrahedral angle (109° 28′) was only about 9°. A small portion of his 440-atom model is shown in Fig. 5-31.

Whereas the shortest of the rings in crystalline silicon are six-membered, the shortest in the Polk model was five-membered. The six-membered rings in the diamond structure all have the "chair" type of configuration (Fig. 5-32b) with staggered (60°) dihedral angles (Fig. 5-33b). In the crystalline wurtzite structure, one-quarter of the six-membered rings have the "chair" configuration and three-quarters have the "boat" configuration (Fig. 5-32a). In the latter, four dihedral angles are staggered, and two are eclipsed (Fig. 5-33a). In the Polk model, the dihedral angles varied as a smooth increasing function from 0° to 60°, indicating the presence of both the chair and boat configurations. The calculated and measured [27] RDFs of *a*-Si are compared in Fig. 5-34. Note the sharpness of the bond length in the Polk model and the apparent disagreement with the measured RDF. One would tend to argue that bond lengths presumably vary more than was assumed by Polk. As a result, the medium-range order and, in particular, the ring statistics of the Polk model are clearly wrong.

In fact, the difficulty lies more in terms of the constraints applied in building stick-and-ball models. Using a Keating-type potential (involving bond-stretching and bond-bending distortions, similar to Eq. (3.3)) in a computer model of as many as 519 atoms, Steinhardt *et al.* [28] obtained an RDF for *a*-Ge that was then relaxed for vibrational, instrumental, and analytical broadening. The resultant RDF was found to be in very good

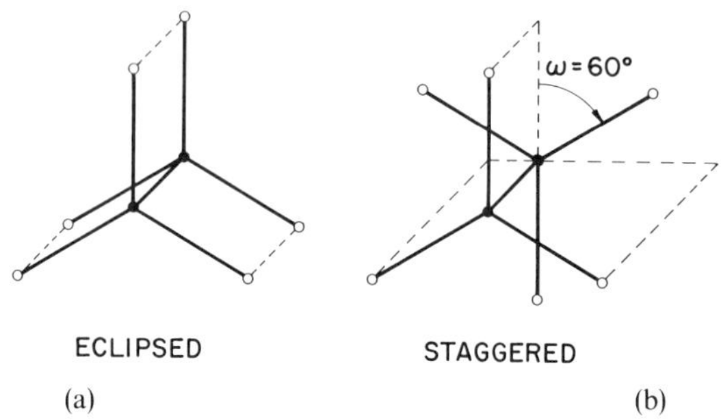

Figure 5-33. The (a) eclipsed and (b) staggered bond orientations in tetrahedrally bonded structures.

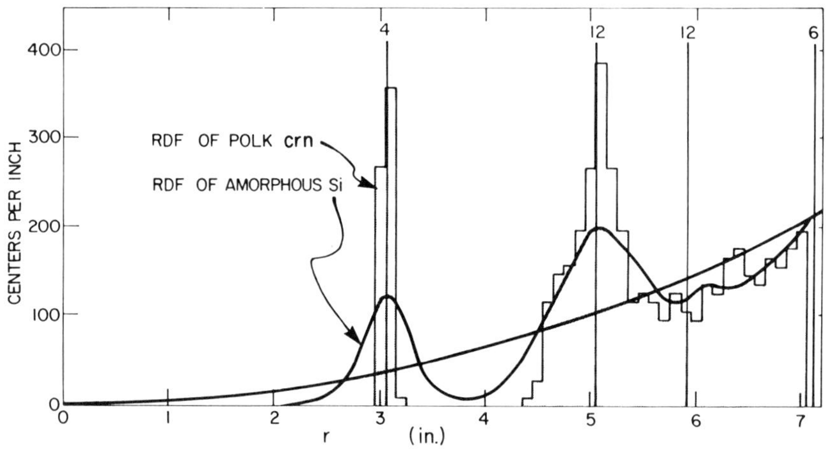

Figure 5-34. Comparison of measured RDF of *a*-Si with that of the Polk CRN model. (Note that the experimental RDF distance was scaled up by a factor of 10^8 to correspond with the model dimensions in inches.) (After Moss and Graczyk [27]. Reproduced with permission of Elsevier Science Publishers.)

agreement with one measured experimentally (Fig. 5-35). Steinhardt *et al.* further showed that the introduction of 1% variation in bond lengths allowed the angular variation to be reduced to 7%, which brought the amorphous structures energetically closer to the crystalline structures while still maintaining good agreement of the RDFs. (There may still be some room to argue that agreement between the Steinhardt *et al.* calculation and the experimental data for the Ge–GeII peaks is far from being excellent.)

On the whole, the Polk model for the structure of amorphous semi-

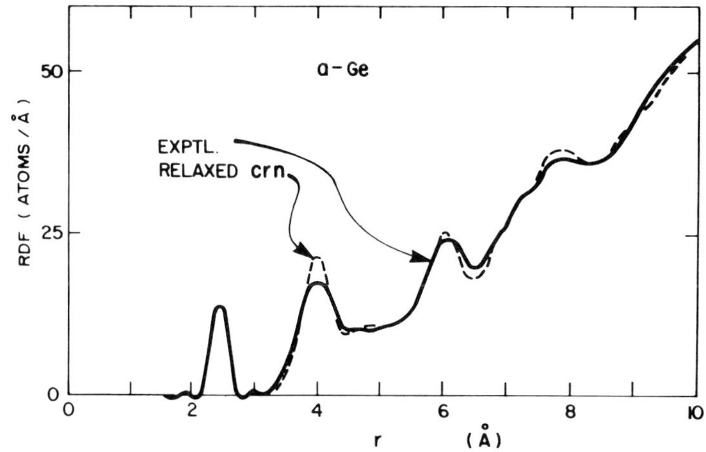

Figure 5-35. Comparison of measured RDF of *a*-Ge with that of the relaxed CRN model. (After Steinhardt *et al* [28]. Reproduced with permission of Elsevier Science Publishers.)

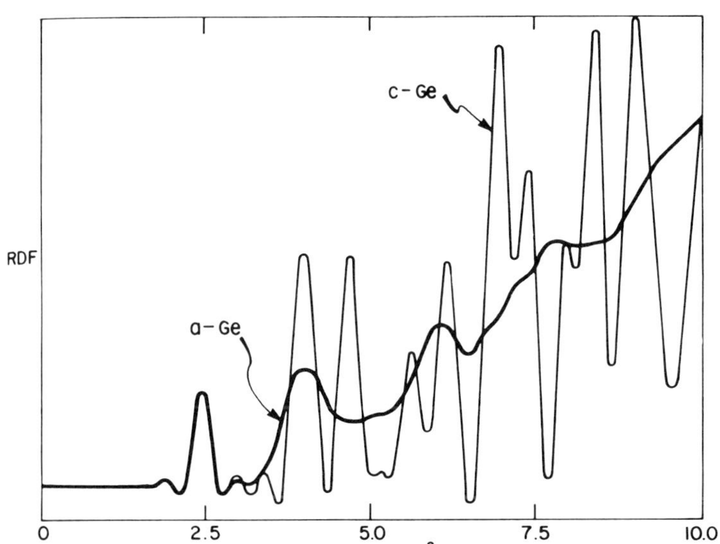

Figure 5-36. Comparison of x-ray derived RDFs of *a*- and *c*-Ge. (After R. J. Temkin, W. Paul, and G. A. N. Connell, *Adv. Phys.* **22**, 581 (1973). Reproduced with permission of Taylor & Francis, Inc.)

conductors appeared successful. The most interesting feature was the third near-neighbor peak in the crystal, which was completely absent in the amorphous material (e.g., see Fig. 5-36 for Ge). More importantly, though, it brought out the following:

1. It is possible to arrange atoms in a fairly tight short-range order, but not have long-range order typical of crystals.
2. Stick-and-ball models generally do not represent the medium-range order well enough to give accurate ring statistics.
3. Computer simulations need to be carried out carefully to include the different relaxation procedures; otherwise, the comparison of the calculated RDF with experimentally measured RDF is not meaningful. In colloquial terms, it is called "GIGO" or "garbage in, garbage out."

One should bear in mind that the Polk model applies to purely covalent solids. In many ways, the structure is quite simple indeed (in fact, the simplest of all considered so far). Next, we consider an even simpler structure: that of glassy metals, where there are no covalent bonds, no fixed coordination, and the metallic bond allows more variation on the bond distance.

5.14. Glassy Metals

Glass formation in metallic systems occurs in four basic binary systems: (a) the transition metal–metalloid (abbreviated as T-M) system; (b) the early and late transition metals (TE, TL) system; (c) the alkaline earth (AE) with rare earths (RE) and simple metals (S) system; and (d) the transition metal–actinides (T-AC) system. Table 5-5 is a summary of the basic binary systems. Of these, the first system, where the content of the transition metal is generally about 70–80% and the metalloid up to about 30%, is perhaps the most studied one.

Models to describe the structure of the glassy metals generally fall within one of the four classes: microcrystallites, random network, dense random packing of spheres, and random packing of trigonal prisms. The first two are quite like the ones proposed for oxide systems, are based on XRD data, and obviously suffer from like criticism. The broadening of x-ray peaks could be fitted to a crystallite size, but such crystallites would often have to be quite small to be meaningful as "ordered" clusters. As in the case of oxide glasses, the random network model fits mostly single-component amorphous

Table 5-5. Binary Glass-Forming Metallic Systems

Group	Class[a]	Typical composition of glasses	Most stable intermetallic compounds
a	T–M	$Pd_{80}Si_{20}$	Pd_3Si, Fe_3P (cementite)
		$Ni_{80}P_{20}$	Pd_2Si, Fe_2P
		$Fe_{40}Ni_{40}P_{14}B_6$ (Metglas 2826)	
b	TE–TL	$Nb_{60}Ni_{40}$	"NbNi" disordered phase
		$W_{45}Fe_{55}$	Fe_7W_6 phase
		$Zr_{76}Fe_{24}$	
	TL–RE	$Co_{33}Gd_{67}$	$CdCo_2$ Laves phase
		$Ni_{30}Gd_{70}$	$GdCo_5$, Gd_2Co_{17} } Frank–Kasper phase
c	AE–AE	$Ca_{67}Mg_{33}$	$CaMg_2$ Laves phase
	AE–S	$Mg_{70}Zn_{30}$	$MgZn_2$ Laves phase
	AE–T	$Ca_{65}Pd_{35}$	$CaPd_2$ Laves phase
	S–RE	$Al_{30}La_{70}$	$LaAl_2$ Laves phase
d	AC–T	$U_{70}Cr_{30}$	

[a] T = transition metal; M = metalloid; TE = early transition metal; TL = late transition metal; RE = rare earth; AE = alkaline earth; S = simple metal; AC-actinide.

solids (*a*-Si and *a*-Ge, discussed earlier). The third and the fourth models describe the structure of most of the glassy metal systems equally well, and they are worth considering here in more detail.

The concept of **dense random packing (DRP)**, sometimes called **random close packing (RCP)**, of single-sized hard spheres to describe liquid structures had been advanced by Bernal [29] and was adopted later for the description of glassy metal structures. The idea is simple conceptually: take monosized hard spheres and place them in a flexible bag; squeeze on the bag till the volume does not shrink significantly. The locations of the centers of the spheres simulate a glassy metal. These simulations have now been tried using 7,934 ball bearings by Finney [30], and as computer experiments by Bennett [31]. The RDFs measured from these simulations are generally in good agreement with those measured using XRD (Fig. 5-37). The packing fraction obtained in both the physical models is about 0.64, which is less than the closest packing fraction (0.74) of the fcc and the hcp structures. This difference in the packing fraction is undoubtedly due to the presence of holes in the amorphous structures that are large enough to accommodate a smaller hard sphere (but not enough for another equisized hard sphere). The centers of the spheres themselves are vertices of one of five types of polyhedra (Fig. 5-38) (allowing for some distortion of the edges). The holes enclosed by the

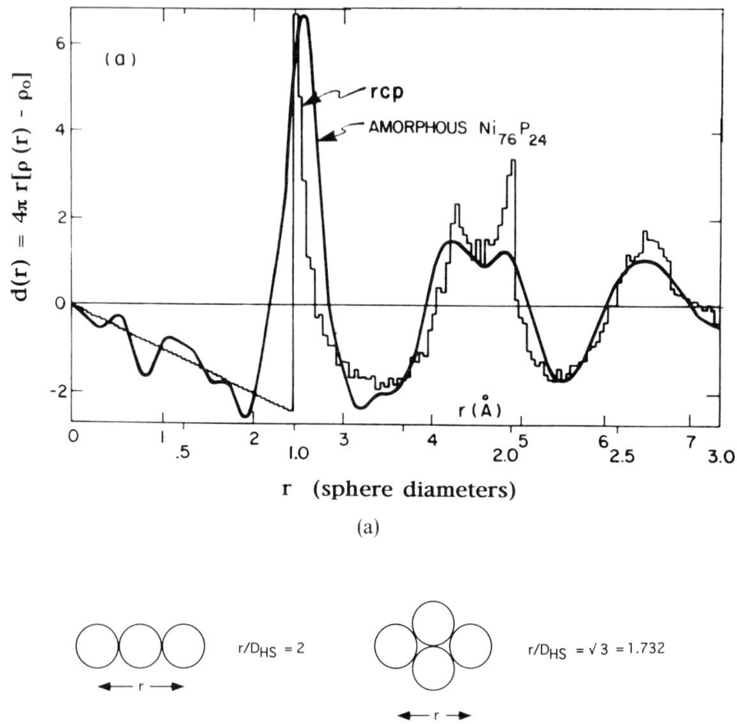

(b)

Figure 5-37. (a) Comparison of the differential correlation function $d(r)$ for Finney's DRP of monosized spheres and the experimental curve for a-$Ni_{76}P_{24}$. Inset (b) the origin of the double peak at second coordination. (After G. S. Cargill, *Sol. St. Phys.* **30**, 227 (1975).)

polyhedra are called **Bernal's canonical holes**. The canonical holes share faces and fill the space. Of the five canonical polyhedra, the smallest, i.e., the tetrahedron, occurs most frequently; there are about 2.9 tetrahedral holes per sphere. This is about 25 times more the next most frequent [32] (the trigonal prism and the tetragonal dodecahedron).

Another way of filling the space in the DRP structures would be through **Wigner-Seitz cells**, which in disordered systems are actually *Voronoi polyhedra* (Fig. 5-39). These are the polyhedra whose faces are the planes perpendicular to, and passing through the midpoint of, the lines joining the sphere centers. (About 40% of the faces are pentagons.) If a large number of fresh peas were placed in a pot, shaken well till volume did not decrease much further, water was added to cover them, and a sufficiently heavy lid

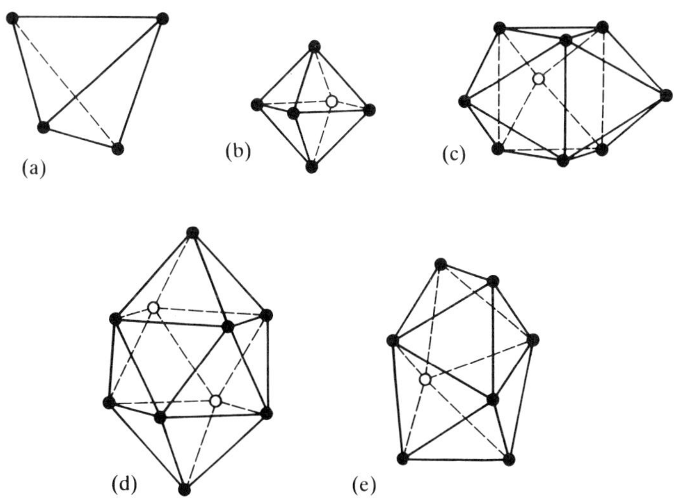

Figure 5-38. The "canonical holes" of Bernal. (a) Tetrahedron. (b) Octahedron. (c) Trigonal prism (shown capped with half-octahedra at each of the three rectangular faces). (d) Archimedean anti-prism (shown capped with half-octahedra on the bottom and the top faces). (e) Tetragonal dodecahedron.

closed off the volume (without any air above the peas), then the peas would essentially dilate into Voronoi polyhedra (experiment of botanist Stephen Hales, 1727).

The **random trigonal prisms packing (TPP)** model was suggested by Gaskell [32B] to describe the structure of the transition metal–metalloid glasses. Apparently, MT_6 trigonal prisms found in T_3M crystalline alloys appear to persist in glasses of close compositions such as T_4M. The T atoms are at the apices of the trigonal prism (the T_I sites), and the M atom is at the center of the prism (Fig. 5-40a). In the cementite (Fe_3C) structure, where the radius ratio of M to T is small, there are three other T atoms that surround the M atom on the equatorial plane (placed in the octahedral site of the rectangle ABED), but at a larger distance from M (the T_{II} sites). The location of the second prism is obtained by rotating the first prism 215° clockwise about the axis AB (Fig. 5-40b). The second prism then shares the edge AB (two T_I sites) with the first prism, and G (a T_{II} site) to make up its triangular base. The small radius ratio allows the trigonal prisms to have little distortion; as a result, the M–T_{II} distances are significantly larger than the M–T_I distances. On the other hand, when the M to T radius ratio is large, as in the Fe_3P structure, the T_{II} sites are placed in the tetrahedral site above the ABC triangle, and again the second prism shares AB (two T_I sites) and G (one T_{II} site); see Fig. 5-40c. Because of the large size of the metalloid

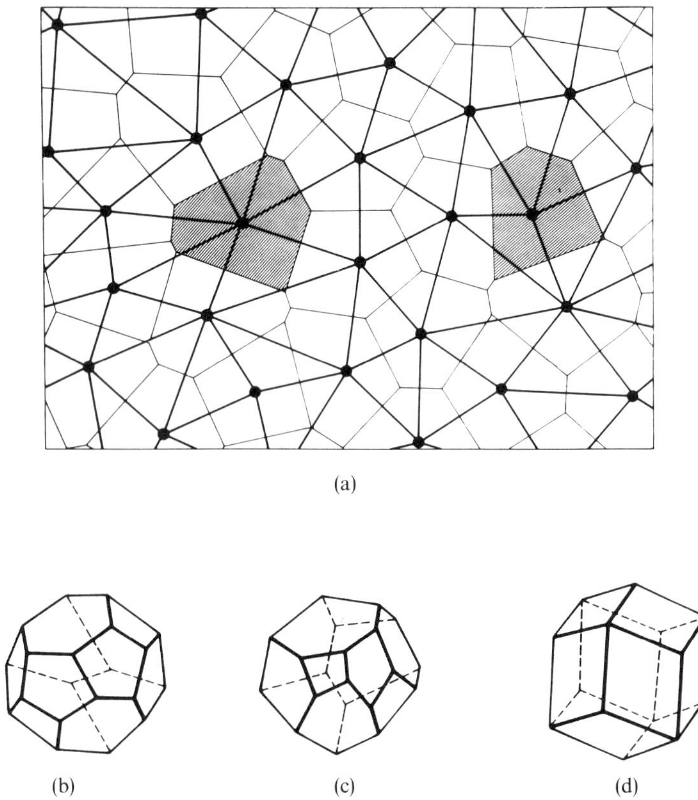

Figure 5-39. (a) 2-D projection of the spaces (Voronoi polyhedra or Wigner–Seitz cells) occupied by dense random packing of atoms. Note that the atom centers are joined, and the perpendicular bisector planes are constructed. The different planes are the faces of the Voronoi polyhedra. (After R. Zallen, *The Physics of Amorphous Solids*, Fig. 2.5, p. 47. Wiley & Sons, New York, 1983. Reproduced with permission of the publishers.) (b) and (c) Two possible shapes of the Voronoi polyhedron when peas are allowed to swell to occupy space. Note the predominance of pentagonal faces. (d) Rhombic dodecahedron cell of the ordered cubic close packing.

atom, the M–T_I distances are larger in the Fe_3P structure than in the cementite structure, and the M–T_{II} distances are not much greater than the M–T_I distances. Randomness in the sharing of edges in both these motifs then allows varying chain or 3-D structures over a variety of composition ranges. For instance, when on an average two edges are shared, the composition would be T_4M.

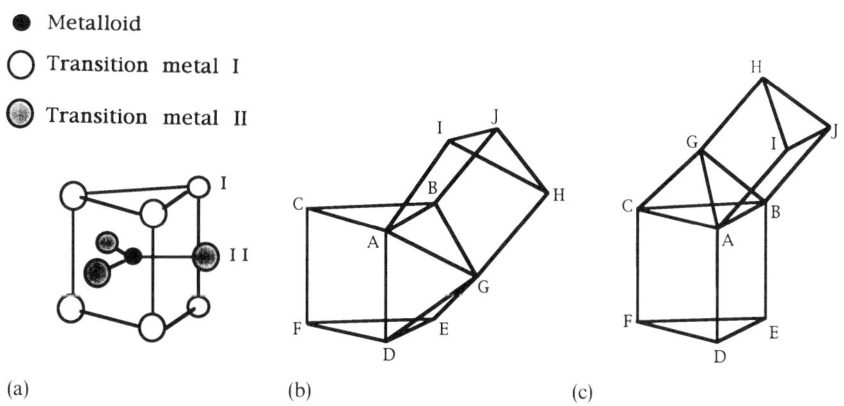

Figure 5-40. (a) Trigonal prismatic packing of transition metals around a metalloid atom. (b) Edge-sharing arrangements in the Fe_3C (cementite) structure. (c) Edge-sharing arrangements in the Fe_3P structure. Note the distance between the metalloid atom M and the second nearest neighbor transition metal (T_{II}) is smaller here than in (b).

5.15. Chalcogenide and Chalcohalide Glasses

The chalcogen elements (S, Se, and Te) can form glasses by themselves upon cooling from a melt. Glasses made by combining chalcogen elements with group IV and V elements are termed chalcogenides. When combined with the halogen group of elements, the glasses are called chalcohalides. Glasses consist of disordered rings (molecules), chains, sheets, and three-dimensional networks. The bonding is generally covalent, with weak van der Waals attraction between the molecules, chains, etc. The coordination number of various elements in the chalcogenide glasses is given by the "8-N" rule, where n is the number of the outer shell electrons (see Section 3.1.4). Thus, the coordination number of the chalcogen elements is 2; that for As and Sb is 3, and for Ge it is 4. In a chalcogenide/chalcohalide glass, the average coordination number, $\langle m \rangle$, may be defined as the atom-averaged coordination number of the constituent atoms. Thus, a 28Ge·12Sb·60Se glass has an $\langle m \rangle$ value of $[\{28(4) + 12(3) + 60(2)\}/100] = 2.68$. The pure element chalcogen glasses ($\langle m \rangle = 2.0$) are essentially chains that polymerize to varying lengths. For instance, S melts at 115°C to a liquid containing roughly an S_8 ring structure. Upon further heating, the rings open up and polymerize to spirals that may have as many as 10^6 atoms. This structure continues into the glassy form when the stiffened liquid is rapidly cooled. Apparently, the Se_8 rings are easier to break up, and hence, a pure Se glass is easier to obtain.

Ge–Sb–Se

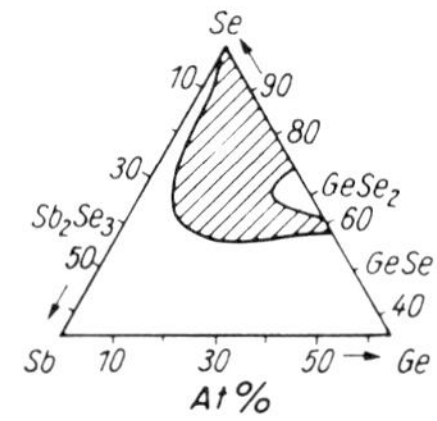

Ge–As–Te

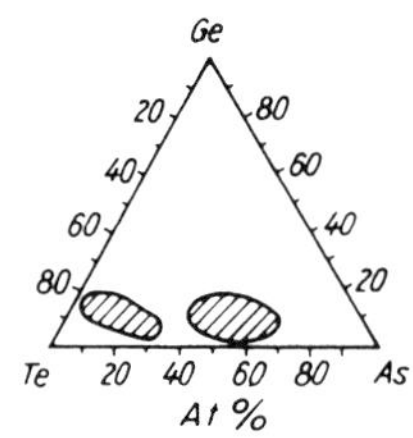

Ge–Se–Te

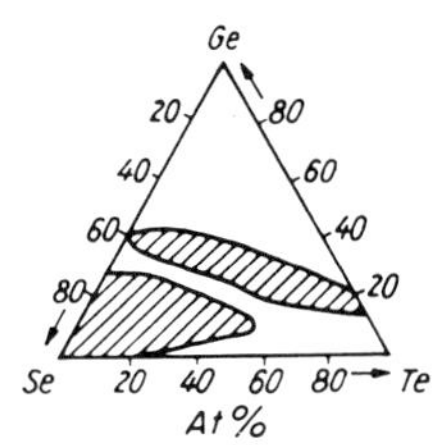

As–S–Se

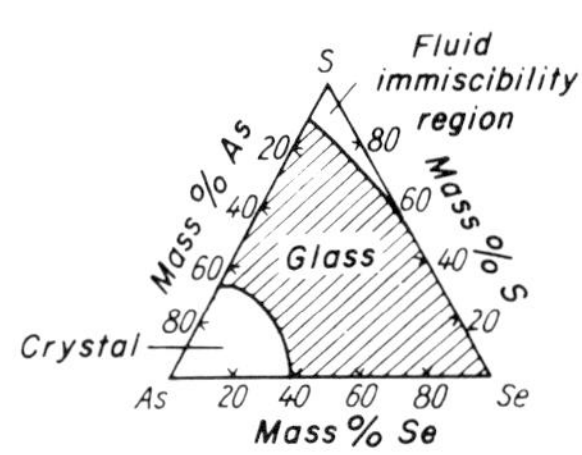

As–Se–Te

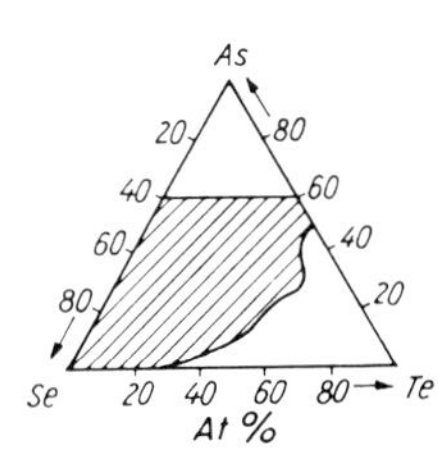

Figure 5-41. Glass formation region shown shaded in the Ge-Sb/As-S/Te ternary diagrams. (From W. Vogel, *Chemistry of Glass*, pp. 184–190. Amer. Ceram. Soc., 1985. Reproduced with permission of the publisher.)

Pure Te glass, on the other hand, is not very readily obtained and requires very rapid quenching from just above the melting point of Te crystal.

The addition of group IV and V elements causes the chains to begin to crosslink because of the increased coordination requirement (4 for group IV and 3 for group V). Since the bonds are covalent, the crosslinking reduces the freedom for disorder. Nevertheless, glass formation is possible over a large range of compositions with various elements. Binary glasses such as As–Se, As–S, Ge–Se, and Sb–Te are not too difficult to obtain. Likewise, large composition ranges in ternary and other multielements can be brought to the glassy state. For instance, the glass formation range is shown shaded on the Ge–Sb–Se, Ge–As–Se, and Si–As–Te ternary diagrams in Fig. 5-41. At low crosslink densities, Ge acts as if it ended four selenium chains; As ends three chains. A 2-D projection ("pancaked version") of one such structure is shown in Fig. 5-42. At moderate to high crosslink densities, the question of crosslink atom bonding is not well understood. On each side of the ternaries, binary crystalline stoichiometric compounds exist. Tie lines joining these binary compounds are called "stoichiometric tie lines." The structure of the glasses changes radically as one crosses these tie lines. Some researchers believe that crosslink atoms can bond randomly (**random covalent network** or the **RCN** model). In other words, in a binary chalcogenide glass $A_{1-x}B_x$, the probability of A–A, A–B, and B–B bonds is determined by random statistics. On the other hand, others believe that heteropolar bonds are always preferred (**chemically ordered covalent network** or the **COCN** model). In this second case, only A–B bonds exist in the stoichiometric crystalline compound (which presumably carry into the glass). On the A-rich

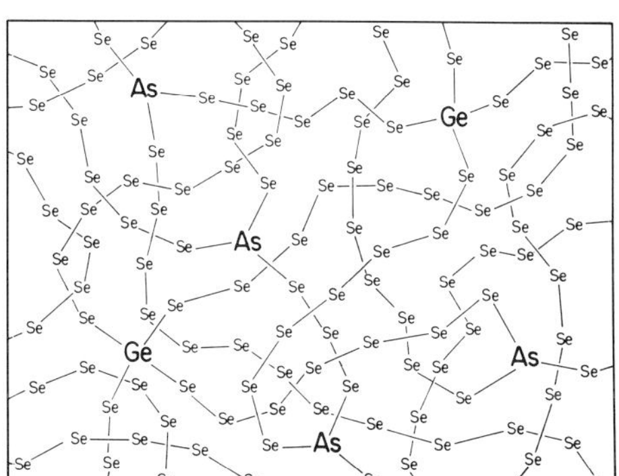

Figure 5-42. 2-D representation of the bonding in Ge–As–Se glass. Note crosslinking of linear chains by As (3-coordination) and Ge (4-coordination). (After R. Zallen, *The Physics of Amorphous Solids*, Fig. 3.7, p. 98. Wiley & Sons, New York, 1983. Reproduced with permission of the publishers.)

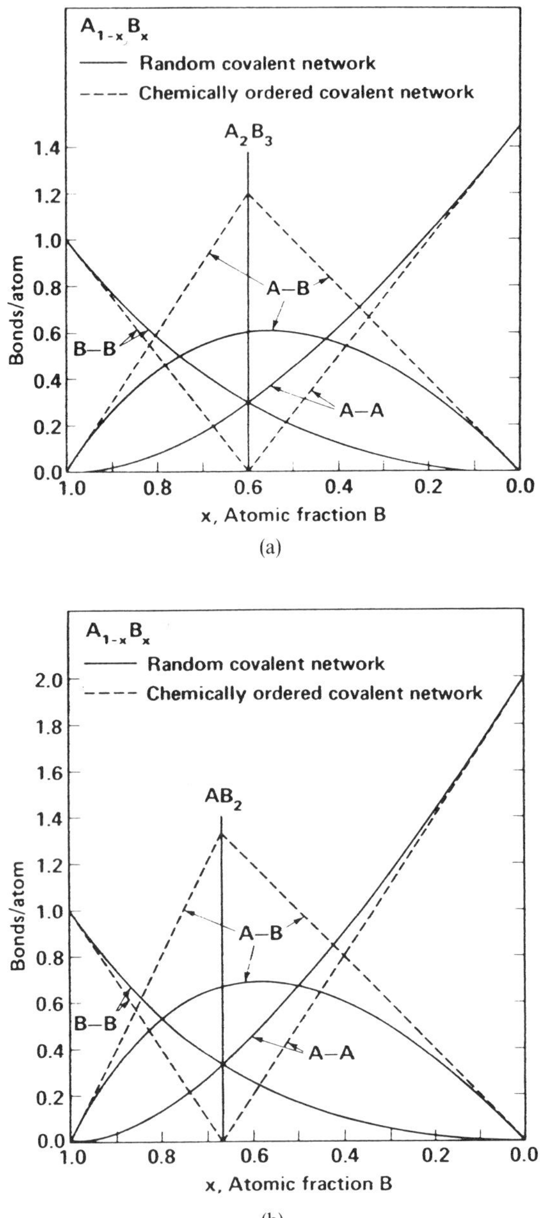

Figure 5-43. Bond statistics in (a) a 3:2 network (e.g., As_2Se_3), and (b) a 4:2 network (e.g., SiO_2). (After Lucovsky and Hayes [33]. Reproduced with permission of Springer-Verlag.)

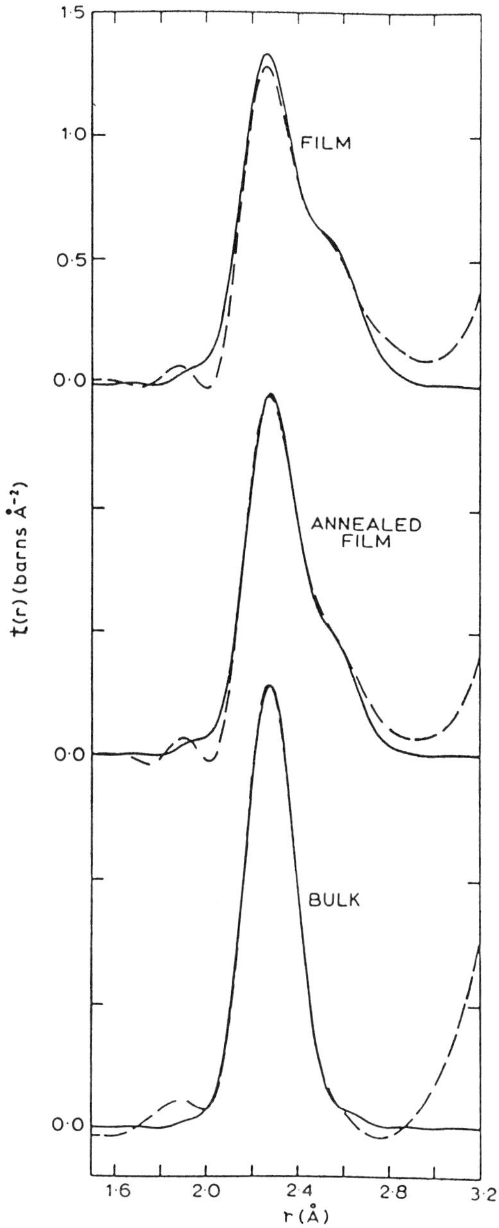

Figure 5-44. Neutron total correlation function for As–S films and bulk. The bulk sample only has one type of bond, presumably heteropolar. This peak clearly has a shoulder in the film indicating the presence, perhaps, of the homopolar ("wrong" type) of bond. (After Daniel *et al.* [34]. Reproduced with permission of Elsevier Science Publishers.)

side of the stoichiometric compound, only A–A and A–B bonds exist, whereas on the B-rich side only B–B and A–B bonds exist. The bond statistics for these two different models are shown [33] for a compound such as Sb_2Se_3 in Fig. 5-43. The characters of the different structural groups present would then dictate the properties of the glasses.

There is some evidence that different preparation techniques may lead to different short- and medium-range order. An example [34] of this is shown in Fig. 5-44, where the neutron diffraction-derived total correlation function, $t(r)$, of As–S films prepared by bulk melting appears to have only one type of bond distance (presumably the heteropolar bond), whereas the vapor-deposited material has two distinct sets of bond lengths (which do not change even upon annealing).

Chalcohalides are formed [35] by melting together the chalcogen elements (Te, Se, and S), halogens, and other group II–IV elements. The glass-forming range is very wide. Compositions such as Te_3Cl_2, Te_3Br_2Se, and Te_4I_5S are typical of this family (also called "TeX" glasses). Chalcohalides are of considerable interest because of their greater transparency in the IR relative to the chalcogenides. In the chalcohalides, the halogens require only one-coordination, and hence act as chain-ending units. As a result, the fluidity of the structure at comparable temperatures greatly increases in comparison to the corresponding chalcogenides. The T_g's of most chalcohalides are in the 60–160°C range.

5.16. Heavy Metal Fluoride Glasses (HMFG)

Suggested reading: M. Drexhage, *Treatise on Mat. Sci. & Tech.*, Vol. 26, *Glass IV* (M. Tomozawa and R. H. Doremus, eds.), pp. 151–243 (1985).

The bonding in the oxide structures is known to be partly ionic and partly covalent. On the other hand, the HMFG are made up of non-oxide ionic salts. The covalency content in these glasses is quite small.

Glass formation in the HMFG system was discovered only as recently as 1974 at the University of Rennes, France. However, because of their possible use as a fiber material for long-distance telecommunication, a vast amount of research effort has been poured into the study of HMFG glasses. The glass formation range is very wide—virtually any metal fluoride can be incorporated into the structure. It is now known that, in addition to the fluorides, other halides can also be included in a wide range of glasses. HMFG compositions are divided into two classes: those based on ZrF_4 or HfF_4 as glass formers, and those excluding ZrF_4 or HfF_4. In the former category, called fluorozirconates or fluorohafnates, the glasses generally have

Table 5-6. Heavy Metal Fluoride Glass Systems and Some of Their Properties

Composition (mole %)	Acronym	T_g (°C)	T_x (°C)	Density (gm/cm³)	n_D
$64ZrF_4$–$36BaF_2$	ZB	300	352	4.66	1.522
$53ZrF_4$–$47ThF_4$	ZT	490	572	5.72	1.551
$50ZrF_4$–$25BaF_2$–$25NaF$	ZBN	240	300	4.50	1.50
$63ZrF_4$–$33BaF_2$–$4GdF_3$	ZBG	310	390	—	1.529
$62ZrF_4$–$33BaF_2$–$5LaF_3$	ZBL	306	380	4.79	1.523
$62HfF_4$–$33BaF_2$–$5LaF_3$	HBL	312	395	5.78	1.514
$57.5ZrF_4$–$33.75BaF_2$–$8.75ThF_4$	ZBT	320	400	4.80	1.523
$57.5HfF_4$–$33.75BaF_2$–$8.75ThF_4$	HBT	319	396	6.19	—
$56ZrF_4$–$15BaF_2$–$6LaF_3$–$4AlF_3$–$20NaF$	ZBLAN	275	405	4.27	1.499
$55ZrF_4$–$30BaF_2$–$15UF_4$	ZBU	320	400	5.01	—
$50ZrF_4$–$43ThF_4$–$7YF_3$	ZTY	465	559	5.41	1.537
$45ZrF_4$–$36BaF_2$–$11YF_3$–$8AlF_3$	ZBYA	344	425	4.54	1.507
$57ZrF_4$–$36BaF_2$–$3LaF_3$–$4AlF_3$	ZBLA	310	390	4.61	1.516
$57HfF_4$–$36BaF_2$–$3LaF_3$–$4AlF_3$	HBLA	312	400	5.88	1.504

50–70% ZrF_4 or HfF_4, although compositions containing as little as about 30 mol % can form glasses. The primary network modifier in these glasses is generally BaF_2 in amounts of about 30 mol %. In the second category, the glasses include large amounts of BaF_2, ThF_4, AlF_3, and ZnF_2. Selected glass compositions of interest and their physical properties (glass transition temperature T_g, crystallization temperature T_x, density, and refractive index) are listed in Table 5-6. Glasses are often called by their acronyms: for instance, glass $53ZrF_4 \cdot 20BaF_2 \cdot 4LaF_3 \cdot 3AlF_3 \cdot 20NaF$ is called "ZBLAN."

Because of the multicomponent nature of these glasses, structural studies have been very difficult. The interpretation of x-ray scattering experiments has often been accomplished using complementary techniques such as molecular dynamics [36]. For the simplest binary system, ZrF_4–BaF_2, the schematics of the glass structure are shown in Fig. 5-45. Unlike other glass network formers where the coordination around the NWF cation is low, the coordination of fluorine around the tetravalent Zr is 7.6 on average. The structure is made up of ZrF_7 and ZrF_8 polyhedra not only joined to each other at corners, but also sharing edges and faces. Fluorine may be of three types: the edge-bridging ("E"), the corner-bridging ("C"), and the nonbridging terminal ("T"). Polyhedral arrangements found in glasses are shown in Fig. 5-46. Some arrangements (Fig. 5-46a) are typical of those found in crystals, while others (Fig. 5-46b) are closely related. (Here, the reader may note that the HMFGs violate just about every criterion of glass formation according to Zachariasen.) The charge neutrality of the polyhedra is observed by "modifying" cations, such as Ba^{2+}, occupying the interstitial spaces. The

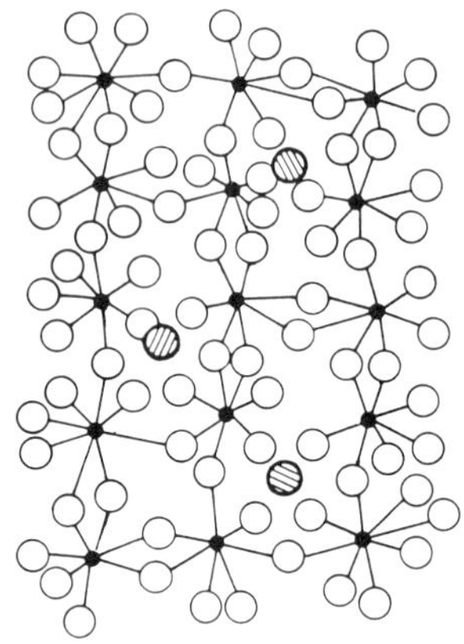

Figure 5-45. MD simulation of $67ZrF_4{\cdot}33BaF_2$ glass. (After Phifer and Angell [36]. Reproduced with permission of Elsevier Science Publishers.)

primary network forming Zr–F bonds in these glasses are believed to be mostly ionic, as opposed to roughly 50–50 ionic and covalent as in SiO_2 and B_2O_3 (see Stanworth's electronegativity criterion for glass formation [22].

In those glasses where AlF_3 is present in large quantities, MD simulations suggest that Al is present as AlF_6 octahedra sharing corners and making a complex network in which Ba, Ca, etc., occupy the interstitial positions. The coordination of fluorine around Ba is found to be between 8 and 9. Interestingly enough, the densities of some of these glasses are higher than those of the corresponding composition crystals [38] (see Section 2.2).

Baldwin and Mackenzie [39] have attempted to classify fluoride glass network formers and modifiers in a manner similar to that for the oxide glasses by Sun (see Section 3.1.2). The dissociation energy of the molecule MF_x was divided by the expected coordination number to obtain value for the B_{M-F} (single M–F bond strength), and classified according to whether or not the bond strength was sufficiently high to discourage crystalline rearrangements. Like the oxides, those elements whose single bond strength with F exceeds about 80 kcal/mol ($\sim$335 kJ·mol^{-1}) are classified NWF, those less than 60 kcal/mol are NWM, and the ones between 60 and 80 kcal/mol are the intermediates. Mackenzie [40] later incorporated Rawson's criterion (Section 3.1.2) of dividing B_{M-F} by the melting temperature of the compound MF_x to assess whether the available energy was sufficient to break the bonds. The concepts so generated were used to examine glass formation in the presence of other halogen ions.

Poulain [41] suggested that glass formation in the HMFG systems would

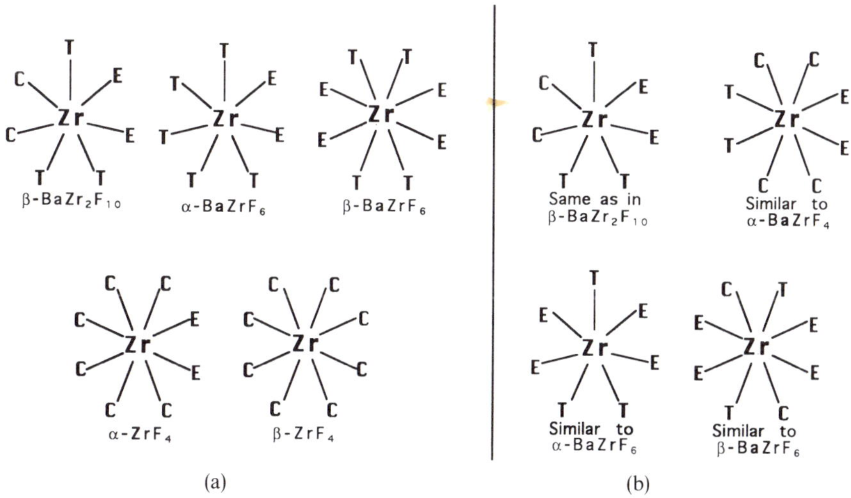

Figure 5-46. Proposed Zr-F polyhedra with F^- characterized as E (edge-bridging), C (corner-bridging), and T (terminal or nonbridging). (a) Polyhedra that may occur in glass and are as found in fluorozirconate crystals. (b) Additional types that may occur in glass and are similar to those found in crystals. (After Phifer and Angell [36]. Reproduced with permission of Elsevier Science Publishers.)

occur if the anionic distribution in the glass is such that the number of potential cationic sites exceeds the available number of cations. This can be possible for high-field-strength cations that, by virtue of having a small ionic radius, can be fitted into a large number of possible sites. However, large repulsion from extremely high-field-strength cations could lead to the disappearance of the ionic bond. Poulain used the ratio of the cationic to anionic field strengths, F_c/F_a, as a parameter to determine glassforming capability. It appears that glass network former cations were those for which $10.5 > F_c/F_a > 2.5$. (The limits were established empirically.) Large-radius cations such as Ba^{2+} and La^{3+} tended to stabilize the structure by precluding highly ordered anionic close packing.

Exercises

1. What is the representative formula for the following glass network?

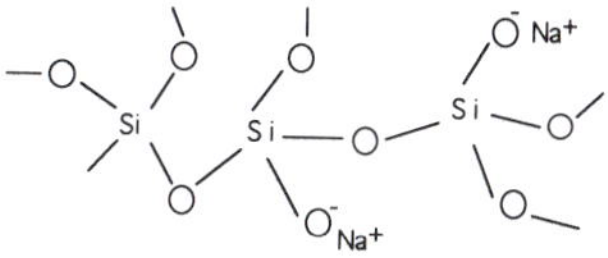

Answer: There are three silicons, two sodiums, two nonbridging oxygens, counted full, two bridging oxygens (shared between tetrahedra shown), counted full, and six bridging oxygens (including missing oxygen on lower left) shared between tetrahedra (not shown), counted halves. Thus, the formula ought to be $Na_2Si_3O_7$ or $Na_2O{\cdot}3SiO_2$.

2. Convert the following to wt. %:
   (a) $15Na_2O{\cdot}10CaO{\cdot}75SiO_2$ (mol %)

   [Answer: $15.53Na_2O{\cdot}9.35CaO{\cdot}75.12SiO_2$]

   (b) $Na_2O{\cdot}2B_2O_3{\cdot}6SiO_2$

   [Answer: $11.05Na_2O{\cdot}24.8B_2O_3{\cdot}64.15SiO_2$]

3. Convert the following wt. % composition to mol %:
   $14BaO{\cdot}10CaO{\cdot}14Al_2O_3{\cdot}62SiO_2$

   [Answer: $6.34BaO{\cdot}12.4CaO{\cdot}19.53Al_2O_3{\cdot}71.73SiO_2$]

4. A glass tank usually melts $15Na_2O{\cdot}10CaO{\cdot}75SiO_2$ (wt. %) continuously using $NaCO_3$, $CaCO_3$, sand, and 25% cullet (broken glass of the same composition) as raw materials. An engineer by mistake happened to mix 1,000 kg of the batch on a 15–10– 75 mol % basis. Calculate the shortest corrective action (one raw material additive).

5. Calculate the f_{NBO} and Y from first principles in a $45Na_2O{\cdot}55SiO_2$ (mol %) glass. Compare your results of Y with Q_n shown in Fig. 5-14. Sketch a representative structure of this glass.

6. Calculate the f_{NBO} and Y from first principles in a $15K_2O{\cdot}10SrO{\cdot}$ $75SiO_2$ (mol %) glass.

7. Calculate the f_{NBO} in a fused silica specimen containing 2 ppm Na_2O and 10 ppm Al_2O_3 impurity. Assume that none of the Na ions are associated with the Al. (Answer: 32 ppm.)

8. Assume that $AlPO_4$ can form a glass isomorphic to fused silica. Sketch the expected RDF if, for simplicity, the ionic radii are assumed to be $Al^{3+} = 0.06$ nm, $P^{5+} = 0.03$ nm, and $O^{2-} = 0.12$ nm. Sketch a representative structure of this glass.

9. Calculate the average chain length for Se chains having chain ending units such that the average coordination number is 1.66.

References

1. R. L. Mozzi and B. E. Warren, *J. Appl. Cryst.* **2**, 164 (1969).
2. R. J. Bell and P. Dean, *Nature (London)* **212**, 1354–1356 (1966).
3. D. L. Evans and S. V. King, *Nature (London)* **212**, 1353–1354 (1966).
4. T. F. Soules, in *Glass Science and Technology* (D. R. Uhlmann and N. J. Kreidl, eds.), Vol. 4a, pp. 267–388. Academic Press, New York, 1990.
5. F. Wooten and D. Weaire, *Solid State Phys.* **40**, 1–42 (1987).
6. S. C. Cherukuri, "Structural Studies of Vitreous Silica and Cristobalite," Ph.D. thesis, Alfred University, 1983.
7. F. Stillinger and T. Weber, *Phys. Rev.* **B31**, 5262 (1985).
8. R. L. Mozzi and B. E. Warren, *J. Appl. Cryst.* **3**, 2512 (1970).
9. T. F. Soules and A. K. Varshneya, *J. Am. Ceram. Soc.* **64**(3), 145–150 (1981).

10a. R. Dupree, D. Holland, P. W. McMillan, and R. F. Pettifer, *J. Non-cryst. Sol.* **68**(2–3), 399–410 (1984).

10b. H. Maekawa, T. Maekawa, K. Kawamura, and T. Yokokawa, *J. Non-cryst. Sol.* **127**, 53–64 (1991).

10c. S. R. Elliott, *J. Non-cryst. Sol.* **123**, 149–164 (1990).

11. A. Dietzel, *Sprechsaal* **62**, 506 (1929).
12. E. J. Gooding and W. E. S. Turner, *J. Soc. Glass Tech.* **18**, 32T (1934).
13. J. Biscoe and B. E. Warren, *J. Am. Ceram. Soc.* **21**, 287 (1938).
14. B. E. Warren, *J. Appl. Phys.* **13**, 602 (1942).
15. P. J. Bray and J. G. O'Keefe, *Phys. Chem. Glasses* **4**, 37–46 (1963).
16. C. M. Kuppinger and J. E. Shelby, *J. Am. Ceram. Soc.* **68**(9), 463 (1985).
17. P. K. Gupta, *Proc. Intl. Cong. on Glass, XIV New Delhi, Part I*, pp. 1–10, 1986.
18. M. E. Milberg, J. G. O'Keefe, R. A. Verhelst, and H. O. Hooper, *Phys. Chem. Glasses* **13**(3), 79–84 (1972).
19. E. D. Lacy, *Phys. Chem. Glasses* **4**(6), 234–238 (1963).
20. J. O. Isard, *J. Soc. Glass. Technol.* **43**, 113–123T (1959).
21. G. J. Bair, *J. Am. Ceram. Soc.* **13**, Part I, 339–347 and Part II, 347–358 (1936).
22. M. F. Mydlar, N. J. Kreidl, J. K. Hendren, and G. T. Clayton, *Phys. Chem. Glasses*, **11**(6), 196–204 (1970).
23. A. E. R. Westman, in *Modern Aspects of the Vitreous State*, Vol. 1 (J. D. Mackenzie, ed.), pp. 63–91. Butterworths, London, 1960.
24. A. K. Varshneya, R. F. Busbey, and T. F. Soules, *J. Non-cryst. Sol.* **69**, 381–385 (1985).

25. K. Kamiya and S. Sakka, *Phys. Chem. Glasses* **20**(3), 60 (1979).
26. D. E. Polk, *J. Non-cryst. Sol.* **5**, 365 (1971).
27. S. C. Moss and J. F. Graczyk, *Phys. Rev. Lett.* **23**, 1167 (1969).
28. P. Steinhardt, R. Alben, and D. Weaire, *J. Non-cryst. Sol.* **15**, 199 (1974).
29. J. D. Bernal, *Nature (London)* **183**, 141 (1959); *Proc. Roy. Soc. (London)* **A280**, 299 (1964).
30. J. L. Finney, *Proc. Roy. Soc. (London)* **A319**, 479 (1970).
31. C. H. Bennett, *J. Appl. Phys.* **43**, 2727 (1972).
32. See S. R. Elliott, *Physics of Amorphous Solids*, 2nd Ed., p. 154. Longman Scientific and Technical, Harlow, Essex, 1990.
32b. P. H. Gaskell, *Nature (London)* **289**, 474 (1981).
33. G. Lucovsky and T. M. Hayes, in *Amorphous Semiconductors* (M. H. Brodsky, ed.), p. 215. Springer-Verlag, Berlin, 1979.
34. M. F. Daniel, A. J. Leadbetter, A. C. Wright, and R. N. Sinclair, *J. Non-cryst. Sol.* **32**, 271–293 (1979).
35. J. Lucas and X. H. Zhang, *J. Non-cryst. Sol.* **125**, 1–16 (1990).
36. C. C. Phifer and C. A. Angell, *J. Non-cryst. Sol.* **94**, 315–335 (1987).
37. See H. Rawson, *Inorganic Glass-Forming Systems*, p. 20. Academic Press, London, 1967.
38. T. Tanba, H. Inoue, Y. Arai, H. Hasegawa, M. Misawa, and I. Yasui, *Proc. Vth Intl. Symp. on Halide Glasses, Shizuko, Japan*, pp. 293–298. (1988).
39. C. M. Baldwin and J. D. Mackenzie, *J. Am. Ceram. Soc.* **62**, 537 (1979).
40. J. D. Mackenzie, Paper 26 at the 2nd Int. Symp. on Halide Glasses, Rensselaer Polytechnic Inst., Troy, New York, 1983.
41. M. Poulain, *Nature (London)* **293**, 279 (1982).

Chapter 6

Composition–Structure–Property Relationship Principles

So far we have studied the range of glass compositions and how the compositions affect the structure. The properties of a substance are controlled by the structure — how the various atoms are organized (or disorganized) to make up the particular substance. This is the chapter where we shall make a gradual transition in the form of basic principles to understanding glass properties — to examine the interrelationship of composition, structure, and properties. The details of specific property variations are saved for specific chapters later.

6.1. General Principles

In the previous chapters, we have emphasized one particular aspect of the glass structure: the lack of long-range periodicity in the arrangement of atoms. Whether the bonding is ionic, covalent, van der Waals, metallic, or a mixture of these, this "*more-or-less randomness*" is the most important feature of a glass. Such an arrangement of atoms in general gives rise to slightly more open packing than that found in the corresponding crystal (although there are exceptions). It follows that whereas small displacements of atoms from their regular periodic locations in a crystal might be higher-energy positions called "defects," the use of such a word in glass leads to mass confusion and is best avoided. Likewise, it is best to look sideways when someone makes a reference to "dislocations" in glass.

A property that is affected by the lack of "dislocations" or "defects" in glass is the strength. Unlike metals where ductility occurs readily through the motion of dislocations and grain boundary sliding, glasses have no such mechanism for deformation. As will be seen later, although plastic flow does occur in glass under extremely high shearing rates, the magnitude of such motion is still small. At ordinary temperatures, therefore, glasses are expected to display mostly a brittle failure mechanism. The randomness is clearly responsible for the liquidlike isotropicity of the glass properties. There are no cleavage planes.

One may readily argue why, unlike crystals, glasses do not have a sharp melting point. The concentration of vacant sites in crystals is generally so low that any large-scale movement of atoms first requires a large number of surface vacancies to be created. As soon as surface atoms occupy the surface vacancies, deeper atoms successively occupy sites nearer to the surface. In other words, the crystal "melts." On the other hand, because of the randomness and the openness of the glassy structure, interstitial displacements within the glass are easily allowed: the magnitude of such displacements increases with temperature. Thus, instead of showing a sharp melting point, glasses become increasingly fluid. As we shall see in the chapter on viscosity, this indeed is the manner in which all glasses behave. One analogy often quoted in this regard is that of a fully packed versus less-than-fully packed elevator. Crystals are much like the former, where people from the interior can only move when the "surface" people come out. Glasses are much like the latter, where people in the interior can move around at will.

Because of the gradual increase of fluidity with temperature, the transition from the glassy state to the supercooled liquid state is smooth and generally broad enough to present ample opportunities for stress relaxation in a "seemingly solid" condition (controllable over physical times). Such is not possible in crystals, simply because the number of defects and dislocations is still small at temperatures well below the melting point of the crystal.

Viscosity of the structure may be expected to increase with the degree of connectivity. B_2O_3 glass, where the polyhedra are triangles and connect at three corners only, is a far more fluid structure than silica glass, where the basic building blocks are tetrahedra connected to each other at all four corners. Viscosity would be expected to increase when the triangularly coordinated boron is converted to a tetrahedrally coordinated boron. Likewise, the creation of nonbridging oxygens (NBOs) leads to increased fluidity. Hence, the higher the amount of the network modifier added to silica glass, the higher the fluidity at a given temperature.

For covalent structures such as the chalcogenide glasses, it may be readily argued that, with increasing degree of crosslinking (i.e., increasing the average coordination number, $\langle m \rangle$), most of the rigidity properties should increase. We shall see later that the glass transition temperature and the room-temperature hardness do increase with $\langle m \rangle$, at least in the chalcogen-rich region (Sreeram *et al.* [1,2]). In chalcohalides, the halogens are chain-ending units. This, clearly, would lead to increased fluidity of the structure at comparable temperatures.

The openness of the glasses over that of corresponding crystals is the "free volume" that the glasses have. By application of very high hydrostatic pressures, it is possible to reduce this free volume, i.e., to make the glasses compact. High-pressure compaction studies in SiO_2 and B_2O_3 show that because the mobility is low at ordinary temperatures, the release of pressure does not cause the atoms to regain their original positions readily: the compaction is "irreversible" at ordinary temperatures.

An increase of NBOs in the structure generally leads to an increase in the thermal expansion coefficient. This is because the average atomic separation between the NBO and the ionically bonded interstitial cation can increase with temperature more readily.

There has been a suggestion [3] that the Young's modulus should be proportional to the number of bridging bonds per unit volume of glass (the nonbridging bonds do not contribute much). This, unfortunately, is far too much of a generalization. For the covalently bonded chalcogenides, an increase in the crosslinking may, in general, be expected [4] to result in an increase of the elastic moduli (and a decrease in the Poisson's ratio).

The interstices in glass act as sieves through which mass can move. Thus, in glasses, conduction of charge occurs predominantly by those ions that can move through the interstices. Network-modifying ions, particularly the alkali ions, invariably carry the bulk of the charge conduction.

One may also suspect that those properties that depend upon motion through the interstices would clearly be enhanced by how open the glass structure is. Thus, the migration of an inert gas atom such as He is expected to be much faster through silica glass than through a soda lime silicate, because the network-modifying cations Na^+ and Ca^{2+} act to block the available pathways in the silica structure. A larger cation in the interstice plays a more effective blocking role than a smaller modifying cation.

It may also be argued that those thermally activated properties that depend upon an atomic/ionic jump from one site to another should, in general, display a distribution of activation energies, because the separation distance of sites in glass is distributed. (Most experiments may, however, only measure an average value, as if the jump distances were all equal.)

The more-or-less randomness, the openness, and the varying degree of connectivity allow the glass structure to accommodate large variations in composition. In other words, glass acts like a solution. Unlike the crystals, there are no stoichiometry requirements in glasses containing more than one component. The entire body of glass may act as a unit cell without displaying repeat units. (The use of periodic boundary conditions in the molecular dynamic and Monte Carlo calculations of the glass structure is merely a tool of computational convenience, and not a reality.) Multicomponent glasses with components having simple chemical ratios are rare: more often, the composition has to be described on a mole percent or a weight percent basis. If we can imagine a $15Na_2O{\cdot}11CaO{\cdot}74SiO_2$ glass to form about as readily as a $16Na_2O{\cdot}11CaO{\cdot}73SiO_2$ glass, then we may also imagine a smooth transition of properties from one composition to the other. The unit addition or substitution of a component can then be deemed a contribution characteristic of that component to the overall property. This notion gives rise to the additivity relationships in glass that are described next.

6.2. Additivity Relationships

An additive property P obeys a linear relation of the type

$$P = \sum_i p_i a_i, \ (i = 1, 2, 3, \ldots, n),$$

where p_i are the additivity factors or the property coefficients for a given component i and a_i are the wt. % or the mol % composition of that component in the glass. The factors p are the contribution to the property made by each 1% of the component. Note that since $\sum a_i = 100\%$, the preceding relation could be rewritten by eliminating the dependent component (usually silica for a silicate glass). The factors would then imply the change in property by substituting 1% of the component i with the dependent component.

The underlying assumption behind the additivity relationships is that of an "ideal mixture/solution" where there are no nonlinear interactions between the different components. Many of the properties of glass have been shown to obey additive relationships: density and refractive index are perhaps the best-known examples. Both these properties are extremely sensitive to the composition and can be measured readily to better than 0.1% accuracy. Additivity factors for each of these have been established for a large number of constituent oxides. For this reason, quality control in large-scale glass manufacture is often carried out using density and refractive index. (Later, we shall show that density and refractive index follow each other quite closely; hence, it is only beneficial to measure one or the other of the two.) Density follows additivity primarily because the volume of an oxide glass is mostly determined by the volume occupied by the oxygen "sublattice," the oxygen ion being about the largest of the ions. The cations are generally small and occupy essentially the interstices within the oxygen sublattice. This is another way of saying that the molar volume of oxide glasses per gram mole of oxygen does not vary very much from composition to composition. The density factors are then based upon the mass contribution per added/substituted atom. Although not very sensitive to composition, factors for elastic properties are also relatively interaction-free.

Nonlinearities, on the other hand, tend to appear when the various constituents interact with each other. Thermal expansion coefficient, viscosity, and electrical conductivity are examples of properties where nonlinearities of the system make the range of validity of the additivity properties quite narrow. For instance, we have already noted in Section 5.6 that, upon the addition of 13 to 25 mol % alkali to a B_2O_3 glass, several of the properties such as the thermal expansion coefficient and the viscosity show extremum behavior (the so-called **boron anomaly**, Section 5.7). Likewise, later on we will show that the presence of two or more different types of alkalis leads to a dramatic departure from additivity of the electrical conductivity. (This is generally known in glass science literature as **the mixed-alkali effect**, Section 14.7.) Clearly, the usefulness of additivity coefficients in such cases is limited to a narrow range of compositions that avoid the extremum behavior. In some cases, efforts have been made to obtain additivity factors including interactive terms and prescribing the range of compositions for the application. Viscosity and thermal expansion are examples where such efforts have met some success.

A word of caution should be noted: there are more reasons in glass science why additivity relations should not work than why they should work. There are enough fine differences between two glasses that what works for one doesn't work for the other for an identical property. In order not to be too disappointed, *it is best to assume that additivity relations work over a narrow range of compositions and that the additivity coefficient of a given oxide may differ from system to system.*

References

1. A. N. Sreeram, D. R. Swiler, and A. K. Varshneya, *J. Non-cryst. Sol.* **127**, 287–297 (1991).

2. A. N. Sreeram, A. K. Varshneya, and D. R. Swiler, *J. Non-cryst. Sol.* **130**, 225–235 (1991).
3. R. J. Charles, *Prog. Ceram. Sci.* **1**, 1–38 (1961).
4. A. N. Sreeram, A. K. Varshneya, and D. R. Swiler, *J. Non-cryst. Sol.* **128**, 294–309 (1991).

Chapter 7

Density and Molar Volume

7.1. Definitions

Density of a substance is defined as the mass per unit volume, and is an intensive property. Appropriate units are g/cm^3 or g/cc (in cgs system) and kg·m^{-3} in the SI system. The dependence of density upon temperature is through the volume thermal expansion coefficient.

Since glasses may, in general, be regarded as solutions, a more useful property is the molar volume V, defined as the volume of one gram mole of glass, and then proceed to define partial molar volumes of the various structural units constituting the glass. The partial molar volume v_i of a species i in a solution is defined by

$$v_i = (\partial V/\partial n_i)_{n_j, T, P} \tag{7.1}$$

and hence, $V = n_i v_i + n_j v_j$.

In essence, the total molar volume is treated as an extensive property in terms of the partial molar volumes of individual species or structural groups. The partial molar volume information can be extracted from the density data of glasses where the constituents have been systematically varied. If a Na^+ is replaced by K^+ in the glass and occupies the same interstice (without changing its size), it is clear that, although the glass density would increase because of the higher atomic weight of K^+, the partial molar volume of the alkali ion would not change. Any increase in the molar volume would be an increase in the partial molar volume of K^+ relative to Na^+. This is likely to have some relation to the relative ionic sizes of the two ions. One may readily note that, although the measurement of density changes in a family of glasses may provide only the trends, the extraction of the partial molar volumes from the density data provides further insight into structure while canceling out the effect of atomic masses.

7.2. Methods of Measurement

Density is measured most commonly by weighing a suitably selected piece of glass in air and dividing this weight by the buoyancy (the reduction in weight) when the sample is suspended in water. According to Archimedes' principle, the buoyancy equals the weight of the displaced fluid, which, for water, equals the volume. Thus, if w_a is the weight of the specimen in air, and w_b in water, then the buoyancy is $w_a - w_b$ and the density $\rho = w_a/(w_a - w_b)$.

Occasionally, the glass may be affected by water immersion, in which case a suitable inert liquid such as kerosene may be selected. The specimen volume in this case is given by $(w_a - w_m)/\rho_m$, where ρ_m is the density of the immersion medium. A kit supplied by scientific balance manufacturers can be utilized to obtain density measurements accurate to 0.001 g/cc.

Automatic measurements of density can be carried out using pycnometers that utilize He gas as the displacement fluid. A small sample, of the order of a few milligrams, is all that is needed to make the measurement. The accuracy of such instruments is generally about 0.002 g/cc. In a manufacturing environment, a comparative density (easily accurate to 0.0002 g/cc) is more rapidly obtained using a gradient column. The gradient column consists of a vertically held long (often 1–1.5 m), glass tube with one end closed. The tube contains a liquid with a linear density gradient. To prepare the density gradient, a heavy liquid such as *sym*-tetrabromoethane (=2.96 g/cc) or methylene iodide (=3.32 g/cc) is first poured into the tube, and then a lighter liquid such as isopropyl salicylate ($\rho = 1.1$ g/cc) is gently poured to float above the heavy liquid.* With time, the liquids diffuse into each other, establishing a gradient with the help of gravity. The gradient is calibrated by allowing density standards to gently float in the column at different heights. The density of an unknown glass is obtained by reading its flotation level against a precalibrated scale placed adjacent to the column.

* Many of these liquids may pose health hazards. The reader is advised to check with the suppliers regarding their handling procedures.

Another method also used commonly is the sink–float method, where the unknown specimen is gently dropped into a tube containing a slightly denser solution of organic liquids such as the ones just named. The temperature around the tube is gradually changed until the previously floating specimen begins to sink. Calibration of the liquid's density against temperature yields the density of the unknown. (See ASTM standard C729-75.)

7.3. Dependence upon Cooling Rate, Temperature, and Composition

As stated in Section 2.2, the volume variation upon cooling is fundamental to the transformation of a supercooled liquid into a glass. Except for a few anomalous cases, such as silica, a fast-cooled glass generally has a higher volume, and hence a lower density, relative to a slow-cooled glass of the same composition. Because density is inversely proportional to the volume, the fractional change in density with a ΔT change in temperature is given by $-3\alpha\Delta T$, where α is the linear thermal expansion coefficient. Later, in Chapters 10 and 13, we shall learn that the thermal expansion coefficient of most glasses increases greatly in the glass transition region and beyond. Thus, the effects of temperature variation on the density of a glass are usually small in the glassy state.

The densities of some single-component glasses are shown in Table 7-1.

Of the various crystalline forms of silica, the density of fused silica is the closest to that of β-cristobalite, which is 2.25 g/cc. The significant drop in the density of fused silica relative to that of α-quartz, which is the normal starting raw material, is indicative of the large structural changes that occur in the fusion process. The twisted hexagonal close-packed structure of α-quartz apparently "unwinds" and opens up. Densities of silica glasses derived from other processes are generally slightly lower, primarily because of impurities and voids.

The density of B_2O_3 glass changes nonlinearly as a function of tempera-

Table 7-1. Densities (g/cm^3) of Primary Glass Formers at Room Temperature

SiO_2	2.20
B_2O_3	1.812–1.84
GeO_2	3.628–3.65
P_2O_5	>2.23

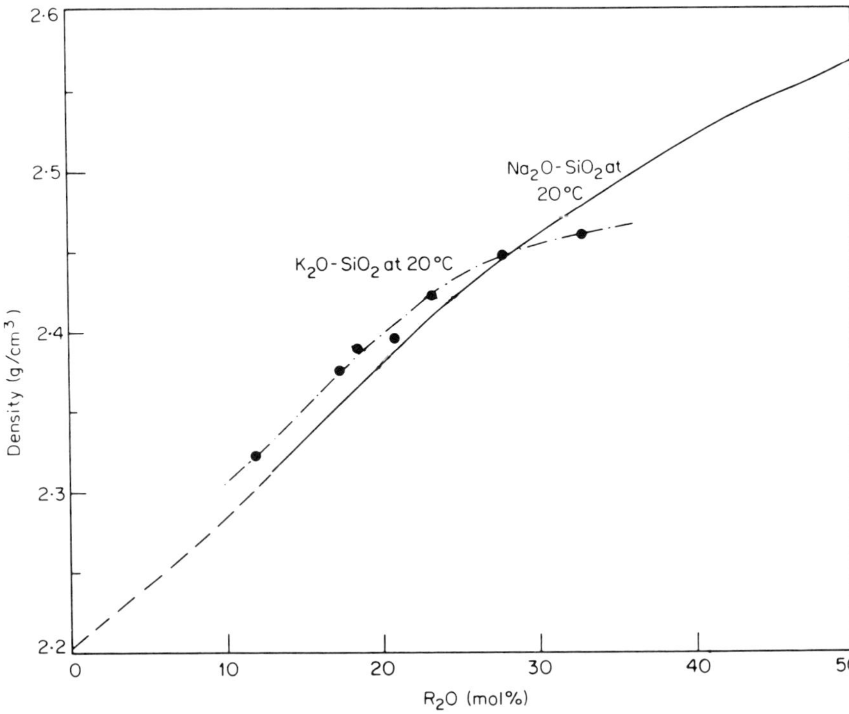

Figure 7-1. Density of alkali silicate glasses.

ture, primarily because of the changes in boron coordination and structural groups.

Densities of a series of B_2O_3–SiO_3 glasses, where both the components are network-forming individually, were measured by Cousen and Turner [1] and Jabra *et al.* [2]. Comparison of the partial molar volumes of B_2O_3 and SiO_2 relative to the molar volume values in the pure state is left to the reader as an exercise.

The addition of network-modifying components increases the density. This is to be expected, since the NWM ions attempt to occupy the interstices within the network. The variation of density with composition in alkali silicate glasses is shown in Fig. 7-1. Apparently, the density increases somewhat rapidly with initial additions of the alkali, and not as rapidly with continuing additions. The partial molar volumes of the alkali, shown in Fig. 7-2, indicate a sharp minimum at around 10 to 12 mol % added alkali. According to Bockris *et al.* [3], the initial additions of the alkali ions are completely "enclosed" within the Si–O–Si interstitial "cages." At 10–12%, the alkali ions begin to "open up" these "cages."

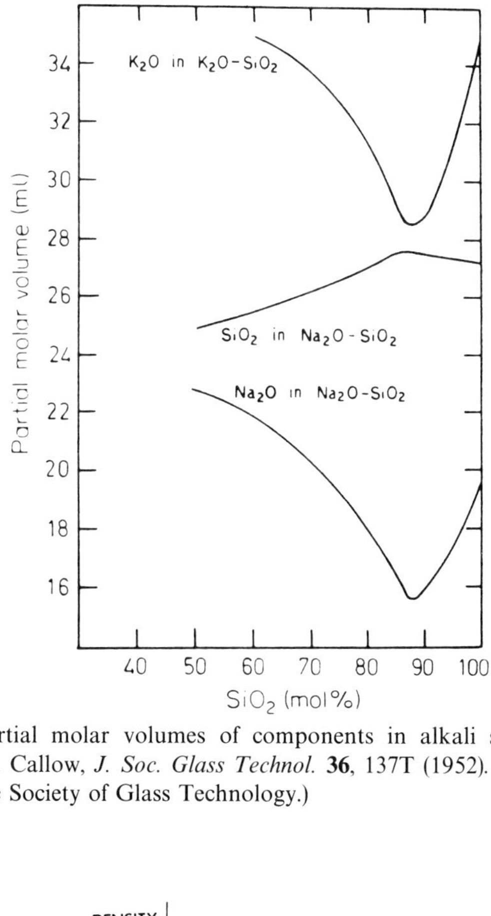

Figure 7-2. Partial molar volumes of components in alkali silicate glasses at 25°C. (After R. J. Callow, *J. Soc. Glass Technol.* **36**, 137T (1952). Reproduced with permission of the Society of Glass Technology.)

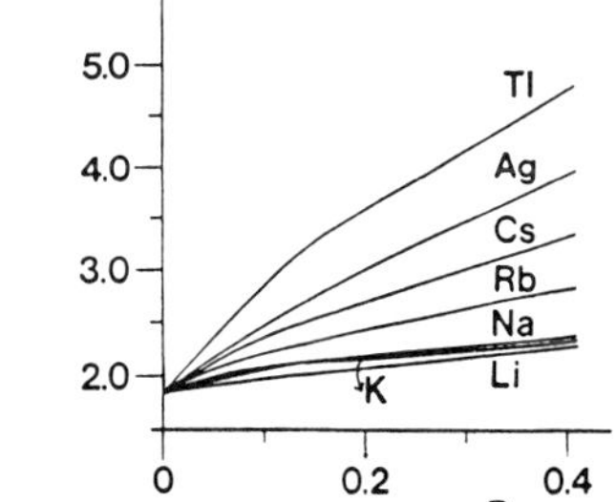

Figure 7-3. Density of various alkali borate glasses at 25°C. (After H. C. Lim and S. Feller, *J. Non-cryst. Sol.* **94**, 36–44 (1987). Reproduced with permission of Elsevier Science Publishers.)

The addition of alkalis to B_2O_3 also results [4] in an increase of the density. Room temperature data show a monotonic increase (Fig. 7-3, from Lim and Feller, Ref. 5) in density. Assuming that the addition of the alkali changes the fraction N_4 of the tetrahedrally coordinated boron (Section 5.6), Lim and Feller [5] have suggested that the volume of the tetrahedrally coordinated boron unit increases with the alkali ion size, while the volume of the triangularly coordinated boron unit stays relatively constant. In borate melts (temperatures in excess of about 900°C), there is some evidence that high-temperature data [6] show density maxima at 30 to 40 mol % added

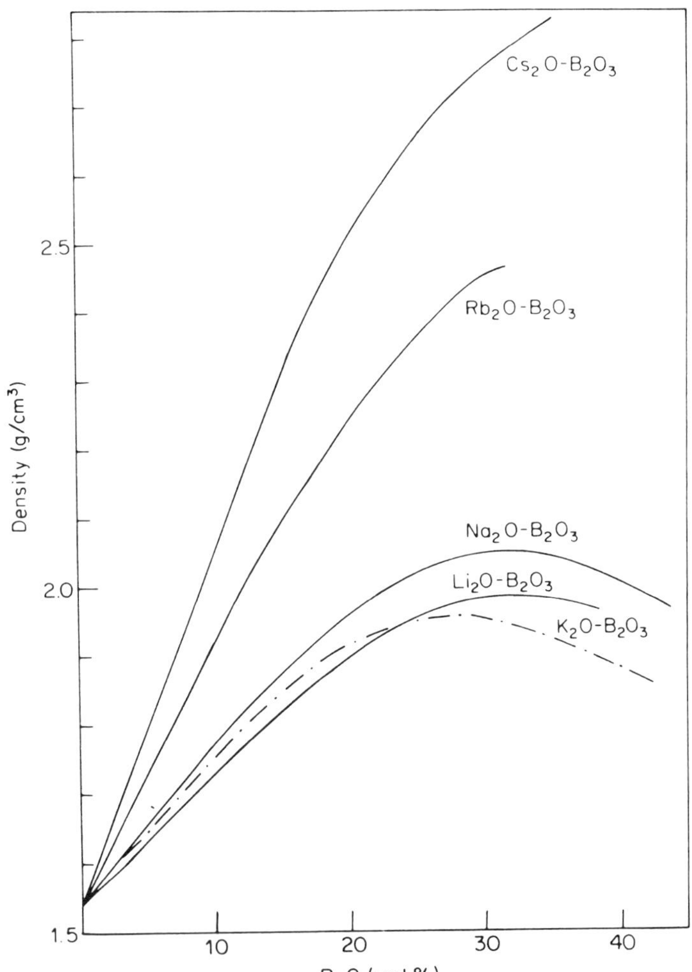

Figure 7-4. Density of various alkali borate melts at 900°C. (Ref. 6.)

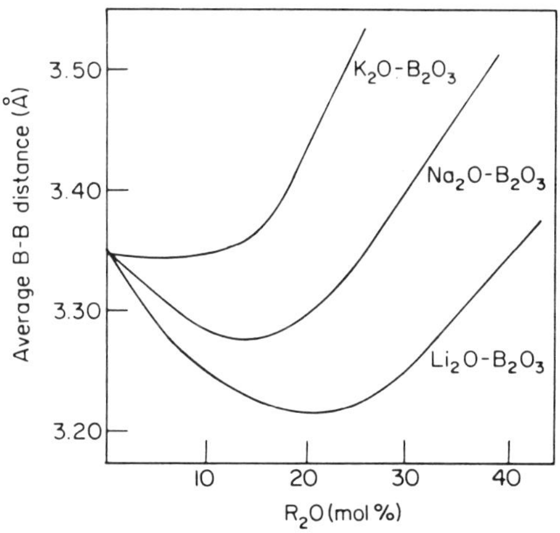

Figure 7-5. Average B–B distance in alkali borate melts at 1,000°C. (After Shartsis *et al.* [4]. Reproduced with permission of the American Ceramic Society.)

alkali (Fig. 7-4). Again, by extracting the partial molar volumes of each of the components, it is suggested [4] that the average B–B distance undergoes a minimum at varying alkali level (Fig. 7-5).

Densities of several glasses with three or more components, including commercial systems, have been measured. Some of these are tabulated by Morey [7], Bansal and Doremus [8], and Mazurin *et al.* [9].

Molar volume of glasses in the system Ge–Sb–As–Se–Te have been measured by Sreeram *et al.* [10]. The values range between 17.6 and 18.5 cm^3/mol. When the values are plotted against the average coordination number $\langle m \rangle$, it is observed that there is a distinct minimum at $\langle m \rangle = 2.40$, suggesting an optimized packing at this value, which is in agreement with Phillips's topological constraints model (Section 3.1.4). The molar volume increases with $\langle m \rangle$ greater than 2.40, and appears to maximize at $\langle m \rangle \cong$ 2.67.

Densities of a large number of heavy metal fluoride glasses have been measured. Some of these are tabulated by Drexhage [11]. The densities are generally in the 4.5–6.0 g/cm^3 range.

7.4. Densification of Glasses by High Pressure or Irradiation

Bridgman and Simon [12] have shown that pressures higher than a threshold 100 kbar (=10 GPa) caused fused-silica glass to undergo compaction or irreversible densification. Densities increased about 7.5%, although a specimen with a 17.5% density increase closely approaching the density of α-quartz was also found. More recently, Susman *et al.* [13] have obtained a 20% increase in the density of fused silica at 16 Gpa pressure (slightly higher than that of α-quartz!). Apparently, the recovery of the density after pressure release is very sluggish for fused silica. On the other hand, for B_2O_3, no pressure threshold has been observed: density increases steadily about 6.6%, and pressure release causes a rapid steady recovery. The behavior of binary and multicomponent silicates was intermediate to that of silica and B_2O_3. Density recovery is faster for all glasses when they are heated at temperatures close to their respective glass transition. There has been considerable discussion of the exact mechanism of the densification and recovery, particularly the question of the role of shear stresses. (See Susman *et al.* [13]; Mackenzie [14]; Cohen and Roy [15]; Poch [16]; Mizouchi and Cooper [17]; Arndt and Stöffler [18].) It is believed that the application of pressure causes a cooperative rotation of ionic polyhedra to move the atoms in a relatively open structure, such as that of a glass, closer to occupy some of the available interstitial volume. The rotation of the polyhedra causes the bonds such as Si–O–Si to bend without changing the near-neighbor separation. The presence of shear accelerates the pressure densification process.

Irradiation of glasses by high-energy particles, such as fast neutrons, is also known to cause densification. Vitreous silica when irradiated with a neutron dosage of 2×10^{19} n/cm^2 densifies 2–3%. Further irradiation to 3×10^{20} n/cm^2 causes a gradual dilatation whereby about 10% of the densification is recovered [19]. The densified structure shows a small increase in Young's modulus and a small decrease in rigidity. The width of bond distance distribution (as measured by x-rays) increases slightly, implying the possibility of strains in the network. Gradual heating effects complete recovery of the normal structure.

7.5. Calculation of Density

Because of the nature of glass as a "solution" of the various component oxides, ideas concerning the use of additivity factors for the calculation of the density ρ of glasses have been studied in great detail. The most successful of these are those by Huggins and Sun [20], based upon the notion that the specific volumes of a component oxide are additive. The scheme of the calculation suggested by them is as follows:

Let f_M be the weight fraction of each component oxide M_mO_m in the glass. First obtain $\Sigma s_M f_M$, where s_M is the ratio of the number of oxygen atoms in the molar formula to the formula weight and may be read from Table

Table 7-2. Factors for the Calculation of Density

		v_M			v'_M
Component, M_mO_n	$s_M \times 10^2$	(A) (N_{Si} = 0.27–0.345)	(B) (N_{Si} = 0.345–0.40)	(C) (N_{Si} = 0.40–0.435)	(D) (N_{Si} = 0.435–0.50)
Li_2O	3.347	0.452	0.402	0.350	0.261
Na_2O	1.6131	0.373	0.349	0.324	0.281
K_2O	1.0617	0.390	0.374	0.357	0.329
Rb_2O	0.53487	0.266	0.258	0.250	0.235
Cs_2O	0.35484				
Tl_2O	0.23542	0.122	0.118	0.115	0.108
BeO	3.997	0.348	0.289	0.227	0.120
MgO	2.480	0.397	0.360	0.322	0.256
CaO	1.7832	0.285	0.259	0.231	0.184
ZnO	1.2288	0.205	0.187	0.168	0.135
SrO	0.96497	0.200	0.185	0.171	0.145
CdO	0.77876	0.138	0.126	0.114	0.0935
BaO	0.65206	0.142	0.132	0.122	0.104
PbO	0.44801	0.106	0.0955	0.0926	0.0807
FeO	1.3918				
$B_2O_3(B_3)$	4.3079	0.590	0.526	0.460	0.345
$B_2O_3(B_4)$	4.3079	0.791	0.727	0.661	0.546
Al_2O_3	2.9429	0.462	0.418	0.373	0.294
Fe_2O_3	1.8785	0.282	0.255	0.225	0.176
	1.5165				
	1.0291				
Bi_2O_3	0.6438	0.105	0.0958	0.0858	0.0687
	4.544				
SiO_2	3.330	0.4063	0.4281	0.4409	0.4542
TiO_2	2.5032	0.319	0.282	0.243	0.176
	1.9120				
ZrO_2	1.6231	0.222	0.198	0.173	0.130
Ta_2O_5	1.1318	0.164	0.147	0.130	0.0997

[a] After Huggins and Sun [20].

7-2, column 2. (For B_2O_3, for instance, $s_M = 3/(2 \times 10.8 + 3 \times 16) = 0.043$.) The fraction N_{Si}, which is the number of gram atoms of Si per gram atom of oxygen in the glass, is then calculated by $N_{Si} = f_{Si}/60.06\Sigma s_M f_M$. Next, look up the values of v_M, the specific volume of 1.0 g of the component oxide, from the column appropriate for a given N_{Si}. The total specific volume v of the glass is obtained by $v = \Sigma v_M f_M$. The density in grams per cubic centimeter is the reciprocal of v. The last column in Table 7-2 is used for the specified N_{Si} range and for poorly annealed glasses.

Table 7-3. Density Difference Factors[a]

Component oxide	Density change (g/cm^3 per 1 wt. % oxide change)
SiO_2	−0.0024
Al_2O_3	+0.0018
CaO	+0.0106
MgO	+0.0050
BaO	+0.0173
Na_2O	+0.0050
K_2O	+0.0028
B_2O_3	+0.0036
ZnO	+0.0145
PbO	+0.0192

[a] After Elliott [21].

In their calculations, Huggins and Sun also attempted to include the effect of boron coordination changes in glass. In Table 7-2, B′ denotes the 3-coordinated boron, and B″ the 4-coordinated boron. Because the boron coordination changes were not precisely known until much later, their procedure to account for the coordination changes can only be considered approximate. Nonetheless, Huggins and Sun's density factors have been found to be remarkably accurate in many cases. Note that their tables also contain additivity factors for the refractive index and the dispersion.

For soda-lime–silica glasses, Elliott [21] has converted Huggins and Sun's scheme to calculating the density changes upon 1% addition of a given oxide. These density difference factors are shown in Table 7-3. The maximum error claimed is ± 0.0005 g/cm^3. It may be interesting to note that, in a typical soda-lime silicate glass, the substitution of 1 wt. % Na_2O by 1 wt. % K_2O leads to a decrease in density.

Exercises

1. Show that when Elliott's factors are converted to a mole basis, the density increase caused by a 1 mol % addition of Na_2O is still about 20% larger than that caused by a 1 mol % addition of K_2O, despite the fact K^+ is nearly double the size of Na^+. What may qualitatively be concluded regarding the relative partial molar volumes of Na_2O and K_2O?

2. Calculate the density of $15Na_2O{\cdot}10CaO{\cdot}75SiO_2$ (mol %) glass using Huggins and Sun's factors. Hence, calculate its molar volume. (Answer: $\rho = 2.479$ g/cm^3. $V = 24.16$ cm^3.)

3. Calculate the change in density expected using Elliott's factors by changing the glass composition in Q.(2) to a $5Na_2O{\cdot}10K_2O{\cdot}10CaO{\cdot}75SiO_2$ (mol %) glass. (Answer: $\Delta\rho = -0.006$ g/cm^3.)

4. Imagine a glass plate of the composition $15Na_2O{\cdot}10CaO{\cdot}75SiO_2$ (mol %) glass where 10 mol % of the surface Na_2O has been replaced by K_2O on an ion-for-ion basis (without electrostatic charge building up) well below the glass transition range. Would you expect any stresses to develop in the glass plate? What if Elliott's density change factors for Na_2O and K_2O were exactly equal on a mole basis?

(Note: Questions 2–4 form the basis of stress development by ion exchange in glass, detailed in Section 14.8.1.)

References

1. A. Cousen and W. E. S. Turner, *Glastech. Ber.* **6**, 393 (1928); *J. Soc. Glass Technol.* **12**, 169T (1928).
2. R. Jabra, J. Pelous, and J. Phalippou, *J. Non-cryst. Sol.* **37**, 349 (1980).
3. J. O'M. Bockris, J. W. Tomlinson, and J. L. White, *Trans. Farad. Soc.* **52**, 299 (1956).
4. L. Shartsis, W. Capps, and S. Spinner, *J. Am. Ceram. Soc.* **36**, 35 (1953).
5. H. C. Lim and S. Feller, *J. Non-cryst. Sol.* **94**, 36–44 (1987).
6. E. F. Riebling, *J. Am. Ceram. Soc.* **50**, 46 (1967).
7. G. W. Morey, *Properties of Glass.* Reinhold, New York. 2nd Ed., 1954.
8. N. P. Bansal and R. H. Doremus, *Handbook of Glass Properties*, Chapter 4. Academic Press, New York, 1986.
9. O. V. Mazurin, M. V. Streltsina, and T. P. Shvaiko-Shvaikovskaya, eds. *Handbook of Glass Data*, Vols. A, B, C, and D. Elsevier, Amsterdam, 1983.
10. A. N. Sreeram, A. K. Varshneya, and D. R. Swiler, *J. Non-cryst. Sol.* **128**, 294–309 (1991).
11. M. G. Drexhage, in *Treatise on Mat. Sci. & Tech.*, Vol. 26, *Glass 4.* (M. Tomozawa and R. H. Doremus, eds.), Academic Press, New York, pp. 151–243. 1985.
12. P. W. Bridgman and I. Simon, *J. Appl. Phys.* **24**(4), 405–413 (1953).
13. S. Susman, K. J. Volin, R. C. Liebermann, G. D. Gwanmesia, and Yanbin Wang, *Phys. Chem. Glasses* **31**(4), 145–150 (1990).
14. J. D. Mackenzie, *J. Am. Ceram. Soc.* **46**, 461 (1963).
15. H. M. Cohen and R. Roy, *Phys. Chem. Glasses* **6**(5), 149–161 (1965).
16. W. Poch, *Phys. Chem. Glasses* **8**(4), 129–131 (1967).

17. N. Mizouchi and A. R. Cooper, *J. Am. Ceram. Soc.* **56**(6), 320–323 (1973).
18. J. Arndt and D. Stöffler, *Phys. Chem. Glasses* **10**(3), 117–124 (1969).
19. W. Primak, *J. Phys. Chem. Sol.* **13**, 279–286 (1960).
20. M. L. Huggins and K.-H. Sun, *J. Am. Ceram. Soc.* **26**, 4 (1943).
21. R. M. Elliott, *J. Am. Ceram. Soc.* **28**(11), 303–305 (1945).

Chapter 8

Elastic Properties and Microhardness of Glass

8.1. Introduction

Elastic properties differentiate a solid from a liquid. The application of a shearing force to solids is met with considerable resistance. The magnitude of the resulting deformation is proportional to the force and remains constant with time if the force is constant. The deformation is instantaneously recovered upon the removal of the force. On the other hand, liquids flow or deform at a constant rate over time if a constant shear force is applied, or the shear force relaxes over time if a constant strain is applied. The terms "instantaneous" and "deform over time," and hence the terms "solid" and "liquid," need to be recognized, keeping time in proper perspective. Elastic properties of glass are important because most glass products critically depend upon the solidlike behavior of glass.

At temperatures well below the glass transition range (see Fig. 2-1), glass may be considered a linear elastic solid obeying Hooke's law when the applied stress magnitudes are low relative to the fracture strength. This means that upon application of a stress (force per unit area), glass undergoes instantaneous deformation such that the ratio of the stress to the resulting strain (change in length per unit length) is a constant called the "modulus of elasticity" (SI units: Pa = N/m^2; cgs units: dyn/cm^2; English units: psi), which is independent of the magnitude of the strain. An example of non-

Hookean elastic behavior observed commonly is the stretching of a rubber band, where the modulus of elasticity varies significantly with the deformation. Non-Hookean behavior such as the nonconstancy of the elastic moduli in glass is often observed when the applied stresses, are very high, perhaps about two-thirds of the fracture strength. This situation is reached, for instance, when glass is loaded by an indenter in a microhardness test, which is a measure of the abrasion resistance. Because of the small area of contact, the applied load translates to very large local stresses that cause yielding and plastic deformation of the glass network (see Section 8.3). Other examples of plastic behavior, such as irreversible compaction under hydrostatic compression, and elastic–viscous (viscoelastic) behavior such as delayed elasticity, also exist in glass. The magnitude of such nonlinearity or nonelastic behavior generally increases with the magnitude of applied stresses (see, for example, Mazurin [1]; Schlapp [2]; and Douglas *et al.* [3]) and with the temperature approaching the glass transition range.

8.2. Elastic Properties of Glass

8.2.1. The Terminology of Elasticity

The semiempirical approach to elasticity of solids is based primarily upon the conclusions from the various experiments conducted on most engineering materials. The solids are subjected to three types of stressing conditions: uniaxial stress, triaxial stress, and pure shear. These are shown schematically in Fig. 8-1. It is assumed that a normal stress σ_x does not produce any shear strain e_{xy} on the x, y, or z planes, and a shear stress σ_{xy} only produces one shear strain e_{xy}. For the situation where a uniaxial stress σ_x (tensile, for instance) is applied on a body producing an elongation or an extensional strain of e_x in the same direction (Fig. 8.1a), the ratio of the linear stress to the linear strain is called the **Young's modulus** (generally denoted as E, sometimes as Y), i.e.,

$$\sigma_x = Ee_x. \tag{8.1}$$

A stress is tensile or positive if it produces an elongation in the applied direction. Accompanying this elongation in the x direction, there will be contraction in the y and z directions such that

$$e_y = e_z = -\nu e_x. \tag{8.2}$$

The ratio ν of the transverse strain to that of the longitudinal strain is called the **Poisson's ratio**.

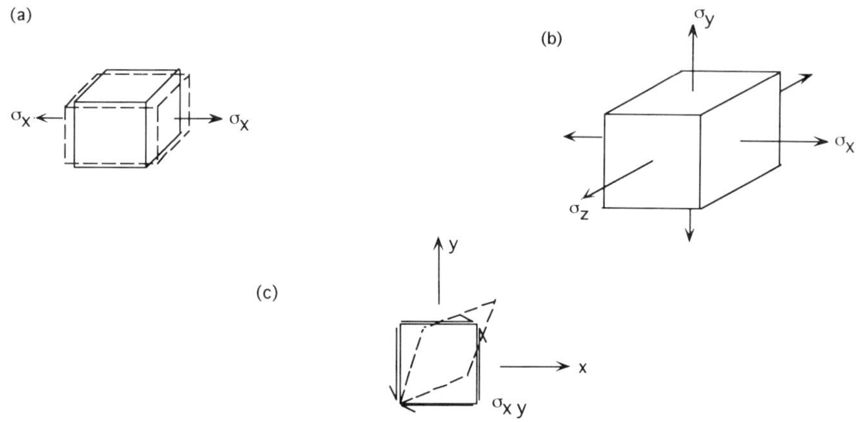

Figure 8-1. State of stress for an element under (a) uniaxial stress, (b) triaxial stress, (c) pure shear.

When the body is subjected to a triaxial system of stresses (Fig. 8.1b), the total strain in the x direction, for instance, is the sum of the contributions due to the individual stress components, i.e.,

$$e_x = (1/E)\,[\sigma_x - \nu(\sigma_y + \sigma_z)]. \tag{8.3}$$

Under a two-dimensional state of pure shear (Fig. 8.1c), the ratio G of the shear stress σ_{xy} to the shear strain e_{xy} is called the **shear or the rigidity modulus**, and its reciprocal, **the shear compliance** J, i.e.,

$$e_{xy} = (1/G)\sigma_{xy} = J\sigma_{xy}. \tag{8.4}$$

There are three equations of the form (8.3) and three of the form (8.4). Together, these are called the **constitutive equations**.

A special situation arises when the three normal components of the triaxial stress system are exactly equal (Fig. 8-2), i.e.,

$$\sigma_x = \sigma_y = \sigma_z = -p, \tag{8.5}$$

which is the "hydrostatic" condition (as if the body were immersed in pressurized water). In this situation, a volumetric strain (dilatation) $e_{\mathrm{m}} = e_x + e_y + e_z$ develops, where extensional strains $e_x = e_y = e_z$ and shear strains $e_{xy} = e_{yz} = e_{xz} = 0$. The ratio of the hydrostatic stress $-p$ to the volumetric strain e_{m} is called the **bulk modulus** (generally denoted as K), i.e.,

$$-p = Ke_{\mathrm{m}}. \tag{8.6}$$

Note that in this definition, the negative sign is used to acknowledge the

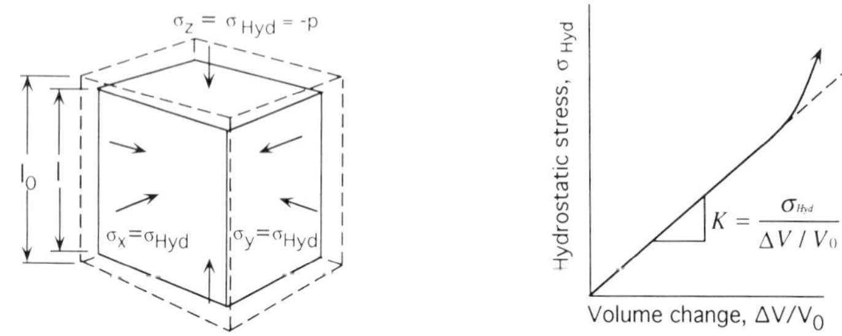

(a) (b)

Figure 8-2. The hydrostatic stress. (a) dimensional changes of the element. (b) plot of the hydrostatic stress versus volume change.

compaction observed when a compressional stress (pressure) is applied. The inverse of K is called the **compressibility** and is generally denoted by κ.

For isotropic systems such as glass, the following relations between the four constants E, K, G, and ν reduce the number of independent constants to only two:

$$G = E/2(1 + \nu); \qquad K = E/3(1 - 2\nu); \qquad \nu = (E/2G) - 1;$$
$$E = 9KG/(3K + G). \tag{8.7}$$

Since the elastic moduli should be greater than 0, Eq. (8.7) yields $-1 \leqslant \nu \leqslant 0.5$. In particular, when $\nu = 0.5$, $K \to \infty$ and $E \to 3G$. Such a material is considered incompressible.

The isotropicity of glass is a very special property. Even the highest-symmetry crystals, such as the cubic crystals, do not conform to the preceding relations: for them, a minimum of three constants are required to describe the elastic behavior. Because much of the behavior of materials is described in terms of generalized equations, it is best to briefly review the formalism of linear elasticity discussed in Appendix 1.

8.2.2. Methods of Measuring Elastic Moduli

Common techniques for measuring the elastic moduli of glasses fall in three categories: those based on measurement of the stress–strain curve, those based upon the propagation of ultrasonic waves, and those based upon the estimation of natural frequencies.

Among the first, the Young's modulus is measured by applying a uniaxial tension to a suitably prepared rod specimen that has been gripped at the

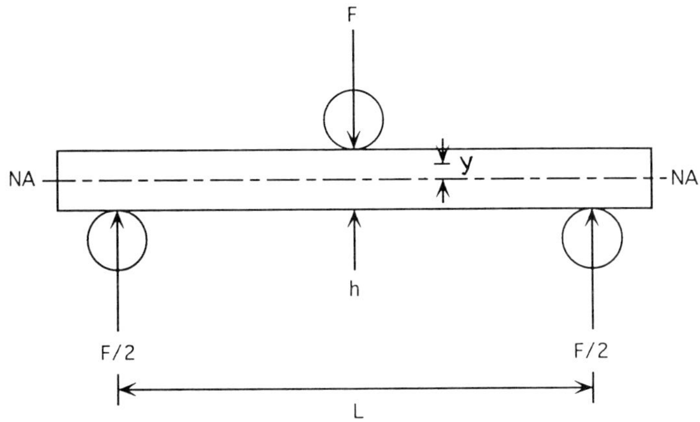

Figure 8-3. Scheme of three-point bending.

ends in a mechanical testing machine, and recording the change in length as the load on the specimen increases. The method sounds simple, yet there are many sources of errors (besides slippage at the grips). The original test length (gage length) of the specimen is somewhat uncertain. Likewise, because of possible misalignment, the applied tension may not be purely uniaxial (presence of some shear is likely). Some of these problems are reduced in a beam-bending or flexural technique. For a centrally loaded and end-supported beam (three-point bend technique, shown in Fig. 8-3), the center deflection δ is determined from

$$\delta = FL^3/48EI, \tag{8.8}$$

where F is the central force (load) on the beam, L is the length of the beam between the end-supports, and I is the geometric moment of inertia ($I = bh^3/12$ for a rectangular beam of width b and height h, and $I = \pi a^4/4$ for a circular rod with radius $= a$). When the end-supported beam is symmetrically loaded at two points (four-point bend technique, shown in Fig. 8-4), the center deflection is given by

$$\delta = F(L_1 - L_2)(2L_1^2 + 2L_1L_2^2 - L_2^2)/96EI, \tag{8.9}$$

where L_1 is the distance between the outer support pegs and L_2 is the distance between the inner loading pegs.

The flexural test methods are not without their problems, either. Standardized procedures are described in ASTM D790-81 and MIL STD 1942 (MR) 1983. Some of the errors have been analyzed by Baretta *et al.* [4]. A round

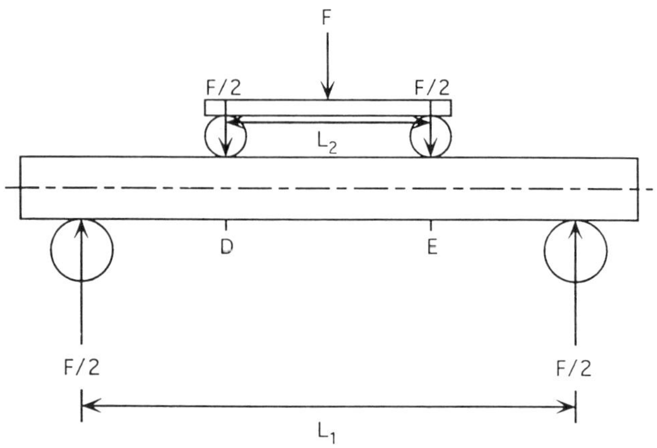

Figure 8-4. Scheme of four-point bending.

robin test, including the schematics of some test configurations, has been reviewed by Quinn [5] recently. Perhaps the most serious problem is that the preceding flexure formulae are valid only for linear beam theory, i.e., when the center deflection is small (say, less than the beam height). For large deflections, corrections to the formulae are required. An approximate technique has been described by West [6].

The bulk modulus must be measured in a hydrostatic pressure chamber. The common equipment is the Bridgman type apparatus described by Birch *et al.* [7]. Fluids such as a mixture of pentene and isopentene are used [8] to transmit the pressure. Corsaro [9] has described an improved device where the sample is surrounded with mercury and the entire chamber is pressurized by air. Changes in the sample volume result in the displacement of mercury and a consequent reduction in the height of the mercury column in an attached capillary. The change in the height of the mercury column is determined precisely by measuring the propagation time of an ultrasonic wave. The assurance of a purely hydrostatic compression in all these devices is critical. One important difficulty in the measurement of the bulk modulus of glass is the irreversible compaction that follows under the action of high pressures. The compaction results simply from a network collapse. Thus, measurements at small values of the pressure may represent the bulk modulus; however, a nonlinear relation between the pressure and volume decrease is likely at high pressures approaching a few hundred megapascals (Section 7.4).

Shear modulus may be measured by applying a twist at the lower end of a rod whose upper end is rigidly held. If θ is the angle of twist, and T the applied torque, then

$$G = 2TL/\pi\theta a^4, \tag{8.10}$$

where L is the length and a the radius of the rod.

This method was used quite successfully by Kurkjian [10]. A relaxometer designed by Rekhson and Mazurin [11] used a sample in the form of a spring, thereby improving the measurement accuracy.

In the ultrasonic wave velocity techniques, an ultrasonic pulse is propagated by a suitable transducer through the specimen, reflected by the back surface and received as an echo [12]. Two types of waves are propagated, each coming from a different transducer. One of these is a longitudinal wave transducer; the other is a transverse (shear) wave transducer. Generally, a 5 to 10 MHz frequency for the former and a 2.5 to 10 MHz frequency for the latter are adequate. The pulse-echo transit is recorded on an oscilloscope. The Poisson's ratio ν is calculated using the formula

$$\nu = [1 - 2(V_T/V_L)^2]/[2 - 2(V_T/V_L)^2], \tag{8.11}$$

where V_T and V_L are the transverse and longitudinal wave velocities, respectively.

The shear modulus G is calculated from

$$G = V_T^2\rho, \tag{8.12}$$

where ρ is the density.

Ultrasonic wave propagation measurements can be quite accurate. Certainly, the sample preparation (generally in the form of a 2 mm to 1 cm thick plate or disk) is relatively simple. The errors often are due to internal porosity or cracks that act to reflect the waves prematurely.

The natural frequency measurements are made by gently tapping a suitable specimen and picking up the natural frequency of vibration with a transducer manually held against the specimen [13, 14]. By tapping the specimen in different ways, both the longitudinal and the shear waves can be excited. A typical test setup is shown in Fig. 8-5. The elastic moduli may be calculated from the various vibration modes. For instance, for a rod supported between two pegs distance L apart, the natural frequency n_T of the transverse vibrations is given by

$$n_T = \frac{\pi}{2L^2}[EIg/w]^{1/2}, \tag{8.13}$$

where I is the geometric moment of inertia, g is the acceleration due to gravity, and w is the weight of the specimen per unit length.

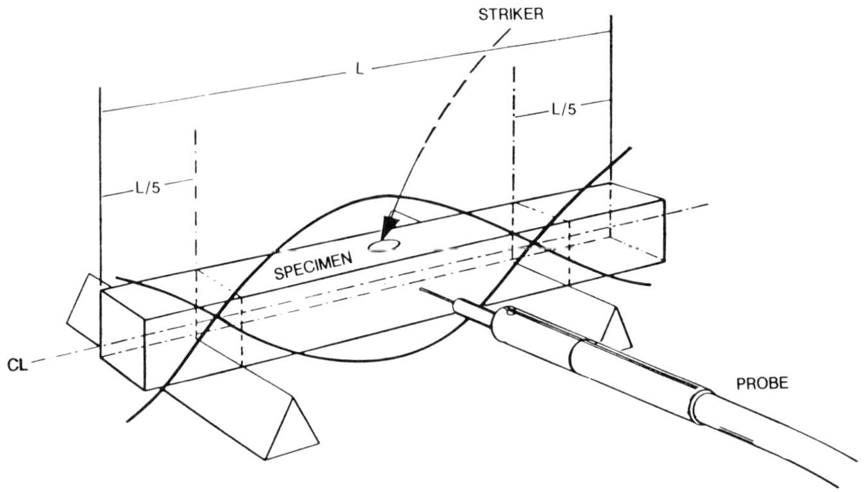

Figure 8-5. Schematic of elastic moduli measurement using the resonance method. (Courtesy J. W. Lemmens, Inc., St. Louis, Missouri.)

The natural frequency of the torsional vibration for a solid round rod fixed at one end is given by

$$n_G = [GAg/w]^{1/2}/4L. \tag{8.14}$$

Here A is the cross-sectional area of the rod.

It is helpful to utilize the standard procedures ASTM C-623 and ASTM C-885, which give procedures and values of correction factors for given specimen geometries. (See also Ref. 15.)

8.2.3. Elastic Moduli versus Glass Composition

The elastic constants of silica glass, summarized by Sosman [16], are as follows:

$$E = 700\ \text{kbar},$$

$$K = 370\ \text{kbar},$$

$$G = 305\ \text{kbar},$$

$$\nu = 0.17.$$

(Note: 1 kbar $= 10^9$ dyn/cm^2 $= 10.20$ kg/mm^2 $= 14.5 \times 10^3$ psi $= 0.1$ GPa.) The variation of E, K, G, and ν in some binary alkali silicates is shown in Fig. 8-6. Note that E and K decrease as the size and the concentration of

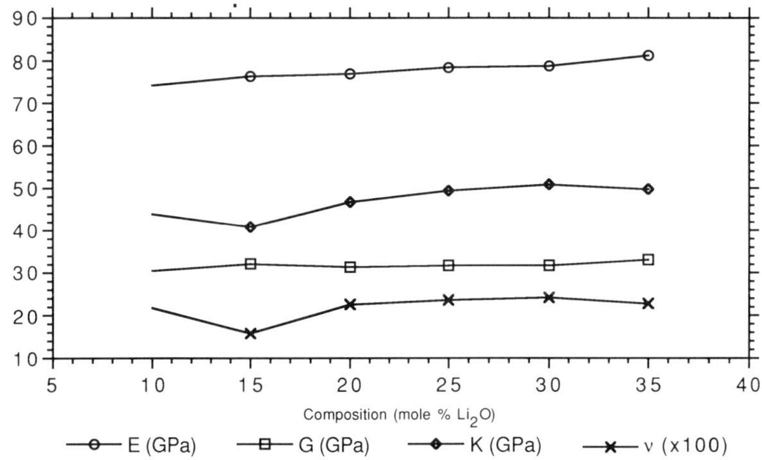

(a)

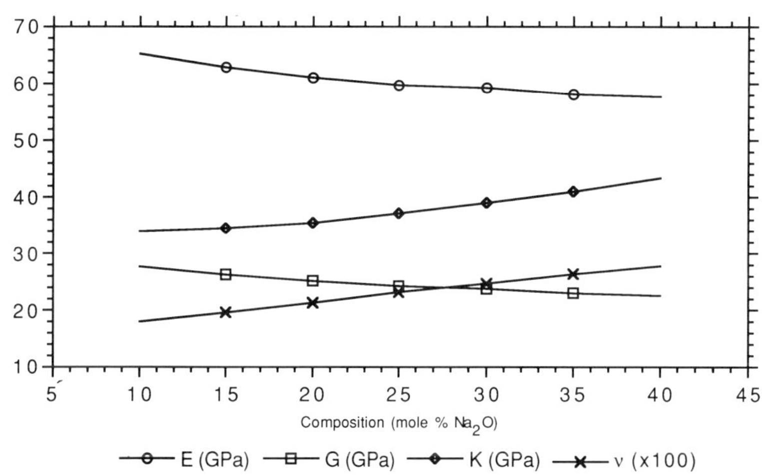

(b)

Figure 8-6. Elastic constants of (a) lithium silicate (data of R. R. Shaw and D. R. Uhlmann, *J. Non-cryst. Sol.* **5**, 237 (1971); (b) sodium silicate (data of M. H. Manghnani and B. K. Singh, *Proc. Xth Int. Cong. on Glass, Kyoto, N11*, p. 104, 1974).

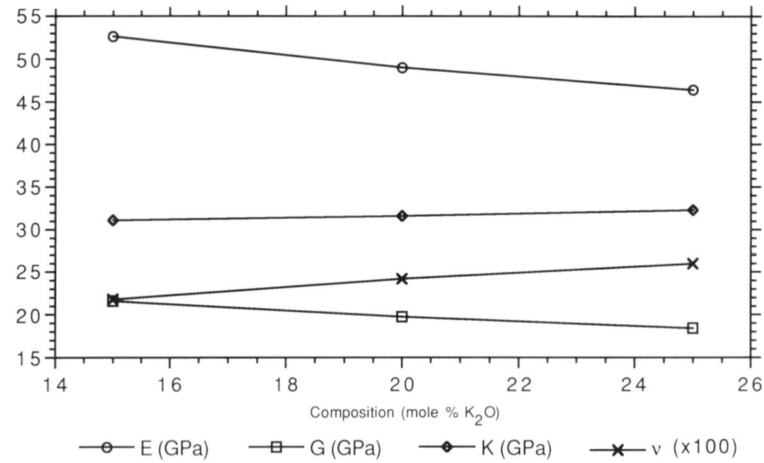

(c)

Figure 8-6. (c) potassium silicate (data of M. H. Manghnani and B. K. Singh, *Proc. Xth Int. Cong. on Glass, Kyoto, N11*, p. 104, 1974).

the alkali cations increases. The clear exception is Li, where *E*, but not *K*, actually increases with the alkali addition. The composition dependence of the elastic moduli in the system of Na_2O–xMO–$5SiO_2$ glasses is shown in Fig. 8-7. Except for PbO, *E* increases with MO addition.

In the Na_2O–B_2O_3–SiO_2 system, the data of Imaoka *et al.* show that at low (up to 10 mol %) alkali content, *E* shows a monotonic decline with B_2O_3 addition. However, in higher-alkali glasses, continuing substitution of SiO_2 by B_2O_3 results in a maximum in *E* (Table 8-1). The elastic constants of some Li_2O–Al_2O_3–SiO_2 glasses are shown in Fig. 8-8. Note that the values of *E* and *K* are clearly higher than those in silica.

Elastic properties in the Ge–Sb–As–Se–Te chalcogenide glasses have been measured by Sreeram *et al.* The Young's modulus showed a general increasing trend with the amount of crosslinking. The Poisson's ratio decreased linearly with the average coordinate number $\langle m \rangle$. The Young's moduli of heavy metal fluorides are in the 55–70 GPa range and are comparable to those in the silicates.

The reader is advised to refer to Bansal and Doremus [17]; Frischat [18]; Mazurin *et al.* [19]; Sreeram *et al.* [20]; and Drexhage [21] for more tabulation of elastic properties data.

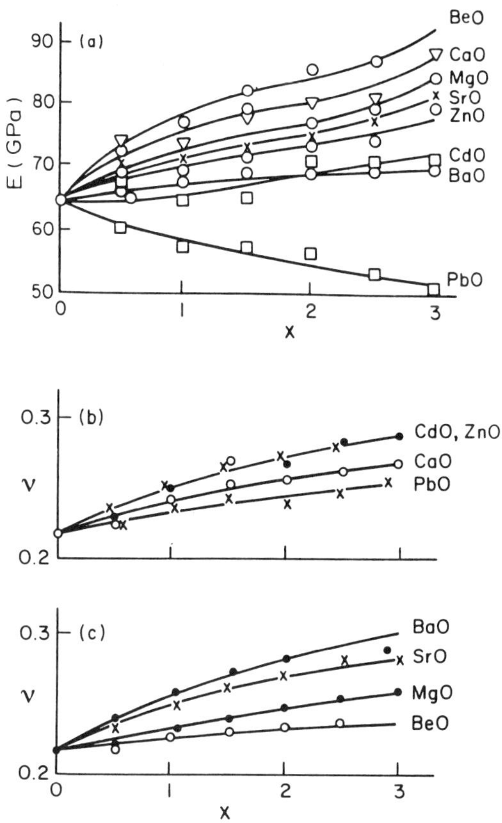

Figure 8-7. Young's modulus and Poisson's ratio of Na_2O–xMO–$5SiO_2$ glasses. M = Be, Mg, Ca, Sr, Ba, Zn, Cd, Pb. (Data of F. K. Alienkov, P. B. Paulavichus and V. A. Slizhis, *Trans. Akad. Nauk SSR Ser.* **B2**, 69 (1962); F. K. Alienkov, V. A. Slizhis, P. B. Paulavichus and P. V. Dundis, *Opt. Mekh. Prom.* no. **9**, 38 (1962); *Stekloobraznoe Sostoyanie* **3(2)**, 30 (1963).)

8.2.4. Temperature and Pressure Dependence of Elastic Moduli

Elastic properties of silica-rich glasses are anomalous at low temperatures (0–60 K) in that extrema are observed. The magnitude of the anomaly is proportional to the silica content. As an example [22], Fig. 8-9 shows the compressibility, κ, of vitreous silica, which has a maximum at around 60 K. The Young's and shear moduli of silica decrease with temperature over the range 0–60 K.

The application of high pressures also causes the elastic properties of

Table 8-1. Young's Moduli of Na_2O–B_2O_3–SiO_2 Glasses[a]

Composition (mol %)			E	Composition (mol %)			E
Na_2O	B_2O_3	SiO_2	(GPa)	Na_2O	B_2O_3	SiO_2	(GPa)
10	10	80	81.5	20	60	20	61.1
10	15	75	78.6	20	65	15	56.2
10	20	70	76.4	20	70	10	50.9
10	25	65	73.9	20	75	5	51.3
10	30	60	63.1	25	50	25	71.8
10	35	55	60.3	25	55	20	73.2
10	40	50	54.1	25	60	15	68.8
10	45	45	50.3	25	65	10	65.1
10	50	40	47.1	25	70	5	67.4
10	55	35	47.4	30	5	65	73.5
10	60	30	46.3	30	10	60	75.1
10	65	25	44.9	30	15	55	75.8
10	70	20	38.6	30	20	50	77.2
10	75	15	37.8	30	25	45	82.4
10	80	10	36.5	30	30	40	84.5
10	85	5	32.8	30	35	35	78.9
20	5	75	76.2	30	40	30	73.5
20	10	70	78.0	30	45	25	79.9
20	15	65	82.8	30	50	20	75.1
20	20	60	83.8	30	55	15	72.4
20	25	55	83.5	30	60	10	70.9
20	30	50	84.2	30	65	5	70.5
20	35	45	76.1				
20	40	40	75.5				
20	45	35	73.3				
20	50	30	70.0				
20	55	25	67.2				

[a] After M. Imaoka, H. Hasegawa, Y. Hamaguchi, and Y. Kurotaki, *Yogyo Kyokai shi*, **79**, 164 (1971).

silicates to become anomalous. The compressibility of α-quartz and most other solids decreases with pressure. On the other hand, Bridgman and Simon [23] showed that the application of low to moderate pressures cause the compressibility κ of vitreous silica to increase (Fig. 8-10). Pressures higher than about 40 kbar (=4 GPa) reverse the trend to become "normal." It appears that the glass network "softens" at low pressures. Vukcevich [22] explained the anomalous properties of silica and silicates as arising from two minima in the Condon–Morse potential energy function, corresponding to a bimodal distribution of Si–O–Si bond angles peaked at 138° and 145°. Changes in pressure or temperature cause a shift in the equilibrium population of the two states.

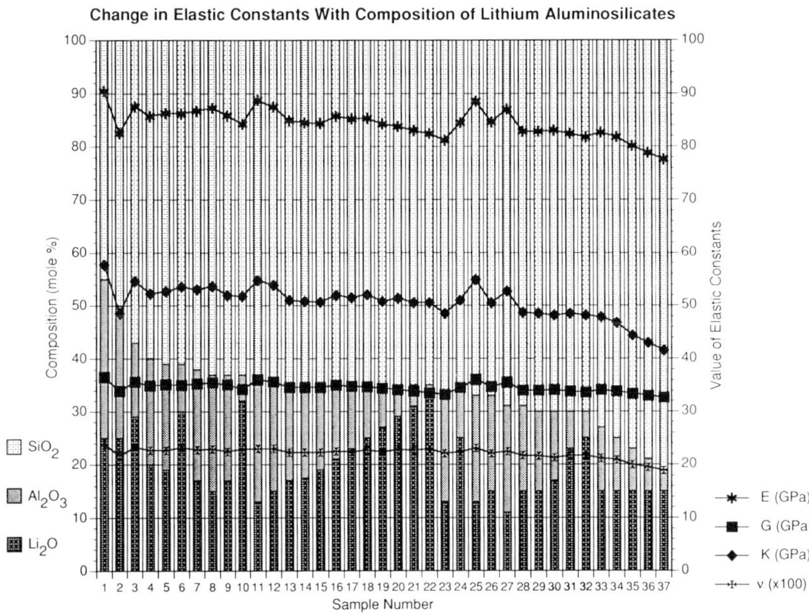

Figure 8-8. Elastic constants of some Li_2O–Al_2O_3–SiO_2 glasses. The composition for each glass may be read by the length of the vertical bars for each constituent oxide. (Data tabulated by N. P. Bansal and R. H. Doremus, *Handbook of Glass Properties*, p. 327. Academic Press, New York, 1986.)

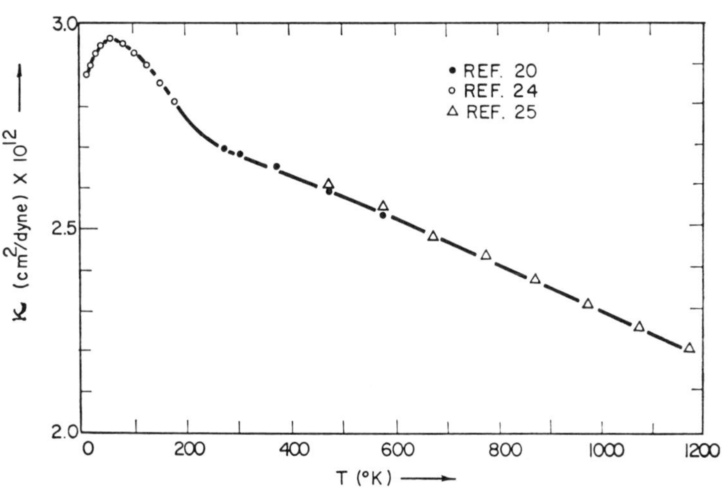

Figure 8-9. Temperature dependence of the compressibility, κ, of vitreous silica at zero pressure. (After Vukcevich [22]. Reproduced with permission of Elsevier Science Publishers.)

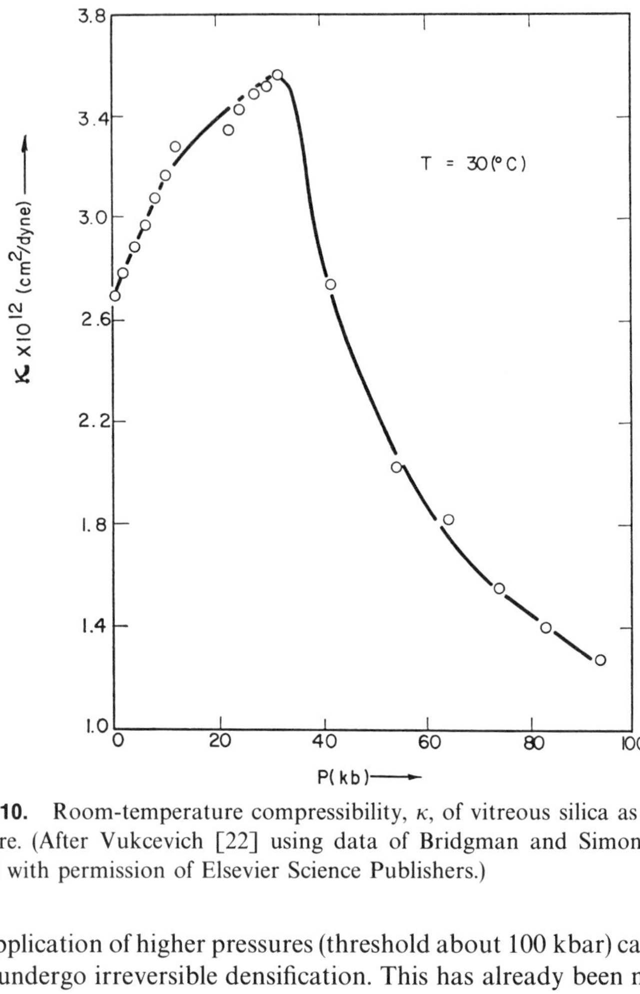

Figure 8-10. Room-temperature compressibility, κ, of vitreous silica as a function of pressure. (After Vukcevich [22] using data of Bridgman and Simon [23]. Reproduced with permission of Elsevier Science Publishers.)

The application of higher pressures (threshold about 100 kbar) causes silica glass to undergo irreversible densification. This has already been mentioned in Section 7.4.

8.2.5. The Atomistic Approach to Elastic Behavior

Atomic separation in solids is usually discussed in terms of the Condon–Morse potential energy curves. The interatomic force, F, equals $-dU/dr$, where U is the potential energy and r is the interatomic distance. F is the sum of (i) an attractive force of the form $-a/r^N$, where a is a constant and $N = 2$ for Coulombic attraction, and (ii) a steep repulsive force b/r^M due to Pauli's exclusion principle. At the equilibrium separation, r_0, the force is zero (see Fig. 8-11). To pull the atoms apart, an applied force must act

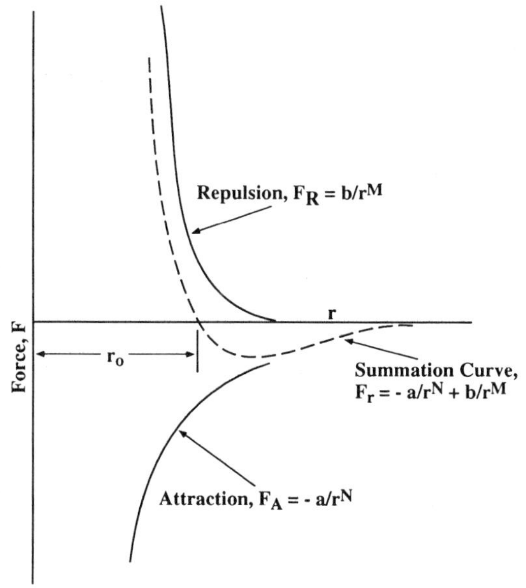

Figure 8-11. Condon–Morse curves showing qualitative variation of force with interatomic separation r.

opposite to F. Hence, the microscopic elastic modulus, E, is $-\partial(F/A)/\partial r$ at $r = r_0$, where A is an effective area that the ions present to the applied force. According to Gilman [24], the area term itself is roughly r_0^2, and hence,

$$E \approx 2\alpha e^2/r_0^4 \approx 2U_m/r_0^3, \tag{8.15}$$

where α is the Madelung constant, e the electronic charge, and U_m is the Madelung energy. In other words, E should be roughly equal to twice the Madelung energy per cubic volume of r_0^3. Note that E is predicted to decrease as the separation between the ions increases, which is as noted earlier.

Makishima and Mackenzie [25] suggested that the Madelung energy U_m per unit volume of a polycomponent glass may be approximated by a sum of the dissociation energy g_i contribution for each constituent oxide per unit volume times a packing density, so that from Eq. (8.15),

$$E = 2V_t \sum_i g_i X_i, \tag{8.16}$$

where

$$V_t = \frac{\rho}{M} \sum_i V_i X_i.$$

M is the effective molecular weight, ρ is the density, X_i is the mole fraction of the component i, and V_i is a packing factor defined for an oxide A_mO_n having ions A and O with radii r_A and r_O, respectively, by the equation

$$V_i = 6.023 \times 10^{23} \times (4\pi/3)\,[mr_A^3 + nr_O^3]. \tag{8.17}$$

Makishima and Mackenzie used Sun's dissociation energies (Section 3.1.2) per mole to calculate g_i, and Pauling's ionic radii to calculate V_i. The computed E was in excellent agreement with the measured values in many compositions belonging to the alkali–alkaline earth–aluminosilicate system, sodium borates, sodium borosilicates, and soda-lime borosilicates. Makishima and Mackenzie were able to predict a ~160 GPa value for a glass containing BeO, Al_2O_3, and ZrO_2 as perhaps the highest obtainable Young's modulus glass.

In an earlier paper, however, Charles [26] suggested that E should be proportional to the number of Si–O–Si bridging bonds per unit volume, and that the modifier oxides should contribute little to E, since they only fill up the voids. As may be noted in the preceding section, although the addition of Na_2O and K_2O does decrease E, such is not true with Li_2O addition to the SiO_2 network. Likewise, the addition of CaO to a sodium silicate increases the modulus, in contrast to Charles's argument.

One must recognize that, in contrast to the case of ionic crystals, any theoretical consideration of the elastic moduli in glasses, where bonding is at least partially ionic, must include the possibility that not all the force applied externally is expended in pulling the ions apart. A portion of this force attempts to straighten the Si–O–Si bond angles between two adjacent tetrahedra. This problem makes the estimation of E using expressions such as those of Gilman, or Makishima and Mackenzie, quite difficult. Likewise, the Si–O–Si bending component in Charles's argument is also missing. Taking into account the atomic geometry such as bond length, bond angles, density, and the nonuniform strains produced by relaxation of atoms, particularly of the O atom in an Si–O–Si unit, under the local interatomic force field, Bell and Dean [27] have attempted to describe analytical expressions for the elastic constants of silica. The agreement has been found satisfactory. Some molecular dynamic calculations made by Soules and Busbey [28] on stretching silica structures show stress–strain curves that yield reasonable values (35 GPa) of the elastic modulus.

8.3. Microhardness of Glass

8.3.1. Microhardness and Its Measurement

The microhardness (or simply the hardness measured over a micron distance scale) is seemingly another property characteristic of solids. One thinks of a harder material as being able to abrade or scratch a less hard material when mechanistically rubbed against it. Thus, the hardness should be a measure of the resistance of a material to being penetrated and eroded by another material's sharp projections. The scratching capability of various minerals is ranked using the Mohs scale of hardness spanning 1 to 10, 10 being diamond and rated the hardest. Most oxide glasses fall between 5 (apatite) and 7 (crystalline quartz). For metals, the hardness is often determined (Rockwell and Brinell methods) by using steel ball indenters (from 1/16 inch up to 10 mm diameter) thrust into the specimen under loads (from 100 to 3,000 kg), and measuring the diameter of the impression. The more acceptable measurement for most materials, including glasses, is the diamond microindenter where either a **square pyramid shape (Vickers)** or an **elongated pyramid shape (Knoop)** small diamond indenter (see Fig. 8-12) is pushed into the material under a known load. The size of the indentations is generally of the order of a few microns under 5 to 1,000 g loads; hence the term "microhardness." For the Vickers indenter,

$$\text{Vickers hardness number (VHN)} = \frac{\text{force}}{\text{area}} = \frac{2F\sin(\theta/2)}{D^2}, \qquad (8.18)$$

where F is the force in kg, θ is the included angle of the Vickers pyramid, and D is the average diagonal of the impression in mm. Since $\theta = 136^\circ$ for the Vickers pyramid, VHN may be written as

$$\text{VHN} = 1.8544\, F/D^2 \qquad (\text{kgf/mm}^2). \qquad (8.19)$$

(To get SI units, note that 1 kgf/mm^2 = 1 kp/mm^2 = 9.81 MPa.)

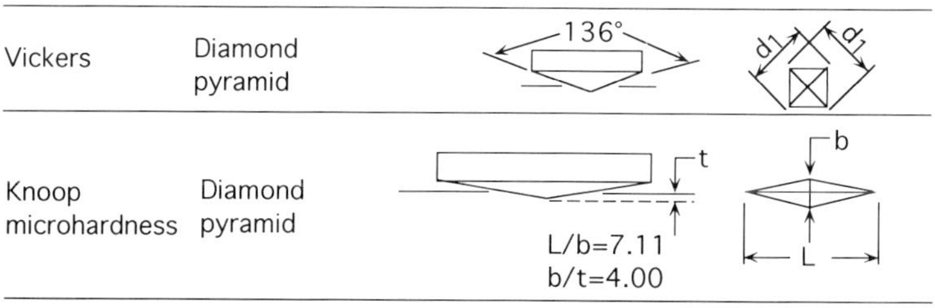

Figure 8-12. The geometry of microhardness indenters.

For the standard Knoop indenter, the long diagonal L is 7 times the width and 30 times the depth. The Knoop hardness number (ASTM C730-85) is given by

$$\text{KHN} = \frac{14.23F}{L^2} \qquad (\text{kgf/mm}^2). \tag{8.20}$$

Generally, a total time of 15–25 seconds is used for either measurement, which is sufficient for the load to gradually descend onto the surface and reach an equilibrium penetration depth.

When an indentation is made in a glass, initially the glass is elastically compressed (by the hydrostatic component) and sheared (by the shear component). Next, after reaching the elastic limit, the balance of the hydrostatic stress component permanently densifies the glass under the indentation, whereas the balance of the shear stress component causes the glass to undergo plastic deformation. The evidence of plastic flow is shown in Fig. 8-13 (after Peter [29]). Values exceeding about 200 kgf/mm^2 (=2 GPa) are suggested [2] for the elastic limit of soda-lime silica glasses. When the indenter is withdrawn, the elastic portion recovers, i.e., a slight shrinkage of the impression size is expected. Clearly, the hardness number cannot be independent of the applied load. Also, when large loads are used, cracking from the diagonals is likely to develop. The cracking, of course, implies the attainment of the fracture strength. In Chapter 18 we shall see how the cracked impressions are utilized to infer strength parameters quantitatively.

Apparently, in the case of Knoop indentation, the elastic recovery occurs primarily along the width diagonal. In addition, the length L of the long diagonal in the Knoop case is much larger than that for the Vickers indentation under similar loads. Lesser shrinkage of the impression and a better reading accuracy are claimed to be two advantages of the Knoop indenter over the Vickers. In a recently developed technique, the microhardness is measured [30, 31] dynamically. Instead of measuring the length of the impression diagonal, the depth of penetration, h, is recorded as a function of load (F) during loading or unloading. The F vs. h plot is fitted to the polynomial

$$F = a_1 h + a_2 h^2 = F_1 + F_2, \tag{8.21}$$

where the constants, a_1 and a_2, are computed from the fit. The first term is the work to create the indentation surface, and the second the work to perform elastic and plastic deformation. One may therefore define a new hardness number "F_2HN" given by

$$\text{F}_2\text{HN} = F_2/\text{area} = a_2 h^2/K^* h^2 = a_2/K^*, \tag{8.22}$$

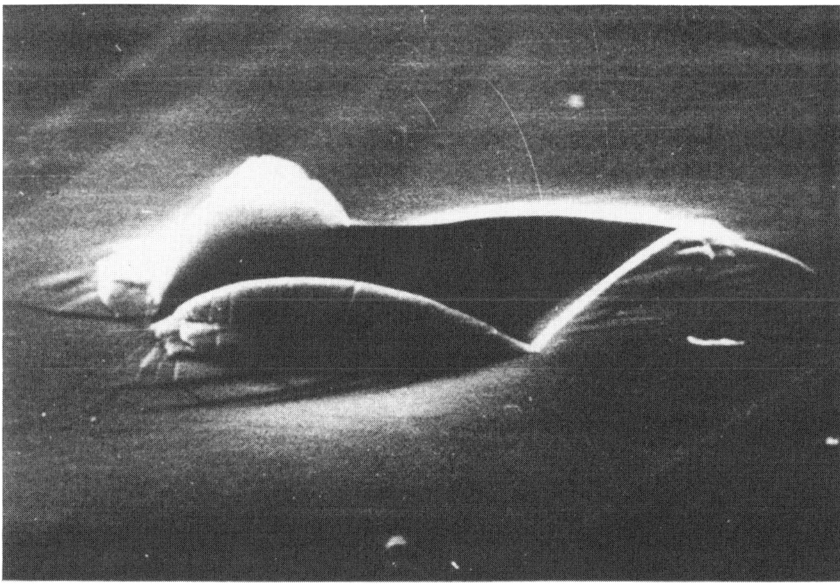

Figure 8-13. Plastic flow of glass under a microhardness indentation. 7,000 ×. (After Peter [29]. Reproduced with permission of Elsevier Science Publishers.)

where K^* is an indenter geometry constant. Thus, the F_2HN is **load-independent hardness.**

The dynamic method gives an instantaneous response of the material under the indentation load, and hence can yield quite useful information such as the elastic recovery (difference between the loading and the unloading) and the load at the instant of crack initiation.

8.3.2. Microhardness vs. Glass Composition

The VHN of SiO_2 glass is generally quoted at about 630 kgf/mm^2 (= 6.18 GPa), and that of B_2O_3 at about 205 kgf/mm^2. The addition of alkalis decreases the hardness of silica presumably because the connectivity of the structure decreases. On the other hand, the addition of alkalis to B_2O_3 glass, causing conversion to four-coordinated borons, increases the connectivity, and hence the hardness. Typical VHN values for some of the binary glasses are shown in Table 8-2 (see Ref. 17).

As in the case of the elastic moduli, the addition of alkaline earths increases the hardness. Most of the common soda-lime silica glasses have a VHN of about 550 kgf/mm^2. The KHN values for corresponding compositions are generally smaller than the VHN values. Additivity factors for the computation of VHN have been proposed by Yamane and Mackenzie [32].

Table 8-2. Vickers Microhardness of Some Binary Glasses

$0.33Li_2O{\cdot}0.67SiO_2$	535 kgf/mm^2
$0.3Na_2O{\cdot}0.7SiO_2$	413
$0.33K_2O{\cdot}0.67SiO_2$	330
$0.3Na_2O{\cdot}0.7B_2O_3$	480

The microhardness of chalcogenide glasses in the Ge–Sb–As–Se–Te family varies [33] between about 100 and 180 kgf/mm^2 and is directly proportional to $(\langle m \rangle - \langle m_0 \rangle)$, where $\langle m \rangle$ is the average coordination number of the constituent atoms and $\langle m_0 \rangle$ is the value of $\langle m \rangle$ for the onset of rigidity. Empirically, $\langle m_0 \rangle$ was found to vary between 1.83 and 1.68, which corresponds to 7- to 8-atom chain segments (a value of 2.00 corresponds to an infinite chain of Se atoms).

Hardnesses of heavy metal fluorides range between 230 and 300 kgf/mm^2, considerably less than those of the silicates, but greater than those of the chalcogenides.

Again, the reader is referred to Bansal and Doremus [17]; Frischat [18]; and Drexhage [21] for more tabulation of the properties. Note that Frischat's tabulated values in gigapascals are a factor of 10 too high.

Exercises

1. Estimate the value of E for fused silica using the Gilman equation. Compare this with the measured value.

2. Estimate the error in the measurement of E by the four-point bending technique if the beam of glass has a uniform distribution of fine bubbles covering 10% of the cross-sectional area.

3. Estimate the percent error in the measurement of E by the pulse-echo technique for the preceding glass. (Hint: See D. Ashkin *et al.*, *J. Amer. Ceram. Soc.* **73**(11), 3376 (1990).)

References

1. O. V. Mazurin, in *Glass Sci. & Tech.*, Vol. 3 (D. R. Uhlmann and N. J. Kreidl, eds.), pp. 119–179. Academic Press, New York, 1986.

2. D. M. Schlapp, *Phys. Chem. Glasses* **6**(5), 168–170 (1965).
3. R. W. Douglas, P. J. Duke, and O. V. Mazurin, *Phys. Chem. Glasses* **9**(6), 169–178 (1968).
4. F. I. Baretta, W. T. Matthews, and G. D. Quinn, *U.S. Army Mat. Technical Report No. TR 87-35*, July 1987.
5. G. D. Quinn, *J. Am. Ceram. Soc.* **73**(8), 2374–2384 (1990).
6. D. C. West, *Experimental Mech.* **21**(2), 185–190 (1964).
7. F. Birch, E. C. Robertson, and S. P. Clark, Jr., *Ind. Eng. Chem.* **49**(12), 1965–1966 (1957).
8. N. Mizouchi and A. R. Cooper, *J. Am. Ceram. Soc.* **56**(6), 320–323 (1973).
9. R. D. Corsaro, *Phys. Chem. Glasses* **17**(1), 13–22 (1976).
10. C. R. Kurkjian, *Phys. Chem. Glasses* **4**(4), 128–136 (1963).
11. S. M. Rekhson and O. V. Mazurin, *Glass Technol.* **18**(1), 7–14 (1977).
12. M. H. Manghnani, *J. Am. Ceram. Soc.* **55**, 360 (1972).
13. S. Spinner and W. E. Tefft, *Proc. ASTM* **61**, 1221–1238 (1961).
14. S. Spinner, T. W. Reichard, and W. E. Tefft, *J. Res. Nat. Bur. Stds.* **64A**(2), 147–155 (1960).
15. G. Picket, *Proc. ASTM* **45**, 846–865 (1945).
16. R. Sosman, *The Properties of Silica*, p. 437, Chemical Catalog Company, Reinhold Publishing Co. New York (1927).
17. N. P. Bansal and R. H. Doremus, *Handbook of Glass Properties*, Chapter 4. Academic Press, New York, 1986.
18. G. H. Frischat, in *Strength of Inorganic Glasses* (C. R. Kurkjian, ed.), p. 609. Plenum Press, New York, 1985.
19. O. V. Mazurin, M. V. Streltsina and T. P. Shvaiko-Shvaikovskaya, *Handbook of Glass Data*, Parts A–D Elsevier Science Publishers, Amsterdam, 1983.
20. A. N. Sreeram, A. K. Varshneya, and D. R. Swiler, *J. Non-cryst. Sol.* **128**, 294–309 (1991).
21. M. G. Drexhage, in *Treatise on Mat. Sci. & Tech.*, Vol. 26, *Glass 4* (M. Tomozawa and R. H. Doremus, eds.), Academic Press, New York, pp. 151–243. 1985.
22. Fig. 2 from M. R. Vukcevich, *J. Non-cryst. Sol.*, **11**(1), 25–63 (1972).
23. P. W. Bridgman and I. Simon, *J. Appl. Phys.* **24**(4), 405 (1953).
24. J. J. Gilman, in *Prog. in Ceram. Sci.* (J. E. Burke, ed.), Vol. 1, pp. 146–199. Pergamon Press, Elmsford, New York (1961).
25. A. Makishima and J. D. Mackenzie, *J. Non-cryst. Sol.* **12**(1), 35–45 (1973).
26. R. J. Charles, *Prog. Ceram. Sci.* **1**, 1–38 (1961).
27. R. J. Bell and P. Dean, *Phys. Chem. Glasses* **18**(6), 105–111 (1971).
28. T. F. Soules and R. F. Busbey, *J. Chem. Phys.* **78**, 6307 (1983).
29. K. W. Peter, *J. Non-cryst. Sol.* **5**(2), 103–115 (1970).
30. F. Frohlich, P. Grau, and W. Grellman, *Physica Stat. Sol.* **A42**, 79 (1977).

31. G. H. Frischat, in *Strength of Inorganic Glasses* (C. R. Kurkjian, ed.), p. 135–145. Plenum Press, New York, 1985.
32. M. Yamane and J. D. Mackenzie, *J. Non-cryst. Sol.* **15**, 153–164 (1974).
33. A. N. Sreeram, A. K. Varshneya, and D. R. Swiler, *J. Non-cryst. Sol.* **130**, 225–235 (1991).

Chapter 9

The Viscosity of Glass

9.1. Introduction

Viscosity is a liquid state property. The application of a shear force causes the atoms and molecular groups to undergo displacement with respect to each other that continues with time so long as the application of force continues. Viscosity is the inverse of fluidity and is a measure of the resistance to shear deformation with time. The time rate of deformation $\dot{e}_{xy}$ is linearly related to the shearing stress σ_{xy} through Newton's law of viscosity:

$$\sigma_{xy} = \eta \dot{e}_{xy}, \tag{9.1}$$

where η is the coefficient of viscosity. (The reader should note the use of "·" to denote the derivative with respect to time.)

The coefficient of viscosity, or simply the viscosity, has the dimensions $ML^{-1}T^{-1}$. When stress is written in the units of Pa, the appropriate unit for η is Pa·s (since strain is unitless). When the stress is written as dyn/cm^2, the unit is called the poise. As it happens, the most commonly used unit is the poise (often abbreviated as P). However, the Pa·s unit is the accepted SI unit. (For conversion, 1 Pa·s = 10 poise.)

Notice the similarity between Eq. (9.1) and Eq. (8.4), which is $\sigma_{xy} = Ge_{xy}$. *When a shear stress acts on an elastic body, the body undergoes instantaneous equilibrium shape change*, the shear strain being related to the shear stress through Eq. (8.4). *On the other hand, when a shear stress acts on a viscous body, the body starts flowing or deforming with time* at a rate given by Eq. (9.1). It is therefore reasonable to suspect that the solutions to many viscosity problems can be found simply by obtaining a solution to an analogous elasticity problem, replacing the shear strain by shear strain rate and writing η in place of G. This "viscosity–elasticity analogy" happens to be true for highly viscous bodies; however, it requires that the body be assumed to be isotropic and *incompressible*. For an incompressible solid, the Poisson's ratio $v = 0.5$, and the Young's modulus $E = 3G$ (*cf.* Eq. (8.7)). A formal statement of this analogy has been made by Goodier [1] as follows:

> Let there be two isotropic and homogeneous bodies of the same shape, one purely viscous, the other purely elastic and incompressible. The two are either loaded in the same way, or, if certain displacements are prescribed as given functions in the elastic body, then the corresponding velocities in the viscous bodies are prescribed as the same functions. Let the *displacements* of the incompressible elastic body be expressed as certain functions of x, y, z, and the modulus of rigidity G. Then the *velocities* of the viscous body are expressed by the same functions when the coefficient of viscosity η is substituted for G as long as the deformation makes no significant change in the boundaries. The converse will also be true.

The reader may note that the analogy is not confined to just the shear stresses. *Elastic equations involving any type of stress can be converted to viscous equations by substituting η for G and* 0.5 *for v, and writing strain rates in place of strains.* Trouton [2] was perhaps the first to note this analogy for the simple case of the extension of pitch under uniaxial load. Trouton defined λ, the **coefficient of viscous traction**, through the equation

$$\sigma_x = \lambda \dot{e}_x \tag{9.2}$$

and showed that $\lambda \cong 3\eta$, which is also predicted by analogy to the elastic situation described by $\sigma_x = Ee_x$.

Like the elasticity situation, Eq. (9.1) represents linear viscosity where η does not depend upon the strain rate. Fluids that obey (9.1) are termed **Newtonian fluids**. Because of the linearity, the **Boltzmann superposition principle** discussed in Chapter 13 and at the end of Appendix 1 is also applicable.

On an atomic scale, the fluidity could be ascribed to the motion of interstitial spaces or "holes" of the size of the molecules. A hole may be thought of as a portion of empty space that passed into the liquid (concurrent with a molecule that detached itself from the interior of the liquid and came to the surface occupying the empty space). Since the molecule has a large entropy in the empty space, we may also argue that so does the hole in a liquid. The high entropy of the hole implies a large number of rearrangements within the liquid, and hence the entropy contribution of the holes is termed **configurational entropy**. In a liquid at a given temperature, the holes have a certain distribution of energies that enables them to overcome potential barriers with a certain probability distribution. In other words, the holes are able to move around with random velocities with various orientations, which are representations of the molecular velocity at that temperature. The lack of large holes in a crystal gives rise to a well-defined melting point. The surface atoms must move significantly before releasing underlying atoms, sequentially, out of their well-defined potential. Glass and glass-forming melts, on the other hand, are presumed to have a large number of holes, collectively called the **free volume**, already in the structure. With increasing temperature, the holes can move, and grow significantly (hence the "shape" of the mass changes) without bringing an appreciable change in the neighborhood potentials. The analogy here to a packed and a less-filled elevator is close. Crystals act like a packed elevator, where people cannot move until the people at the door have moved out at the instant of opening the door. Glass and glass-forming melts act like a less-filled elevator, where individual motion is possible at all times, and more so when the door opens.

9.2. Viscosity — Temperature Dependence

As pointed out in connection with the *V–T* diagram (Section 2.2), a glass-forming melt acts as a liquid at high temperatures and turns into a solid upon cooling. This η–T relationship for a typical soda lime container glass is shown in Fig. 9-1. At around 1,450°C, where the glass is "melted," the viscosity is around 10 Pa·s. (For those who might have been in the moonshine business—this is roughly the viscosity of hot molasses. The melt is a thick syrup at this point. Also, for comparison, the room-temperature viscosity of water is 0.001 Pa·s, and that of glycerine about 1 Pa·s). The melt can flow, but not readily. As the temperature decreases, the viscosity increases several orders of magnitude monotonically. At around 750°C, the viscosity is about $10^{6.5}$ Pa·s, at which the mass would slump under its own weight. Below about 450°C, the viscosity of glass is higher than about 10^{15} Pa·s, at which point it becomes difficult to distinguish the glass from a solid form. It should be noted that there are no discontinuities in this curve, which is the key to continuous glass forming discussed in Chapter 20.

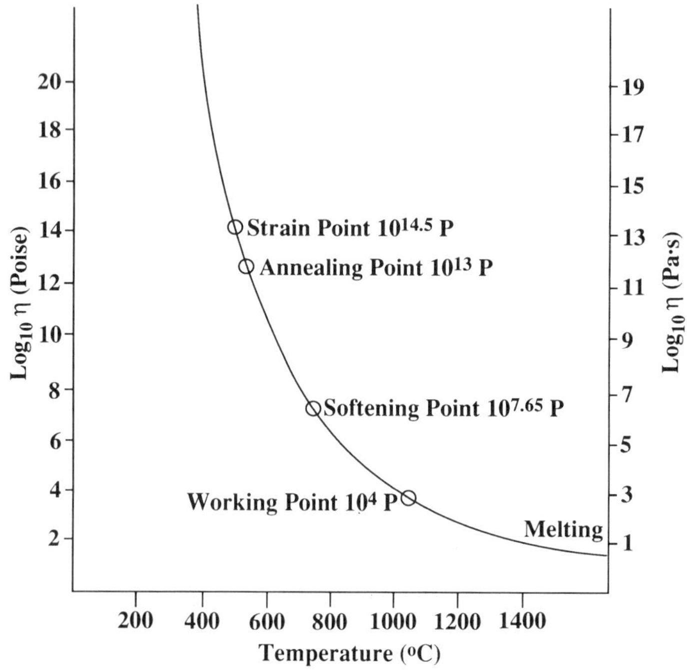

Figure 9-1. Variation of the viscosity of a common soda lime silica glass with temperature.

The temperature dependence of the glass-forming melt viscosity can be examined using Eyring's approach involving motion of the holes under the influence of shear stresses. Consider a latticelike liquid, Fig. 9-2, where a shear stress is applied on the atomic layer C causing the development of an x-velocity gradient in the y direction. The shape of the liquid changes only if there is significant relative motion between the atoms and the holes (which can happen only if a velocity gradient exists between the different layers). The probability of an atom jumping into a hole located adjacent to it is the product of the probability that it can successfully overcome a potential

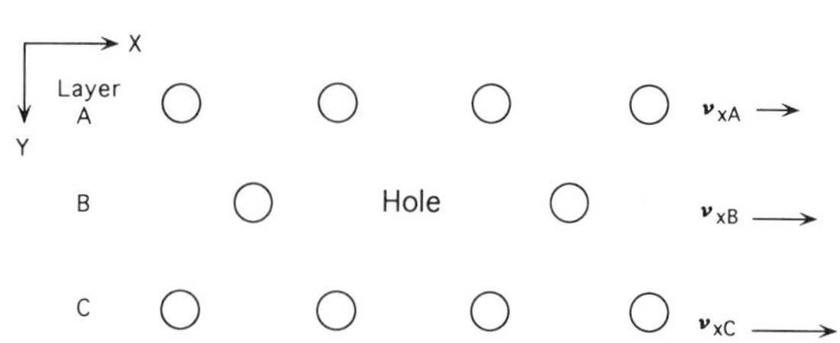

Figure 9-2. Hole motion under the action of shearing stress.

barrier and the probability P_h that a suitably-sized hole exists adjacent to it. In the absence of a shear stress, the former probability is proportional to a Boltzmann probability factor, $\exp[-\Delta G_0/k_B T]$, where ΔG_0 is the height of the potential barrier (Fig. 9-3a). The frequency of jumping from the left ν_0 is identical to the frequency of jumping from the right, and is given by

$$\nu_0 = [k_B T/h] \exp[-\Delta G_0/k_B T] P_h, \tag{9.3}$$

where k_B is the Boltzmann constant, h is Planck's constant, and T is the absolute temperature. A shear stress σ_{yx} (acting in the x-direction on the y-plane) biases the potential energy function by an amount $\sigma_{yx} V_a/2$ on each side (Fig. 9-3b), where V_a is the atomic volume. Hence, the frequency ν_+ of forward jump may be written as

$$\nu_+ = [k_B T/h] \exp[-(\Delta G_0 - \sigma_{yx} V_a/2)/k_B T] P_h = \nu_0 \exp[\sigma_{yx} V_a/2k_B T], \tag{9.4}$$

and the frequency ν_- of backward jump is given by

$$\nu_- = [k_B T/h] \exp[-(\Delta G_0 + \sigma_{yx} V_a/2)/k_B T] P_h = \nu_0 \exp[-\sigma_{yx} V_a/2k_B T]. \tag{9.5}$$

The net forward velocity is given by

$$\upsilon_{xB} - \upsilon_{xA} = (\nu_+ - \nu_-)\delta x,$$

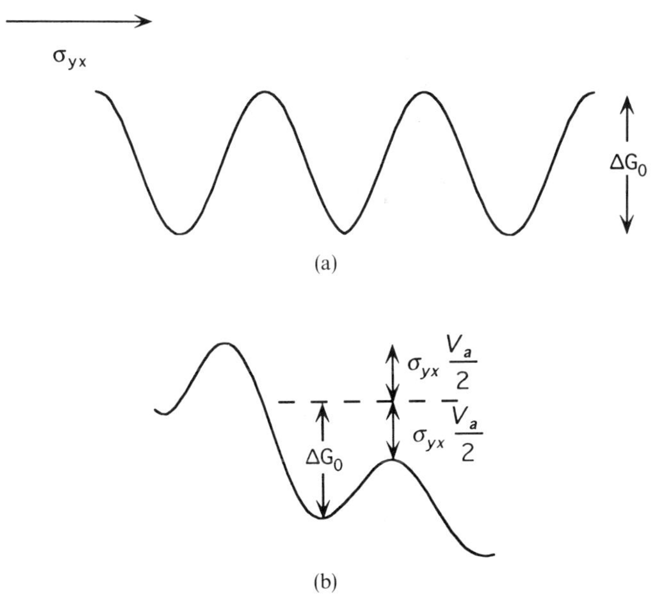

Figure 9-3. Potential diagram for atoms (a) when no shear stress exists, and (b) biased by the presence of shear stress.

which yields

$$\partial v/\partial y = (v_+ - v_-)\delta x/\delta y \cong (v_+ - v_-)$$
$$= 2v_0 \sinh(\sigma_{yx} V_a/2k_B T) \cong v_0 \sigma_{yx} V_a/k_B T. \quad (9.6)$$

The shear strain rate $\dot{e}_{yx} = \partial v/\partial y$, and hence

$$\eta = \sigma_{yx}/[v_0\sigma_{yx}V_a/k_BT] = [h/V_a][\exp(\Delta G_0/k_BT)](P_h)^{-1}. \quad (9.7)$$

This is the general equation describing the temperature dependence of the viscosity. The primary issue now is to determine P_h. One possibility would be to assume that, if ΔE_h is the energy required to form a hole, then P_h is given by a Boltzmann probability function

$$P_h = \exp[-\Delta E_h/k_BT]. \quad (9.8)$$

Substitution of (9.8) in (9.7) yields

$$\eta = [h/V_a][\exp(\Delta G_0 + \Delta E_h)/k_BT)], \quad (9.9)$$

which is an Arrhenius rate-type equation:

$$\eta = \eta_0 \exp(\Delta H/RT), \quad (9.10)$$

where R is the gas constant, η_0 is a pre-exponential factor, and ΔH is the activation energy for viscosity. This would imply that a plot of $\ln \eta$ or $\log \eta$ vs. $1/T$ would be a straight line, the slope of which gives $\Delta H/R$. Note also that, at high temperatures, ΔE_h could be quite small, since the holes are abundant in the structure. Thus, *the activation energy for viscosity in the low-temperature regime is higher than that in the higher-temperature regime.*

In addition to (9.10), there are other expressions used to describe the η–T relationship:

$$\eta = \eta_0 T \exp(\Delta H/RT) \qquad \text{(modified Arrhenius equation)}, \quad (9.11)$$

$$\eta = \eta_0 \exp[\Delta H/R(T - T_0)], \quad (9.12)$$

or, more commonly,

$$\log \eta = -A + B/(T - T_0) \qquad \text{(Vogel–Fulcher–Tammann equation)}, \quad (9.13)$$

where A, B, and T_0 are constants. Also,

$$\log\left\{\frac{\eta(T)}{\eta(T_0)}\right\} = \frac{[a_1(T - T_0)]}{[a_2 + (T - T_0)]}\left\{\frac{T\rho}{T_0\rho_0}\right\} \qquad \text{(Williams–Landel–Ferry equation,} \quad (9.14)$$

where T_0 is a reference temperature and a_1 and a_2 are constants.

Equation (9.13), abbreviated as the VFT relation, is perhaps the most useful expression relating the glass viscosity to the temperature in practical applications. Equation (9.14), abbreviated as the WLF equation, was developed to describe the viscosity of polymers. It has been found that if T_0 is taken as T_g (the glass transition temperature, defined in Section 13.3), then for most polymeric glasses, $a_1 = 17.44$ and $a_2 = 51.6$ K. The relationship of Eqs. (9.13) and (9.14) to the concept of holes is better understood in light of the free volume theories of the glass transition behavior, and is hence, deferred till Chapter 13.

Adam and Gibbs [3] developed an Arrhenius-type expression for the temperature dependence of viscosity where the activation energy for viscous flow was related to the configurational entropy of small structural units within the liquid. Since the size of these units changed with temperature, so did the activation energy. Again, because these thermodynamic concepts are more useful to the understanding of glass transition, a detailed discussion of the Adam–Gibbs relationship is deferred till Chapter 13.

9.3. Viscosity Reference Points

Because of the importance of viscosity to commercial processes, there are at least four standard viscosity reference points utilized in common practice. These are (see Fig. 9-1):

The **working point** 10^3 Pa·s $= 10^4$ poise

The **softening point** $10^{6.65}$ Pa·s $= 10^{7.65}$ poise

The **annealing point** 10^{12} Pa·s $= 10^{13}$ poise

The **strain point** $10^{13.5}$ Pa·s $= 10^{14.5}$ poise

In addition, melting point (viscosity = 10 Pa·s) and flow point (viscosity = 10^4 Pa·s) are also used occasionally.

As one would expect, the working point is the viscosity at which a mass of glass is delivered to a machine to be worked on, and the softening point is the minimum viscosity that the worked mass of glass must have before being allowed to stand on its own (or it will deform under its own weight). The temperature interval between the working point and the softening point is called the **working range**.

The usefulness of the viscosity reference points lies in the specification of temperatures at which the viscosity is a certain value. The softening point was defined by Littleton [4] as the temperature at which a glass fiber

0.55–0.75 mm diameter and 23.5 cm long, of which the top 10 cm are heated at a rate of 5°C/min when suspended vertically in a furnace of specified characteristics, elongates under its own weight at a rate of 1 mm/min.

The annealing point and the strain point approximately bracket the middle and the lower end, respectively, of the glass transformation range used for practical annealing of glass products. (In glass technology, the term "annealing" refers to releasing stresses that develop during the glass-forming process; it is discussed in detail in Section 13.9.) A useful definition of the annealing point is "the temperature at which a glass would release 95% of its stresses within 15 minutes." At strain point, the stress release occurs over roughly six hours. It is indicated later in Section 13.9 that these definitions are best stated as "stress release occurs within a matter of minutes at the annealing point" and "within a matter of hours at the strain point."

9.4. Measurement of Viscosity

There is no single technique that can be utilized to measure the viscosity of glass over the entire range, spanning roughly 14 orders of magnitude. The measurements have to be made using various techniques applicable for limited ranges. These are summarized in the table.

Range	Method	Viscosity values
Melting	Stokes' falling sphere/bubble rise	$<10^4$ Pa·s
	Margules rotating cylinder	$<10^6$ Pa·s
Softening and Annealing	Parallel plate	10^5 Pa·s $< \eta < 10^9$ Pa·s
	Penetration viscometer	10^5 Pa·s $< \eta < 10^9$ Pa·s
	Fiber elongation	10^5 Pa·s $< \eta < 10^{15.5}$ Pa·s
	Beam bending	10^7 Pa·s $< \eta < 10^{12}$ Pa·s
	Disappearance of stress	10^{11} Pa·s $< \eta < 10^{14}$ Pa·s

Measurement of viscosities higher than about 10^9 Pa·s requires a waiting period for glass stabilization. (The origin of this problem is also discussed in Chapter 13.) The waiting time is generally of the order of $3\eta/G$ (recall, G is the shear modulus). Since $G \approx 30$ GPa, viscosities larger than about $10^{15.5}$ Pa·s cannot be accurately measured as a practicality.

In the method that uses the rise of a fluid bubble, a bubble of a known gas is introduced in the molten glass, and the velocity of rise is monitored. The velocity of rise v is given by

$$v = [2r^2(\rho - \rho_0)g/9\eta][(3\eta + 3\eta_0)/(2\eta + 3\eta_0)], \tag{9.15}$$

where ρ and η are the density and the viscosity, respectively, of the glass, ρ_0 and η_0 are the corresponding values for the fluid inside the bubble, r is the radius of the bubble, and g is the acceleration due to gravity.

In the falling sphere method, a platinum ball of radius r is allowed to fall freely in the molten glass, and its velocity v of fall is measured. For $\eta_0 \to \infty$, Eq. (9.15) reduces to

$$v = 2r^2(\rho - \rho_0)g/9\eta. \tag{9.16}$$

There are two techniques adopted for the rotating cylinder method. In the original Margules viscometer, a platinum cylinder is rotated at a constant velocity in the molten glass. Glass viscosity is calculated from the torque

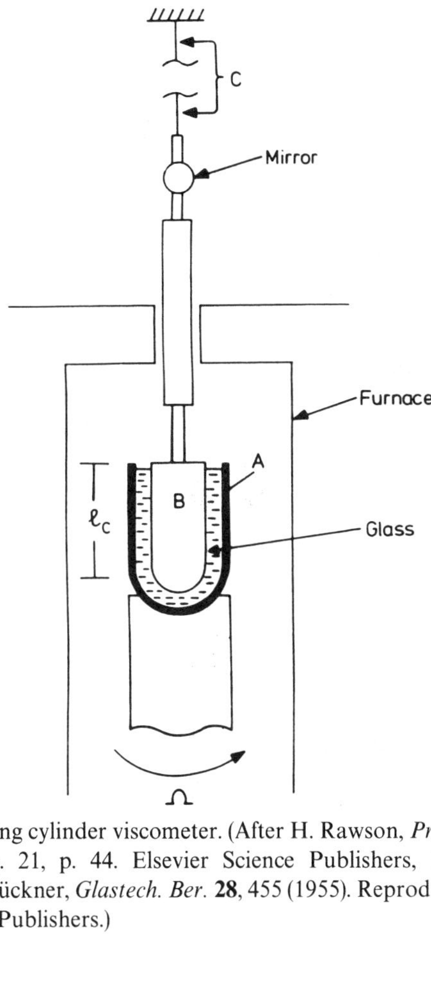

Figure 9-4. Rotating cylinder viscometer. (After H. Rawson, *Properties and Applications of Glass*, Fig. 21, p. 44. Elsevier Science Publishers, 1980. Adapted after A. Dietzel and R. Brückner, *Glastech. Ber.* **28**, 455 (1955). Reproduced with permission of Elsevier Science Publishers.)

applied to the cylinder to maintain a given velocity of rotation. In the Dietzel and Brückner version, shown in Fig. 9-4, the molten glass is filled in the space between two concentric cylinders. The outer cylinder is rotated at a constant angular velocity, and the twist angle of the inner cylinder is measured. This method is described in detail by Rawson [5].

A schematic of a parallel-plate viscometer is shown in Fig. 9-5. A disk of

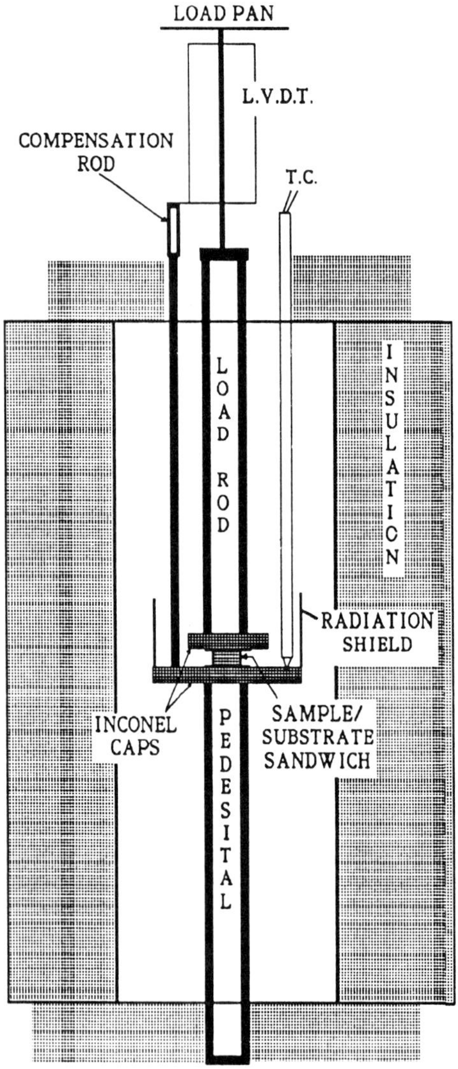

Figure 9-5. Schematic of a parallel-plate viscometer.

glass, roughly 6–8 mm dia × 3–5 mm thick, is sandwiched between two parallel plates inside a well-insulated furnace as shown. Direct contact with the plate material is avoided by employing thin Pt foil or alumina substrates. The upper pedestal is loaded, and the rate of sagging is recorded as a function of time. This method is described in detail by Varshneya *et al.* [6].

In the penetration viscometer method [7, 8], a ball of defined geometry is loaded and allowed to penetrate a given distance into the glass. Viscosity is calculated from the rate of penetration.

The fiber softening method is perhaps the most important of all the viscosity measurement techniques. A 0.55 to 0.75 mm diameter glass fiber is drawn. One end is fused to make a ball. The fiber is cut to 23.5 cm length, and then suspended inside a specified furnace, which only covers the top 10 cm (Fig. 9-6). The furnace is heated at 5°C/min. The sagging of the lower end is sighted using a telescope and measured as a function of time. If p is the length of the fiber in the furnace and L its length outside, then

$$\eta = [(Lp + p^2/2)\rho g - \gamma p/r]/3(dp/dt), \tag{9.17}$$

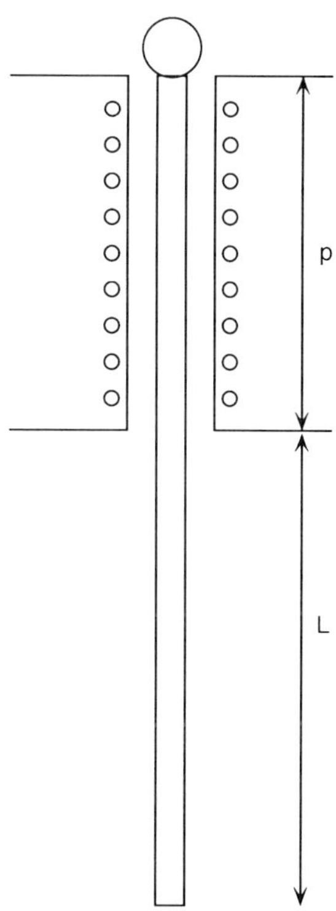

Figure 9-6. Schematic of the Littleton softening-point method.

where r is the average radius, ρ and γ are the density and the surface tension of the glass, and g is the acceleration due to gravity. (It may be noted that the term $\gamma p/r$ arises because of surface energy and attempts to shrink the fiber. Hence, L and p may be chosen such that the fiber neither elongates nor shrinks. This would then allow a measurement of γ of the glass.) One plots the rate of elongation vs. the temperature. The temperature at which $dp/dt = 1$ mm/min is identified as the softening point. Because of the varying ρ and γ, the actual value of the viscosity varies somewhat from glass to glass. For a soda lime glass, $dp/dt = 1$ mm/min corresponds to roughly $10^{6.65}$ Pa·s viscosity. (See also ASTM standard C338-73 for the measurement of the softening point.)

To measure high viscosities, one may need to load the fiber by placing weights in a pan attached to the lower end. Viscosities in this manner have been measured to the practical highest limit of about $10^{15.5}$ Pa·s.

In the beam bending method, a simply supported beam is centrally loaded and its rate of sag at the center is recorded. The equation to calculate the viscosity is derived by extending viscoelastic analogy (Section 9.1) to the three-point elastic beam bending method (Eq. 8.8)) and is being left as an exercise for the reader.

9.5. Viscosity vs. Composition and Temperature Data

A plot of log η vs. $1/T$ for various glass-forming oxides is shown in Fig. 9-7. The measured activation energies are shown in Table 9-1. The large difference between the viscosities of silica and boric oxide glasses is understandable because silica is a fully connected tetrahedral network, whereas boric oxide is only triangularly connected. The difference between the tetrahedrally coordinated germania and silica viscosities is not easily understood. Presumably there are subtle differences between the two; for instance, one of the four oxygens in germania is farther away from Ge relative to the other three. We might also imagine that, in single-component oxide glasses, ΔH might involve the cation–oxygen bond-breaking. Since the single bond strengths of the glass-forming cation and oxygen are $>380\ \mathrm{kJ{\cdot}mol^{-1}}$ ($\sim$90 kcal/mol) (recall Sun's single bond strength criterion for glass formation), one would expect the activation energies to be of the same order. Yet, the data shown in Table 9-1 appear to vary from oxide to oxide. It is clear that fluidity may not involve a single bond breaking; there may be a double bond breaking in some cases, while in others, only bond-bending may be adequate. Mott [9] has suggested that the viscosity of SiO_2 near the glass transition involves the motion of charged defects comprising positively charged threefold-coordinated silicon together with a negatively charged

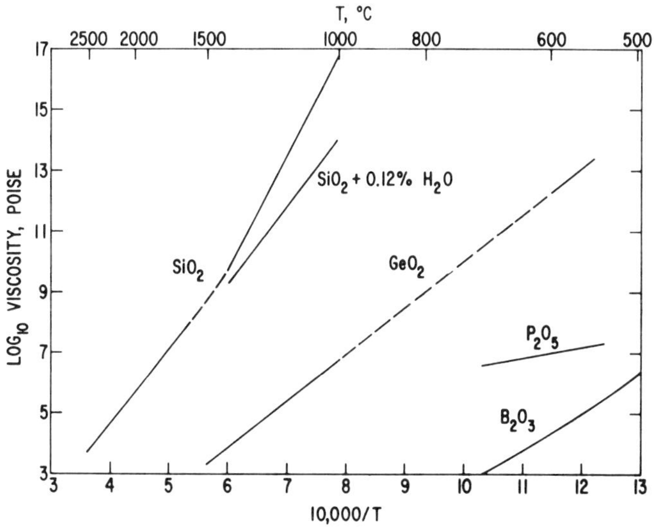

Figure 9-7. Variation of log (viscosity) vs. 10,000/T for simple glass-forming oxides. (After R. H. Doremus, *Glass Science*, Fig. 2, p. 105. Wiley & Sons, New York, 1973. Reproduced with permission of the publishers.)

NBO resulting from a broken bond. At slightly higher temperatures, a higher activation energy involving defects with two broken bonds was postulated in agreement with the data of Hetherington *et al.* [10].

The observed activation energy does vary with temperature. In B_2O_3 glass, the ΔH in the transformation range is about 325 kJ·mol^{-1} and decreases to only about 57 kJ·mol^{-1} at fairly high temperatures. As indicated in Section 9.2, the difference is presumably due to the energy required to create holes within the structure.

Table 9-1. Activation Energies for Viscous Flow in Some Simple Glasses

Glass	Activation energy (kcal mol^{-1})	Temperature range (°C)
Vitreous silica	170	1,100–1,400
	123	1,600–2,500
Vitreous germania	75	540–1,500
Vitreous P_2O_5	41.5	545–655
Vitreous B_2O_3	83–12	26–1300

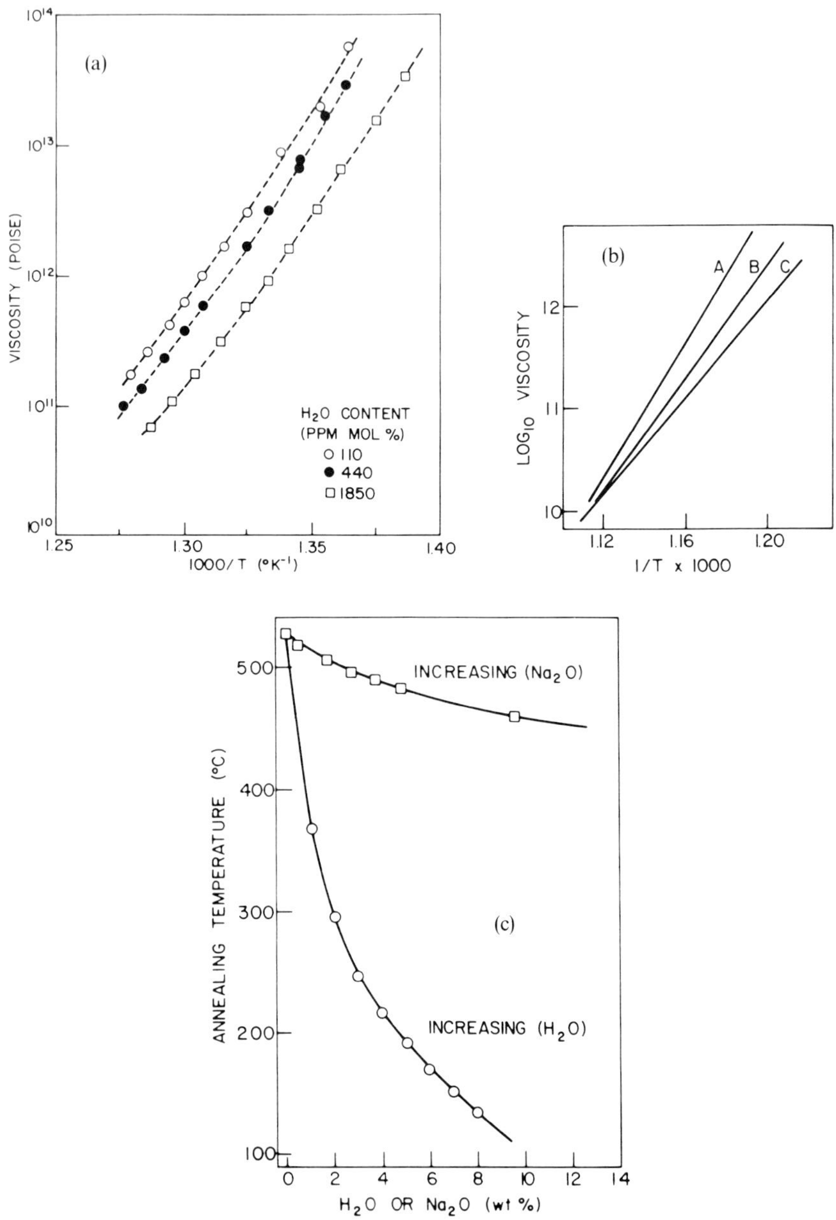

Figure 9-8. Effect of water on the temperature dependence of the viscosity of glasses. (a) Na_2O–SiO_2 glasses. (After J. E. Shelby and G. L. McVay, *J. Non-cryst. Sol.*, **20**, 439 (1976).) (b) Na_2O–CaO–SiO_2 glasses. —OH content: A = 110 ppm, B = 356 ppm, and C = 763 ppm. (After P. W. McMillan and A. Chlebnik, *J. Non-cryst. Sol.*, **38/39**, 509 (1980).) (c) Comparison of changes in T_g of Na_2O–K_2O–ZnO–Al_2O_3–SiO_2 glass with the addition of H_2O or Na_2O. (After C. K. Wu, *J. Non-cryst. Sol.*, **41**, 381 (1980).)

Shown also in Fig. 9-7 is the influence of water impurity on silica glass. The rather dramatic effect is true of nearly all oxide glasses [see Fig. 9-8 and Ref. 11]. Since viscosity property is of the utmost importance to glass-forming processes, needless to say, the practical significance of water impurity in glass is overwhelming. Water in glass is primarily derived from raw materials as surface-adsorbed moisture and as chemically bound water of crystallization. It may also be derived from the combustion products of the fuel used to melt the glass.

Additions of alkali greatly reduce the viscosity of the tetrahedrally connected networks. Such an effect, however, is nonlinear with respect to the composition. The initial additions reduce the viscosity much more than the later ones (Fig. 9-9a). (The observation implies that additivity factors for viscosity apply over only a narrow range of compositions.) When the alkali additions to an alkali silicate glass are plotted on a mol/cc basis, the data for Li_2O, Na_2O, and K_2O fall on the same plot (Fig. 9-9b).

Mixing alkalis generally lowers the viscosity of alkali silicate glasses at a given temperature, if the viscosities of the end members are relatively

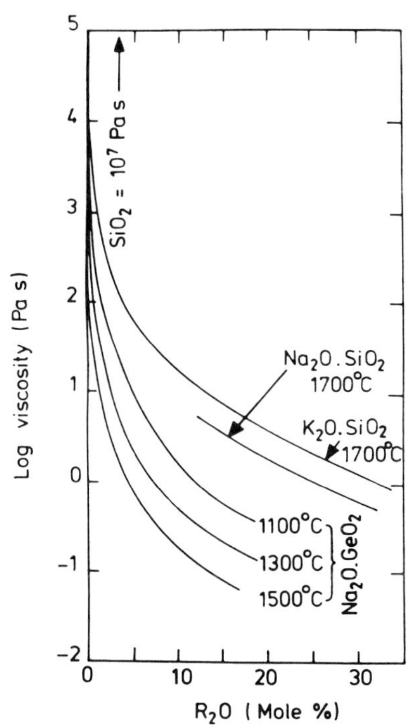

(a)

Figure 9-9. (a) Viscosity of melts in binary silicate and germanate systems.

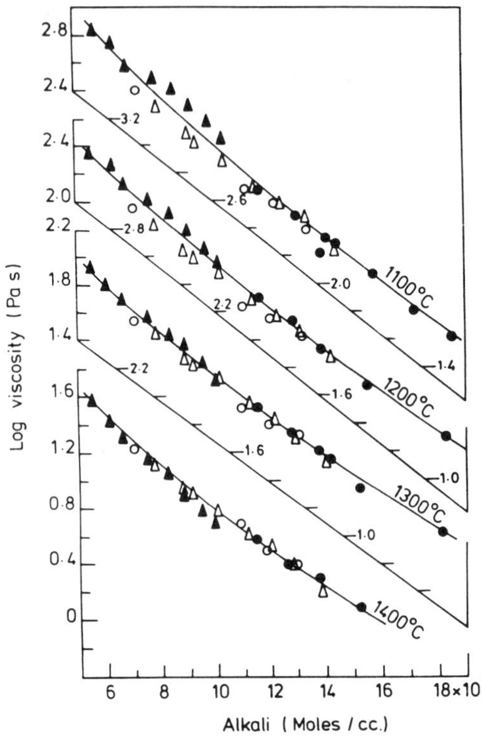

(b)

Figure 9-9. (b) Viscosity of alkali silicate glasses as a function of the added mole/cc alkali ▲ K_2O, ○ and △ Na_2O, ● Li_2O. (After H. Rawson, *Properties and Applications of Glass*, Fig. 31, p. 64. Elsevier Science Publishers, Amsterdam, 1980. Reproduced with permission of the publishers.)

close. Figure 9-10 plots the isokom (equal viscosity) temperatures for $x Na_2O \cdot (20 - x) K_2O \cdot 80SiO_2$ glasses (data of Nemilov (12)). This behavior, which is attributed to the phenomenon called the **mixed-alkali effect** (Chapter 14), is less pronounced in lower total alkali, or mixed Li_2O/K_2O-containing glasses. Nearly all other third-component additions to alkali silicates increase the viscosity, as is shown, for instance, in Fig. 9-11 for the gradual replacement of Na_2O by CaO, MgO, or Al_2O_3 in a $2Na_2O \cdot 6SiO_2$ glass.

In alkali borate melts, the viscosity–alkali content plots show pronounced maxima at low temperatures (Fig. 9-12a, taken from Visser and Stevels [13]). The magnitude of these maxima tends to decrease with increasing temperature. This observation is one of the two primary features of the **boron anomaly**

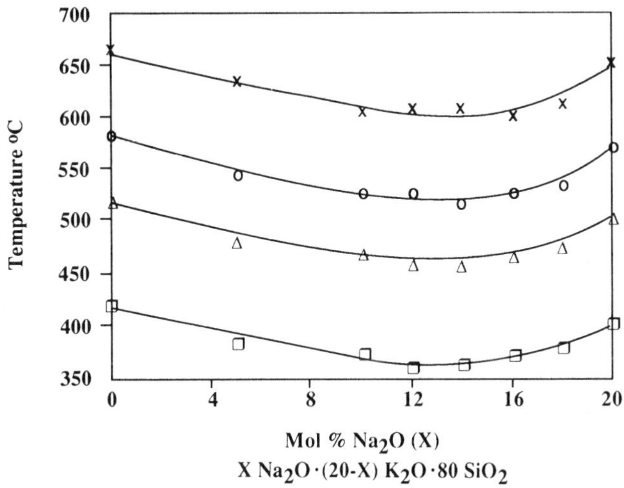

Figure 9-10. Isokom temperatures for $x\mathrm{Na_2O}\cdot(20-x)\mathrm{K_2O}\cdot 80\mathrm{SiO_2}$ glasses. Log η (poise) = 8.0 (×); 10.0 (○); 12.0 (△); 16.0 (□). (Data of Nemilov (12).)

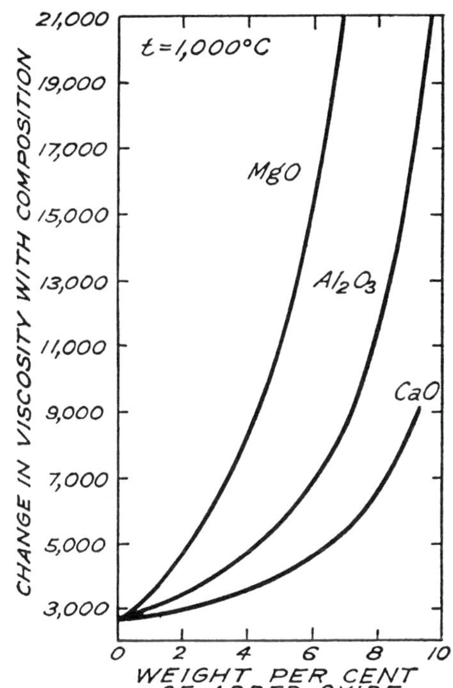

Figure 9-11. Effect on the viscosity (poise) at 1,000°C of molecular replacement of Na_2O by CaO, MgO, or Al_2O_3 in $Na_2O\cdot 3SiO_2$ glass. (After S. English, *J. Soc. Glass Technol.* **8**, 205 (1924). Reproduced with permission of The Society of Glass Technology.)

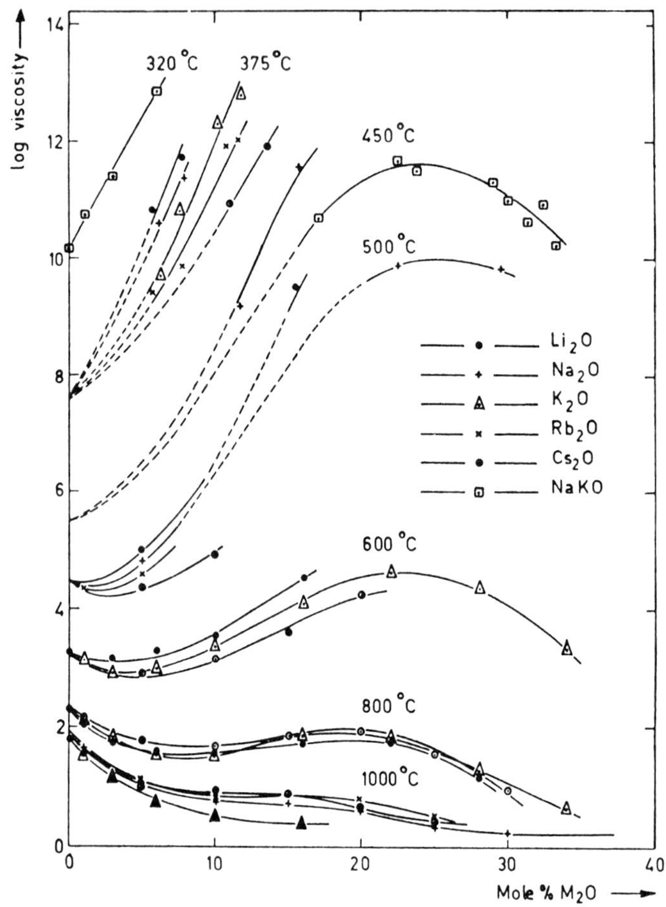

(a)

Figure 9-12. (a) Effect of the composition on the viscosity (poise) of alkali borate melts. (After Visser and Stevels [13]. Reproduced with permission of Elsevier Science Publishers.)

(Section 5.7). Like the silicates, mixed-alkali borate glasses also show negative deviation relative to that displayed by the single-alkali borates. However, the mixed-alkali effect does not interfere with the appearance of boron anomaly concurrently [12]. The apparent independence of the two phenomena is shown in Fig. 9-12b (after Kuppinger and Shelby [14]). It has been suggested [14] that the increase in viscosity is related to the replacement of the planar boroxol groups containing boron in 3-coordination, first, by

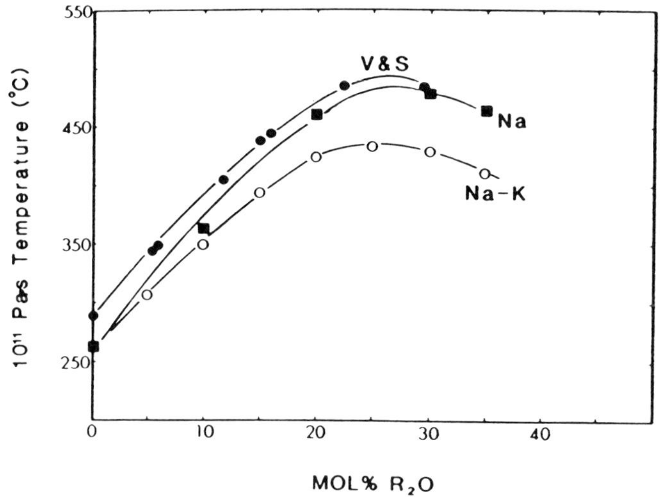

(b)

Figure 9-12. (b) Effect of the composition on the 10^{11} Pa·s isokom in mixed-alkali borate glasses compared with that of single-alkali borates. Curve marked V&S is the data of Visser & Stevels (13). (After Kuppinger and Shelby [14]. Reproduced with permission of The American Ceramic Society.)

three-dimensional tetraborate groups having boron in 4-coordination, and later, by nonbridging oxygens.

The variation of viscosity with temperature in some compositions of commercial importance is shown in Fig. 9-13. Note that the soda lime silicate and the borosilicate glasses have similar annealing points, yet the latter has a much higher working point (and a higher melting point). The relative difficulty associated with the working and the melting of borosilicates has been used to call them "hard" glasses as opposed to "soft" glasses for the soda limes and the lead silicates. (The usage has nothing to do with the hardness property discussed in Section 8.3.) Note also the curves for the soda lime and the alkali lead silicate glasses. The working points for both are around 1,000°C; however, the lead silicates have a much lower softening point in comparison to the soda lime. The temperature interval between the working point and the softening point, otherwise called the **working range**, is hence considerably larger for the alkali lead silicates. As a result, alkali lead silicate glasses are preferred for making intricate shapes. A glassblower or a machine can work on the lead silicates considerably longer before the need to reheat the workpiece arises. It happens that the thermal expansion coefficient of the alkali lead silicates is a close match to that of the soda limes (see Section 10.4), a fact that is important in allowing direct sealing of

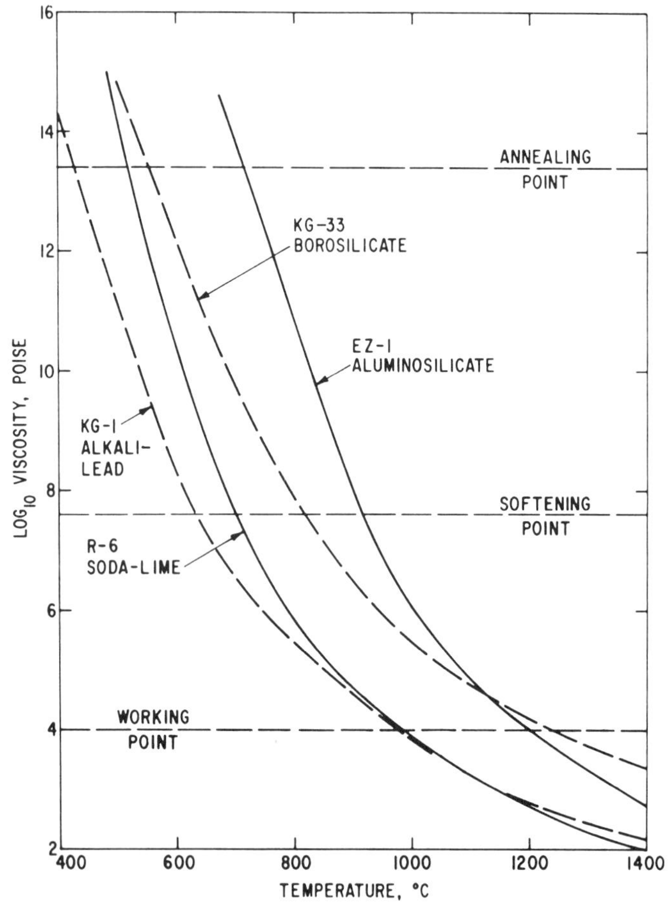

Figure 9-13. Viscosity of some commercial silicate glasses. (After R. H. Doremus, *Glass Science*, Fig. 1, p. 103. Wiley & Sons, New York, 1973. Reproduced with permission of the publisher.)

intricate parts made out of the alkali lead silicates with simple shapes of inexpensive soda limes for many products. The household incandescent lamp is an excellent example where the outer shell is made of soda lime glass and the electrical lead mount structure is made of alkali lead silicate glass.

Additivity factors for viscosity of the soda lime silicate commercial glasses have been worked out by Lakatos *et al.* [15] and by Oksoy [16]. They provide the values of the constants A, B, and T in the Vogel–Fulcher–Tammann equation (9.13), and they apparently work quite well.

Viscosity data for the chalcogenide systems are confined mostly to a measurement of the T_g. As expected, the presence of crosslinking atoms increases the T_g. By modifying the Gibbs–DiMarzio relation for the T_g of polymers, Sreeram *et al.* [17] were able to show that the T_g of Ge–Sb–As–Se–Te glasses could be written in the form

$$T_g = T_0/[1 - \beta(\langle m \rangle - 2)], \tag{9.18}$$

where T_0 is the T_g of the parent Se–Te chain, $\langle m \rangle$ is the average coordination number, and β is a system constant. The term $(\langle m \rangle - 2)$ represents the average number of crosslinking bonds.

Viscosity of heavy metal fluoride glasses has been studied extensively and is reviewed by Drexhage [18]. The T_g of most HMFG glasses of interest varies between 240 and 340°C. The viscosity decreases rapidly upon further

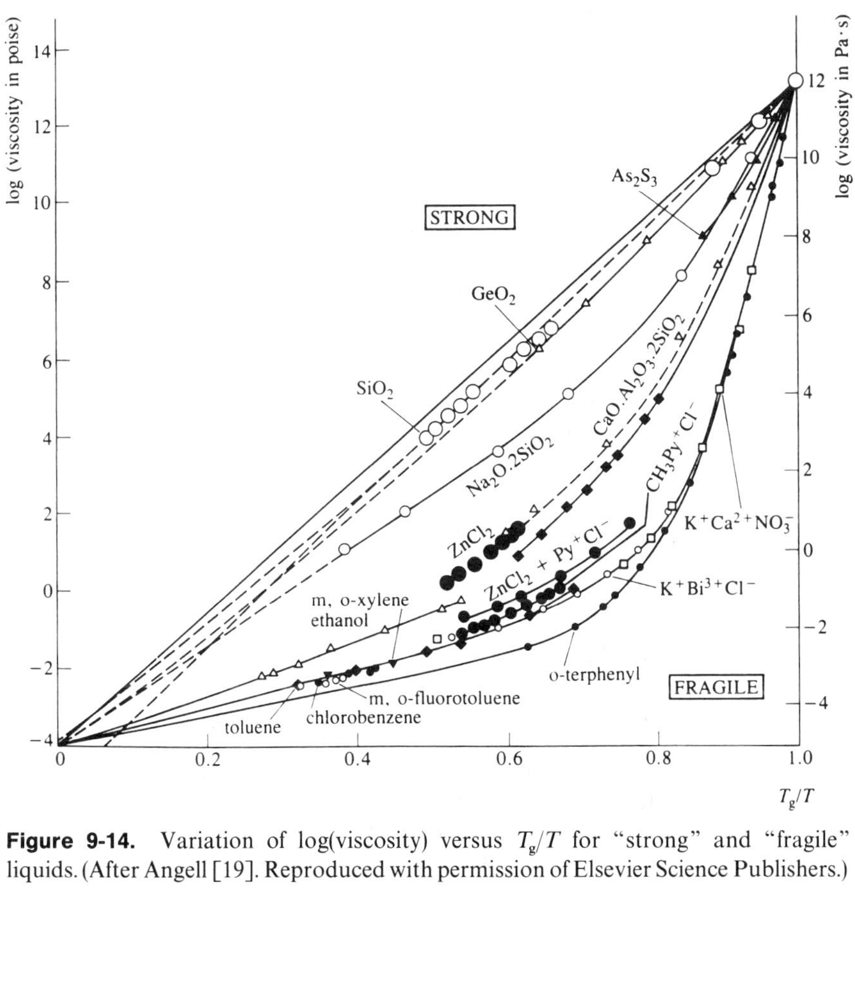

Figure 9-14. Variation of log(viscosity) versus T_g/T for "strong" and "fragile" liquids. (After Angell [19]. Reproduced with permission of Elsevier Science Publishers.)

heating until, at around 700–800°C, their viscosity is of the order of only 0.5 P.

The observed viscosity–temperature dependence of a large number of glasses has been utilized by Angell and co-workers [19, 20] to classify glass-forming liquids into **strong** and **fragile** liquids. When the viscosity is plotted against the normalized temperature T_g/T (Fig. 9-14), "strong" liquids are those that fall closer to a straight-line (i.e., Arrhenius) behavior. Examples of strong liquids are fused silica and germania, which are well-connected networks with minimal structural changes upon heating. "Fragile" liquids, e.g., *o*-terphenyl and heavy metal fluoride glasses, display a significant departure from Arrhenius behavior: they have high effective activation energies close to the T_g, with significant lowering at increased temperatures. Apparently, the fragile liquids have relatively unconstrained structures that are able to assume many configurations as the temperature is increased in the vicinity of T_g. This would result in increased configurational entropy S_c^*, and the effective activation energy in the Adam–Gibbs relationship (Eq. (13.58)).

9.6. Non-Newtonian Viscosity

Any fluid where η is a function of the shear strain rate $\dot{e}_{xy}$ in Eq. (9.1) is called a non-Newtonian fluid. Common classifications of non-Newtonian fluids are discussed next. Their shear stress (σ_{xy}) vs. shear strain rates ($\dot{e}_{xy}$) and η vs. $\dot{e}_{xy}$ are schematically shown in Fig. 9-15 and 9-16.

1. Bingham plastic: This material shows no deformation with time until it exceeds a yield point. Thereafter, the relationship between the shear stress and the shear strain rate is linear. (Some paints or slurries may approximate this model over a certain range of temperatures.)

2. Pseudoplastic (shear thinning): The apparent viscosity decreases with the magnitude of the shear stress. Glass at high shearing stresses may approach this behavior.

3. Dilatancy (shear thickening): This is the inverse of the pseudoplastic behavior. The apparent viscosity increases with increasing shear stress. Concentrated ceramic slips often show dilatant behavior.

4. Thixotropy: The material shows a reversible decrease in the apparent viscosity with time at a constant shear rate. Unlike shear thinning, the viscosity depends on time, and not on the magnitude of the shear stress. Thixotropy occurs because of a reversible structural change with time.

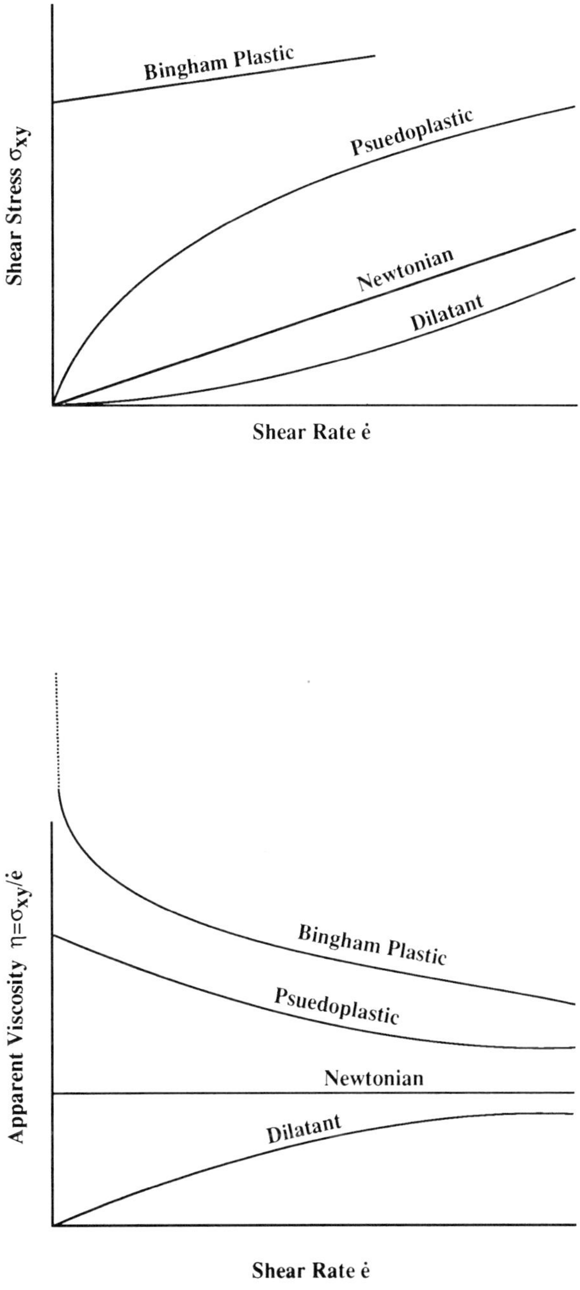

Figure 9-15. Classes of non-Newtonian behavior.

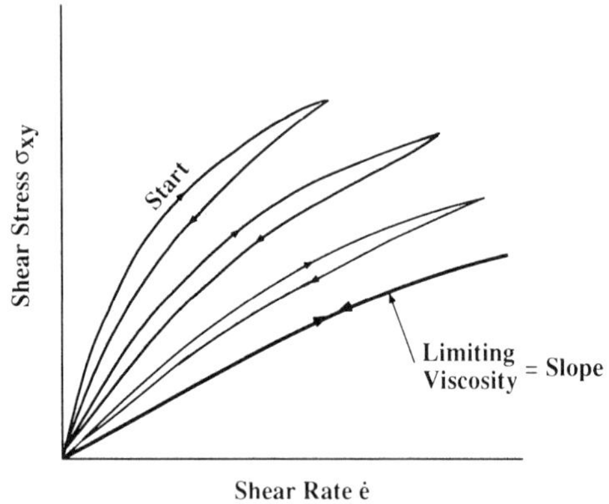

Figure 9-16. Thixotropy. (Note the apparent decrease of viscosity upon cycling.)

Hysteresis curves develop (Fig. 9-16), the apparent viscosity decreasing and approaching a limiting value with repeated cycling.

The pseudoplastic behavior or the shear thinning is of considerable interest in glass-forming. During high-speed pressing operations, glass may have sufficiently high shearing rates to cause departure from Newtonian behavior. Shear thinning may lead the glass to flow out of a controlled shape during the forming process, particularly during high-speed pressing and blowing operations. In addition, the lowered viscosity could also cause increased ionic mobility (Stokes–Einstein relationship), leading to increased phase separation that, in turn, would cause **workability** problems during glass-forming.

Table 9-2. Shear Strain Rates Marking the Onset of Shear Thinning[a]

Temperature (°C)	Newtonian viscosity (poise)	Inverse relaxation time (s^{-1})	Critical strain rate[b] for onset of non-Newtonian flow (s^{-1})
530	10^{14}	5×10^{-3}	5×10^{-6}
570	10^{12}	4×10^{-1}	2×10^{-4}
630	10^{10}	2×10^{1}	6×10^{-3}
690	10^{8}	1×10^{2}	2×10^{-1}
720	10^{7}	1×10^{3}	1.2×10^{0}
800	10^{6}	1×10^{4}	7×10^{0}
900	10^{5}	1×10^{5}	4×10^{1}

[a] After Simmons and Simmons (23).

[b] Onset of shear thinning is defined by calculating the shear strain rate causing a viscosity decrease of 10%.

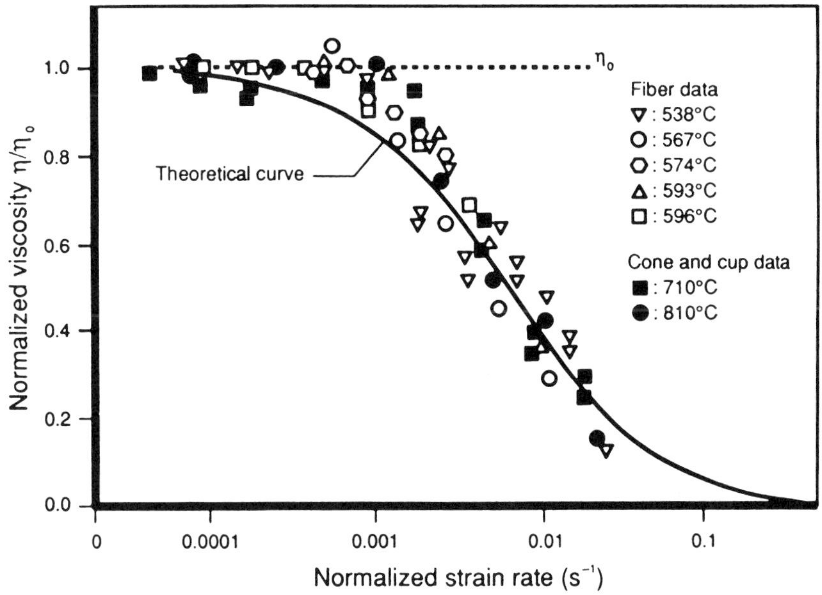

Figure 9-17. Normalized viscosity vs. normalized strain rate at various temperatures and for various methods. (After Simmons and Simmons [23]. Reproduced with permission of The American Ceramic Society.)

Much of the research in the area of non-Newtonian viscosity of glass has been performed by Brückner and co-workers [21, 22] and Simmons and Simmons [23]. Table 9-2, taken from Simmons and Simmons, shows that the onset of non-Newtonian behavior in a soda lime glass may occur at a strain rate as low as $5 \times 10^{-6}\,\mathrm{s}^{-1}$ at 530°C, whereas the viscosity remains Newtonian up to a strain rate of $40\,\mathrm{s}^{-1}$ at 900°C. A fit of the normalized viscosity η/η_0 (where $\eta_0 =$ the Newtonian viscosity) to the equation

$$\eta/\eta_0 = 1/[1 + (\dot{e}_{xy})\eta_0/\sigma_\infty], \tag{9.19}$$

where σ_∞ is the high-temperature cohesive shear strength, for two different types of experiments is shown in Fig. 9-17. The estimated values of σ_∞ are plotted in Fig. 9-18. As Brückner [21] points out, not only does the viscosity become nonlinear at high shearing rates, the shear stress also builds up rapidly as a result of an inadequate rate of stress relaxation. If the shear stress exceeds the cohesive shear strength σ_∞, then it is possible to have fracture in a "seemingly" liquid state (see Question 1 in Chapter 2, regarding the distinction between a solid and a liquid). Thus, high-speed pressing and blowing processes require a consideration of non-Newtonian flows and high-temperature fracture of glass.

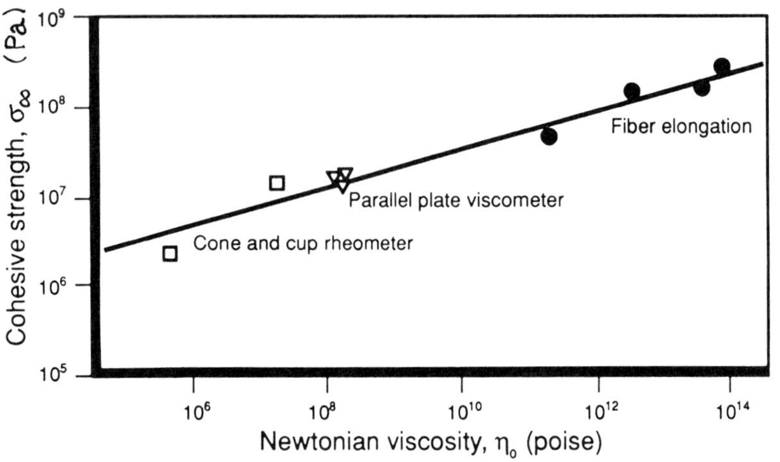

Figure 9-18. Variation of the cohesive strength vs. Newtonian viscosity. (After Simmons and Simmons (23). Reproduced with permission of The American Ceramic Society.)

9.7. Volume Viscosity

So far we have discussed shear-associated viscosity, which leads to a shape change by a rearrangement of the interstitial volume, but not to a change in volume itself. The volume viscosity η_v is defined by

$$-(p - p_0) = \eta_v \dot{V}, \tag{9.20}$$

where p_0 is the pressure necessary to maintain the material at the same volume. The equation is obtained by replacing shear stress by $-(p - p_0)$ and shear strain rate by volume strain rate in Eq. (9.1). Volume viscosity influences the interstitial volume presumably by the changing of bond angles and a consequent cooperative readjustment of the packing. The concept of volume viscosity has been discussed in relation to the isothermal relaxation of volume during glass transition by Davies and Jones [24], who suggested that the kinetics of glass transition involved a relaxation time that was related to, more appropriately, the volume viscosity and not the shear viscosity. (The influence of viscosity on glass transition is discussed in more detail in Section 13.4.) Davies and Jones further showed that, at least in the glass transition range, the activation energies for the shear viscosity and the volume viscosity were about the same. (For more discussion, see Mazurin [25].) The same may not be true below glass transition temperatures. For instance, Mizouchi and Cooper [24] showed that the rates of density recovery at room temperature in pressure-densified B_2O_3 were measurable and depended upon the density, as well as on the extent of initial densification (Section 7.3). Thus, whereas the shear fluidity at room temperature would be negligible, by contrast, the volume fluidity at low temperatures appears significant. Likewise, volume viscosity may be a major factor determining stress relaxation during ion exchange stuffing (Section 18.12). Unfortunately, our understanding of the exact mechanisms underlying the concept of the volume viscosity at temperatures well below the glass transition range is not satisfactory.

Exercises

1. A fiber of uniform radius r and length L is suspended in a furnace holding a constant temperature. A mass M (much greater than the mass of the fiber} is attached to the bottom of the fiber. Develop an expression to determine the glass viscosity using the measured elongation rate.
2. Develop the formula for the computation of viscosity by measuring the sag rate at midpoint in a three-point beam bending method.
3. Develop the formula for the standardized measurement of the softening point with and without the surface tension correction.
4. Calculate the rate of extension under its own weight of a clad-core composite glass fiber of length L, having radii r_1 (for the core) and r_2 (for the clad), and suspended in a uniform temperature furnace such that the corresponding viscosities are η_1 and η_2, respectively. Hence, estimate the change in the extension rate when $(r_2 - r_1)/r_2 \cong 0.1$ and $\eta_2 = 100\eta_1$. (Note: Because of alkali volatilization and surface silica enrichment, the high "skin" viscosity is typical and affects many glass-forming processes.)
5. Calculate the rate of rise of spherical bubbles of different sizes in a molten glass tank having glass of density 2.5 g/cm^3 and viscosity 10^2 P. Show that micron-sized bubbles may take months to rise from the tank bottom to the surface 125 cm above. (Hence, fine bubbles in glass tanks, otherwise called "seeds," must generally be removed by chemical means.)

References

1. J. N. Goodier, *Phil. Mag.* **22**, 678 (1936).
2. F. T. Trouton, *Proc. Roy. Soc.* (*London*) **A77**, 426 (1906).
3. G. Adam and J. H. Gibbs, *J. Chem. Phys.* **43**, 139 (1965).
4. J. T. Littleton, *J. Soc. Glass Technol.* **24**, 176T (1940).

5. H. Rawson, *Properties and Applications of Glass*, p. 44. Elsevier, Amsterdam, 1980.
6. A. K. Varshneya, N. H. Burlingame, and W. H. Schultze, *Glastech. Ber.* **63K**, 447–459 (1990).
7. R. W. Douglas, W. L. Armstrong, J. P. Edward, and D. Hall, *Glass Technol.* **6**(2), 52–56 (1965).
8. L. Chen, L. Zhang, and H. Zhang, *Cer. Bull.* **69**(4), 677–679 (1990).
9. N. F. Mott, *Phil. Mag. B* **56**(2), 257–262 (1987).
10. G. Hetherington, J. Jäckle, and J. C. Kennedy, *Phys. Chem. Glasses* **5**, 130 (1964).
11. R. F. Bartholomew, in *Treatise on Mat. Sci. & Tech.*, Vol. 22, *Glass III* (M. Tomozawa and R. H. Doremus, eds.), pp. 75–127. Academic Press, New York, 1982.
12. S. V. Nemilov, *J. Appl. Chem. USSR, Eng. Trans.* **42**, 46 (1969).
13. T. J. M. Visser and J. M. Stevels, *J. Non-cryst. Sol.*, pp. 376–394, 395–400, 401–409 (1972).
14. C. M. Kuppinger and J. E. Shelby, *J. Am. Ceram. Soc.* **68**(9), 463 (1985).
15. T. Lakatos, L.-G. Johansson, and B. Simminsköld, *Glass Technol.* **13**, 88–95 (1972).
16. D. Oksoy, L. D. Rye, and E. N. Boulos, *Glastech, Ber*, In press, 1993.
17. A. N. Sreeram, D. R. Swiler, and A. K. Varshneya, *J. Non-cryst. Sol.* **127**, 287–297 (1991).
18. M. G. Drexhage, in *Treatise on Mat. Sci. & Tech.*, Vol. 26, *Glass IV*, pp. 151–243. Academic Press, (M. Tomozawa and R. H. Doremus, eds.) New York, 1985.
19. C. A. Angell and W. Sichina, *Ann. NY Acad. Sci.* **484**, 241 (1986).
20. C. A. Angell, *J. Non-cryst. Sol.* **102**, 205 (1988).
21. R. Brückner, in *Advances in the Fusion and Processing of Glass III* (A. K. Varshneya, P. P. Bihuniak, and D. F. Bickford, eds.) Ceramic Trans. Volume 29, p. 3–24, American Ceramic Society, Columbus, OH (1993).
22. P. Manns and R. Brückner, *Glastech. Ber.* **61**(2), 46 (1988).
23. J. H. Simmons and C. J. Simmons, *Cer. Bull.* **68**(11), 1949 (1989).
24. R. O. Davies and G. O. Jones, *Adv. Phys.* **2**, 370–410 (1953).
25. O. V. Mazurin, *J. Non-cryst. Sol.* **87**, 392–407 (1986).
26. N. Mizouchi and A. R. Cooper, *J. Am. Ceram. Soc.* **56**(6), 320–323 (1973).

Chapter 10

Thermal Expansion of Glass

10.1. Introduction

When a glass is heated, it generally expands. If the temperature over the body of glass is equal everywhere, and the body is not restrained, then there will be no development of stress in the body. If, on the other hand, there is nonuniform heating over the body, then the different layers of glass will attempt to expand differently. The enforcement of the compatibility criteria (Appendix 1) results in a different set of dimensional changes, and consequently, stresses develop. The tensile stress components could lead to glass failure (which is discussed in detail in Chapter 18). The magnitude of the stresses so generated is related to the thermal expansion. The expansion due to heat flow into, or the contraction due to heat flow out of, the body is therefore an important consideration in product design. Thus, architectural glass window panes need to tolerate a temperature gradient across their thickness created by the difference between room temperature and the temperature on the outside. A lighted automobile headlamp should be able to stand the sudden cooling ("thermal shock") created by rainfall. It may be realized that thermal compatibility principles apply not only to the various parts of the glass body but also to other materials that have been joined to the glass. Thus, the success of glass-to-metal seals in a variety of today's products critically hinges upon the thermal expansion of glass (and of the metal).

A consideration of the thermal expansion is also important in the annealing of glassware mentioned briefly in Section 9.3. Later, in Section 13.8, we will show that this annealing refers to the release of thermal stresses generated from the nonuniform cooling of glass through the glass transition range during glass-forming. For a similar reason, the thermal compatibility of glass with other materials, such as metals, is an important issue.

In this chapter, we will review the thermal expansion property of the glasses.

10.2. Definitions

Over the temperature range T_1 to T_2, the mean or the average volume expansion coefficient β_m is given by

$$\beta_{\mathrm{m}} = \frac{V_2 - V_1}{V_1(T_2 - T_1)}, \tag{10.1}$$

where V_i is the volume at temperature T_i. The change in volume is to be measured at constant pressure and physical state.

The **true or instantaneous volume expansion coefficient** β is given by

$$\beta = \frac{1}{V}\left(\frac{\partial V}{\partial T}\right)_p. \tag{10.2}$$

The corresponding linear expansion coefficients (α) are obtained by replacing volume V by length L. Appropriate units for thermal expansion coefficient are /°C in the cgs system and K^{-1} in the SI system. Also, for small expansion coefficients,

$$\beta = 3\alpha. \tag{10.3}$$

10.3. Methods of Thermal Expansion Measurement

In most practical applications, the linear thermal expansion is of primary interest. It is generally measured by three methods: pushrod dilatometers, interferometers, and standard glass seals.

The experimental setup for a typical commercial pushrod dilatometer is shown in Fig. 10-1. The specimen, which is butted against the flat wall of the measuring system, pushes on a horizontally spring-loaded pushrod upon heating. This change relative either to the system tube itself or to a second pushrod against a standard specimen is measured using an LVDT (linear

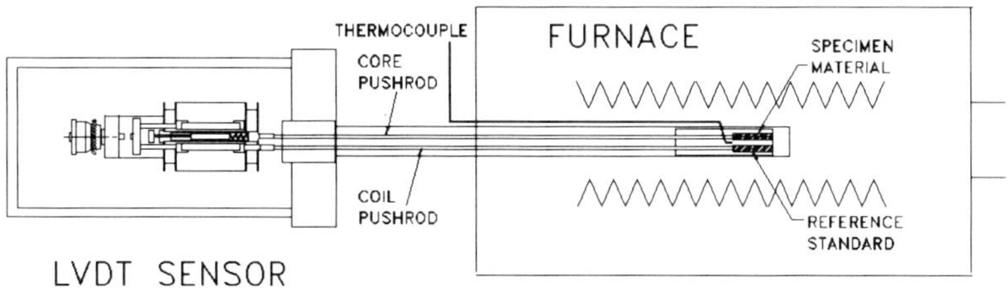

Figure 10-1. Schematic of a double pushrod differential dilatometer. (Courtesy: Theta Industries, Inc., Port Washington, New York.)

variable differential transducer). The measuring system and the pushrods are generally made out of alumina or fused silica. An equal-length platinum sample is generally a good standard to use. The key to getting good data is the uniformity of temperature over the specimen and the standard, both of which should have flat, parallel ends. The differential accuracy improves if the unknown and the standard are close in expansion characteristics. Some commercial differential pushrod dilatometers can be accurate to as little as 0.5%, which is generally quite adequate. A typical plot from a dilatometer (after adding the expansion of the standard specimen) is shown in Fig. 10-2. Note the rapid rise in expansion that marks the onset of the glass transition range. The curved hook downwards is an artifact of the measuring technique itself. At this temperature, the expansion of the specimen is countered by an

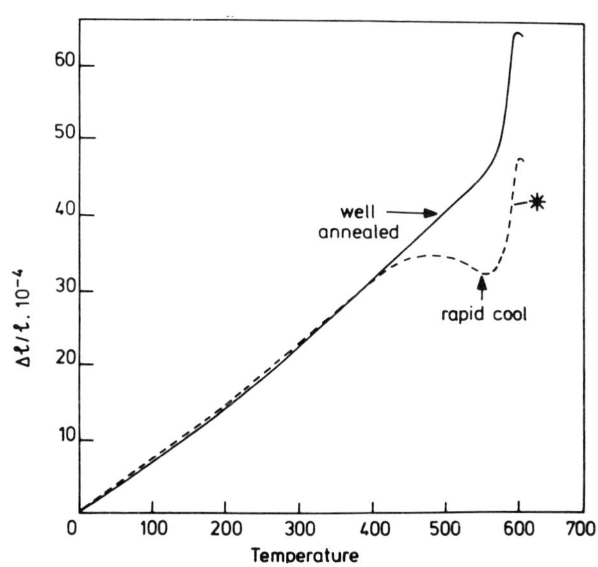

Figure 10-2. Effect of the initial cooling rate on the expansion of a glass.

apparent contraction due to specimen softening (pushrods begin to penetrate into the specimen, or the specimen begins to sag because the viscosity is about 10^9–10^{10} Pa·s). This temperature is called the **dilatometric softening temperature** T_d. The measurements should be stopped soon after. (Failure to stop the furnace from heating could result in specimen getting much too soft, and possible damage to the system due to glass adherence.)

The return (contraction) path is rarely parallel to the expansion path in the glass transition range. This is explained further in Chapter 13. For now, it suffices to suggest that *for many applications, it is the contraction curve that is important* in the consideration of stress generation, and not the expansion curve.

In the interferometric method, three carefully ground and polished pieces, two of a known and one of the unknown glass, separate two optically flat and parallel surfaces. As the temperature increases, the separation also increases, which can be monitored by interferometric methods [1, 2].

When a glass is to be employed for sealing purposes, a more accurate method of determining the expansion is by employing the unknown as a component in a standard seal. Convenient seal forms are the sandwich seal (described in ASTM F144-73) and the trident seal [3]. A sandwich seal is formed by fusing three plates to each other in such a manner that the outer pieces may be a known glass, and the inner piece is the unknown glass. The trident seal is a variation of the sandwich seal. Here the pieces need not be plates, nor need they be fused to each other along the entire length. They merely have to be joined together at the ends. One then has to make a measurement of the optical retardation (generally at the center) and convert this optical retardation to a stress. (See "Photoelastic Properties," Section 19.2.6.) Once the stress is known, the mismatch is calculated [4] from standard formulae. A thermal contraction coefficient may be extracted [5] from the mismatch if the temperature interval between the set-point and room temperature is known.

10.4. Thermal Expansion vs. Composition and Temperature

Before studying various thermal expansion data, it should be noted that most expansion coefficients in glass literature are quoted in terms of 10^{-7}/°C. On the other hand, metallurgists quote the metal expansion data at 10^{-6}/°C or ppm. (This, of course, is important during a conversation.) Yet another issue, often, is the range of temperatures over which the value of the expansion coefficient is quoted. Glass manufacturers generally quote the mean expansion coefficient over 20–300°C, which is the assumed range for most of the expansion coefficient values cited unless specified otherwise.

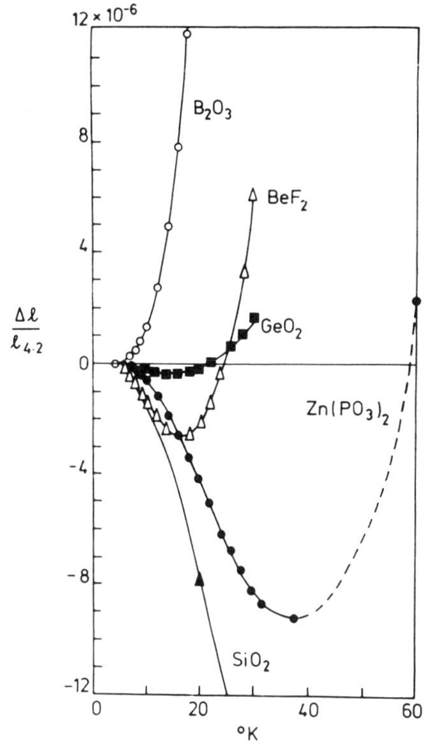

Figure 10-3. Thermal expansion curves of some single-component glasses at low temperatures. (After J. T. Krause, and C. R. Kukrjian, *J. Am. Ceram. Soc.* **51**, 226 (1968).)

Of all the single-component glasses, silica glass happens to have perhaps the lowest thermal expansion coefficient. The average value between 20 and 1,000°C is roughly $5.5 \times 10^{-7}/°C$. (Because of this rather low value of α, silica glass is preferred for products that require a high thermal shock resistance—see later discussion.) In marked contrast to the expansion of silica glass, the mean linear expansion coefficient from 20°C for cristobalite increases from about $55 \times 10^{-7}/°C$ to about $200 \times 10^{-7}/°C$ at 200°C, and undergoes a large expansion increase at the $\alpha \rightarrow \beta$ inversion to an effective mean expansion coefficient of $550 \times 10^{-7}/°C$ at about 270°C; subsequently, it decreases to that of silica glass at temperatures above 1,000°C. (The mean expansion coefficient values stated here are $\frac{1}{3}$ of the measured mean volumetric expansion coefficient. See Ref. 6.) The contrast between the expansions of silica and those of other glass-forming oxides such as B_2O_3 ($150 \times 10^{-7}/°C$) and GeO_2 ($77 \times 10^{-7}/°C$) is also noticeable. In the vicinity of 0 K, vitreous silica and several other tetrahedral network formers display negative

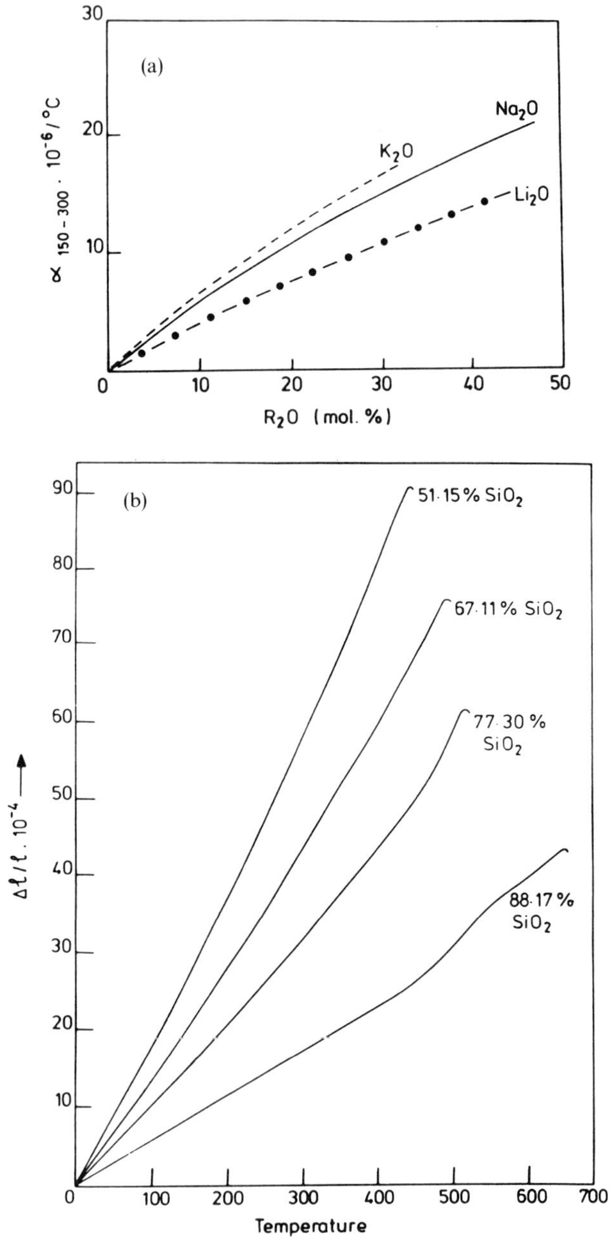

Figure 10-4. (a) Thermal expansion coefficients of $R_2O–SiO_2$ (R = Li, Na, K) glasses as a function of composition. (b) Thermal expansion curve for $x Na_2O \cdot (100 \times x) SiO_2$ glasses.

expansion (Fig. 10-3). For vitreous silica, the anomaly continues [7] from 0 K to about room temperature. This anomalous behavior, similar to that for the elastic properties of silica, has been explained by Vukcevich [8] in terms of changing population of Si–O–Si bond angle distribution (Section 8.2.4). The anomalous behavior of silica at low temperatures is discussed further in the following chapter on heat capacity.

The addition of alkali to the fully connected silica network breaks up the oxygen bridges, resulting in a monotonically increasing thermal expansion coefficient as a function of the alkali content (Fig. 10-4a): the higher the alkali content, the larger the expansion (Fig. 10-4b). In the alkali borate glasses, however, pronounced minima in the thermal expansion coefficient are observed as a function of the alkali concentration (Fig. 10-5). This observation, otherwise called the boron anomaly, has already been discussed in Section 5.7. B_2O_3 additions to SiO_2 also increase the expansion coefficient monotonically (Fig. 5-17) despite the absence of nonbridging oxygens. Here, the expansion increases as a result of the freedom allowed by the triangular boron connections.

The addition of TiO_2 to silica lowers [9] the expansion coefficient. A $7.5TiO_2$–$92{\cdot}5SiO_2$ (wt. %) glass made by the CVD process (Section 20.3.2) has an expansion coefficient of $0 \pm 0.3 \times 10^{-7}/°C$ over the 5–35°C range (Corning "Ultra Low Expansion" ULE glass). Such zero-expansion glasses are used as honeycomb substrates for space-based lightweight mirrors.

Table 10-1 lists the expansion coefficients of some commercial glasses. The soda lime silica glasses generally have an expansion coefficient of 85–95 ×

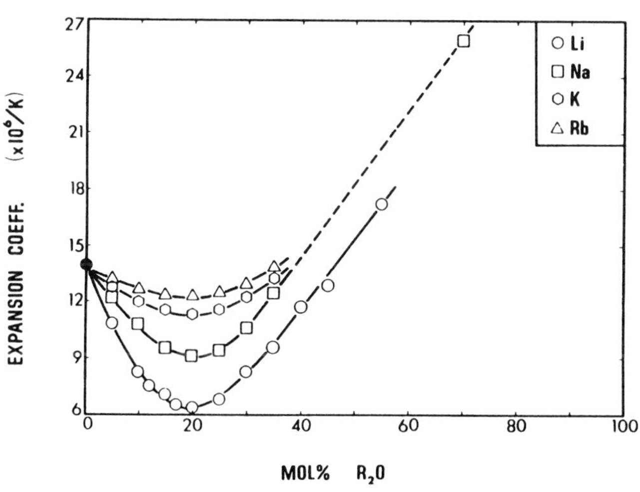

Figure 10-5. Thermal expansion coefficients of R_2O–B_2O_3 glasses as a function of composition. (After J. E. Shelby, *J. Am. Ceram. Soc.*, **66**(3), 225–227 (1983).)

Table 10-1. Thermal Expansion Coefficients of Some Commercial Glasses

		Expansion coefficient $10^{-7}/°C$	
Glass Code	Type	0–300°C	Set point—25°C
GE 001	Potash soda lead	93.5	101
GE 008	Soda lime	93.5	105
GE 012	Potash soda lead	89.5	97
Corning 1720	Aluminosilicate	42	52
GE 706	Borosilicate	48	55
GE 725	Borosilicate	35.5	39
Corning 7720	Borosilicate	36	43
Corning 7740	Borosilicate	32.5	35
Corning 7913	96% silica	7.5	5.5
Corning 7940	Fused silica	5.5	3.5

$10^{-7}/°C$. Lying between the soda lime silicate glasses on one hand, and vitreous silica on the other, are the alkali borosilicate and the alkaline earth aluminosilicate glasses. The expansions of the alkali borosilicates range from $10 \times 10^{-7}/°C$ (Vycor) to about $55 \times 10^{-7}/°C$ (GE 706, Corning 7052), and those of the alkaline earth aluminosilicates are around $45 \times 10^{-7}/°C$ (GE180).

Table 10-2. Additivity Factors, p_i, for Calculating Thermal Expansion Coefficients[a]

Material	Winkelmann and Schott[b]	English and Turner[c]	Gilard and Dubrul[d]
SiO_2	0.267	0.050	0.04
B_2O_3	0.033	−0.653	$-0.4 + 0.01w$
Na_2O	3.333	4.16	$5.1 - 0.0333w$
K_2O	2.833	3.90	$4.2 - 0.0333w$
MgO	0.033	0.45	0
CaO	1.667	1.63	$0.75 + 0.035w$
ZnO	0.60	0.70	$0.775 - 0.025w$
BaO	1.0	1.40	$0.91 + 0.014w$
PbO	1.0	1.06	$1.15 - 0.005w$
Al_2O_3	1.667	0.14	0.2

[a] $\alpha = \sum p_i \cdot w_i \times 10^{-7}/°C$; w_i = wt. %.
[b] Winkelmann, A., Schott, O., 1984, *Ann. Phys. (Leipzig)*, 51: 735.
[c] English, S., Turner, W. E. S., 1929, *J. Am. Ceram. Soc.*, 12: 760
[d] Gilard, P., Dubrul, L., 1934, *Verre Silic. Ind.*, 5: 122

Additivity factors for the various components in the glasses have been obtained by various workers. Some of these are listed in Table 10-2. For commercial soda-lime silicate compositions, Lakatos *et al.* [10] have suggested the use of the following expression:

$$\alpha = [51{\cdot}3 + 210{\cdot}864\ Na_2O + 275{\cdot}584\ K_2O + 13{\cdot}8887\ CaO - 23{\cdot}93\ MgO - 88{\cdot}638\ Al_2O_3] \times 10^{-7}/^{\circ}C,$$

where the oxide formulae represent the number of moles of the oxide per mole of SiO_2 in the glass composition.

10.5. Concepts of Glass Expansion

In ionically bonded crystalline materials, one thinks of the thermal expansion as arising from the asymmetric shape of the Condon–Morse potential diagram. The asymmetric shape is due to the anharmonic terms in the expression for the potential. When the energy is raised by adding thermal energy, the ion is able to vibrate between separations r_a and r_b. The time-averaged location of the ion appears to be r_e, which increases with temperature (Fig. 10-6). (In this instance, the reader should note that the dissociation (or sublimation) energy of ions is related to the depth of the

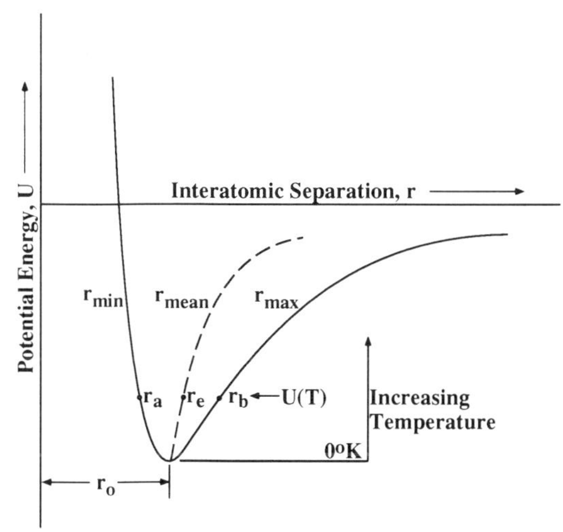

Figure 10-6. Change in the atomic separation, r, with temperature. At 0 K, the equilibrium separation is r_0. Thermal energy at T_1 causes oscillations between r_a and r_b; the mean position r_e is greater than r_0 because of the asymmetry of the potential function.

potential well, the Young's modulus is related to the curvature of the trough, and the thermal expansion is related to the asymmetry of the potential function.)

However, much like the argument presented for elasticity in glass (Chapter 8), network structures can also expand or contract by a bond-bending mechanism. Thus, in effect, the expansion of a glass may have very little resemblance to the thermal expansion of a crystal of identical composition.

Bond-bending, specifically the increasing population of the Si–O–Si bond angles from 145° to 138° with temperature, is thought [8] to be responsible for the negative expansion of fused silica over 0–200 K.

10.6. Thermal Stresses and the Thermal Shock Resistance

As stated in the Introduction, the development of a nonuniform temperature distribution in a glass (or any solid) results in the development of strains and stresses. The constitutive equations of the type of (8.3) governing the relation of the normal component of the strains to the normal components of stresses must be modified as

$$e_x = (1/E)\,[\sigma_x - \nu(\sigma_y + \sigma_z] + \alpha T. \tag{10.4}$$

In the simple case of a bar that is restrained in the x-direction and is free to move in the y- and z-directions, the solution of the modified constitutive equations (see Section 8.1) leads to the only nonvanishing stress component,

$$\sigma_x = -E\alpha T, \tag{10.5}$$

and the nonvanishing strains,

$$e_z = e_y = \alpha T(1 + \nu). \tag{10.6}$$

The normal stress components in an infinite plate (z = thickness direction) cooled such that temperature is a function only of z, are given by

$$\sigma_x = \sigma_y = -[E/(1 - \nu)]\alpha(T - \bar{T}) \tag{10.7}$$

and

$$\sigma_z = 0. \tag{10.8}$$

Application of (10.7) and (10.8) means the development of a biaxial tension (positive sign of the stress) in glass wherever the temperature is lower than the average and a compression (negative sign of the stress) wherever the situation is opposite. The biaxial stress distribution has the same mathematical form as that of the temperature distribution. It follows that if T were of

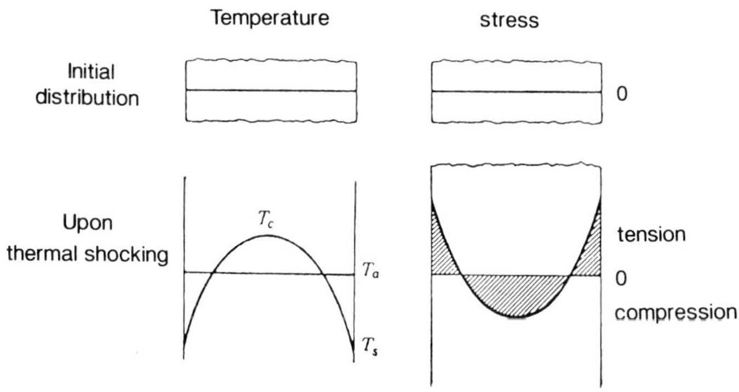

Figure 10-7. Stress production in glass due to thermal shock.

a parabolic form with the outside cooler than the inside (temperature profile produced by normal cooling means), then the stress profile will also be parabolic—tension on the outside and compression on the inside (Fig. 10-7). It may be readily shown that if $T_i - T_s = \Theta$, where T_i and T_s are the inside and the surface temperatures, then $(T_i - \bar{T}) = \Theta/3$, and $\bar{T} - T_s = 2\Theta/3$. Hence, for the interior,

$$\sigma_x = \sigma_y = -[E/(1 - \nu)]\alpha(\Theta/3), \quad (10.9)$$

and for the surface,

$$\sigma_x = \sigma_y = [E/(1 - \nu)]\alpha(2\Theta/3). \quad (10.10)$$

It may also be shown that, for a plate cooled symmetrically at a rate of Φ°C/s, the planar stresses are given by

$$\sigma_x = \sigma_y = [E/3(1 - \nu)]\Phi L^2\alpha(\rho C_p/k), \quad (10.11)$$

where L is the half-thickness of the plate, and ρ, C_p, and k are the density, specific heat, and thermal conductivity of the glass, respectively. (Note that, in Eq. (10.11), Φ is positive for cooling and negative for heating.) Surface and interior stresses for several other standard geometries have been tabulated by Kingery *et al.* [11].

The outside surface tension developed upon cooling can lead to glass fracture. Exposing a glass product to a sudden change in temperature and observing whether the glass fails or not is the basis of the **thermal shock** or "thermal endurance" test. Typically, a number of glass specimens are heated in an oven, and then plunged suddenly in an ice–water mixture. The survivors are put back in the oven at a higher temperature and plunged back into the cold bath. The process is carried until all the samples have failed. A plot of the number of survivors versus the interval of temperature change ΔT

(= oven temperature minus the ice–water temperature) readily yields that value of the interval at which the probability of survival is 50%. This temperature interval is termed the **thermal shock resistance** or the **thermal endurance** of the glass. For glass articles of similar dimensions but of different-composition glasses, the thermal endurance is estimated by

$$\Delta T = m[S/\alpha E][k/\rho C_{\mathrm{p}}]^{1/2}, \tag{10.12}$$

where S is the tensile strength of glass and m is a constant [12]. The properties E and $[k/\rho C_p]$ are relatively insensitive to the composition, and hence the thermal endurance is approximately inversely proportional to the thermal expansion coefficient. For articles of the same composition and shape, the variability clearly arises from the variability of the tensile strength of the individual specimen, which in turn reflects the magnitudes of permanent stresses and surface damage due to the manufacturing process. For many situations where the length dimension varies, it has been found that the thermal endurance is approximately given by the relation

$$\Delta T = nS/\alpha L^{1/2}, \tag{10.13}$$

where n is another constant. Thus, in addition to the inverse proportionality to the thermal expansion coefficient, the thermal endurance decreases as $L^{1/2}$, i.e., thinner glass articles can withstand higher temperature shock.

Since the stresses in the interior of the glass are of opposite sign, it is possible to test the endurance of the interior by "upshocking" of the temperature (for instance, sudden immersion in hot water or in molten salt).

The chief problem in carrying out thermal shock tests is the rate at which the outside surface temperature is changed. This, of course, depends upon how fast the article is transferred from one bath to the other, and the rate of heat transfer between the glass and the immersion bath. It should be recognized that different liquids do not give the same quenching effect (for instance, immersion in liquid N_2 is not the most effective way of quenching because of the formation of a vapor cushion). Likewise, immersion from a 100°C oven to 4°C water does not impart the same thermal shock as immersion from a 160°C oven to 64°C water. Another problem is the effect of the immersion fluid on the strength of glass. Water generally has a corrosive effect compared to oils (see Chapter 15). Hence, immersion from a 160°C oven to 64°C water will not give the same thermal endurance results as immersion from a 160°C oven to 64°C oils. Despite these experimental problems, thermal shock resistance tests are widely accepted quality control procedures for accelerated life testing in severe environments.

Exercises

1. Describe how you would measure the density of a $15Na_2O{\cdot}10CaO{\cdot}75SiO_2$ (mol %) glass from 150°C to 400°C.
2. Some 'bead' seals (cylindrical glass sealed over a round metal wire) are notorious for having excessive hoop stresses in the glass near the metal interface. What thermal shock procedure would you recommend to assure the quality of the seals?
3. A semi-infinite glass plate, thickness $2L$, expansion coefficient α, has been cooled such that the temperature profile through the thickness is parabolic (the temperature is the highest in the middle). Develop an expression for the stress profile. Show that the depth of the tension in the surface is 0.216 of the total plate thickness.
4. Write typical values of α_m for a common soda-lime container glass measured using a pushrod dilatometer when (i) heated from 20 to 300°C, (ii) cooled from 300 to 20°C, (iii) heated from 20 to 550°C, (iv) cooled from 550 to 20°C.
5. How are the preceding measurements affected when the glass has bubbles covering about 10% of the cross-sectional area?
6. It is proposed to modify the composition of a $15Na_2O{\cdot}10CaO{\cdot}75SiO_2$ (mol %) glass such that it would have a slightly higher annealing point, but a slightly lower thermal expansion coefficient. What "minor" composition change would you recommend?
7. Under what conditions will a specimen of soda-lime silica window glass display a negative thermal expansion coefficient regardless of the measurement technique?
8. A composite glass fiber of radius r is made by fusing two equal cross-section area strips together, "beading" them in the middle, and then gently drawing the bead. The fiber solidifies at the lower of the two T_g's and generally assumes a curved shape. Assume that the elastic properties of the two glasses are identical. Develop a simplified formula to calculate $\Delta\alpha$ between the two glasses by measuring the midpoint deflection h of a segment, length L of the curved fiber. (Note: This is based upon the bimetallic strip temperature sensor. You may need to consult *Roark's Formulas for Stress and Strains* by W. C. Young, McGraw-Hill, New York (1989).

   [Answer: $\Delta\alpha = 10hr/[\Delta T\{L^2 + 4h^2\}]$, where ΔT is the difference from the lower T_g to the room temperature. See A. Roth, *Vacuum Sealing Techniques*, p. 109, Pergamon Press, Oxford, 1966.]

9. Why do you think the thermal expansion coefficient of fused silica is nearly zero?

References

1. R. N. Work, *J. Res. Natl. Bur. Stand.* **47**, 80–86 (1951).
2. See also W. A. Plummer and H. E. Hagy, *Appl. Opt.* **7**, 825–831 (1968).
3. H. E. Hagy, *J. Am. Ceram. Soc.* **62**(1–2), 60–62 (1979).
4. See A. K. Varshsneya, in *Treatise in Mat. Sci. & Tech.*, Vol. 22, *Glass III*, pp. 241–305. Academic Press, New York, 1982. eds. M. Tomozawa and R. H. Doremus.
5. A. K. Varshsneya, *J. Amer. Ceram. Soc.* **63**(5–6), 311–315 (1980).
6. R. B. Sosman, *The Properties of Silica*, p. 367. Chemical Catalog Co., Reinhold Publishing Company, New York, 1927.
7. D. F. Gibbons, *Phys. Chem. Sol.* **11**, 246 (1959).
8. M. R. Vukcevich, *J. Non-cryst. Sol.* **11**(1), 25–63 (1972).
9. See H. A. Miska, in *Ceramics and Glasses*, Vol. 4, *Engineered Materials Handbook*, pp. 1016–1020. ASM Publ., 1992.
10. T. Lakatos, L.-G. Johansson, and B. Simminsköld, *Glastek. Tidsk.* **28**, 69–73 (1973).
11. W. D. Kingery, H. K. Bowen, and D. R. Uhlmann, *Introduction to Ceramics*, 2nd Ed., page 819. John Wiley, New York, 1975.
12. See F. V. Tooley, *Handbook of Glass Manufacture*, Vol. 1, p. 18. Ogden, New York, 1953.

Chapter 11

Heat Capacity of Glass

11.1. Introduction

If a material is heated, its temperature rises. Heat capacity, generally denoted by C, is the amount of heat Q (calories or joules) required to raise the temperature of the body by 1°C and is reported on a unit mass or mole basis. Thus, the units of C are cal/(g·°C) or cal/(mol·°C) (cgs units), and $J{\cdot}kg^{-1}{\cdot}K^{-1}$ or $J{\cdot}mol^{-1}{\cdot}K^{-1}$ (SI units). (As a guide, 1 calorie ≡ 4.18 J.) Often the term "specific heat" is also used. It refers to the ratio of the heat capacity to that of an equal mass of water at 15°C. Specific heat is, therefore, unitless. Occasionally, the term mean heat capacity is used to refer to an average heat capacity over a prescribed range of temperature.

Raising the temperature of the body involves adding energy to the atoms to cause them to vibrate more vigorously. The true heat capacity at temperature T is given by $\delta Q/dT$, where the use of δQ instead of dQ denotes an imperfect differential, i.e., a path-dependent quantity (see Section 13.5). When pressure is kept constant, the measured heat capacity is denoted by C_p, and when volume is kept constant, the measurement refers to C_v. Laboratory measurements are usually performed at constant (atmospheric) pressure and, hence, yield C_p. Scientific treatments, on the other hand, often include C_v. Fortunately, the difference between C_p and C_v is often small and can be neglected below room temperature. For solids, the heat capacity is the energy of phonons associated with allowed lattice vibrations. (The reader is referred to C. Kittel, *Introduction to Solid State Physics*, John Wiley & Sons, 1971, for a better understanding of lattice dynamics and their relationship to the heat capacity of solids.) The heat capacity near absolute zero is vanishingly small (arises from zero-point oscillations), but rises rapidly with increasing T, approaching the classical **Dulong and Petit limit** of $3R$/g·atom for solids, where R is the gas content. Hence, for SiO_2 in the solid state, the limiting heat capacity is 3 atoms/(mole of SiO_2) $\times$ $3R$/gram mole, which is about 0.3 cal/(g·°C). For B_2O_3, the limit would be $5 \times 3R$/g·mol, i.e., about 0.43 cal/(g·°C). The limit is rarely reached: glasses transform into supercooled liquid where additional contribution of the configurational entropy to the heat capacity must be considered, and Dulong and Petit's law ceases to apply. (From thermodynamics: $C_p = T(\partial S/\partial T)_p$, where the entropy S is equal to $k_B \ln W$; k_B is the Boltzmann constant; W is the number of distinct configurations. W increases considerably with increasing fluidity.)

11.2. Measurement of Heat Capacity

Traditionally, the heat capacity is measured using a calorimeter. A heated specimen is gently dropped in a well-insulated container having a known quantity of water, and the rise in temperature is measured. A simple heat balance equation yields the heat capacity of the unknown solid. In modern practice, the heat capacity is determined using a differential scanning calorimeter (DSC), where a small quantity of the unknown glass (as little as about 30–200 mg) placed in an Al (or Pt) pan is heated such that its temperature rises at a predetermined rate. The instrument adjusts the rate of heat input (dQ/dt) to obtain the required dT/dt and plots dQ/dt as a function of T. The displacement of dQ/dt relative to the "baseline" obtained by running an empty pan equals $C_p \cdot dT/dt$ (see Section 13.6.3 for more interpretation).

11.3. Composition Dependence

The heat capacity of most silicate and borate glasses at 25°C is generally about 900 J·kg^{-1}·K^{-1} and is not very sensitive to compositional changes. Additivity factors (Table 11-1) for calculating mean heat capacity C_m from composition data have been given by Sharp and Ginther [1]. The mean heat capacity is given by

$$C_m = (aT + c_0)/(0.00146T + 1). \qquad (11.1)$$

Table 11-1. Factors for Calculating Mean Heat Capacity[a]

Oxide	a	c_0
SiO_2	0.000468	0.1657
Al_2O_3	0.000453	0.1765
CaO	0.000410	0.1709
MgO	0.000514	0.2142
K_2O	0.000335	0.2019
Na_2O	0.000829	0.2229
B_2O_3	0.000635	0.198
SO_3	0.00083	0.189
PbO	0.000013	0.049

[a]After Sharp and Ginther (1).

The factors work reasonably well, except for glasses that contain a large amount of PbO.

11.4. Temperature Dependence

Suggested reading: W. A. Phillips, ed. *Amorphous Solids: Low-Temperature Properties.* Springer-Verlag, Berlin, 1981. (In particular, see article by R. O. Pohl, pp. 27–52.)
A. J. Leadbetter, *Phys. Chem. Glasses* **9**(1), 1–13 (1968).

Table 11-2 shows the variation of C_p for fused silica as a function of T (after Sosman [2]). A plot of C_p for boric oxide glass upon heating is shown in Fig. 11-1 and is typical of silicate glasses. The behavior of C_p at the onset of the glass transition range is a topic best deferred till Chapter 13.

At very low temperatures, the heat capacity of electrically insulating solids is given in terms of the Debye T^3 law as

$$C_v = \frac{12\pi^4}{5} N k_B \{(T/\theta_D)^3\}, \tag{11.2}$$

where N is the number of atoms in the specimen and θ_D is the Debye temperature. Higher terms in T^5, T^6, etc. arise from dispersion and may be neglected. The Debye temperture is given by

$$\theta_D = (\hbar/k_B)[6\pi^2 v_D^3 N/V]^{1/3}, \tag{11.3}$$

where V is the volume and the velocity v_D is the weighted average of the

Table 11-2. True (C_p) and Mean (C_m) Heat Capacities of Fused Silica as a Function of Temperature[a]

t, °C	C_p	C_m	t, °C	C_p	C_m
−255	4.8		400	260	222
−250	7.7	93.6	450	266	227
−243	12.7		500	270	231
−233	19.8		550	273	235
−223	26.7		600	275	238
−200	43.4	110.1	700	281	244
−150	79.5	126	800	286	249
−100	112.8	141	900	289	253
−50	142.2	154.5	1,000	292	257
0	165.7	165.7	1,100	294	260
50	185	176	1,200	303	263
100	201	185	1,300	316	267
150	215	193	1,400	328	270
200	227	200	1,500	339	275
250	236	206	1,600	346	279
300	245	212	1,700	352	283
350	254	212			

[a]After Sosman [2].
* Values are 10^3 C_p or C_m in cal/(g·°C).

longitudinal and transverse sound waves in the medium:

$$\frac{3}{v_D^3} = \frac{1}{v_L^3} + \frac{1}{v_T^3}. \tag{11.4}$$

Only long-wavelength acoustic modes are excited at low temperatures. The wavelength is much larger than the atomic spacings, and hence for these modes the solid may be treated as an elastic continuum with macroscopic

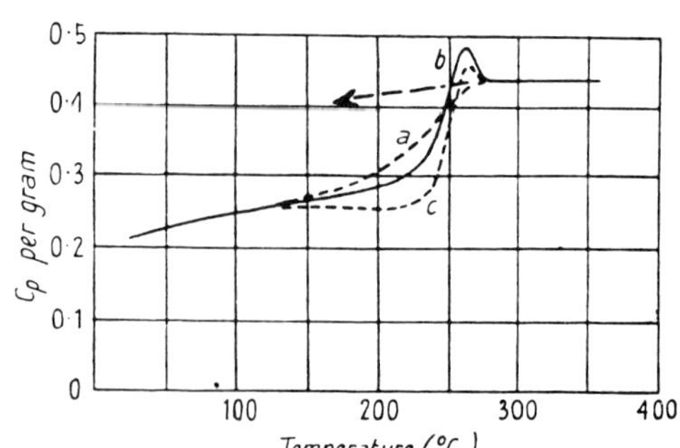

Figure 11-1. Heat capacity of boric oxide glass.

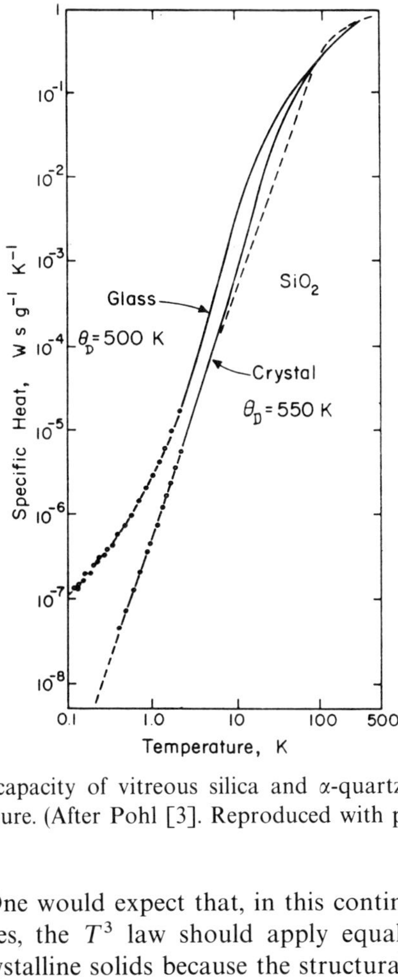

Figure 11-2. Heat capacity of vitreous silica and α-quartz at low temperatures. θ_D = Debye temperature. (After Pohl [3]. Reproduced with permission of Springer-Verlag.)

elastic constants. One would expect that, in this continuum approximation at low temperatures, the T^3 law should apply equally well to both the amorphous and crystalline solids because the structural irregularities of the disorder should become less and less important as the phonon wavelength increases. Figure 11-2 (from Pohl [3]) shows the behavior of heat capacity of vitreous silica and α-quartz at low temperatures. Two areas of disagreement may be noted:

1. The heat capacity of glass decreases more slowly than that of the crystal, and varies more like

$$C_p = c_1 T + c_2 T^3, \tag{11.5}$$

   where c_1 and c_2 are constants, instead of the T^3-dependence; and
2. the glass appears to have excess heat capacity.

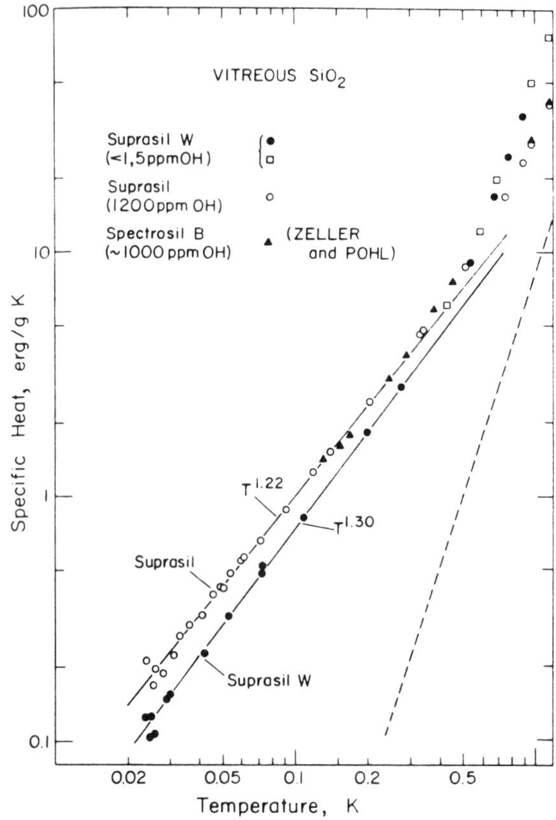

Figure 11-3. Low-temperature heat capacity of silica glasses with varying OH content showing the T^n ($n = 1.2$–1.3) dependence. (After Pohl [3]. Reproduced with permission of Springer-Verlag.)

The first of these is highlighted in Fig. 11-3 (from Pohl [3]), where the heat capacity for several forms of silica glass is shown to have T^n dependence; $n = 1.2$–1.3. In addition, there are some noticeable differences due to the –OH content. The second issue is highlighted by plotting the normalized heat capacity ($=C_{exp}/C_{Debye}$) against T. This is shown in Fig. 11-4 (from Pohl (3)) for silica, Se, and polyethylene glasses and crystals. The heat capacity appears to extrapolate to values of C_{exp}/C_{Debye} greater than 1, suggesting that the thermal property is not in agreement with the elastic property. A peak is observed for all amorphous solids, though not exclusively. See also Ref. 4 for C_v data for superionic and chalcogenide glasses.

These disparities can be rationalized by the presence of excess low-frequency vibration modes in amorphous solids. The low excitation energies

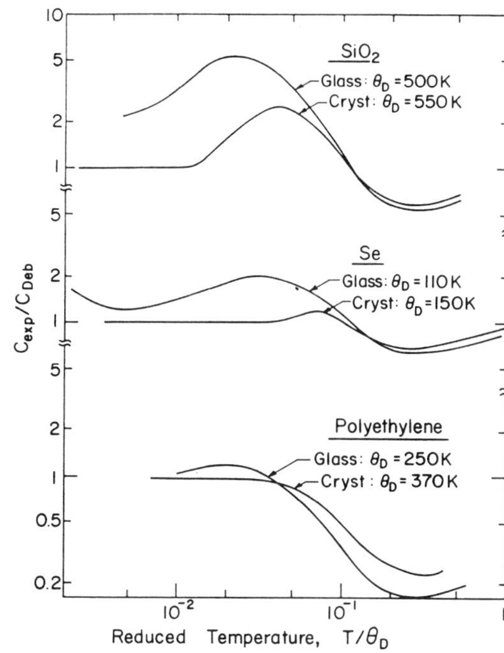

Figure 11-4. Excess heat capacity of glasses at low temperatures relative to their crystalline forms. (For silica, the crystal form used was α-quartz.) (After Pohl (3). Reproduced with permission of Springer-Verlag.)

could be generated by tunneling of states whose density is essentially constant because of the spread of asymmetries and barrier heights between neighboring potential wells. If these modes are treated as harmonic modes, then their characteristic temperatures would be in the 10–60 K range. As Fig. 11-3 suggests, some of these could arise because of impurities. That could explain why some crystals also show a peak (see Fig. 11-4). However, it is doubtful whether the impurities account for the entire anomalous behavior. Mechanisms to explain the apparent discrepancy between the elastic and the thermal properties have yet to be understood.

Exercise

1. Calculate the limiting heat capacity of $Na_2O{\cdot}3SiO_2$ glass in the solid state.

[Answer: 0.298 cal/(g·°C) or 75.2 J·mol^{-1}·K^{-1}.]

References

1. D. E. Sharp and L. B. Ginther, *J. Am. Ceram. Soc.* **34**, 260 (1951).
2. R. B. Sosman, *The Properties of Silica.* Chemical Catalog Co. (Reinhold Publ. Corp.), New York, 1927.
3. R. O. Pohl, in *Amorphous Solids: Low-Temperature Properties* (W. A. Phillips, ed.), p. 27. Springer Verlag, Berlin, 1981.
4. A. Avogadro, S. Aldrovandi, and F. Borsa, *Phil. Mag. B* **56**(2), 227–236 (1987).

Chapter 12

Thermal Conductivity and Heat Transfer in Glass

12.1. Introduction

Thermal conductivity, denoted generally by K, is defined by **Fourier's first law** as the amount of heat transmitted per unit cross-sectional area per unit time through the body in the presence of a unit temperature gradient. Thus, for the linear flow of heat in the x direction,

$$Q = -KA(dT/dx), \tag{12.1}$$

where Q is the heat current (flux per unit time), A is the cross-sectional area, and T is the temperature. The negative sign implies a positive flow of heat along decreasing temperatures. Appropriate units for K are $\text{cal}\cdot\text{cm}^{-1}\cdot{}^{\circ}\text{C}^{-1}\cdot\text{s}^{-1}$ in the cgs system, and $\text{W}\cdot\text{m}^{-1}\cdot\text{K}^{-1}$ in the SI system. (For conversion, $1\ \text{W}\cdot\text{m}^{-1}\cdot\text{K}^{-1} = 2.388 \times 10^{-3}\ \text{cal}\cdot\text{cm}^{-1}\cdot{}^{\circ}\text{C}^{-1}\cdot\text{s}^{-1}$.) For situations involving transient flow and no internal generation of heat, the application of the conservation of energy principle yields

$$dT/dt = d[\kappa dT/dx]/dx = \kappa d^2T/dx^2, \tag{12.2}$$

where κ is called the **thermal diffusivity**, and is equal to $K/\rho C_p$, where ρ is the density and C_p is the heat capacity. The use of κ in place of D (mass diffusivity, see Chapter 14) enables one to convert solutions of mass-transfer problems to equivalent heat-transfer problems. Solutions to heat flow problems in a variety of situations are given by Carslaw and Jaeger [1].

Thermal conduction in opaque non-metallic solids occurs by a transfer of energy between lattice waves or phonons. As an approximation, one may write

$$K = C_v \omega \Lambda / 3, \tag{12.3}$$

where ω is the velocity of acoustic waves in the solid and Λ the mean free path for phonons. The phonon mean free path is a function of two processes: geometric scattering and interaction with other phonons. For perfectly harmonic oscillators, the second mechanism does not exist, and hence Λ is determined simply by the boundaries and defect sites. It has been shown that the inclusion of anharmonicity causes Λ to be proportional to $1/T$ at high temperatures. One would, therefore, expect K to decrease with increasing temperature.

In the temperature range where glasses become transparent to the emitted or incident radiation (**diathermanous**), *heat may also be transferred through glass by a photon conduction mechanism.* The hemisperhical radiation conductivity K_R in a semi-infinite body is given by the equation

$$K_R = 16n^2 \sigma T^3 / 3\alpha, \tag{12.4}$$

where n is the refractive index, σ is Stefan's constant, and α is the spectral absorption coefficient. As one may note, K increases with T^3. For most clear glasses, the transparency up to about 4 to 5 μm wavelength yields a sufficiently low α, such that K_R is of the same order as K at about 400°C. It rapidly begins to take over as the dominant heat transfer mechanism with increasing temperature. At 1,100°C, the K_R of clear glass is approximately 0.175 $\text{cal·cm}^{-1}\text{·°C}^{-1}\text{·s}^{-1}$, whereas $K = 0.0022$ $\text{cal·cm}^{-1}\text{·°C}^{-1}\text{·s}^{-1}$ (Fig. 12-1). Oxide glasses are essentially opaque (very high α) to all radiation beyond 4.5 μm. For this reason, radiation pyrometers that have wavelength pass windows beyond 4.5 μm, such as 4.6–5.3 μm, receive the radiation from a top fraction of the surface layer. They are thus excellent for measuring the surface temperature of glass: they do not "look" into the depth. By the same token, the efficient transfer of heat by the photon conduction method at wavelengths less than 4.5 μm indeed allows glasses to be readily melted to a depth of 1.5 meters in continuous melting tanks without the risk of having solid stagnant glass at the tank bottom. Briefly, the top layer of glass heats up by conduction from the hot flames, and absorption of all emitted radiation beyond 4.5 μm by the hot crown. This layer then emits radiation in a spherically symmetrical manner. The layer below heats due to (a) conduction from the upper layer, (b) complete absorption of radiation $>$4.5 μm from the layer above, and (c) partial absorption of radiation $<$4.5 μm from the

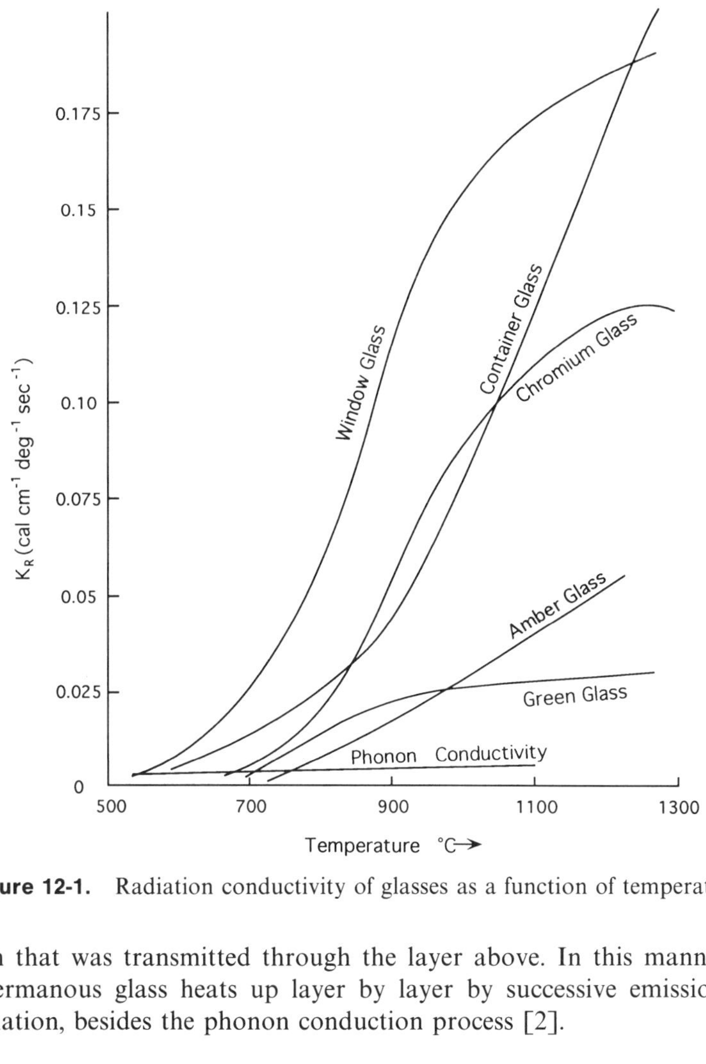

Figure 12-1. Radiation conductivity of glasses as a function of temperature.

crown that was transmitted through the layer above. In this manner, the diathermanous glass heats up layer by layer by successive emission and reradiation, besides the phonon conduction process [2].

12.2. Measurement of Thermal Conductivity

The measurement of *K* requires the establishment of an equilibrium temperature gradient across a specimen of known thickness, and the measurement of the heat flow across it. One could generate heat by passing electrical current on one side of the specimen and waiting to obtain a steady flow of heat. By determining the rate of heat flow (for instance, by calorimetry) and the temperature differential across a certain width, the application of Eq. (12.1) yields K. As indicated earlier, more often the measurement of the thermal diffusivity κ is required. Here, one would measure the rise in temperature across a glass slab with time and apply the solution of (12.2) appropriate for the boundary conditions of the experiment. Equipment based upon diffusivity of heat generated by a laser flash can be obtained commercially.

In addition to the T^3 dependence of K_R indicated by Eq. (12.4), the effective radiation conductivity in a general non-hemispherical medium also depends upon specimen dimensions on account of the probability of the absorption of photons of various wavelengths. These complications make the measurement of K_R in a hot glass difficult (besides being difficult to interpret).

12.3. Composition and Temperature Dependence

The thermal conductivity of vitreous silica [3] at 0°C is 0.0033 $\text{cal}\cdot\text{cm}^{-1}\cdot{}^\circ\text{C}^{-1}\cdot\text{s}^{-1}$ (1.38 $\text{W}\cdot\text{m}^{-1}\cdot\text{K}^{-1}$), and 0.0046 $\text{cal}\cdot\text{cm}^{-1}\cdot{}^\circ\text{C}^{-1}\cdot\text{s}^{-1}$ (1.92 $\text{W}\cdot\text{m}^{-1}\cdot\text{K}^{-1}$) at 100°C. In many binary silicates, borates and aluminates, the values of K at 100°C are (4) in the vicinity of 0.0025 ± 0.0003 $\text{cal}\cdot\text{cm}^{-1}\cdot{}^\circ\text{C}^{-1}\cdot\text{s}^{-1}$.

The thermal conductivity of several glasses has been measured by Salama *et al.* [5] and by Ratcliffe [6]. Values for some soda-lime silica, sodium borosilicate, and sodium aluminosilicate glasses are shown in Table 12-1 (data quoted by Bansal and Doremus [7]). Additivity factors for calculating thermal conductivity from the known weight percent glass composition have been provided by Sharp [8].

Stephens [9] found that K for a commercial "Pyrex" glass could be given as function of T over 90–525 K by the relation

$$K(\text{cal}\cdot\text{cm}^{-1}\cdot{}^\circ\text{C}^{-1}\cdot\text{s}^{-1}) = -0.003523 + 0.002454 \log_{10} T \qquad (12.5)$$

As shown in Fig. 12-2, the thermal conductivity of glasses is generally much smaller than that of the crystals. Whereas the $1/T$ dependence appears true for the crystals, the same is not true for the glasses. One may hypothesize that the phonon mean free paths in glasses are relatively unchanged with temperature, being determined primarily by geometric scattering from crystallite surfaces, and that K reflects the changes in the heat capacity [10].

The low-temperature behavior of thermal conductivity in glasses is typically that shown in Fig. 12-3 (Fig. 5.1 from Ref. 11) for vitreous silica. In region A, the thermal conductivity decreases slowly with decreasing temperature. Near ~10 K, the thermal conductivity has a plateau. In region C, the conductivity varies as T^2. The behavior is understood in terms of the intrinsic

Table 12-1. Thermal Conductivity (due to Phonons) of Some Ternary Glasses

A. Thermal Conductivity of Na_2O–CaO–SiO_2 Glasses

Composition (mol %)			K($W\,m^{-1}\,K^{-1}$) at		
Na_2O	CaO	SiO_2	0°C	30°C	100°C
18	5	77	0.933		1.043
18	10	72	0.910		1.016
18	15	67	0.905		0.972
39.27	2.50	58.23		0.644	
39.51	4.52	55.98		0.634	
39.30	7.49	53.12		0.625	
39.29	10.02	50.69		0.657	

B. Thermal Conductivity of Na_2O–B_2O_3–SiO_2 Glasses

Composition (mol %)			K ($W\,m^{-1}\,K^{-1}$) at	
Na_2O	B_2O_3	SiO_2	0°C	100°C
18	5	77	1.017	1.118
18	10	72	1.010	1.075
18	15	67	0.985	1.075
18	20	62	0.974	1.059
18	30	52	0.941	1.038

C. Thermal Conductivity of Sodium Aluminosilicate Glasses

Composition (mol %)			K ($W\,m^{-1}\,K^{-1}$) at	
Na_2O	Al_2O_3	SiO_2	0°C	100°C
18	3	79	0.972	1.081
18	6	76	0.964	1.072
18	9	73	0.960	1.031
18	12	70	0.947	1.026

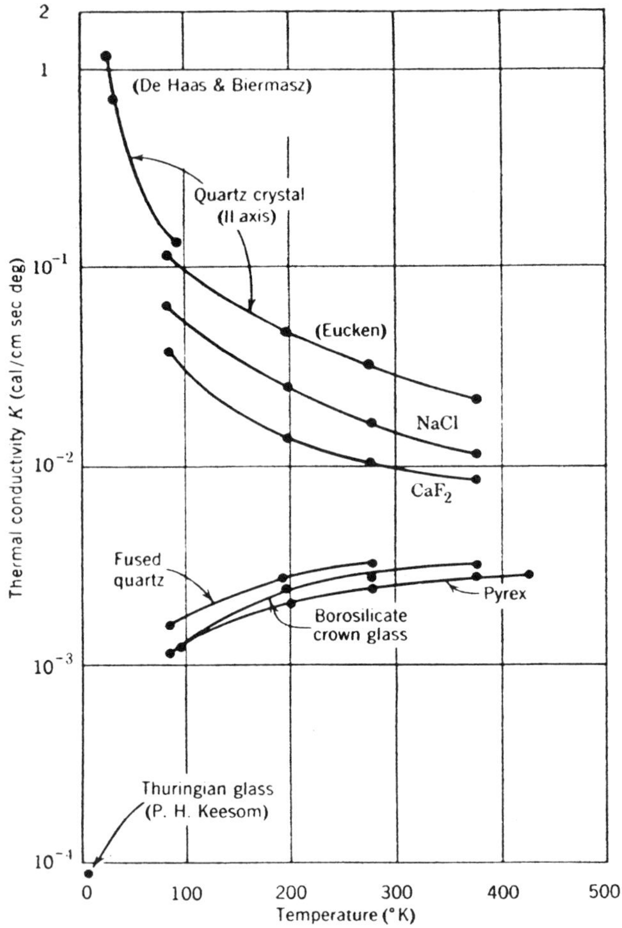

Figure 12-2. Temperature dependence of the phonon thermal conductivity of various glasses and crystals. (After C. Kittel [10], Fig. 27, p. 231.) Elsevier Science Publishers. Reproduced with permission of the publisher.

scattering of long waves in a disordered structure. It is found that thermal phonons alone contribute to the thermal conductivity at low temperatures in amorphous solids, that the mean free path for the longitudinal and the transverse waves are about the same, but more importantly, that the additional excitation states that contribute to the excess heat capacity must be localized (making no contribution towards the phonon transport).

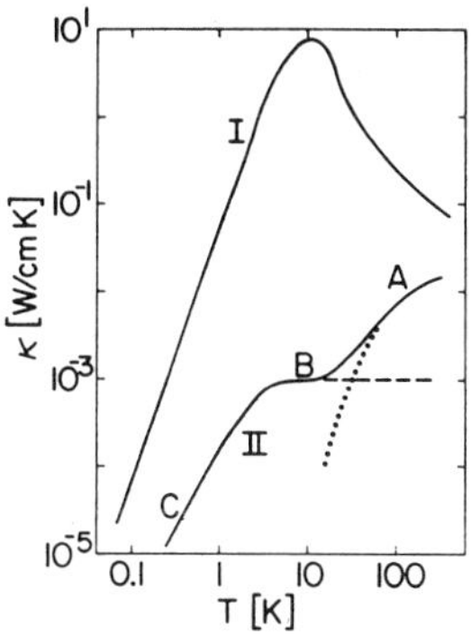

Figure 12-3. Low-temperature thermal conductivity of α-quartz (I) and fused quartz (II). (After A. C. Anderson [11]. Reproduced with permission of Springer-Verlag.)

Exercises

1. Calculate K_R for a soda-lime glass at 1,400°C. Assume $\alpha = 0.2\ \text{cm}^{-1}$, and compare with K for copper $= 0.92\ \text{cal}\cdot\text{cm}^{-1}\cdot{}^{\circ}\text{C}^{-1}\cdot\text{s}^{-1}$).
2. Why are colored glass tanks generally shallower than the clear glass tanks?

   (Answer: Colored glasses have a larger absorption coefficient, hence they have smaller K_R than the clear glasses (see Fig. 12-1), which, in turn, causes the temperature gradients in a colored glass tank to be steeper through the depth than in a clear glass tank.)

References

1. H. S. Carslaw and J. C. Jaeger, *Conduction of Heat in Solids*. Oxford at the Clarendon Press, 1947.
2. R. Gardon, *J. Am. Ceram. Soc.* **44**, 200 (1961).
3. A. Euchen, *Ann. Physik* **34**, 185 (1911).
4. A. Russ, *Sprechsaal* **61**, 887 (1928).
5. S. N. Salama, S. M. Salman, and S. Gharib, *J. Non-cryst. Sol.* **93**, 203–214 (1987).
6. E. H. Ratliffe, *Phys. Chem. Glasses* **4**(4), 113 (1963).
7. N. P. Bansal and R. H. Doremus, *Handbook of Glass Properties*. Academic Press, New York, 1986.

8. D. E. Sharp, tabulated by G. W. Morey, *Properties of Glass*, 2nd Ed., Reinhold Publications Co. New York, p. 219 (1960).
9. R. W. B. Stephens, *Phil. Mag* **14**, 897 (1932).
10. C. Kittel, *Introduction to Solid State Physics*, 4th Ed. Wiley, New York, 1971.
11. A. C. Anderson, in *Amorphous Solids — Low Temperature Properties* (W. A. Phillips, ed.), p. 66. Springer-Verlag, Berlin, 1981.

Chapter 13

Glass Transition Range Behavior

13.1. Introduction

Glass transition (or transformation) range (or, abbreviated, the T_g range) marks the temperature interval over which a given system gradually transforms upon cooling from a supercooled liquid state into the glassy state. If we examine Fig. 13-1 (which is the same as Fig. 2-1), we note that the failure to crystallize *below* point "b" results in the volume V of the material following the extension of the equilibrium liquid state to temperatures below the melting point T_m. This extension, called the supercooled liquid state, has a higher internal energy with respect to the corresponding crystalline state. For this reason, *the supercooled liquid is in "metastable" equilibrium with respect to the crystal.* The kinetics of change to the crystalline state, however, are sluggish because of the energy barriers involved in nucleation/crystallization. The temperature is high enough, and the viscosity is low enough, that a small temperature decrease near T_m causes a spontaneous rearrangement of the atomic structure characteristic of the liquid at the changed temperature. Thus, *for a thermodynamic treatment, the supercooled liquid line may continue to be treated as an "equilibrium" liquid line.* Upon continued cooling, the situation could not go on forever. As we noted in Chapter 9, cooling brings about contraction of the "free volume." The free volume generally shrinks considerably more in proportion than does the rest of the atomic

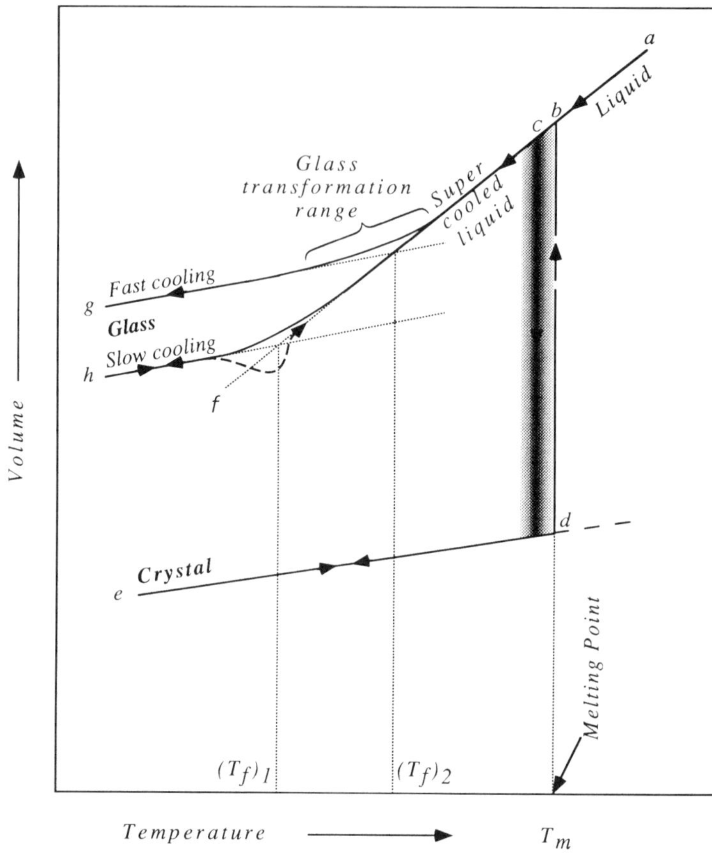

Figure 13-1. Volume–temperature diagram for a glass-forming melt.

volume. Qualitatively, we could argue that this rapid contraction of the free volume causes an enormous increase in the viscosity. Since the viscosity η is related to the mobility β (through the Stokes–Einstein relation $\beta = 1/6\pi\eta r$, where r is the atomic radius), it is clear that the "spontaneity" in the readjustment of the structure becomes increasingly difficult with decreasing temperatures. Hence, the volume begins to depart from the supercooled liquid line. The temperature corresponding to this situation is the beginning of the glass transition range. A stage comes when the ionic (or atomic) mobilities are so low that no perceptible rearrangement of the structure occurs over practical times. The temperature here is the end of the glass transition range. For all practical purposes, the material has fully transformed into a (solid) glass. Despite the fact that glass appears to be in "equilibrium," the truth is that it is continuously attempting to approach the "equilibrium" liquid line. *Glass is, therefore, unstable with respect to the liquid.* The intersection of the extrapolated liquid and glassy state lines is called T_f, **the fictive temperature** (already introduced in Section 2.1).

How would the glass respond if a mechanical or a temperature impulse were applied in or near the transition range?

If a mechanical impulse, say a constant shear stress, is applied to the glass in the solid state, then a shear strain would be generated instantaneously characteristic of an elastic solid. The magnitude of the shear strain is proportional, linearly or nonlinearly, to the shear stress through the shear modulus, and will remain constant so long as the applied stress remains constant. If, on the other hand, the mass is a liquid, then we know that the body will flow, the rate of deformation with time being related to the applied stress — again, linearly or nonlinearly — through the viscosity property. The flow will continue so long as the shear stress is applied. Clearly, the response of the glass to an applied mechanical impulse in or near the transition range is a time-dependent structural change representing a combination of solidlike and liquidlike behaviors. This is called **viscoelastic behavior**. Since the temperature is unchanged, the physical properties of the glass, such as thermal expansion and elastic moduli, remain nearly unchanged.

If a temperature impulse is applied, structural rearrangements occur as a result of which properties are likely to change with time. *In the glassy state, the structural rearrangements are so slow that properties do not change as a function of time. In the supercooled liquid state, the properties change too rapidly to be detected by a measuring instrument. It is only in the glass transition range that the structure relaxes at a perceptible rate, and consequently, the properties change while we are watching them.*

The behavior of glass and its properties in and near the glass transition range is the subject of this chapter. We shall first study the viscoelastic behavior of glass, which is the time-dependent response to a mechanical impulse, usually at constant temperature, and then turn to study the time-dependent structural relaxation when a temperature impulse is applied.

13.2. Viscoelastic Properties of Glass

Suggested reading: S. M. Rekhson, *J. Non-cryst. Sol.* **95 & 96**, 131–148 (1987). G. W. Scherer, *Relaxation in Glass and Composites*. John Wiley, New York, 1986.

The viscoelastic behavior of any substance includes both elastic (instantaneous) and viscous (time-dependent) response. Because of the generally low magnitudes of stresses involved in glass near the transition range, the elastic response is Hookean (elastic moduli are independent of the magnitude of the applied stress), and the viscous response is Newtonian (viscosity coefficient is independent of the magnitude of the shear stress). For this reason, it is often convenient to apply the principles of linear viscoelasticity to any relaxation phenomena in glass. The formalism of relaxation and linearity is discussed next.

13.2.1. About Relaxation

The term **relaxation** is used to describe the approach to equilibrium of a system driven out of equilibrium by a perturbation in thermodynamic state variables. Let us consider the application of an incremental shear strain of which the deviatoric component is $\Delta\varepsilon_{12(1)}$ at time t_1 (see equations A1.30–A1.35 for a description of the deviatoric components). This would generate an instantaneous incremental shear stress whose deviatoric component is ΔS_{12} according to the relation (A1.35),

$$\Delta S_{12} = 2G\Delta\varepsilon_{12(1)}.$$

If G varies with time as $G_t = G_0\Psi(t)$, where the function $\Psi(t)$ varies between 1 and 0 as time t increases from zero to infinity, then the shear stress would vary with time t according to

$$\Delta S_{12(t)} = [2G_0\Psi(t - t_1)]\Delta\varepsilon_{12(1)}, \qquad t_1 \leqslant t. \tag{13.1}$$

Now, we add another incremental shear strain $\Delta\varepsilon_{12(2)}$ at time t_2 and watch the shear stress at t ($t_2 \leqslant t$):

$$\Delta S_{12(t)} = [2G_0\Psi(t - t_1)]\Delta\varepsilon_{12(1)} + [2G_0\Psi(t - t_2)]\Delta\varepsilon_{12(2)}. \tag{13.2}$$

Here, we assumed that the first term on the right-hand side continued to change with time independent of whether the second term existed or not. Basically, we assumed the process to be linear, and simply superimposed the two events. In essence, *the effect of the combined strains applied at different times is the sum of the effects each would have produced alone.* This notion, shown schematically in Fig. 13-2, is termed the **Boltzmann superposition principle** (also mentioned at the end of Appendix A1). Linearity also implies that Ψ should not depend upon the magnitude or the sign of $\Delta\varepsilon$. (If Ψ depended upon $\Delta\varepsilon$, then upon the application of $\Delta\varepsilon_{12(2)}$, the magnitude of $\Psi(t - t_1)$ at $t+$ would be different from that an instant before, resulting in a totally different contribution from the first term in Eq. (13.2).) For most materials, as long as the deformations involved are small, the assumption of

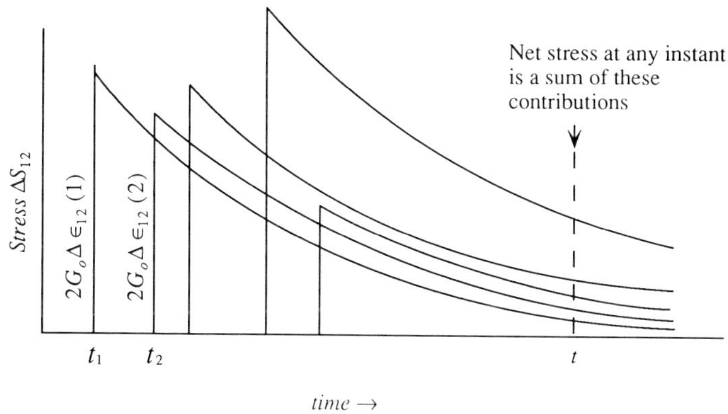

Figure 13-2. Principle of Boltzmann superposition.

linearity is reasonable. Generalization of (13.2) yields

$$\Delta S_{12(t)} = \sum_k [2G_0\Psi(t - t_k)]\Delta\varepsilon_{12(k)}, \qquad t_k \leqslant t, \qquad (k = 1 \text{ to } n). \quad (13.3)$$

If $\Delta\varepsilon_{12}$ is continuous, then $\Delta\varepsilon_{12(k)} = (d\varepsilon_{12}/dt)dt$, which enables us to write

$$S_{12(t)} = 2G_0 \int_0^t \Psi(t - t') \frac{d\varepsilon_{12}}{dt'} dt' \qquad (13.4)$$

This is called a memory integral, since the value of the shear stress depends upon the entire shear strain history. To generalize, we may define

$$X(t) = \text{force at time } t,$$

$$Y(t) = \text{constraint at time } t.$$

The force and the constraint would be related through

$$Y(t) = \int_0^t \frac{dX(t')}{dt'} \Phi(t - t')dt', \qquad (13.5)$$

$$X(t) = \int_0^t \frac{dY(t')}{dt'} \Psi(t - t')dt', \qquad (13.6)$$

where $\Phi(t)$ and $\Psi(t)$ are the linear response functions independent of $dX(t')/dt'$ or $dY(t')/dt'$, respectively. Equation (13.5) represents the response of the function to an applied force. Equation (13.6), on the other hand, describes the response to an imposed constraint. Examples of forces, constraint and response functions are shown in Table 13-1.

An experiment of the type (13.5), where we apply a constant force and

Table 13-1. Constraints and Response Functions

Force	Constraint	Equation	Response function
Shear force	Shear strain	(13.5)	Shear compliance
		(13.6)	Shear modulus
Electric field	Polarization	(13.5)	Dielectric permittivity
		(13.6)	Electric modulus
Pressure	Volume	(13.5)	Bulk compliance
		(13.6)	Bulk modulus

watch its influence on the constraint, is called a retardation or creep experiment. As an example, upon the application of a constant shear stress $S_{12}(t)$ ($= 0$ at $t < 0$; $S_{12}(0)$ at $t \geqslant t_0$) (see Fig. 13-3a), an instantaneous elastic strain (constraint) $\varepsilon_{12(E)}$ develops, followed by the development of a delayed strain $\varepsilon_{12(D)}$ and, eventually, viscous flow whose slope is given by $S_{12}(0)/2\eta$. Note that if a constant uniaxial stress σ_{11} were applied, and we measured the extensional rate $\dot{e}_{11}$ as $t \to \infty$, then from (A1.7) and (A1.30–A1.32) we get

$$\sigma_{11} = (\tfrac{3}{2})S_{11} = (\tfrac{3}{2})[2\eta\dot{\varepsilon}_{11}] = 3\eta\dot{\varepsilon}_{11} = 3\eta\dot{e}_{11} \tag{13.7}$$

Thus, the slope of the uniaxial strain with time will be given by $\sigma_{11}/3\eta$. An experiment of the type (13.6), where a constant constraint is applied and we measure the response of a force, is called a relaxation experiment. The application of an instantaneous constraint — for instance, a shear strain $\varepsilon_{12}(t)$ ($=0$ at $t < 0$; $=\varepsilon_{12}(0)$ at $t \geqslant t_0$) — will cause the development of an instantaneous Hookean stress $S_{12}(0) = 2G\varepsilon_{12}(0)$, which will relax to zero with time (Fig. 13-3b). The response functions Φ and Ψ are, respectively, the **retardation function** and the **relaxation function**. The most common type of structural retardation experiment is when the temperature is changed, and we study the changes in the properties such as the expansion coefficient or density as a function of time. (Often, the use of the terms "relaxation" and "retardation" is quite loose in the published literature.)

In addition to the creep and stress relaxation experiments, one other experiment often convenient to conduct [1] is using oscillatory impulses. For such experiments, the stress and strain may be written as

$$S_{12} = \mathbf{Re}[S_{12}(0)\exp\{j\omega t\}], \qquad \text{and } \varepsilon_{12} = \mathbf{Re}[\varepsilon_{12}(0)\exp\{j\omega t\}], \tag{13.8}$$

where **Re** stands for the real part of the complex number defined by the Euler identity $\exp(j\theta) = \cos\theta + j\sin\theta$, where $j \equiv \sqrt{-1}$, and

$$S_{12}(0) = 2G\varepsilon_{12}(0). \tag{13.9}$$

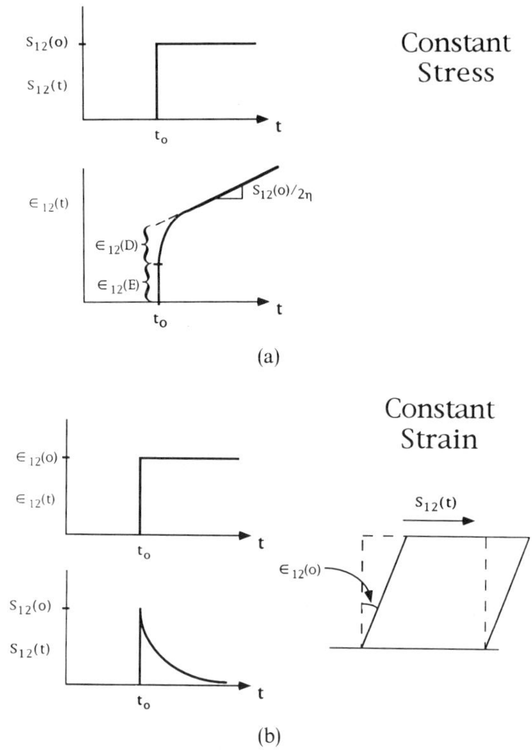

Figure 13-3. (a) Effect of the application of a constant stress on a viscoelastic body (creep experiment). (b) Effect of the application of a constant strain on a viscoelastic body (relaxation experiment).

A **complex shear modulus** G^* is defined by

$$S_{12} = 2G^*\varepsilon_{12}, \tag{13.10}$$

where

$$G^* = G' + jG''. \tag{13.11}$$

G' is referred to as the **storage modulus** and G'' as the **loss modulus**. For a perfectly elastic solid, the stress and the strain are in phase with each other; consequently, $G' = G$ and $G'' = 0$. When a shear strain given by (13.8) is applied to a purely viscous body, the shear stress leads the shear strain by

90° (using Eq. 9.1); consequently, $G' = 0$ and $G'' = G$. All of the energy introduced into such a system will be dissipated into heat. For viscoelastic solids neither of the components is zero, and in general, is frequency dependent. The ratio of the two is called the **loss angle** or the **loss tangent** $\tan\delta$, or Q^{-1}, i.e.,

$$\tan\delta = G''/G' = Q^{-1}, \tag{13.12}$$

where δ is the phase angle difference between the constraint and the force. The value $2\pi\tan\delta$ is called the **internal friction**. Q is called the **quality factor**. (For a review, see Mazurin [2].) The response of the material to an input of sinusoidal shearing strain is shown in Fig. 13-4. Since the shear compliance J is the reciprocal of G (Eq. 8.4), one may define a **complex compliance** J^*:

$$J^* = 1/G^* = J' + jJ''. \tag{13.13}$$

Strictly speaking, $J = 1/G$ is true only for the purely elastic case. In general

$$J' = \frac{G'}{G'^2 + G''^2} \qquad J'' = -\frac{G''}{G'^2 + G''^2}. \tag{13.14}$$

Likewise, a **complex viscosity** η^* may also be defined:

$$\eta^* = \eta' - j\eta'', \tag{13.15}$$

where η' is called the **dynamic viscosity**. For the stress and the strain given by (13.8) and using the definition $S_{12} = 2\eta^*\dot{\varepsilon}_{12}$, which is analogous to Eq. (A1.35), it may be shown that

$$\eta^* = G^*/j\omega; \quad \eta' = G''/\omega; \quad \eta'' = G'/\omega. \tag{13.16}$$

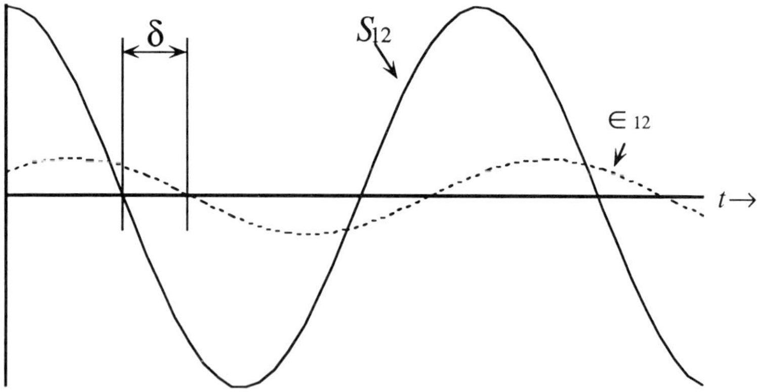

Figure 13-4. Effect of the application of a sinusoidal strain on a viscoelastic body (internal friction experiment).

13.2.2. The Maxwell Element

The simplest model to simulate the behavior of a linear viscoelastic solid is the Maxwell element (Fig. 13-5). It consists of a perfectly Hookean elastic spring in series with a dashpot that simulates a perfectly Newtonian fluid. The deviatoric component of the Hookean strain $\varepsilon_{12(E)}$ in the spring upon the application of a shear stress whose deviatoric component is S_{12} is given by $S_{12} = 2G\varepsilon_{12(E)}$. For the dashpot, the strain rate is given by $\dot{\varepsilon}_{12(V)} = S_{12}/2\eta$. With the assumption of small deformations, linearity allows us to write the total strain ε_{12} as a sum of the contributions from the spring and the dashpot:

$$\varepsilon_{12} = \varepsilon_{12(E)} + \varepsilon_{12(V)}. \tag{13.17}$$

At constant stress, the contribution of the dashpot to the total strain is determined by integrating $\dot{\varepsilon}_{12(V)} = S_{12}/2\eta$ with respect to t, hence

$$\varepsilon_{12} = [S_{12}/2G] + S_{12}t/2\eta. \tag{13.18}$$

The total strain varies with time such that the slope is $S_{12}/2\eta$ — in other words, the specimen displays creep (as shown in Fig. 13-3a, but without the curved portion).

The response of the Maxwell element to a constant applied strain is more interesting (Fig. 13-5a). Since $\dot{\varepsilon}_{12} = \dot{\varepsilon}_{12(E)} + \dot{\varepsilon}_{12(V)}$, differentiating (13.17) we get

$$\dot{\varepsilon}_{12} = (\dot{S}_{12}/2G) + (S_{12}/2\eta). \tag{13.19}$$

At constant strain, $\dot{\varepsilon}_{12} = 0$. Substituting in (13.19) and integrating both sides, we get

$$S_{12}(t) = S_{12}(0)\exp[-t/\tau], \tag{13.20}$$

where $S_{12}(0) = 2G\varepsilon_{12(0)}$ is the deviatoric component of shear stress due to an applied strain $\varepsilon_{12(0)}$ at $t = 0$, and $\tau = \eta/G$ is called the shear stress (or Maxwell) **relaxation time**.

Eq. (13.20) shows that, upon the application of a strain, an instantaneous shear stress results because of the action of the spring, but this shear stress relaxes exponentially with time as a result of viscous flow in the dashpot. The relaxation of the stress in a Maxwell element is shown in Fig. 13-5b. When the time $t = \tau$, the stress $S_{12}(t) = S_{12}(0)\cdot e^{-1} \approx 0.37S_{12}(0)$; i.e., at the end of one relaxation time unit, about a third of the original stress remains.

For an oscillatory strain given by (13.8), we may write the frequency-dependent values of the storage and loss moduli as

$$G'(\omega) = \frac{G(\omega\tau)^2}{1 + (\omega\tau)^2}; \qquad G''(\omega) = \frac{G\omega\tau}{1 + (\omega\tau)^2}; \qquad \delta = 1/\omega\tau. \tag{13.21}$$

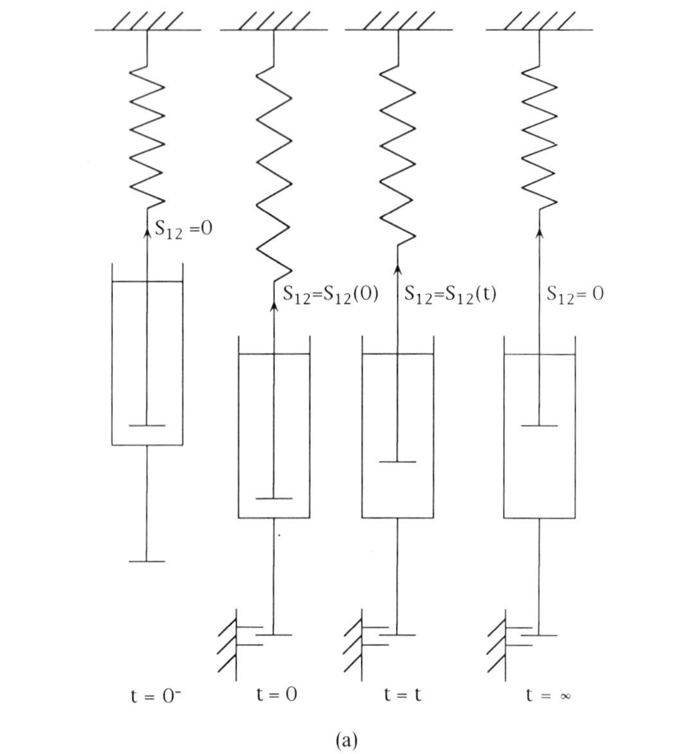

Figure 13-5. Maxwell element. (a) Its working principle. (b) Response after the application of a constant strain.

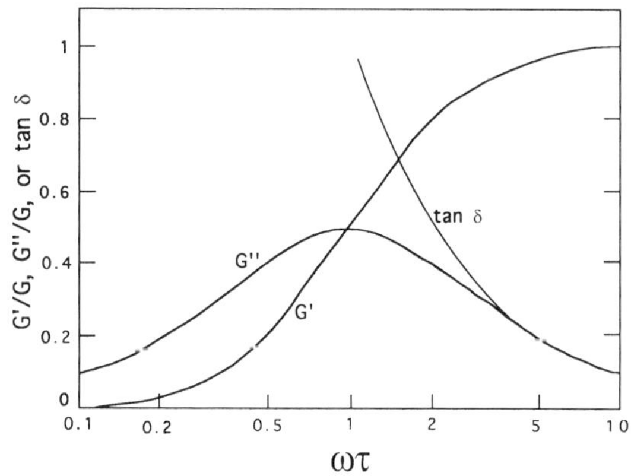

Figure 13-6. Response of a Maxwell body as a function of frequency.

Figure 13-6 shows plots of G', G'', and $\tan\delta$ as a function of ω for the Maxwellian element. As $\omega \to \infty$, $G'' \to 0$ and $G^* \to G'(\infty)$. Since the response to an infinite frequency implies an instantaneous response, we may write the infinite frequency shear modulus $G'(\infty) \equiv G(0)$, the instantaneous shear modulus ($=G$, the frequency-independent shear modulus for a perfectly elastic solid).

We may write $S_{12}(t) = 2G(t)\varepsilon_{12(t)}$, where $G(t)$ is the shear relaxation modulus. Hence, from (13.20),

$$G(t) = G(0)\exp[-t/\tau] = G(0)\Psi(t), \tag{13.22}$$

where $\Psi(t)$ is the stress relaxation function. Integrating both sides of (13.22) with respect to t, we get the Ferry relation

$$\eta = \int_0^\infty G(t)dt = G(0)\int_0^\infty \Psi(t)dt. \tag{13.23}$$

Equation (13.23) is sometimes utilized to define the viscosity η. This is a particularly interesting relationship in that it connects a "solid" property to a "liquid" property through a time-dependent behavior. If $\Psi(t)$ were always 1, the material would appear perfectly elastic solid at all times. *The detection of any significant relaxation by our instruments causes us to interpret that this body is "more or less" liquid.*

13.2.3. The Voigt–Kelvin Element

In this simulation, the spring and the dashpot are in parallel (Fig. 13-7a), and hence the strain is equal in both the components at all times and the

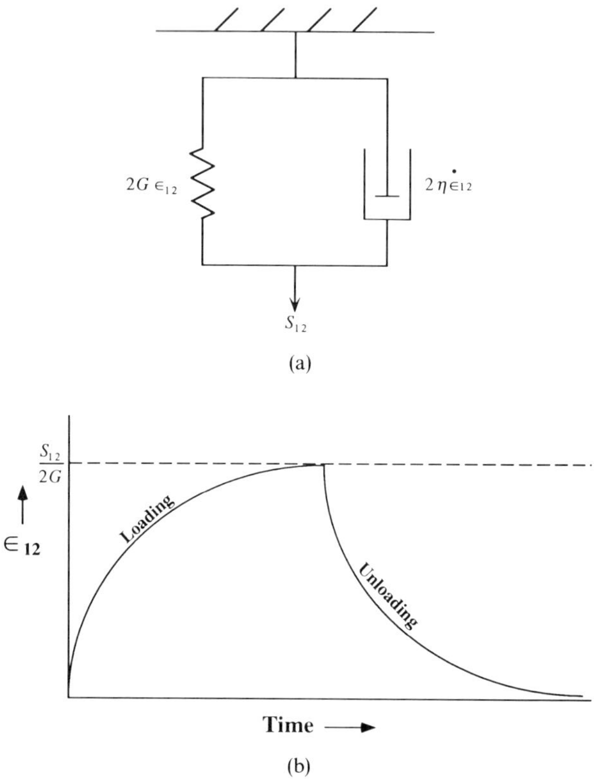

Figure 13-7. Voigt–Kelvin element. (a) Its composition. (b) Response after the application of a constant stress.

stress is distributed. Thus,

$$S_{12} = S_{12(E)} + S_{12(v)} = 2G\varepsilon_{12} + 2\eta\dot{\varepsilon}_{12}. \tag{13.24}$$

Under the condition that S_{12} = constant, the integration of (13.24) yields

$$\varepsilon_{12}(t) = (S_{12}/2G)[1 - \exp(-t/\tau)]. \tag{13.25}$$

Thus, upon the application of a stress in a Voigt–Kelvin element, the strain exponentially builds up to its full value (Fig. 13-7b) and is recovered completely upon unloading.

13.2.4. Delayed Elasticity

Delayed elasticity, sometimes called "anelasticity," refers to the development of the full elastic strain over a period of time. Since the Voigt–Kelvin element

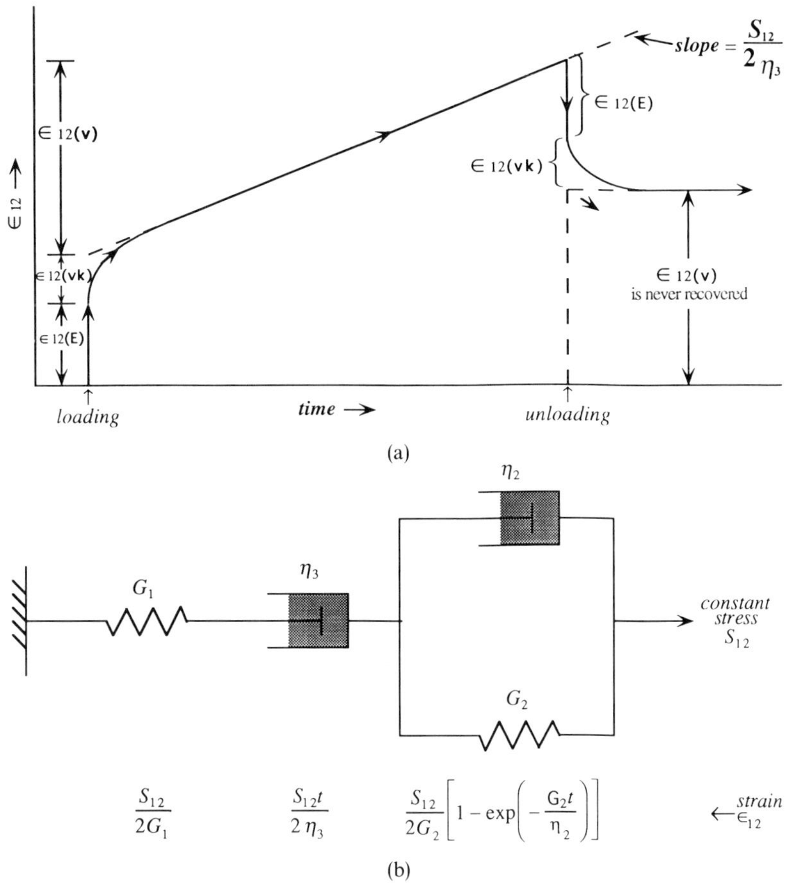

Figure 13-8. Delayed elasticity. (a) Variation of strain with time after loading and unloading. (b) Burger element.

satisfies this condition (Fig. 13-7b), it is the simplest model representing delayed elasticity. More frequently, the behavior of glass under constant stress, and subsequent unloading, is observed to be as shown in Fig. 13-8a, which is an extension of Fig. 13-3a. Upon the application of a load, the specimen displays an instantaneous elastic deformation followed by a delayed buildup of strain, ultimately showing creep behavior typical of viscous flow. *When the load is removed, the elastic portion of the total strain is recovered instantaneously, followed by a delayed recovery of the elastic portion of the remaining strain. The purely viscous deformation is never recovered.* One may simulate this behavior by a spring (representing the initial elastic response), a Voigt–Kelvin element (representing the delayed strain), and a dashpot (representing the final viscous flow under continued application of stress). This model, shown in Fig. 13-8b, is called a **Burger element**. Conceptually, it is a Maxwell element and a Voigt–Kelvin element in series. When the applied stress S_{12} is a constant, the strain $\varepsilon_{12(E)}$ in the spring is given by $S_{12}/2G_1$. The strain in the Voigt–Kelvin element is $\varepsilon_{12(vk)} = (S_{12}/2G_2)[1 - \exp(-t/\tau_2)]$, where $\tau_2 = \eta_2/G_2$. The strain in the dashpot is obtained by integrating both sides of the equation, $\dot{\varepsilon}_{12(v)} = S_{12}/2\eta_3$, which gives $\varepsilon_{12(v)} = S_{12}t/2\eta_3$. Hence, the total strain is given by

$$\varepsilon_{12} = [S_{12}/2G_1] + (S_{12}/2G_2)[1 - \exp(-t/\tau_2)] + [S_{12}t/2\eta_3]. \quad (13.26)$$

It may be recognized that the multiplication by G_1 will lead to the third term on the right containing another relaxation time τ_3 $(=\eta_3/G_1)$. Thus, *the delayed elasticity in a Burger element is controlled by more than one relaxation time.* Real viscoelastic substances may indeed have a multitude of Voigt–Kelvin and Maxwell elements such that the relaxation function is described by a spectrum of relaxation times. However, having a spectrum of relaxation times does not negate obeying linearity. Linearity applies so long as the τ's themselves are not a function of stress (or strain).

13.3. Structural Relaxation Due to Temperature Change

In Fig. 13-1, we have already noted the variation of the volume of glass as a function of the cooling rate. The slower the cooling rate employed, the smaller the final volume. *Hence, properties of glass depend upon its thermal history.* The change in volume upon reheating is equally interesting (shown in Fig. 13-9), because the effect of the rate employed is again noticeable. The volume follows the cooling curve as long as the glass remains "perceptibly solid." Thereafter, the volume departs from the cooling curve. At an infinite heating rate, the volume may be expected to follow the extension of the glass line up to T_f on the equilibrium liquid line, since T_f was defined to be the effective temperature at which the structure of the liquid was instantaneously frozen into the glass (Section 2.2). It will overshoot if heating is continued, and turn upwards to meet the equilibrium line in the manner shown by curve "a." At most practical heating rates, the volume of the glass may contract further (curve "b") or expand approaching the equilibrium line, depending upon the relation of the reheating rate to the cooling rate. A rapidly cooled specimen shrinks significantly upon reheating before catching up with the equilibrium liquid (curve "c"). *In all cases, the volume upon heating never retraces the cooling path in the transition range.* The **glass transition temperature**, T_g, is normally defined by the intersection of the glassy state line with the tangent to the steep portion of the state curve in the transition range.

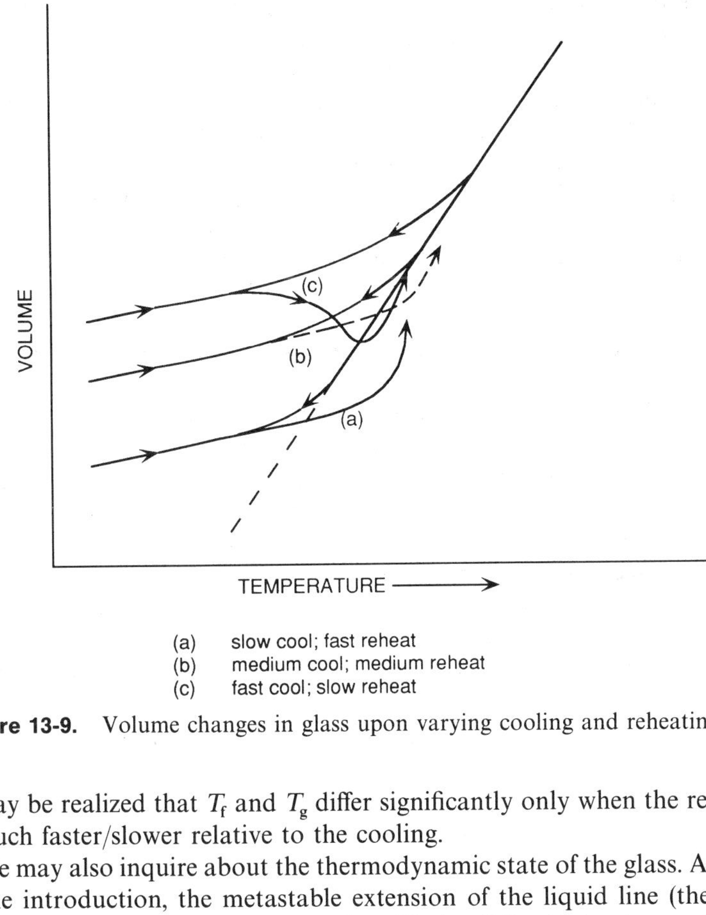

Figure 13-9. Volume changes in glass upon varying cooling and reheating rates.

It may be realized that T_f and T_g differ significantly only when the reheating is much faster/slower relative to the cooling.

We may also inquire about the thermodynamic state of the glass. As stated in the introduction, the metastable extension of the liquid line (the supercooled liquid) may be treated as an "equilibrium" liquid: its thermodynamic state can be uniquely described by any two of the three thermodynamic variables P, V, and T (pressure, volume, and temperature). At a given temperature and pressure, the supercooled liquid has a fixed volume, fixed thermal expansion coefficient, fixed specific heat, etc., independent of the path, all of which are characteristics of a "stable" or "metastable" equilibrium state. On the other hand, we know that variations in the cooling rate can cause the glass line to depart from the equilibrium liquid line at varying temperatures. Hence, fixing two of the three thermodynamic variables (P, V, T) alone does not uniquely specify the third. Because no changes apparently occur in glass over short periods (a few minutes to a few years), it may be possible to introduce a fourth thermodynamic variable to represent the "seemingly frozen structural state" of the glass despite the thermodynamic classification of its state as metastable with respect to the crystal and unstable with respect to the liquid. There may thus be opportunities to apply the laws of thermodynamics using four (or more?) variables.

In studying the structural relaxation, it may be readily understood why three types of theories to explain the glass transition have been advanced: those based on viscosity and free volume, those based on thermodynamics, and finally, those based on purely kinetic arguments. Any theory must explain how the properties vary as a function of temperature, time, thermal history, and the direction of the temperature change.

13.4. Viscosity–Free Volume Theories of Glass Transition

The original concepts of the viscosity-free volume theories of glass transition were advanced by Turnbull and Cohen [3], and Cohen and Grest [4].

As indicated earlier, the underlying idea here is that the increase in viscosity upon cooling results from a contraction of the free volume. In other words, viscosity changes and free-volume changes are interrelated. The changes in free volume approximate the changes in the total volume. We might argue that the initial volume V_i would be "frozen" into the glassy state if the cooling were sufficiently rapid, and that, given sufficient time, the volume would relax to that of the equilibrium liquid. For intermediate times, we could determine the volume V' using an expression such as

$$\frac{V' - V_e}{V_i - V_e} = \exp(-\Delta t/\tau), \tag{13.27}$$

where τ = volume relaxation time, and Δt = available time.

The reader may note that Eq. (13.27) is of the Maxwellian relaxation type. It may be recognized that the farther away one is from the equilibrium, the faster the rate of approach. Now,

$$V_i - V_e \approx V_e \beta_f \Delta T, \tag{13.28}$$

where β_f is the thermal expansion coefficient of thc frcc volumc. Also, the volume relaxation time may be written as

$$\tau = \eta\kappa, \tag{13.29}$$

where η is the "volume viscosity" (Section 9.7) and κ the compressibility. Substituting (13.29) and (13.28) in (13.27), we get:

$$\frac{V' - V_e}{V_e \beta_f \Delta T} = \exp[-\Delta T/\eta B\kappa], \tag{13.30}$$

where $B = \Delta T/\Delta t$ = the cooling/heating rate. (To establish the sign convention, we note that $\Delta T/B = \Delta t$, which is positive regardless of the direction of the temperature change.)

The left side of Eq. (13.30) is the fractional deviation from the equilibrium volume following a temperature change of ΔT. To estimate the total deviation from the equilibrium volume, one needs to iterate for small steps of the temperature change over the entire heat treatment while accumulating the deviation at each step.

As an example, let us consider a soda-lime silica glass that is being cooled at a rate of 10°C/s. We shall attempt to set up Eq. (13.30) in small steps of ΔT, say 10°C, of cooling, calculating at each step the $V' - V_e$ (which is approximately the change in the free volume, neglecting atomic "lattice" expansion) and accumulating it for the entire cooling path. For the glass, $\kappa = 10^{-11}$ Pa^{-1}, and an approximate Arrhenius form of the viscosity (which includes free volume effects) for the soda-lime silica glass may be written as (we assume that the volume viscosity is the same as the shear viscosity)

$$\eta(\text{Pa·s}) = 10^{-43} \exp(90{,}000/T) = 10^{-43} \cdot 10^{(39{,}200/T)}. \qquad (13.31)$$

Also, $\beta_f \cong 10^{-4}/K$ (which is three times the difference between the linear expansion coefficient in the glassy state and in the liquid state from a dilatometer plot).

The values of η and β_f allow the computation of the "relaxation modulus" $\exp[-\Delta T/\eta B\kappa]$ in Eq. (13.30) as shown in Table 13-2, column 5. Column 6 shows the accumulated $(V' - V_e)/(V_e\beta_f\Delta T)$ along the cooling curve, which is obtained by starting at a high temperature, multiplying by the relaxation modulus, and adding 1 for the next lower temperature step. Curve (1) in Fig. 13-10 is a plot of dV'/V_e, which is the fractional expansion of the free volume of glass obtained by adding the accumulated $(V' - V_e)/(V_e\beta_f\Delta T)$ to the fractional expansion dV_e/V_e, as a function of T. (To help with the plotting, the ordinate has been scaled such that the 45° line has a slope of $10^{-4}/K$ and represents the fractional expansion, dV_e/V_e, of the equilibrium liquid. The numbers in column 6 are then marked off this line.) If the cooling were performed at 0.1°C/s (instead of 10°C/s), a recomputation would generate curve (2) on Fig. 13-10 (the actual computation is shown in columns 7–9 of Table 13-2). Note the similarity between the cooling curves shown in Figs. 13-1, 13-9, and 13-10. As expected, the slower cooling allows greater consolidation of the free volume; hence, the apparent departure from the equilibrium volume is at lower temperatures. The fractional difference between the volumes in the glassy state obtained by the two cooling rates is 2.5×10^{-3}, which is approximately the same as the fractional difference between the observed densities of the fast- and slow-cooled soda-lime silica specimens.

Table 13-2. Computation of Volume Changes in Glass after Heat Treatment

			1 Cooling ($B_1 = 10°C/s$)			2 Cooling ($B_2 = 0.1°C/s$)			3 Heating ($B_3 = 0.1°C/s$)	
T	$\frac{39{,}200}{T}$	η Pa·s	$\frac{BT}{\eta B\kappa}$	$\exp - \frac{\Delta T}{\eta B\kappa}$	$\frac{\sum \Delta V}{(V_e \alpha_f \Delta T)}$	$\frac{\Delta T}{\eta B\kappa}$	$\exp - \frac{\Delta T}{\eta B\kappa}$	$\frac{\sum \Delta V}{(V_e \alpha_f \Delta T)}$	$\exp - \frac{\Delta T}{\eta B\kappa}$	$\frac{\sum \Delta V}{(V_e \alpha_f \Delta T)}$
600 K	65.3	$10^{22.3}$	$10^{-11.3}$	1	12.635	$10^{-9.3}$	1	10.130	1	12.635
650 K	60.3	$10^{17.3}$	$10^{-6.3}$	1	7.635	$10^{-4.3}$	1	5.130	1	7.635
660 K	59.4	$10^{16.4}$	$10^{-5.4}$	1	6.635	$10^{-3.4}$	1	4.130	1	6.635
670 K	58.5	$10^{15.5}$	$10^{-4.5}$	1	5.635	$10^{-2.5}$	1	3.130	1	5.635
680 K	57.7	$10^{14.7}$	$10^{-3.7}$	1	4.635	$10^{-1.7}$	0.980	2.130	0.980	4.540
690 K	56.8	$10^{13.8}$	$10^{-2.8}$	1	3.635	$10^{-0.8}$	0.853	1.167	0.853	3.025
700 K	56.0	$10^{13.0}$	$10^{-2.0}$	0.99	2.635	10^{0}	0.368	0.368	0.368	0.744
710 K	55.2	$10^{12.2}$	$10^{-1.2}$	0.939	1.66	$10^{0.8}$	0.0018	0.0018	0.0018	−0.00046
720 K	54.4	$10^{11.4}$	$10^{-0.4}$	0.672	0.765	$10^{1.6}$	0	0	0	0
730 K	53.7	$10^{10.7}$	$10^{0.3}$	0.136	0.136	$10^{2.3}$	0	0	0	0
740 K	53.0	$10^{10.0}$	10^{1}	4.5×10^{-5}	4.5×10^{-5}	10^{3}	0	0	-0	0
750 K	52.3	$10^{9.3}$	$10^{1.7}$	0	0	$10^{3.7}$	0	0	0	0
760 K	51.6	$10^{8.6}$	$10^{2.4}$	0	0	$10^{4.4}$	0	0	0	0

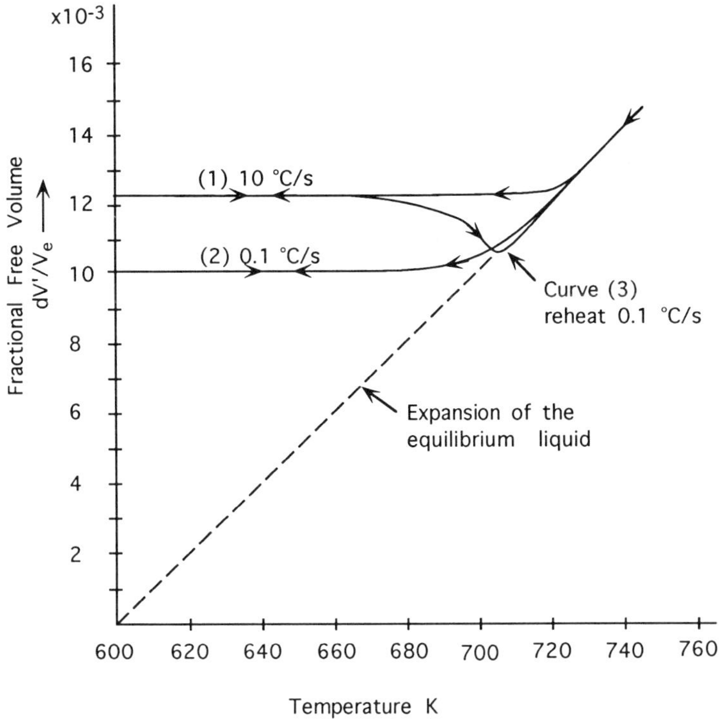

Figure 13-10. Calculated cooling and heating V–T curves. (Note: The overshoot of curve [3] past the dashed line has been exaggerated for clarity.)

Let us turn our attention to the computation of the heating curve. Assume that specimen [1], which was cooled at 10°C/s, is now heated at 0.1°C/s. In Table 13-2, the accumulated deviation 12.635 units is transferred to column 10. At each small step ΔT of heating, the deviation is reduced by a unit free volume expansion ($= V_e \beta_f \Delta T$) and multiplied by the relaxation modulus. The resulting iteration is plotted as the fractional free volume change, curve [3], in Fig. 13-10. Again, we note the remarkable qualitative agreement between our calculation and the experimental observations shown schematically in Fig. 13-9. In particular, the slight overshoot at 710 K and a rapid recovery thereafter should be noted. (To make direct comparison with total volume, we should have added the "solid" expansion coefficient — that observed in the glassy state — to our computations to introduce a slope to the V–T curves in the glassy state. The reader is advised to use a computer to carry out the iterations just shown.)

What we have shown so far is that despite the smoothness of the viscosity versus temperature curve, the V–T curve displays a gradual departure from equilibrium over the 10^{10}–10^{14} Pa·s viscosity range, and thus a "glass transition" develops. The computations utilized an Arrhenius rate-type temperature dependence for the viscosity. As noted in Section 9.2, the Arrhenius dependence is consistent with a Boltzmann distribution probability P_h of finding holes (Eq. (9.8)). In other words, the number of adequately sized holes decreases exponentially. If, instead, we assume that P_h follows a Poisson distribution according to the hole size V_h relative to the average free volume per atom V_f^* as

$$P_h = \exp[-V_h/V_f^*], \tag{13.32}$$

and assume that there exists a temperature T_0 at which V_f^* is zero, then it may be shown that at $T > T_0$,

$$P_h = \exp[-V_h/V_0^*\Delta\beta(T - T_0)], \tag{13.33}$$

where V_0^* is the average lattice volume per atom at T_0 and $\Delta\beta$ is the difference of the volume thermal expansion coefficient between the glass and the supercooled liquid. Substitution of (13.33) into (9.7) leads to the development of the Vogel–Fulcher–Tammann (VFT) equation for viscosity (9.13). If it were assumed that the condition $V_f^* = 0$ is unattainable at practical cooling rates, and that a free volume V_{fg}^* is "frozen" at the glass transition temperature T_g, then the Williams–Landel–Ferry (WLF) equation (9.14) develops. The key outcome in both the VFT and the WLF equations is that *an infinite viscosity is predicted at T_0 that is a hypothetical temperature where the free volume disappears, attainable only by an infinitely slow cooling rate.*

Despite the seemingly enormous success of the free volume theory to explain the glass transition behavior, there are problems. One of the most serious problems with the free volume theory, ironically, is the V–T behavior of the simplest of all the glasses, namely silica. Figure 2-7 clearly shows that the free volume of the glass is less than that of the equilibrium liquid on the left side of the diagram. Note also that slower cooling results in a higher "apparent free volume" than a faster cooling. How could the glass transition be considered as originating in the material's inability to acquire a larger free volume? Some additional objections to the free volume theories of the glass transition have been mentioned by Gupta [5].

13.5. Thermodynamics of the Glass Transition

We begin our discussion with a review of the notation to be used:

N_i = number of moles of species i.

U = internal energy
$dU = U_B - U_A$ in going from state A to state B
V = specific volume
$dV = V_B - V_A$ in going from state A to state B
$v = V/N$ = molar volume
P = pressure
T = temperature
δQ = a small quantity of heat (positive, if it is given to the system)
δw = amount of work done on the system (positive, if it increases U)
S = entropy
$s = S/N$ = molar entropy
H = enthalpy = $U + PV$
$h = H/N$ = molar enthalpy
G = Gibbs function or Gibbs free energy = $U + PV - TS$
$g = G/N$ = molar Gibbs function
F = Helmholtz free energy = $U - TS$
μ = chemical potential
r = number of components

Note that the use of δQ instead of dQ is consistent with the premise that the small amount of heat is an imperfect differential, and hence a path-dependent quantity. Likewise, δw is also an imperfect differential. Some useful definitions to recall are the first derivatives of U with respect to the other thermodynamic variables:

$$\left\{\frac{\partial U}{\partial S}\right\}_{V;N_i\ldots N_r} \equiv T,$$

$$\left\{\frac{\partial U}{\partial V}\right\}_{S;N_i\ldots N_r} \equiv -P,$$

$$\left\{\frac{\partial U}{\partial N_j}\right\}_{S;V;N_k\ldots} \equiv \mu_j, \qquad k \neq j. \tag{13.34}$$

The intensive properties T, P, and μ so defined allow the differential form of U to be written as a product of intensive and extensive properties:

$$dU = TdS - PdV + \mu_1 dN_1 + \cdots + \mu_r dN_r. \tag{13.35}$$

At constant N, there are only three independent second derivatives of U: $\partial^2 U/\partial S^2$, $\partial^2 U/\partial V^2$, and $\partial^2 U/\partial S \partial V$. After the application of Maxwell relations, these second derivatives define three fundamental property coefficients:

$$\text{Coefficient of volume thermal expansion: } \beta = \frac{1}{v}\left\{\frac{\partial v}{\partial T}\right\}_p. \tag{13.36}$$

$$\text{Isothermal compressibility: } \kappa = \frac{-1}{v}\left\{\frac{\partial v}{\partial P}\right\}_T. \tag{13.37}$$

$$\text{Heat capacity at constant pressure: } C_p = T\left\{\frac{\partial s}{\partial T}\right\}_p = \frac{1}{N}\left\{\frac{\delta Q}{\partial T}\right\}_p. \tag{13.38}$$

$$\text{Heat capacity at constant volume: } C_v = T\left\{\frac{\partial s}{\partial T}\right\}_v = \frac{1}{N}\left\{\frac{\delta Q}{\partial T}\right\}_v. \tag{13.39}$$

Let us also recall the Laws of Thermodynamics. According to the Zeroth Law, if A is in thermal equilibrium with B and A is also in thermal equilibrium with C, then B and C are in states which are in thermal equilibrium with each other. The First Law describes the principle of conservation of energy and is stated as

$$\delta Q = dU - \delta w = dU + pdV. \tag{13.40}$$

It forbids the existence of a perpetual machine of the first kind (a perpetual motion machine that creates energy). The Second Law forbids the existence of a perpetual machine of the second kind (a device to perform work but deriving its energy from a constant-temperature reservoir). Some other equivalent statements of the Second Law are:

(a) When any actual process occurs, it is impossible to invent a means of restoring every system concerned back to its original state.
(b) All natural processes are irreversible.
(c) In any natural process, the entropy production is always positive.

The Third Law establishes

$$S = 0 \qquad \text{where} \left\{\frac{\partial U}{\partial S}\right\}_{V;N_i\ldots} = 0,$$

$$\text{i.e., at } T = 0 \text{ for equilibrium states.} \tag{13.41}$$

(The reader is strongly recommended to refer to a textbook on thermodynamics, for instance, *Thermodynamics* by H. B. Callen, (Wiley, 1960), to review topics such as perfect/imperfect differentials, the various laws of thermodynamics, and the Maxwell's equations.)

Let us now turn our attention to the glass transition.

In the introduction earlier, we considered the glassy state as a "metastable equilibrium" state with respect to the extrapolated "equilibrium" supercooled liquid line. We assumed that if infinite time were allowed, glass would gradually move towards the "equilibrium" line. The question is whether the metastable nature of the glassy state exempts glass from being bound by the laws of thermodynamics? One of the early works in this field was that of Simon (6), who plotted the heat capacity of glycerol in the liquid, glassy, and crystalline states as a function of T (Fig. 13-11). There appeared a discontinuity in the molar heat capacity at the glass transition temperature T_g. The entropy may be computed from the experimental heat capacity curve by integrating the expression for C_p from (13.38) as

$$S(T) = \int_{T_0}^{T} \frac{C_p(T')}{T'} dT' + S(T_0), \tag{13.42}$$

where T_0 is a reference temperature just above the freezing point. This computed S–T curve is shown in Fig. 13-12. Like the V–T curve, the S–T curve also shows a change (discontinuous?) in slope at T_g corresponding to the jump in C_p. More importantly, however, the losses of entropy from a point just above the freezing point down to absolute zero may be shown to be 15 cal·mol^{-1}·deg^{-1} for the crystal route, but only 10 via the glass route.

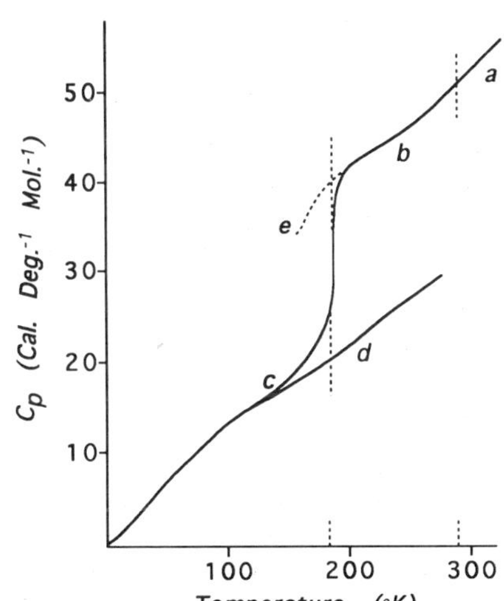

Figure 13-11. Heat capacity of glycerol in the liquid (a), supercooled liquid (b), glassy (c), and crystalline states (d). (e) Results when the supercooled liquid is cooled extremely slowly. (After Simon [6].)

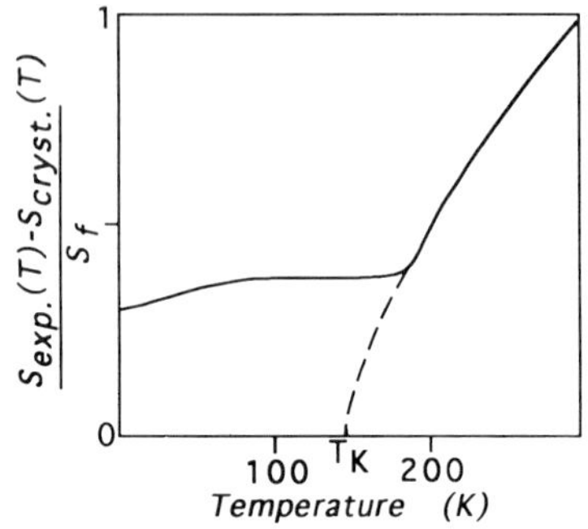

Figure 13-12. Experimental entropy of supercooled and glassy glycerol relative to that of the crystal in the units of entropy of fusion, S_f. (After J. Jäckle, *Physica* **127B**, 79 (1984). Reproduced with permission of Elsevier Science Publishers.)

This would imply that the glass has a residual entropy (in this case, 5 cal·mol^{-1}·deg^{-1}) at absolute zero, which is in apparent contradiction to the Third Law. Simon suggested that the metastability of the glassy state was the underlying reason for it not to follow the laws of thermodynamics, and that if one cooled more and more slowly (at an infinitely slow rate), then the liquid would follow the equilibrium line, resulting in the disappearance of the entropy anomaly. Later work did show that slower cooling would allow the supercooled liquid to continue along line "ab" to "e," which apparently reduced the residual entropy at $T = 0$. This raised a few more intriguing questions. If the liquid were cooled sufficiently slowly, one may expect both the liquid and the crystal to have zero entropy at $T = 0$. What would be the structure of this zero-entropy liquid? If it were different from that of the crystal, then how could there be two different configurations for the same composition, yet with an entropy of zero? (Recall that $S = k_B \ln W$, where k_B is the Boltzmann constant and W is the number of arrangements possible.) If, on the other hand, they were the same configurations, then the implication was a continuous change to crystalline arrangement without the inclusion of a jump involving latent heat of crystallization. Available evidence refuted at least the latter possibility.

The entropy anomaly was given a more dramatic turn by Kauzmann [7], who plotted the difference between the entropies of the supercooled liquid and the corresponding crystal for several glass-making substances. This plot, which looks much like Fig. 13-12, is shown in Fig. 13-13. Kauzmann indicated that the extrapolations of the supercooled curves reached zero entropy at a positive temperature called **Kauzmann temperature** T_K, shown in Fig. 13-12. Hence, the extremely slow-cooled, equilibrium liquid would attain negative entropy at $T = 0$. Since we continued to deal with equilibrium

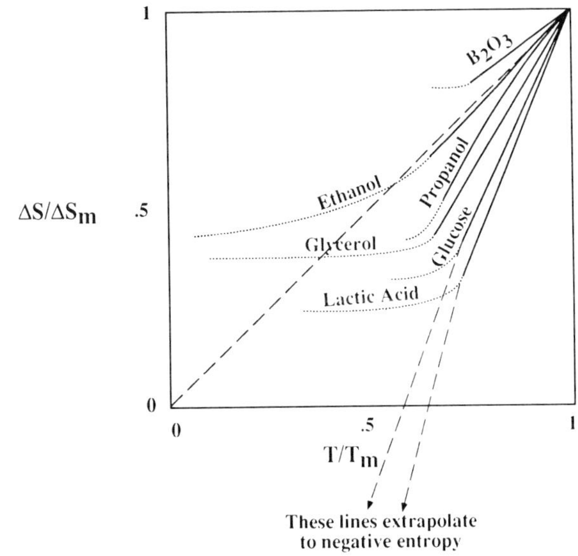

Figure 13-13. Kauzmann paradox. Supercooled liquid lines appear to extrapolate to negative entropy. (Redrawn after Kauzmann [7]. Reproduced with permission of The American Chemical Society.)

states, the exemption of metastability given to glasses having positive residual entropy at $T = 0$ would not apply. Again, this was a case where the Third Law was clearly violated. Kauzmann himself presented an interesting explanation for this so-called **Kauzmann paradox**. He argued that the barrier to crystallization consisted of two parts: the barrier to nucleation and the barrier to flow (activation energy for diffusion). At low temperatures, the barrier to flow dominated. In other words, both the approach to stabilization (to stay on the equilibrium supercooled liquid line) and the crystallization were controlled by similar barriers. At a sufficiently low temperature, crystallization would be inevitable, and the question of the negative entropy was operationally meaningless. Later, however, we shall note that there is a more acceptable explanation (offered by Gibbs and DiMarzio) to the Kauzmann paradox.

It appears that the thermodynamically unstable nature of glass with respect to the "equilibrium" liquid state, or the metastability with respect to the equilibrium crystal state, does not exempt it from obeying the laws of thermodynamics. Indeed, diamond at room temperature is metastable with respect to graphite, yet there are no known anomalies to its thermodynamic behavior. Our argument centers, once again, on the time scale of our experiments. In the supercooled liquid state, the atomic arrangements are rapid enough to obey the laws of thermodynamics. In the glassy state, the "frozen" behavior of the structure in a particular arrangement allows the application of thermodynamics, provided that this particular configuration is also regarded as a thermodynamic variable alongside P, V, and T. These configurational variables are termed **internal order parameters**. This is evident if we examine Fig. 13-9. At any given temperature (and pressure), a particular composition does not have a unique value of volume. We know that the cooling rate variable influences the volume. Hence, as an example, a suitable order parameter Z could be specified as Z_{Tf}, where T_f is the fictive temperature.

Consider the cooling of a liquid from point B to a glass at temperature T (Fig. 13-14). For the sake of discussion, we imagine the glass to be produced by two different cooling rates. The slower one brings the glass along BDCM. The state of the glass at T is given by the point M. Its fictive temperature is $\bar{T}$, which is the point C on the equilibrium curve, and hence, the thermodynamic specification of the state at T is $(P, T, Z_{\bar{T}})$. Using a slightly faster cooling rate, the glass is brought down the path BDLQ such that its fictive temperature is $\bar{T} + d\bar{T}$, represented by the point D. The specification of the state at the point Q is $(P, T, Z_{\bar{T}+d\bar{T}})$. For simplicity, let us assume that the path CM is small, i.e., $\bar{T} = T + dT$, and L has the temperature $\bar{T}$ so that the specification of the state at L is $(P, \bar{T}, Z_{\bar{T}+d\bar{T}})$. Since the points C and D are on the equilibrium line, their fictive temperature is just the physical temperature. In other words, C is $(P, \bar{T}, Z_{\bar{T}})$ and the point D is $(P, \bar{T} + d\bar{T}, Z_{\bar{T}+d\bar{T}})$.

Bringing the glass from state Q to state L is simple: it requires adding a

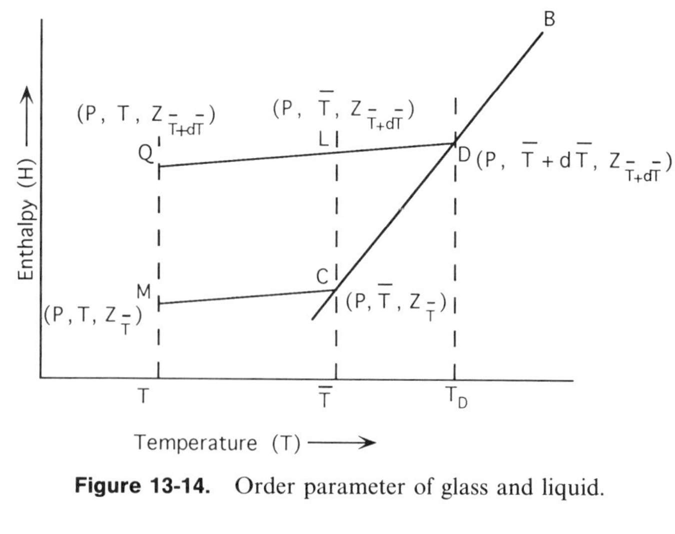

Figure 13-14. Order parameter of glass and liquid.

heat $\delta Q = dH = C'_{pg}dT$, where C'_{pg} is the heat capacity in the glassy state. The entropy change in the process would simply be $\delta Q/T$ $(= C'_{pg}dT/T)$. We could imagine the glass to be in contact with a heat reservoir at constant pressure, and as long as dT is sufficiently small, the path QL could be made reversible. Bringing state M to L, however, is not as easy as adding δQ despite the smallness of the path. To do this, we must take the glass from M to C, then from C to D, and finally from D to L. Each of these paths individually could be made reversible; however, the total path is not reversible, as we now prove:

Entropy change dS_{total} in going from M to L is given by

$$dS_{total} = dS_{MC} + dS_{CD} + dS_{DL} \tag{13.43}$$

$$= C_{pg}[dT/T] + C_{pe}[d\bar{T}/\bar{T}] + C_{pg}[-d\bar{T}/\bar{T}]$$

$$= C_{pg}[dT/T] + \Delta C_p[d\bar{T}/\bar{T}], \tag{13.44}$$

where $\Delta C_p = C_{pe} - C_{pg}$.

To determine the heat requirement δQ, we note that $\delta Q = dH$ at constant P (from the definition of H, and the First Law). Thus,

$$dH_{total} = dH_{MC} + dH_{CD} + dH_{DL} \tag{13.45}$$

$$= C_{pg}[dT] + C_{pe}[d\bar{T}] + C_{pg}[-d\bar{T}] \tag{13.46}$$

$$= C_{pg}[dT] + \Delta C_p[d\bar{T}].$$

The total entropy change can also be written as

$$dS_{total} = dS_{reversible} + dS_{irreversible},$$

where $dS_{reversible} = \delta Q/T = dH_{total}/T = \{C_{pg}[dT] + \Delta C_p[d\bar{T}]\}/T$. Hence,

$$dS_{irreversible} = \{C_{pg}[dT/T] + \Delta C_p[d\bar{T}/\bar{T}]\} - \{C_{pg}[dT] + \Delta C_p[d\bar{T}]\}/T$$

$$= \Delta C_p\{(1/\bar{T}) - (1/T)\}d\bar{T}. \tag{13.47}$$

According to the Second Law, the entropy production for an irreversible reaction must be positive. Hence,

$$\Delta C_p\{(1/\bar{T}) - (1/T)\}d\bar{T} > 0,$$

and since ΔC_p is always positive (shown later), we get

$$d\bar{T} < 0 \text{ when } \bar{T} > T, \quad \text{and} \quad d\bar{T} > 0 \quad \text{when} \quad \bar{T} < T. \tag{13.48}$$

This conclusion describes the location of the point L relative to M when the glass at M is heated. If the physical temperature is less than the fictive temperature (on the left side of the equilibrium line in Fig. 13-14), the glass upon heating should move towards a point L such that the fictive temperature is lowered (it should move downwards in the fourth quadrant toward the equilibrium line). In the limit, an unstabilized glass left isothermally at

L with order parameter $Z_{\bar{T}+dT}$ will move towards point C having a stabilized structure characteristic of a lower fictive temperature $\bar{T}$. When, on the other hand, the physical temperature is higher than the fictive temperature (on the right side of the equilibrium line), the state upon continued heating should move to raise its fictive temperature (i.e., move upwards in the first quadrant to meet the equilibrium line). Both these conclusions are generally true as summarized in Fig. 13-9. In rare cases, however, it is possible for the state on the left side of the equilibrium liquid line to move such that its apparent fictive temperature increases. (The answer to this riddle has to be the inadequacy of the description in terms of a "single" order parameter $\bar{T} = T_f$, as we shall see later). A hysteresis between the cooling curve and the heating curve will still be present, indicating a finite, irreversible production of entropy.

What we have shown is that the inclusion of at least one more thermodynamic variable, such as the order parameter, enables us to apply thermodynamic principles to the glassy state (regardless of its metastability), and that the changes in the order parameter during a heat treatment follow the requirement that entropy has to be produced irreversibly.

In the glass transition range, the order parameter continues to change with time. This makes the precise measurement of properties difficult. The question of precision, or lack thereof, brings us to inquire as to which order of thermodynamic transition closely describes the glass transition. To address this issue, we first review the basics of thermodynamic phase transitions.

The change dG in the Gibbs function associated with a thermodynamic transition can be written as

$$dG = -SdT + VdP + \mu_1 dN_1 + \mu_2 dN_2 + \mu_3 dN_3 + \cdots \tag{13.49}$$

Hence, for a simple system where only one phase exists,

$$dg = -sdT + vdP. \tag{13.50}$$

A first-order thermodynamic transition is defined as the one where g remains continuous, but the first derivatives of g are discontinuous. There are two first derivatives of g: $(dg/dT)_p = -s$, and $(dg/dP)_T = v$. Examination of Fig. 13-1 reveals that the freezing of a liquid at the point 'c' fits this description.

A second-order transformation is one where g is continuous, as are the first derivatives of g; however, the second derivatives of g are discontinuous. In other words, V–T and S–T curves are continuous, but the β–T, κ–T, and C_p (or C_v)–T curves show discontinuities. For an nth-order thermodynamic transition, one expects to have all derivatives of g up to the $(n-1)$th order to be continuous, and the nth derivative to be discontinuous at the transition.

First-order transitions obey the Clausius–Clapeyron equation, which is

$$\frac{dP}{dT} = \frac{L}{T\Delta v}, \tag{13.51}$$

where L = the latent heat of the phase transition.

There are two equations interrelating the changes in the various second derivatives at a second-order transition (in other words, although there are three independent second derivatives β, κ, and C_p [or C_v], the changes in them can not be all independent at the transition):

$$\frac{dP}{dT} = \frac{1}{vT}\left(\frac{\Delta C_p}{\Delta\beta}\right) \tag{13.52}$$

and

$$\frac{dP}{dT} = \left\{\frac{\Delta\beta}{\Delta\kappa}\right\}. \tag{13.53}$$

These are called the Ehrenfest relations for second-order transition. Equating (13.52) and (13.53), one gets

$$\frac{\Delta C_p \Delta\kappa}{Tv\Delta\beta^2} \equiv \Pi = 1. \tag{13.54}$$

The ratio Π is called the Prigogine–de Fay ratio. Its closeness to 1 may be taken as a fingerprint for an ideal second-order transition.

The continuity of V–T and S–T in the T_g range glasses (Fig. 13-1) clearly suggests that the normally observed glass transition may not be described as a first-order thermodynamic transition. A typical plot of C_p vs. T is shown

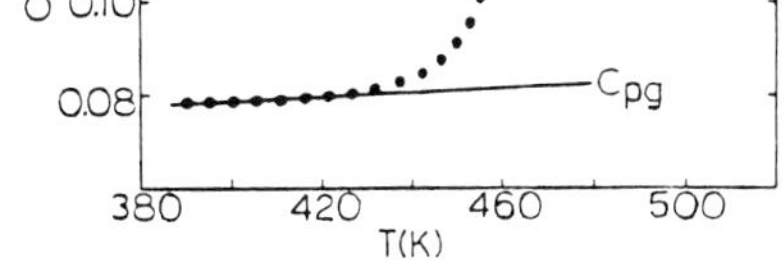

(a)

Figure 13-15. (a) Heat capacity of As_2S_3 glass measured at 10 K/min heating following a −2.5 K/min cooling. (After Esteal *et al.*, *J. Am. Ceram. Soc.* **60**(3–4), 134 (1977). Reproduced with permission of The American Ceramic Society.)

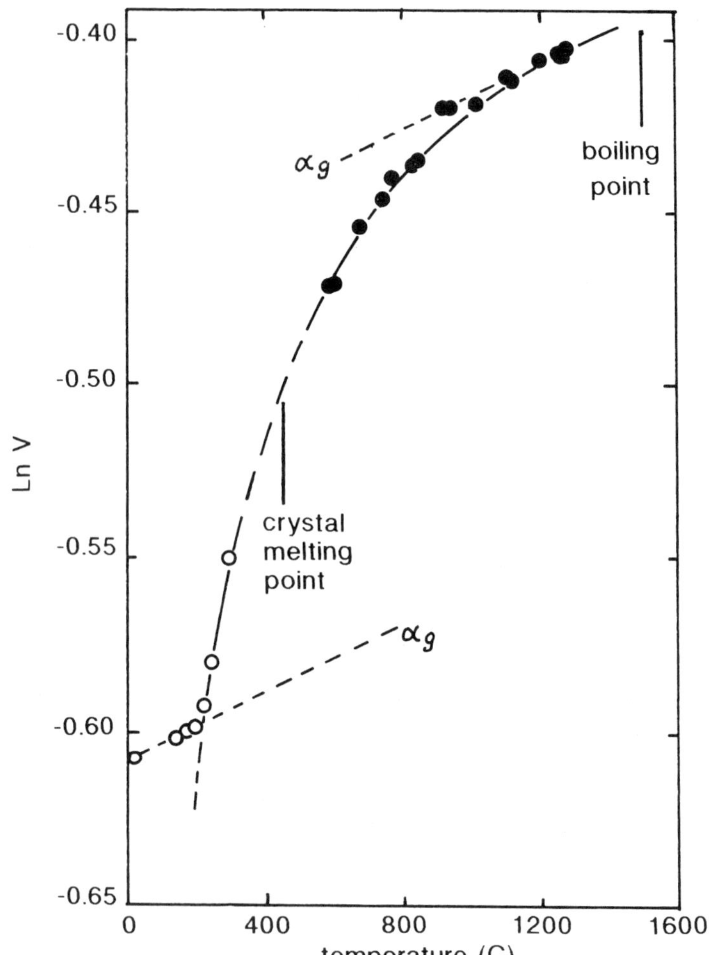

(b)

Figure 13-15. (b) V–T diagram for B_2O_3 glass. (After J. S. Haggerty and A. R. Cooper, *Phys. Non-cryst. Sol.* (J. A. Prins, ed.), p. 346, 1965. North Holland Publishing Co., Amsterdam. Reproduced with permission of Elsevier Science Publishers.)

in Fig. 13-15a. We begin to observe steep changes in the values of the properties at T_g. The V–T data for B_2O_3 shown in Fig. 13-15b also imply a very rapid rise in the expansion coefficient (slope of the $\ln(V)$–T curve) near 210°C. One wonders if the curves of the second derivatives do have discontinuities smoothed out by the precision of the measuring equipment, or whether they are actually continuous. Our approach here would be to examine what is expected on the basis of theoretical arguments. An important contribution on this topic was presented by Gibbs and DiMarzio [8]. They attempted to obtain the configurational entropy of a polymer by calculating the number of ways in which n_x liner chains, each x segments long, could be placed on a lattice where the coordination number was z (generally = 4) along with n_0 holes, all placed in an evacuated box of a known volume. Distinct spatial conformations developed as a result of assumed potential energy minima as a function of the bond angle. Each configurational state constituted $f_i n_x$ conformations of the type i for several values of i ($\sum f_i n_x - n_x$) and n_0 holes. Gibbs–DiMarzio suggested that structural configurations involved increasingly large molecules with decreasing temperatures. The total energy $\Phi(\{\ldots f_i n_x \ldots\}, n_0)$ of the system was a sum Φ_1 of the individual energy of the molecules (intramolecular energy) plus the hole energy Φ_2 (intermolecular energy). For example, for $z = 4$, Φ_1 and Φ_2 could be written as

$$\Phi_1 = f\varepsilon_2(x-3)n_x + (1-f)\varepsilon_1(x-3)n_x$$

and $\Phi_2 = z'\alpha n_0 S'_x/2$.

$$\Phi(\{\ldots f_i n_x \ldots\}, n_0) = \Phi_1 + \Phi_2, \tag{13.55}$$

where ε_1 is the energy of a bond in one of the $(z-1)$ possible orientations, and ε_2 is the energy of the remaining "flexed" bonds whose fraction is f, z' is the coordination number of the "best" lattice characterizing the packing of the segments, α is the energy of interaction (van der Waals bond energy) between a pair of chemically nonbonded nearest neighboring segments, and S'_x is the fraction $[(z'-2)x+2]n_x/\{[(z'-2)x+2]n_x + z'n_0\}$. A partition function Q_T at temperature T could be written as

$$Q_T = \sum_{\substack{W=1 \\ (\{\cdots f_i n_x \cdots\}, n_0)}}^{W=\infty} W(\{\ldots f_i n_x \ldots\}, n_0)\exp[-\Phi(\{\ldots f_i n_x \ldots\}, n_0)/k_B T] \tag{13.56}$$

where W is the total number of configurational states of the system consistent with the requirement that $f_i n_x$ molecules are in conformation i and there are n_0 holes. W was written in the form of Flory's expression (9), which required summation from $W = 0$ to $W = \infty$. To allow this summation, Q_T was rewritten as $Q'_T - Q''_T$, where Q'_T summed from $W = 0$ to $W = \infty$, and Q''_T summed from $W = 0$ to $W = 1$. The internal energy contribution of Q_T to the Helmholtz free energy F could be determined using the relation

$$F = -k_B T \ln Q_T. \tag{13.57}$$

It could be shown that

$$F(T_2+) = F(T_2-),$$

where T_2 was the temperature at which $W_{max} = 1$. Likewise, the first derivatives were also continuous at T_2, i.e.,

$$S(T_2+) = S(T_2-), \qquad \text{and} \qquad U(T_2+) = U(T_2-).$$

However, it could also be shown that the second derivatives α, κ_T, and C_p displayed discontinuities at T_2, implying that T_2 was a true second-order transition.

Thus, in a manner similar to the free volume concepts, Gibbs–DiMarzio were able to demonstrate that *there existed a temperature T_2 at which all the segments and the holes could only be arranged in one particular way; in other words, the configurational entropy disappeared at T_2*. T_2 was thus the limiting value of T_g in a hypothetical experiment where cooling occurred over infinitely long periods. At temperatures higher than T_2, the available energy allowed the system to fill the available volume in a number of arrangements; hence, the configurational entropy was finite in a normally cooled glass.

The strength of the Gibbs–DiMarzio model lay in providing a "natural" explanation to the Kauzmann paradox. It was not necessary to invoke imminent crystallization as an ideally cooled (infinitely slowly cooled) liquid approached the Kauzmann temperature T_K. One merely had to take away the flexibility of configurational arrangements. At T_K, which may also be identified as T_2, the number of configurations reduces to 1 so that the configurational contribution to the entropy is zero between $T = T_K$ and $T = 0$. Upon continued cooling below T_K, the system so formed continues to lose vibrational entropy until $S = 0$ at $T = 0$ in full accordance with the Third Law (see Fig. 13-16, after Stillinger [10]). Thus, *the glass transition of an infinitely slowly cooled liquid at the Kauzmann temperature T_K is believed to*

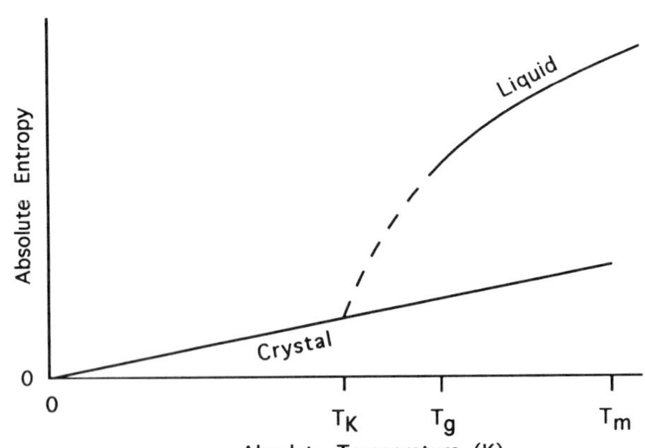

Figure 13-16. Absolute entropy of liquid and crystal with temperature. Note discontinuity of slope at T_K (Kauzmann temperature).

be a true second-order thermodynamic transition. Ordinarily cooled glasses fall short of approaching T_K, and hence the derivative of the entropy is not expected to display a discontinuity at the normally observed T_g. Thus, T_g *obtained by ordinary cooling is not a true second-order transition.* The residual entropy of the ordinarily cooled glasses at $T = 0$ is a manifestation of the disordered liquid structure for which there exist a large number of configurations, all compatible with the same macroscopic state. One of these configurational states is "frozen." With reheating to the supercooled liquid region, and subsequent slow cooling to $T = 0$, the glass may be frozen into yet another configuration; however, this configuration is microscopically indistinguishable from the previous one. (There may be some contribution from vibrational properties in addition to the purely geometric structural considerations; see J. Jäckle [11].) It must be emphasized that the explanations for the Kauzmann paradox remain speculations only. Stillinger [10], for instance, believes that there is no thermodynamic transition at all at T_K. There is no possibility of any experimental evidence ever being provided. Even for the case of $H_2SO_4{\cdot}3H_2O$, $Ca(NO_3)_2{\cdot}4H_2O$, or lithium acetate glasses where the T_K is believed to be particularly close to the normal T_g, the cooling would have to be carried out over many thousand years [12]. Interested reader is referred to articles by Stillinger [10], Jäckle [11, 12], and Angell *et al.* [13] for further reading.

Gibbs–DiMarzio were also able to demonstrate the decrease in configurational entropy with falling temperatures to be so rapid that it could completely vanish 5 to 10% below the T_g obtained ordinarily. This latter observation appeared to hold well for many polymeric glasses (WLF Eq. (9.14)). According to Gibbs–DiMarzio, flow had to be dependent upon the relative rate of the disappearance of the configurational entropy. This argument was subsequently used by Adam and Gibbs [14] to suggest a viscosity–temperature relationship that involved the configurational entropy and not the free volume. Adam and Gibbs assumed that a liquid consists of smaller units that can rearrange. There is a potential energy barrier for rearrangement. The size of the units is a function of the temperature and is determined by the configurational entropy of the system. As temperature decreases, the configurational entropy decreases, and hence the size of the units increases. At zero configurational entropy, the size is essentially the entire system; in other words, there is no further freedom to rearrange (and the viscosity becomes infinite). The Adam–Gibbs equation is

$$\eta = \eta_0 \exp[A/TS_c], \tag{13.58}$$

where $A = \Delta\mu S_c^*/k_B$; $\Delta\mu$ is the potential barrier hindering rearrangement, k_B is the Boltzmann constant, and S_c^* is the configurational entropy of the smallest group of molecules that can undergo rearrangement. S_c^* is given by $k_B \ln W^*$, where W^* is the number of configurations available to that group and is expected to be around 2. S_c is the configurational entropy given by

$$S_c = \int_{T_K}^{T} \frac{\Delta C_p}{T} dT. \tag{13.59}$$

A second argument regarding the expectation from the theory of glass transition was given by Gupta and Moynihan [15]. The equilibrium liquid free energy surface g_e is a unique function of T and P, i.e., all the order

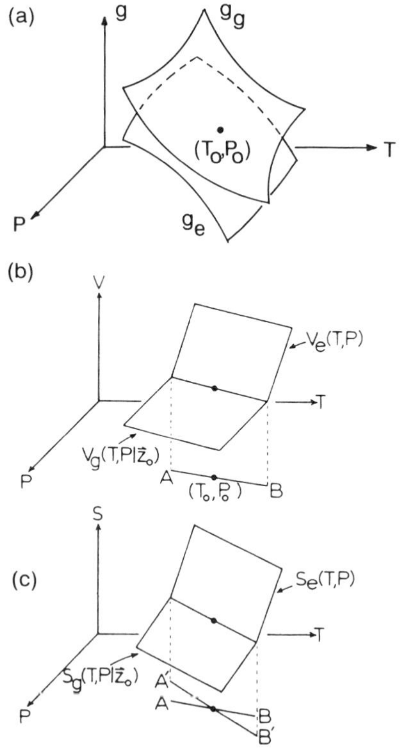

Figure 13-17. (a) Gibbs free energy, (b) volume, and (c) entropy surfaces for equilibrium liquid and glass. Note that the Gibbs free energy surface of glass is always above that of the liquid, and is tangent to it at (T_0, P_0) in the most general case. Lines AB and A′B′ are the loci of points at which the partial derivative of the free energy difference function with respect to P (for volume) and T (for entropy) equal zero. In general, AB and A′B′ are not coincident, intersecting only at (T_0, P_0). (After P. K. Gupta and C. T. Moynihan, *J. Chem. Phys.* **65**(10), 4136 (1976). Reproduced with permission of The American Institute of Physics.)

parameters Z have an equilibrium value characteristic of (T, P). Hence, $g_e = g_e(T, P)$ (see Fig. 13-17). The glass free energy surface g_g, on the other hand, is not determined entirely by (T, P), but also depends upon the values of the order parameter/parameters Z that have been effectively frozen in at (T_0, P_0). Hence, $g_g = g_g(T, P/Z_0)$. The two surfaces must be tangent to each other at least at the point (T_0, P_0) (see Fig. 13-17a). In general, the surface g_g must lie above the surface g_e, since the liquid is never unstable with respect to glass. Depending upon the magnitude of the different order parameters relative to their equilibrium values at (T_0, P_0), the glass and the liquid free energy surfaces may also be tangent to each other along some direction in the T–P plane. The free energy difference function f at temperature T may be written as

$$f = g_g(T, P/Z_0) - g_e(T, P) = 0 \qquad \text{at} \qquad (T_0, P_0),$$

and positive elsewhere. The tangency with respect to T and P at (T_0, P_0) requires

(i) $$\left\{\frac{\partial f}{\partial T}\right\}_P = 0 = \frac{\partial g_g}{\partial T} - \frac{\partial g_e}{\partial T} = -s_g + s_e, \qquad \text{i.e., } s_g = s_e; \tag{13.60}$$

(ii) $$\left\{\frac{\partial f}{\partial P}\right\}_T = 0 = \frac{\partial g_g}{\partial P} - \frac{\partial g_e}{\partial P} = v_g - v_e, \qquad \text{i.e., } v_g = v_e; \tag{13.61}$$

(iii) Curvature of f with respect to T should be positive:

$$\frac{\partial^2 f}{\partial T^2} \geq 0.$$

But

$$\left\{\frac{\partial^2 f}{\partial T^2}\right\}_P = \frac{\partial^2 g_g}{\partial T^2} - \frac{\partial^2 g_e}{\partial T^2} = -\frac{\partial s_g}{\partial T} + \frac{\partial s_e}{\partial T} = -\frac{C_{pg}}{T} + \frac{C_{pe}}{T}, \quad \text{i.e., } \frac{\Delta C_p}{T} \geq 0. \tag{13.62}$$

Note, this implies that the heat capacity of the equilibrium liquid is higher than that of the glass.

(iv) Curvature of f with respect to P should be positive:

$$\left\{\frac{\partial^2 f}{\partial P^2}\right\}_T \geqslant 0.$$

But

$$\frac{\partial^2 f}{\partial P^2} = \frac{\partial^2 g_g}{\partial P^2} - \frac{\partial^2 g_e}{\partial P^2} = \frac{\partial v_g}{\partial P} - \frac{\partial v_e}{\partial P} = -v_g\kappa_g + v_e\kappa_e$$

$$= v\Delta\kappa, \qquad \text{since } v_g = v_e, \tag{13.63}$$

i.e., $v\Delta\kappa \geqslant 0$.

Thus, the compressibility of the liquid is higher than that of the glass.

(v) $$\operatorname{Det}\begin{vmatrix} \dfrac{\partial^2 f}{\partial T^2} & \dfrac{\partial^2 f}{\partial T \partial P} \\ \dfrac{\partial^2 f}{\partial P \partial T} & \dfrac{\partial^2 f}{\partial P^2} \end{vmatrix} \geqslant 0, \text{ for the entire space.}$$

Now, since

$$\frac{\partial^2 f}{\partial T \partial P} = \frac{\partial^2 f}{\partial P \partial T} = \frac{\partial v_g}{\partial T} - \frac{\partial v_e}{\partial T} = v_g \beta_g - v_e \beta_e = -v\Delta\beta.$$

Hence, the condition reduces to

$$\operatorname{Det}\begin{vmatrix} \dfrac{\Delta C_p}{T} & -v\Delta\beta \\ -v\Delta\beta & v\Delta\kappa \end{vmatrix} \geqslant 0,$$

or

$$\frac{\Delta C_p \Delta\kappa}{Tv\Delta\beta^2} = \Pi \geqslant 1. \tag{13.64}$$

It may be shown that the equality in Eq. (13.64) applies, i.e., the Prigogine–deFay condition is obeyed, when the volume and the entropy surfaces (Fig. 13-17b and c) are tangent to each other at a line (lines AB and A′B′ coincide). The inequality applies when the tangency of the free energy surfaces is at a point (T_0, P_0) only, i.e., the lines AB and A′B′ do not coincide except at the point (T_0, P_0). The first of these is satisfied by the glass having either only one order parameter or several order parameters whose gradients in the volume and entropy spaces are identical. For the inequality to hold, there must be two or more order parameters. In this case, it may be recognized that each of the order parameters would equilibrate differently (the "fictive temperature" on the V–T curve may not be the same as that on the S–T curve).

Thus, if the glass had only one order parameter, the ratio Π would come out to be 1 and the glass transition would behave as if it were a second-order transition. On the other hand, the presence of two or more order parameters, whose variations in the V and S spaces are not similar to each other, would cause Π to be larger than 1, and the transition would not be of second order.

Note also from (13.64) that, since ΔC_p and $\Delta\kappa$ are positive, $\Delta\beta$ may have either sign.

Since available experimental data for B_2O_3 and $0.4Ca(NO_3)_2{\cdot}0.6KNO_3$ glasses suggest that $\Pi \sim 4.7$ and 4.5, respectively, we are led to conclude that the practical glass transition in general is not a second-order transition. A similar conclusion was derived earlier from Gibbs–DiMarzio theory. *To avoid confusion, glass transition is best considered as a "frozen-in transition"* without assigning any specific thermodynamic order to it.

13.6. Kinetics of the Glass Transition

13.6.1. "Watching" the Relaxation (Concept of the Deborah Number)

In all our discussion relating to the glass transition so far, we have included the term "relaxation" of order parameters (such as the free volume) to distinguish between a glass and an equilibrium liquid. The kinetics of any relaxation must be described in terms of our concept of time — seconds, hours, months, years, etc. Humans perceive the glass transition at a viscosity $\sim 10^{12}$ Pa·s or so because the properties, such as the volume, relax as we watch a normally cooled glass (cooling rates about 1–10°C/min). Above the glass transition range, the properties are "instantly" in equilibrium. Below this range, the magnitudes of the properties are "essentially" frozen-in. The ratio of the relaxation time to the time for observations is called the **Deborah number (DN)**. The origin of the term Deborah number is interesting. In the Old Testament, Judges 5, verse 5, the prophetess Deborah is claimed to have stated, "... the mountains flowed before the Lord ..." Putting everything in proper perspective, we can safely assume that the mountains do flow — continental drift is a scientifically established fact. However, such flow could only be witnessed by some superpower that could watch the phenomena over geological times. Mortal beings such as humans live far too short a time to watch many of the relaxation behaviors that nature encompasses. In gases and liquids, the time for relaxation is much shorter than the time for any experimental observation to be made; hence, the DN may be as small as 10^{-8} or less, whereas in solids, where most relaxation phenomena are essentially frozen, the DN may be of the order of 10^8 or more. We are tempted to *define the glass transition as one having* DN $\cong 10^{-3}$ to 10^3. Since this range is quite broad (because of the capability of our measuring instruments), and since the physical properties of glass do vary greatly if the cooling rate is varied, the usefulness of DN in scientific discussion of the glass transition is limited.

13.6.2. Tool's Fictive Temperature Theory

The most important systematic study of the dependence of glass properties as a function of temperature covering the glass transition range was carried out by Tool at the U.S. National Bureau of Standards (now called the National Institute for Standards and Technology) [16]. He studied the expansion of two borosilicate glass specimens of identical composition which had been cooled either very rapidly or very slowly. The expansion curves were made to coincide at 615°C, as shown in Fig. 13-18. Next, Tool defined the fictive temperature T_f *of a glass to be the temperature on the equilibrium curve at which the glass would find itself in equilibrium* (the lowest energy state other than the corresponding crystal) if brought suddenly to it. Since the slowly cooled glass was held at 500°C for several weeks, Tool assumed this glass specimen to have a fictive temperature of 500°C (marked E). Line EE′ was then drawn such that it grazed the curves merging at 615°C. Line A intersects EE′ at 653°C, which was then assumed to be the initial fictive temperature of the chilled sample. The slope of the lines A and B is 66×10^{-7}/°C (α_g of glass) and 317×10^{-7}/°C (α_e of the supercooled liquid), respectively. To describe the linear expansion of the specimens, Tool wrote

$$(L_T - L_0)/L_0 = \alpha_g(T - T_0) + (\alpha_e - \alpha_g)(T_f - T_{f0}), \qquad (13.65)$$

where the length L_0 is the length at some reference temperature T_0, and T_{f0} is the initial fictive temperature. The first term on the right-hand side is the change in length due to a change in T at fixed T_f; the second term is the effect due to varying T_f at fixed T.

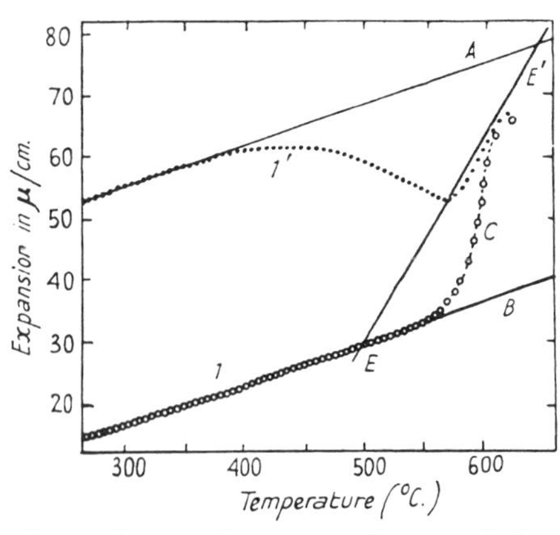

Figure 13-18. Experimental expansion curves for annealed sample (1) and chilled sample (1′). (After Tool [16]. Reproduced with permission of The American Ceramic Society.)

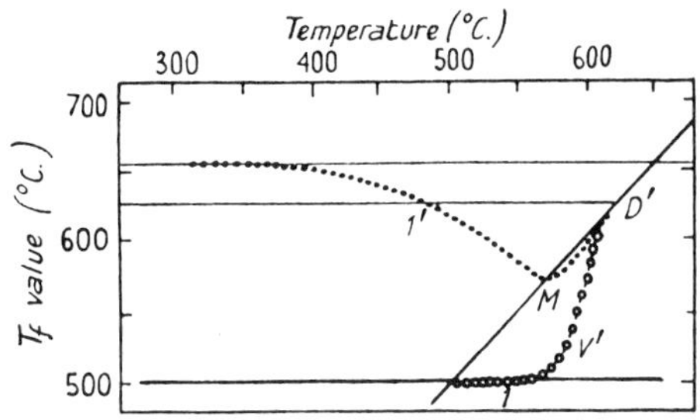

Figure 13-19. Graphical estimation of the fictive temperature for the annealed and chilled specimens. (After Tool [16]. Reproduced with permission of The American Ceramic Society.)

For the chilled sample,

$$(L_T - L_{300})/L_{300} = 66 \times 10^{-7}(T - 300) + 251 \times 10^{-7}(T_f - 653), \quad (13.66)$$

and for the slowly cooled sample

$$(L_T - L_{300})/L_{300} = 66 \times 10^{-7}(T - 300) + 251 \times 10^{-7}(T_f - 500). \quad (13.67)$$

Using the expansion data, it was hence possible to plot T_f vs. T for both the specimens. This is shown in Fig. 13-19. Tool empirically showed that the curves of T_f vs. T fitted the equation

$$\frac{dT_f}{dT} = \pm[K/B]\cdot[T - T_f]\exp(T/g)\exp(T_f/h), \quad (13.68)$$

where B is the constant heating/cooling rate with $K/B = \exp(-45.32)$; $g = 19.78$ K, and $h = 43.22$ K. Apparently, the empirical fitting was good above 500°C, but not so good below 500°C. Now, if we identify T_f with $\bar{T}$ as shown in Fig. 13-14 (i.e., call it an order parameter), then dT_f/dT must be associated with enthalpy (or δQ) changes, as (from Eq. (13.46))

$$\delta Q = dH = C_{pg}[dT] + \Delta C_p[dT_f], \quad \text{or} \quad \delta Q/dT = C_{pg} + \Delta C_p[dT_f/dT]. \quad (13.69)$$

This would bring about an endothermic or exothermic reaction depending upon the sign of dT_f/dT. Tool plotted dT_f/dT computed from Fig. 13-19, and compared it with results from a differential thermal analysis (DTA). The agreement between the two, shown in Fig. 13-20, was remarkable. In the slow-cooled (annealed) sample, a deep endotherm at ~600°C was noted, whereas the "chilled" glass showed a marked exotherm at lower temperatures, followed by a small endotherm at ~600°C.

In summary, Tool suggested that, depending upon the cooling rate, glass

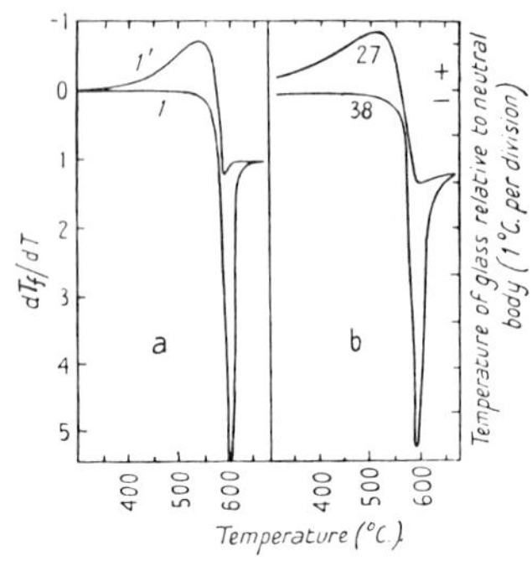

Figure 13-20. Exothermic and endothermic effects. (a) Computed slope of the curves shown in Fig. 13-19. (b) DTA curves for the annealed (number 38) and the chilled (number 27) samples. (After Tool [16].)

has a frozen-in structure that corresponds to some temperature, called the fictive temperature T_f, of an equilibrium liquid at constant pressure. Each nonequilibrium state has a one-to-one correspondence with a single equilibrium state; hence, the thermodynamic state of the glass may be described by any three of the four thermodynamic variables: P, V, T, and T_f. *As we cool a glass, its T_f varies not only as a function of the physical temperature T, but also as a function of itself and the distance from equilibrium ($=T - T_f$). Hence, the relaxation of a property is intrinsically nonlinear.* The rate of change, however, is considered to be independent of the direction of the temperature change.

In a manner analogous to Tool's definition of the T_f, we could also define the **fictive pressure** P_f on a V–P diagram. When pressure is changed on a glass in the transition range, its structure changes. The structure of a nonequilibrium state on the glass line in V–P space would have a one-to-one correspondence with a single equilibrium state at P_f on the extrapolated portion of the equilibrium line. Thus, P_f could be treated as yet another thermodynamic variable in describing the glassy state.

13.6.3. About the DTA and DSC of the Glass Transition

This is a convenient place to understand what we should expect from a differential thermal analysis (DTA) and a differential scanning calorimetry (DSC) of a glass when heated through the transition range.

The reader should recall that, in a DTA, the rate of heat input $\dot{Q}$ is kept constant and the difference in temperature ΔT of the specimen (mass m_s) with respect to a neutral standard is measured against T. This means that

$$m_r C_{pr} \Delta T_r / \Delta t = m_s C_{ps} \Delta T_s / \Delta t, \tag{13.70}$$

where the subscript "r" refers to the reference standard and "s" to the specimen. Hence,

$$[m_r C_{pr} - m_s C_{ps}]/m_s C_{ps} = [\Delta T_s - \Delta T_r]/\Delta T_r = \Delta T/\Delta T_r. \tag{13.71}$$

Thus, a plot of ΔT against T is a plot of $[m_r C_{pr} - m_s C_{ps}]/m_s C_{ps}$ against T. From Eq. (13.46), $C_{ps} = \{C_{pg} + \Delta C_p[dT_f/dT]\}$, where ΔC_p is always positive. The instrument is balanced to obtain $\Delta T = 0$ initially. In other words, the ΔT versus T plot is equivalent to the plot of $\{-\Delta C_p[dT_f/dT]\}/\{C_{pg} + \Delta C_p[dT_f/dT]\}$ versus T. For the "annealed" sample upon heating, $[dT_f/dT]$ starts out being zero, gradually becomes large positive, and finally becomes less positive before settling to a value of unity. Hence, ΔT starts from zero, becomes strongly negative (shows an endothermic behavior), and finally becomes less negative. For the chilled sample upon heating, $[dT_f/dT]$ starts out being zero and becomes negative. Hence, ΔT starts from zero and becomes positive (shows exothermic behavior). When $[dT_f/dT] = 0$, $\Delta T = 0$. As $[dT_f/dT]$ turns to become positive, ΔT begins to show an endothermic behavior, finally settling at a negative value. It may be recognized that ΔT is not exactly proportional to dT_f/dT.

In the "power-compensated" DSC mode, one varies $\dot{Q}$ supplied to the specimen (encapsulated inside a pan) to keep the temperature equal to that of an empty pan that is being heated at a constant $\dot{T}$. (The instrument is adjusted to give a null "baseline" with one empty pan.) We plot $\dot{Q}$ for the unknown specimen inside a pan relative to an empty pan as a function of T. Since,

$$\dot{Q} = (\delta Q/dT)\dot{T} = m_s\{C_{pg} + \Delta C_p[dT_f/dT]\}\dot{T}, \tag{13.72}$$

the plot of $\dot{Q}$ vs. T closely resembles dT_f/dT vs. T. The behavior of $\dot{Q}$ vs. T would basically be similar to that in the DTA, except that a positive $\dot{Q}$ in DSC implies an endotherm as opposed to a positive ΔT in DTA being an exotherm. One may also note that at low temperatures, where $dT_f/dT = 0$, C_{pg} can be directly extracted from $\dot{Q}$ using a DSC (Eq. (13.72)). Measurement of C_{pg} using a DTA is, however, not so straightforward. Quantification of the DTA to measure δQ is utilized in the construction of the "heat flux" DSC instruments.

A value of T_g for the reheated glass is normally obtained by the intersection of the two tangents at the start of the DSC endotherm (Fig. 13-21). The fictive temperature T_f, which is the intersection of the glassy state curve with the equilibrium liquid curve prior to reheating is obtained by the area matching technique described in Section 13.7.

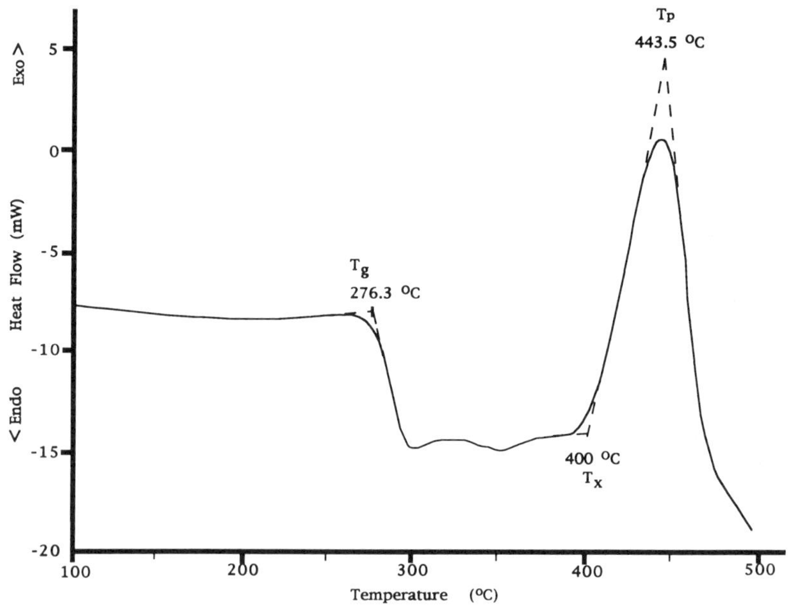

Figure 13-21. Typical DSC curve of a glass. T_g is the onset of glass transition. T_x is the onset of crystallization exotherm. T_p marks the peak crystallization rate.

13.6.4. Difficulties with Tool's Single T_f Concept

Ritland [17] obtained two groups of borosilicate glasses of indentical chemical composition having refractive index ($n_D = 1.51453$) at room temperature and pressure via two separate routes:

(i) Group A samples were soaked in a molten salt bath at 530°C for 24 h and then air-chilled to room temperature.
(ii) Group B samples were rate-cooled at 16°C/h through the glass transition range.

Identical refractive index implied having identical density or identical volume. (This relationship is called the Lorentz–Lorenz relation; see Eqs. (19.24) in Section 19.2.1.) Since P, V, and T of the two groups were identical, the T_f would have to be identical according to Tool. If it is assumed that the air-chilling of group A was "instantaneous," then $T_f = 530°C$ for both.

Ritland, then, placed both the groups in a furnace standing at 530°C. After heat treatment for increasing periods of time, the specimens were withdrawn and air-chilled. Thc room-temperature refractive index of each specimen was then measured. If the two groups were identical, they ought to have

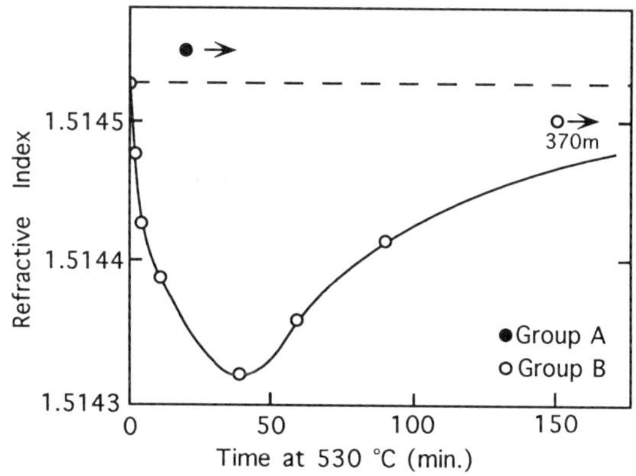

Figure 13-22. Room-temperature refractive index as function of treatment time at 530°C. (After Ritland [17]. Reproduced with permission of The American Ceramic Society.)

retained constant 1.51453 refractive index. Figure 13-22 shows that group A (the chilled specimens) maintained the refractive index; however, the rate-cooled specimens (group B) did not. The group B specimens showed an initial decrease and a slow recovery: they apparently had not approached the equilibrium value even after about 2.5 h in the furnace. Since the samples were small enough to allow the time to physically approach 530°C to be neglected, it had to be concluded that the two groups did not have an identical structure characteristic of a single fictive temperature, and that the entire history had to be included in some way.

More elaborate experiments to challenge Tool's single fictive temperature concept were carried out by Spinner and Napolitano [18] (the next generation at the NBS!). They stabilized a group of 1 cm × 1 cm × 0.1 cm borosilicate glass specimens by heating them at 585°C for a long time and subsequently chilling them. Smaller subgroups were then placed inside furnaces holding various temperatures lower than 585°C. By carefully withdrawing specimens over times up to several days, air-chilling and measuring the room-temperature refractive index n_D, they developed the "approach curves" for each temperature shown in Fig. 13-23. The equilibrium n_D could be plotted as a function of the stabilization temperature, which provided a calibration curve (Fig. 13-24).

In the next phase of experiments, they interrupted the approach of a subgroup to T_1 by switching it to a different furnace holding T_2 at "cross-over" points marked A through E on Fig. 13-23. The measured room temperature n_D of the specimens as a function of time (Fig. 13-25) showed a rapid reversal in the n_D trend, and a subsequent gradual turnaround to

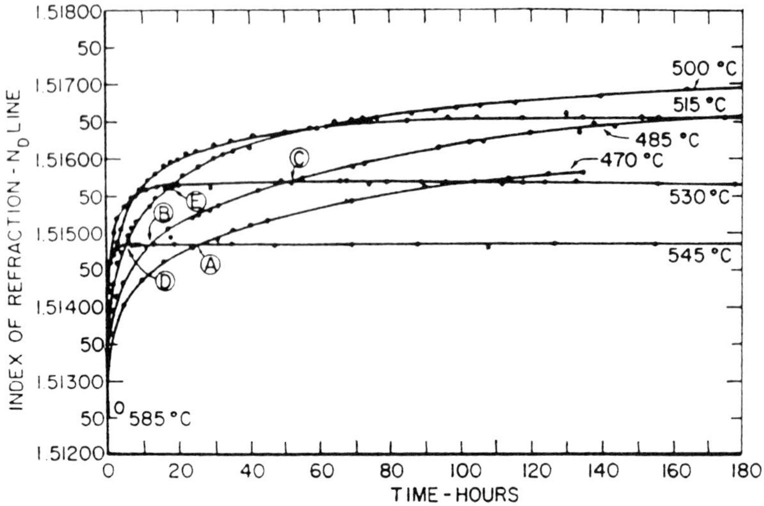

Figure 13-23. Room-temperature refractive index–time "approach curves" for BSC517. (After Spinner and Napolitano [18].)

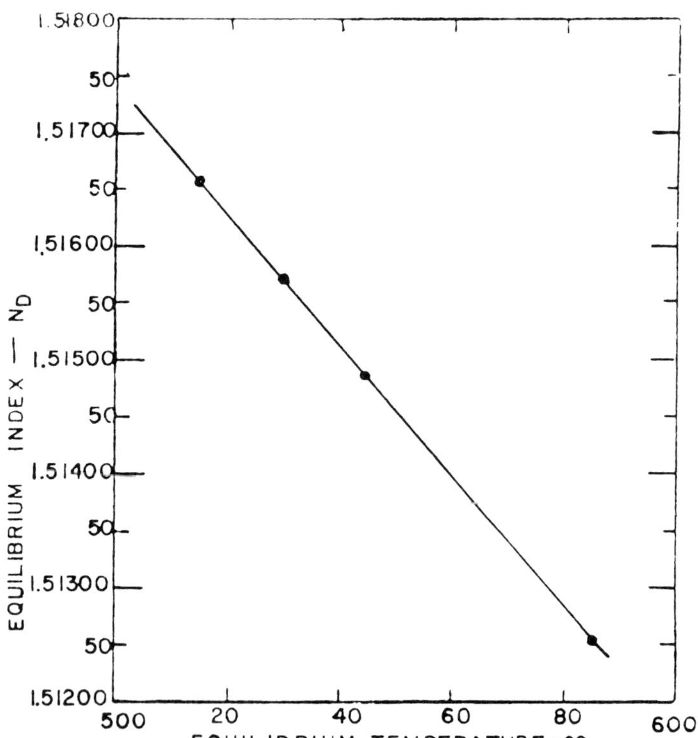

Figure 13-24. Equilibrium room-temperature index as a function of temperature for BSC517. (After Spinner and Napolitano [18].)

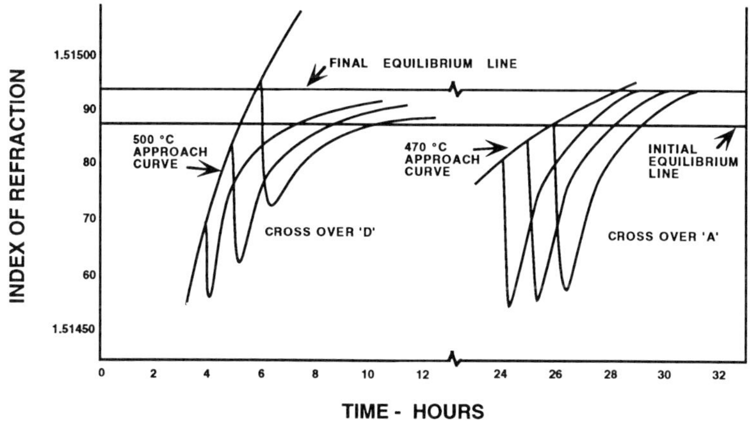

Figure 13-25. Refractive index vs. time at crossovers A and D. (After Spinner and Napolitano [18].)

the equilibrium n_D at T_2 as much as 24 h later for crossover C group (Fig. 13-26). Again, the distinct departure from constancy of n_D could be construed to imply that, in each case, the glass had a unique **thermal history** or a unique **fictive history** by cooling through the glass transition range. An equivalent statement would be to suggest that glass has a **memory** of its *last cooling* through the transition range.

The mathematical maneuver to compute the fictive history was provided by Narayanaswamy [19], who showed how each step of the structural

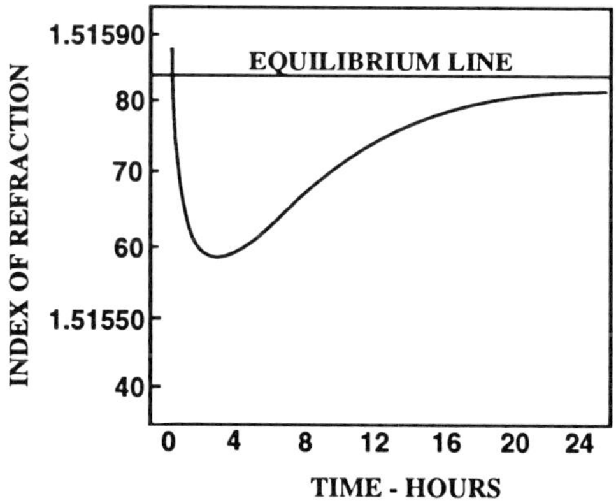

Figure 13-26. Refractive index vs. time at crossover C. (After Spinner and Napolitano [18].)

relaxation could be uncoupled from the previous step. For a simple Maxwellian relaxation function $\Psi(t) = \exp[-t/\tau]$, where τ is the relaxation time, Eq. (13.3) could be rewritten as

$$S_{12(t)} = 2G_0 \int_0^t \exp[-(t - t')/\tau] \frac{d\varepsilon_{12}}{dt'} dt'. \tag{13.73}$$

If $\tau(=\eta/G)$ is a constant, then the integration in (13.73) can be carried out readily for a variety of the strain history. Difficulty arises when τ itself varies with time, particularly because η changes. If η is a function of the temperature alone (single exponential dependence), then the integration is possible using the technique (described later) of Lee *et al.* [20]. Trouble further arises in structural relaxation experiments, where η changes with fictive temperature even at constant temperature. In essence, each incremental step in (13.2) depends upon the previous step. In other words, the process of structural relaxation is intrinsically nonlinear. In describing the fictive temperature concept, Tool had recognized the extent of nonlinearity of the structural relaxation process by suggesting Eq. (13.68), which, since $B = dT/dt$, may be rewritten as

$$\frac{dT_f}{dt} \cong K'_T[T - T_f], \tag{13.74}$$

where K'_T = a factor "inversely proportional to the viscosity." Note that this statement by Tool is approximately correct since from Eq. (9.10), $\eta^{-1} \sim \exp[-\Delta H/RT] \sim \exp[T/g]$ over a limited temperature range. Hence,

$$\frac{dT_f}{\mathrm{dt}} = K''[T - T_f]/\eta, \tag{13.75}$$

which is actually a simple first-order partial differential linear equation. Thus, according to Tool, structural relaxation is "linear" with respect to the driving force ($T - T_f$), with the only "nonlinearity" coming from the fictive temperature dependence of viscosity included as the $\exp(T_f/h)$ term in Eq. (13.68). The crossover experiments have indicated that two glass specimens brought to the same location on the V–T diagram through two different routes relax differently, and hence the relaxation function must have at least two different exponentials. Likewise, viscoelastic relaxation of stress at constant strain experiments by DeBast and Gilard for a soda lime silica glass have shown that, instead of a single Maxwellian relaxation, the uniaxial stress relaxation function is more appropriately given by

$$\Psi_u(t) = \exp[-(t/\tau)^b], \tag{13.76}$$

where $b = 0.5435$. It may be shown that an exponent $b < 1$ approximates a distribution of relaxation times:

$$\Psi_u(t) = \sum_k w_k \exp[-(t/\tau_k)], \tag{13.77}$$

where w_k are the weights associated with each relaxation process. Equation (13.76) with the exponent b $(0 < b < 1)$ is called the **Kohlrausch–Williams–Watts** or **KWW function**, or simply a **stretched exponential**. One may readily imagine that, in structural relaxation experiments, two samples being produced differently would attain a different fictive history at every instant. Tool's approach did not correctly account for the buildup of the "fictive history."

13.6.5. Narayanaswamy's Model of Structural Relaxation

Narayanaswamy retained three basic features of the Tool concept:

(i) The property coefficient in the liquid state α_{pe} is a sum of the contributions α_{pg} from the glassy state (due to the change in physical temperature but with fixed structure) and α_{ps} due to change of structure.
(ii) The structural relaxation is intrinsically linear except for the nonconstancy of the relaxation time arising from the fictive temperature dependence of the viscosity.
(iii) The response function for a property p to a step change in temperature from T_1 to T_2 could be written as

$$M_p^*(t, \Delta T) = \frac{(p - p_{2,\infty})}{(p_{2,0} - p_{2,\infty})} = \frac{(T_{fp} - T_2)}{(T_1 - T_2)} \tag{13.78}$$

The subscripts 0 and ∞ denote the instantaneous and longtime values of p. Thus, $p_{2,0}$ corresponds to the property at the initial temperature T_1, and $p_{2,\infty}$ to that at T_2. M_p^* is normalized to yield 1 at time $t = 0$ and 0 at $t = \infty$. The nonlinearity of M on the magnitude and sign of ΔT is eliminated by taking the limit to $\Delta T \to 0$, i.e.,

$$M_p(t) = M_p^*(t, \Delta T)_{\Delta T \to 0}. \tag{13.79}$$

The mathematical maneuver required to allow an integration of relaxation occurring at different times is the assumption of **thermorheological simplicity (TRS)**, advanced by Lee, Rogers and Woo. By assuming TRS, one assumes that *the functional form of the relaxation does not change with temperature.* Narayanaswamy assumed that TRS could be extended to include variation with T_f. In other words, b in Eq. (13.76) is independent of temperature. (The constancy of b was demonstrated by DeBast and Gilard.) The physical time t is compressed or expanded to ξ, called the **reduced time** such that the extent of relaxation for a time dt at temperature T having relaxation time constant τ is equivalent to relaxation for time $d\xi$ at temperature T_s having a relaxation time constant τ_s. The total reduced time ξ is given by

$$\xi = \int_0^t \phi(T, T_{fp})dt' \tag{13.80}$$

where $\phi(T, T_{fp})$ is given by the shift function,

$$\ln \phi(T, T_{fp}) = \ln(\eta_B/\eta) = \frac{H_g}{R}\left\{\frac{1}{T_B} - \frac{1}{T}\right\} + \frac{H_s}{R}\left\{\frac{1}{T_B} - \frac{1}{T_{fp}}\right\}, \tag{13.81}$$

and the total activation energy

$$H = H_g + H_s. \tag{13.82}$$

(Note: The viscosities are reversed in the middle equality of Eq. (9) in Narayanaswamy's paper.)

Equation (13.81) implies the use of a slightly different functional form of the T_f-dependence of η than that used by Tool. Schematics of changing the time scale and the applicability of the TRS are shown in Fig. 13-27. The response function may now be written as $M_p(\xi)$. The change in property from a temperature T_0 at time $t = 0$ to temperature T at time t is thus given by

$$p(t) - p(0) = p(0)\alpha_{pe}(T - T_0) - p(0)\alpha_{ps} \int_0^\xi M_p(\xi - \xi')\frac{dT}{d\xi'}d\xi'. \tag{13.83}$$

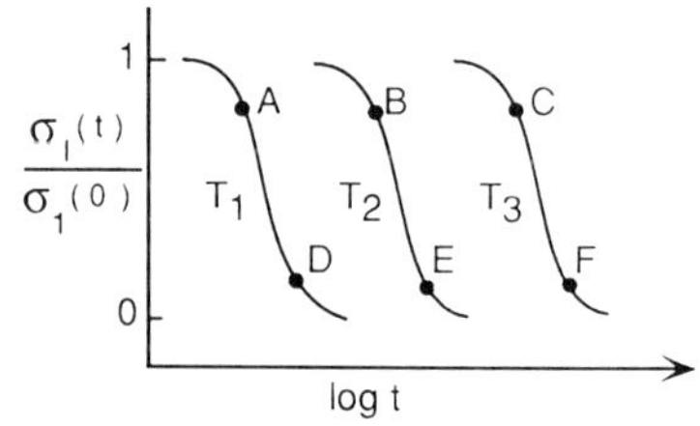

Figure 13-27. Stress relaxation curves at various temperatures ($T_1 > T_2 > T_3$) plotted against log(time). For thermorheologically simple material, distances AB = DE, BC = EF. (After G. Scherer, *Relaxation in Glass and Composites*, Fig. 5.2, p. 53. J. Wiley & Sons, New York, 1986. Reproduced with permission of the publishers.)

Equation (13.83) is now a linear equation where the form of the response function no longer depends upon the temperature (only the magnitude may). The first term gives the total change in p after complete relaxation, and the second term gives the part of p that is yet to be relaxed up to reduced time ξ (or physical time t). The "fictive history," hence, can be written as

$$T_{fp} = T + \int_{\xi(T)}^{\xi(T_0)} M_p(\xi - \xi') \frac{dT}{d\xi'} d\xi'. \tag{13.84}$$

The fictive temperature T_{fp} is thus obtained by adding to the physical temperature T a part that is "retained" or "remembered" by the structure. In the liquid state, the relaxation is essentially instantaneous; hence, $M_p = 0$. This reduces $T_{fp} = T$. On the other hand, if the glass were quenched at infinite rate from the temperature T_0, the response function would be $M_p = 1$ for all times ξ. From (13.84), this gives $T_{fp} = T + (T_0 - T) = T_0$.

13.6.6. Description of the Crossover Experiment in Terms of Narayanaswamy's Model for Relaxation

The first step in the kinetic description of the crossover experiment was to show the applicability of the principle of TRS. This was accomplished by computing the reduced time for each "approach curve" described by Spinner and Napolitano. The response modulus is obtained by the refractive index vs. time data and using the definition of M from Eq. (13.78). The corresponding reduced time was then computed by (13.80)–(13.82). From the experimental data, the reference temperature T_B was 800 K for a strain-point viscosity of $10^{14.5}$ poise. Narayanaswamy showed that if $H_g = 84$ kcal/mol and $H_s = 74$ kcal/mol, then the calculated M vs. ξ curves merged into one another quite well (Fig. 13-28), confirming the argument that the relaxation function did not vary with temperature (TRS). A numerical fit of the merged data to the expression

$$M_n(\xi) = a_1 \exp(-\xi/\tau_1) + a_2 \exp(-\xi/\tau_2), \tag{13.85}$$

where $a_1 + a_2 = 1$, yields

$$a_1 = 0.163, \qquad a_2 = 0.837, \qquad \tau_1 = 1.11 \text{ h}, \qquad \text{and} \qquad \tau_2 = 9.09 \text{ h}.$$

The analysis supports the notion that the response function need not be single relaxation-time-controlled for the applicability of the TRS.

Let us consider the crossover experiment "A" shown in Fig. 13-21. In the first part of the experiment, specimens equilibrated at 858 K are placed in a furnace standing at 743 K. Since $n(858) = 1.51250$ and $n(743) = 1.51920$, Eq.

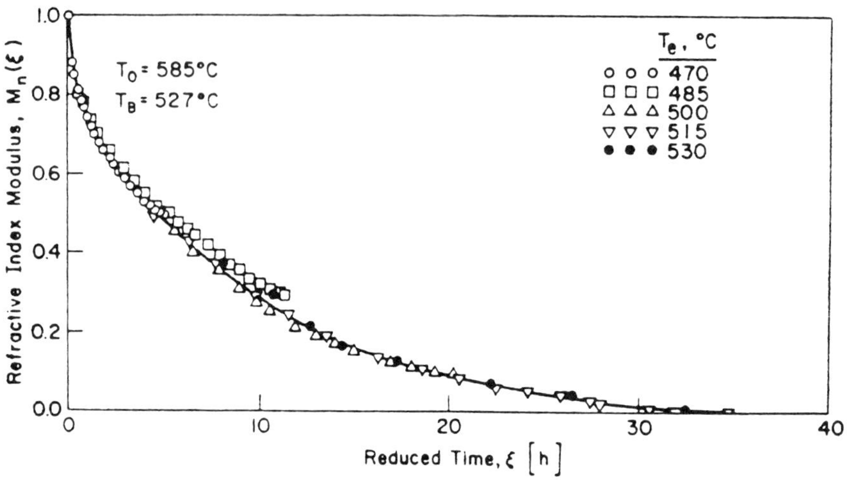

Figure 13-28. Master relaxation curve for BSC517. (After Narayanaswamy [19]. Reproduced with permission of The American Ceramic Society.)

(13.78) gives

$$M_n(\xi) = \frac{(n - 1.51920)}{(1.51250 - 1.51920)} = \frac{(T_{fn} - 743)}{(858 - 743)}. \tag{13.86}$$

(Note: Eq. (15) in Narayanaswamy's paper should read 858 instead of 800.) The second part of (13.86) yields

$$T_{fn} = 115M_n(\xi) + 743. \tag{13.87}$$

Values of n are then read off from Fig. 13-23 as a function of the real time. Table 13-3 shows the computed values of n, M_n, T_{fn}, ϕ, and ξ as a function of the real time t. (A polynomial fit was made to the experimental n–t curve to enable tabulation for the various time values.) The total reduced time at the end of 25 h of real time (at the instant of crossover) is computed to be 2.18 h. The meaning of our computation so far is that 25 h of relaxation at 743 K is equivalent to 2.18 h of relaxation at 800 K.

The specimen is now switched into a furnace holding 818 K. After the crossover, the evolution of the fictive temperature as a function of the reduced time ξ measured since the crossover is given by

$$T_{fn} = [818 - 75M_n(\xi)] + 115M_n(\xi + 2.18). \tag{13.88}$$

Again, we may rationalize the preceding relationship by indicating that at very long times, both the relaxation functions approach zero, i.e., $T_{fn} \to$ 818 K. The second term on the right is the continuation of relaxation from the first step; Eq. (13.88) reduces to Eq. (13.87) for $\xi = 0$, and the first term

Table 13-3. Computation[a] of Reduced Time Buildup for 470°C Approach Curve in Spinner and napolitano Experiment

t (h)	n	M_n	T_{fn} (K)	$\ln(\phi)$	ϕ	$d\xi = \phi dt$
0.00	1.51250	1.000	858	−0.907	0.403	—
0.5	1.51276	0.961	853.6	−1.133	0.322	0.181
1.0	1.51302	0.922	849.1	−1.360	0.257	0.145
1.5	1.51328	0.883	844.7	−1.590	0.204	0.115
2.0	1.51354	0.844	840.3	−1.823	0.162	0.091
2.5	1.51379	0.807	835.9	−2.058	0.128	0.072
5.0	1.51403	0.772	831.8	−2.279	0.102	0.287
7.5	1.51424	0.740	828.1	−2.474	0.084	0.232
10.0	1.51438	0.719	825.7	−2.605	0.074	0.198
12.5	1.51451	0.700	823.5	−2.727	0.065	0.174
15.0	1.41460	0.687	822.0	−2.812	0.060	0.156
17.5	1.51467	0.676	820.8	−2.878	0.056	0.145
20.0	1.51474	0.666	819.6	−2.945	0.053	0.136
22.5	1.51479	0.658	818.7	−2.993	0.050	0.129
25.0	1.51485	0.649	817.7	−2.993	0.047	0.121
					Total	2.182 h

[a] The last column is computed by averaging the ϕ value with the preceding and multiplying by the time step. Some errors due to coarse time step selection are likely.

is the effect of a 75 K jump applied in the physical temperature. Substitution of the M_n description (13.78) along with values (13.85) into (13.88) yields the expression for the evolution of T_{fn} after crossover:

$$T_{fn} = 818 - 9.6 \exp[-\xi/1.11] + 12.9 \exp[-\xi/9.09], \qquad (13.89)$$

which, by differentiation with respect to ξ, may be shown to have a maximum at $\xi \cong 2.3$ h. In fact, after conversion of ξ at 800 K to real time t at 818 K (which is being left to the reader as an exercise), it may be shown that the quantitative agreement of the crossover behavior is excellent with that observed experimentally by Spinner and Napolitano (Fig. 13-25).

It may be realized that if M_n were Maxwellian, i.e., *if* M_n *had a single relaxation time, there would be no crossover phenomena.* In the case of a single structural relaxation time, Narayanaswamy's approach reduces to Tool's approach.

13.7. Properties Data in the T_g Range

The observed properties behavior in the T_g range conforms to the following generalities (Gupta and Huang [21]):

(i) **Non-exponential**: A description of the relaxation requires at least two relaxation times. A stretched exponential function is often useful.
(ii) **Nonlinear**: Each relaxation time depends upon the fictive temperature.
(iii) **Near thermorheologically simple**: For small excursions in the temperature (or pressure), the magnitude of the relaxation as a function of the reduced time can be expressed as a master curve.
(iv) **Property type**: It is possible for the fictive parameter (temperature or pressure) to be different for different properties.

Kurkjian [22] has measured the instantaneous shear modulus (at 15 MHz) of a soda-lime silica glass (for which $\eta = 2.5 \times 10^{13}$ poise at 520°C) from room temperature to about 550°C. The shear modulus decreased from about 3.05×10^{11} dynes/cm^2 to only about 2.68×10^{11} dynes/cm^2. The shear stress relaxation function over this temperature range could be fitted to a master curve (Fig. 13-29) and found to be adequately described by a sum of six exponentials where the longest relaxation time was about 2×10^4 times the shortest. The linear sum of the six exponentials may be shown roughly equivalent to a KWW function (Eq. (13.76)) with $b \approx 0.5$. Volume relaxation function is also found to fit a KWW form, but with $b = 0.68$.

The heat capacity increases rapidly with temperature at the onset of the glass transition, in general agreement with the kinetic and thermodynamic concepts. Since the expected behavior of the enthalpy upon cooling and heating is as shown in Fig. 13-30a, the heat capacity during cooling and heating varies as shown in Fig. 13-30b. A marked hysteresis between the cooling and the heating curve is observed. According to Moynihan *et al.* [23], upon cooling a glass to a temperature T_1 well below the transition range,

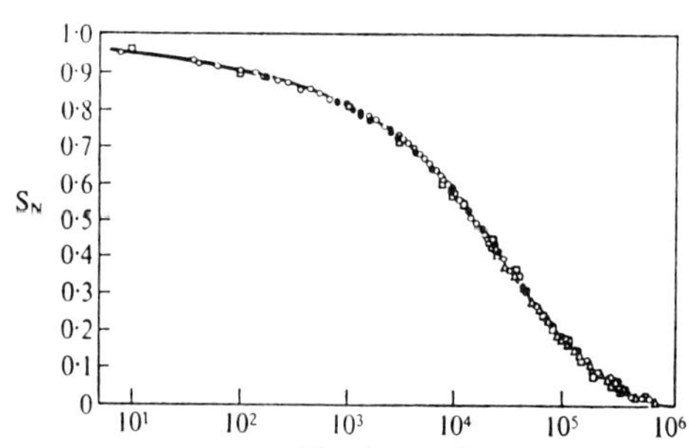

Figure 13-29. Master curve for normalized stress relaxation at 473°C produced from data at various stresses at various experimental temperatures. (After C. R. Kurkjian, *Phys. Chem. Glasses* **4**[6], 128 (1963). Reproduced with permission of The Society of Glass Technology.)

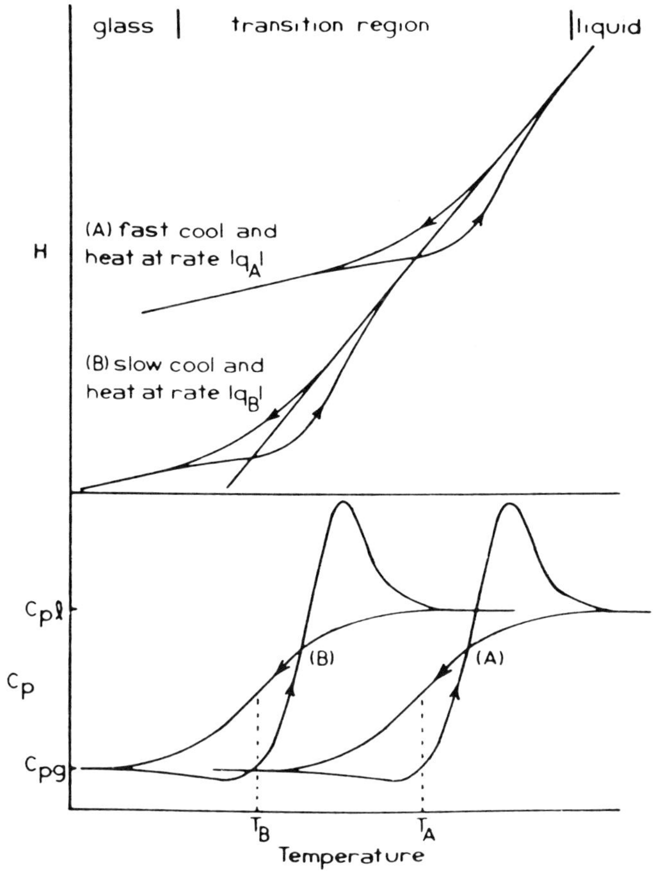

Figure 13-30. Enthalpy and heat capacity vs. temperature for liquid cooled and reheated through the glass transition range at rates q_A and q_B. (After Moynihan *et al.* [23b]. Reproduced with permission of The American Chemical Society.)

its fictive temperature reaches a low limit, identified as T_f, depending upon the cooling rate. Upon reheating, the fictive temperature would rise from this low limit. T_f can be calculated from the equation

$$\int_{T_0}^{T_f} (C_{pe} - C_{pg}) dT_f = \int_{T_0}^{T_1} (C_p - C_{pg}) dT. \tag{13.90}$$

The T_f upon cooling can be graphically computed from the reheating data by equating the areas under the curves bounded by the glass and the equilibrium liquid heat capacity data (Fig. 13-31). By computing the T_f for

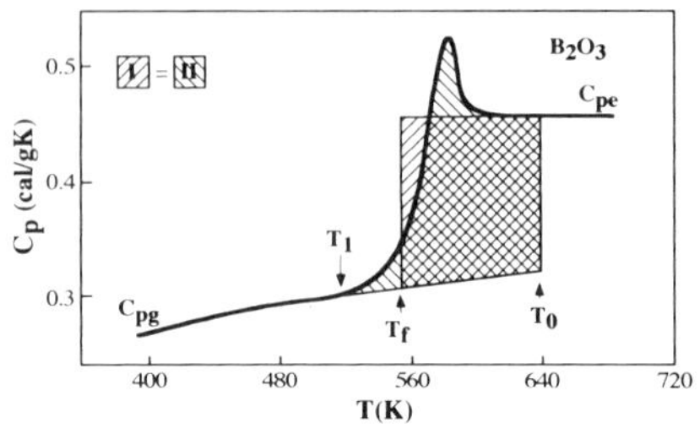

Figure 13-31. Heat capacity vs. temperature for B_2O_3 glass determined at 10 K/min heating rate following cooling at −10 K/min. The fictive temperature is found by equating areas I and II. (After Moynihan *et al.* [23a]. Reproduced with permission of The American Ceramic Society.)

many cooling rates B ($\equiv dT/dt$), Moynihan *et al.* [23] suggested that

$$\frac{d \ln|B|}{d(1/T_f)} = -\frac{\Delta H}{R}, \tag{13.91}$$

where ΔH is the activation enthalpy for shear viscosity assuming an Arrhenius relationship.

The magnitude of ΔC_p observed in several glass-forming liquids apparently correlates well with the classification of **strong** and **fragile** liquids (Section 9.5). Comparison of Fig. 13-32 with Fig. 9-14 reveals that the strong liquids have minimal ΔC_p, whereas the fragile liquids have large ΔC_p. A large number of available configurations for the relatively unconstrained structures of the fragile liquids causes the configurational entropy to change rapidly in the vicinity of T_g, giving rise to a high activation energy (Adam–Gibbs, Eq. (13.58)).

The thermal expansion coefficient of several glasses has been measured from room temperature to beyond the glass transition range. One of the earliest measurements was by Haggerty and Cooper [24] on three silicate glasses. It was found that the thermal expansion coefficient also increased rapidly with temperature at the onset of the glass transition (again, in agreement with thermodynamics). The expansion coefficient in the supercooled liquid region was about three to seven times that in the glassy state.

With caution, the concept of a single fictive temperature may be used to predict the behavior of properties in the T_g range, particularly for glasses having widely varying thermal histories. A fast-cooled glass has a higher fictive temperature, and hence a higher volume, a more open structure, and all the properties that are consistent with a higher temperature or more open structure in comparison to a slow-cooled glass. Thus, *a fast-cooled glass*

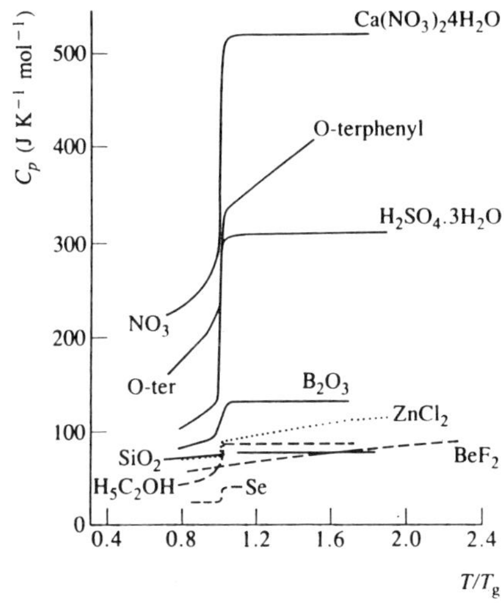

Figure 13-32. Variation of heat capacity as a function of normalized temperature for "strong" and "fragile" liquids. (After S. A. Brawer, *Relaxation in Viscous Liquids and Glasses*, The American Ceramic Society, Columbus, Ohio, 1985. Reproduced with permission of The Publisher.)

shows a lower refractive index, a lower viscosity, and a higher mobility of atoms/ions through the structure initially than does a slow-cooled glass. All the property measurements made to date support these notions. An example of the electrical resistivity behavior of a soda–potash–lead oxide silicate glass is shown [26] in Fig. 13-33, where the chilled glass shows the lowest resistivity, and the well-equilibrated glass has the highest. If one attempts to measure the viscosity of glass in the transition range, experimental data should only be taken when the glass has stabilized. Since the unstabilized glass would most likely have a fictive temperature higher than the physical temperature, glass would asymptotically approach an increased viscosity. At sufficiently low temperatures, the rate of stabilization may be so slow that only an apparent viscosity (which is lower than the real viscosity) is measured even after months (years!) of waiting. (Waiting for a glass to achieve an equilibrium viscosity higher than about 10^{18} Pa·s is not very practical; see Douglas [25].)

The rate of approach to equilibrium clearly depends upon the viscosity of the initial structure. Thus, the densities of two specimens of soda-lime silica glass that have been stabilized previously at 560°C and 500°C approach the equilibrium value at different rates after being introduced into a 530°C furnace (Fig. 13-34, by Hara and Suetoshi [27]). The 560°C stabilized specimen, having a more open, and hence a lower-viscosity, structure,

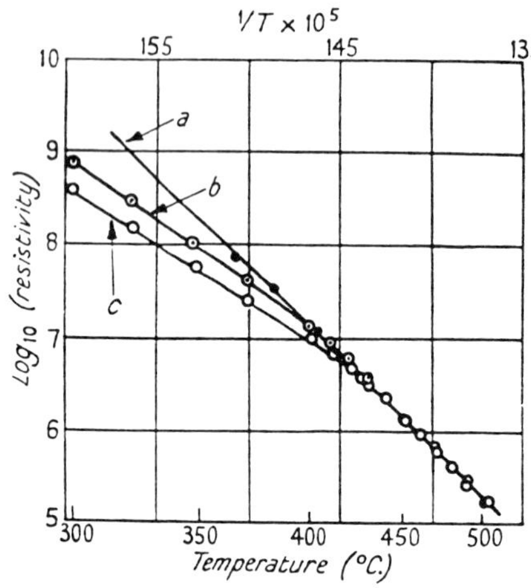

Figure 13-33. Electrical conductivity as a function of temperature of (a) well-equilibrated, (b) normally annealed, and (c) chilled glasses. (After Littleton and Wetmore [26]. Reproduced with permission of The American Ceramic Society.)

accumulates a greater magnitude of reduced time. It therefore relaxes to the 530°C structure at a faster rate.

Caution is advised in two primary areas: description of far-from-equilibrium relaxation, and carrying over the relaxation function measured using one property to describe another. Whereas much of the relaxation data for small excursions in temperature or pressure are consistent with the assumption of the TRS, Gupta and Huang found that TRS could not describe the enthalpy relaxation in rapidly quenched glass fibers. Likewise, Moynihan *et al.* [23] have suggested that the structural relaxation function may be property-type-

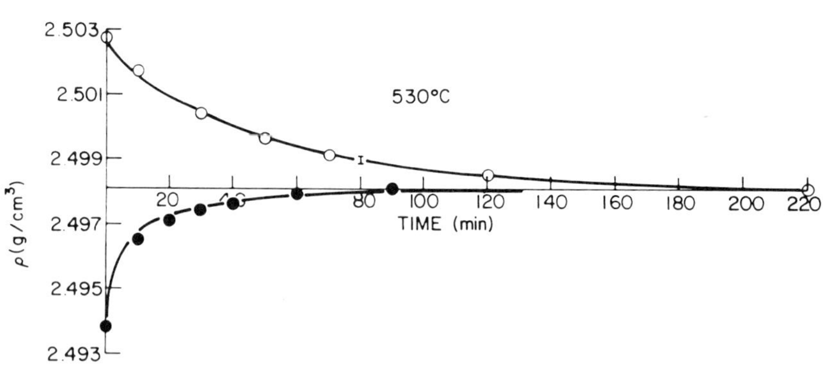

Figure 13-34. Relaxation of density with time at 530°C for soda-lime silica glass previously equilibrated at 560°C (●) and 500°C (○). (Data of Hara and Suetoshi [27].)

sensitive. For instance, the fictive temperature for volume relaxation may be different from that for enthalpy relaxation.

13.8. Development of Permanent Stresses in Glass by Cooling through the Glass Transition Range

Suggested reading: R. Gardon and O. S. Narayanaswamy, *J. Am. Ceram. Soc.* **53**(7), 380 (1970).
R. Gardon, *Glass Sci. & Tech.*, Vol. 5 (D. R. Uhlmann and N. J. Kreidl, eds.), pp 146–217. Academic Press, New York, 1980.
O. S. Narayanaswamy, *J. Am. Ceram. Soc.* **61**(3–4), 146–152 (1978).

When a glass is cooled through the glass transition range, permanent stresses develop in different regions. In the later chapter on the strength of glass, it will be emphasized that glass invariably breaks in tension. Hence, those regions of glass that develop permanent compression upon cooling are strengthened, and those where tension develops are weakened. Since compression and tension go hand in hand, the forming technology of glass products for low scrap rate during process and a safe product in service demands attention in relation to the magnitude and location of the developed stresses in glass.

If glass were perfectly elastic, cooling from the outside would generate tension on the outside and compression on the inside (Section 10.6). In particular, slow or natural cooling generates a parabolic temperature gradient that results in a parabolic stress gradient. Upon the elimination of the temperature gradient at room temperature, all stresses would also disappear. If glass were fluid, any stresses that developed would be relaxed rapidly even in the presence of temperature gradients. It is in the glass transition region that structural rearrangement occurring at different rates at different locations leads to incomplete stress relaxation, and hence to the appearance of permanent stress.

Qualitatively, there are two dominant contributions to the permanent stress development: the viscoelastic contribution and the structural contribution. The latter is further classified depending upon which of the transient or permanent structural heterogeneities cause it. These are discussed next.

THE VISCOELASTIC CONTRIBUTION. This mechanism is schematically shown for a glass plate in Fig. 13-35. In (a), the glass plate has a high uniform temperature, and hence no stress at all. Upon cooling, a temperature gradient develops, cooler on the outside and warmer on the inside, as shown in (b). In a perfectly elastic body, such temperature gradient would cause the

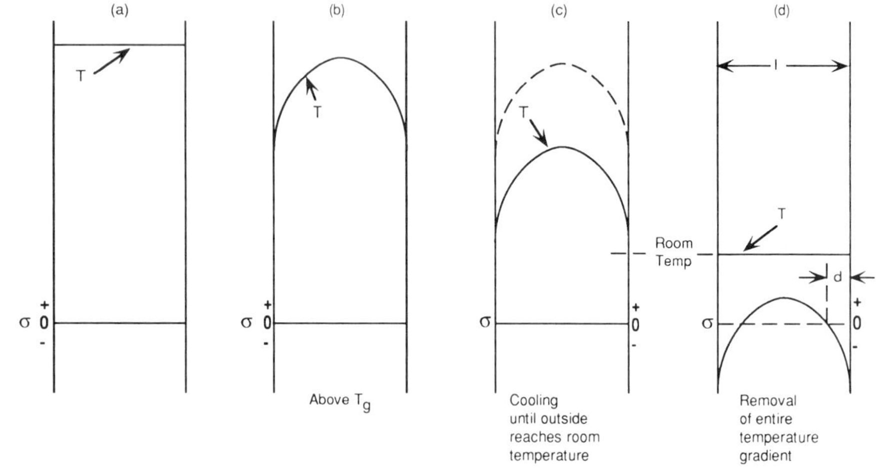

Figure 13-35. Simplified concept of permanent stress production in glass due to a "frozen temperature gradient." (a) Glass with no temperature gradient well above the transition region has no stress. (b) Temperature gradient develops upon cooling; however, no stress develops because of rapid relaxation while the glass remains well above T_g. (c) Cooling to below T_g while maintaining the same "frozen temperature gradient" produces no stress yet. (d) Final removal of the temperature gradient produces stresses that are now permanent.

development of tension on the outside and compression on the inside. However, since glass is still well above T_g into the supercooled liquid region, these stresses relax rapidly. In (c), the plate has been brought down in temperature with the same "frozen" temperature gradient — however, in such a manner that the outside has reached room temperature. If the effective thermal contraction coefficient of the body is assumed to be independent of temperature, then since each layer of glass cooled through exactly the same temperature interval, the various layers of the plate in situation (c) still do not have any dimensional incompatibility. Hence, the plate does not develop any stresses. In the next step, we superimpose a reverse temperature gradient (as in (d)) to make the temperature uniform (room temperature) everywhere. Such a gradient where the outside is warmer than the inside would cause the generation of a stress gradient with compression on the outside and tension on the inside. We may recognize that the final stress distribution is the negative of the stress that relaxed viscoelastically. It really does not matter how many steps we take to arrive at the final state. We are allowed to perform these idealized operations conceptually, because of linearity and the Boltzmann superposition principle.

STRUCTURAL CONTRIBUTION DUE TO TRANSIENT HETEROGENEITY. In the preceding discussion, we ignored the rather large volume change a glass-forming liquid undergoes during and above the glass transition range. For simplicity, the structure, and hence the volume, may be characterized by the fictive temperature, T_f, which is essentially the physical temperature in the liquid range and which lags gradually in the transition range until "frozen" at some value T_{fr} in the "apparently solid" state. When a glass with a temperature gradient is being cooled uniformly through the transition range, the various layers also possess a gradient in T_f, i.e., a structural heterogeneity, which vanishes rapidly as the glass approaches T_{fr} just below the transition range. We noted earlier that the removal of the physical temperature gradient ΔT in the solid state brings about permanent stresses. Likewise, we may expect the development of permanent stresses due to the removal of the fictive temperature gradient. There is one difference, however, which is best understood with the help of Fig. 13-36. The figure shows the temperature, time, and stress development interrelationship. Since the fictive temperature gradient diminishes through the transition range, *stresses begin to appear while the glass is in the transition range*. The physical temperature gradient ΔT, on the other hand, does not usually begin to diminish until well below the transition range. Hence, the viscoelastic contribution does not appear until

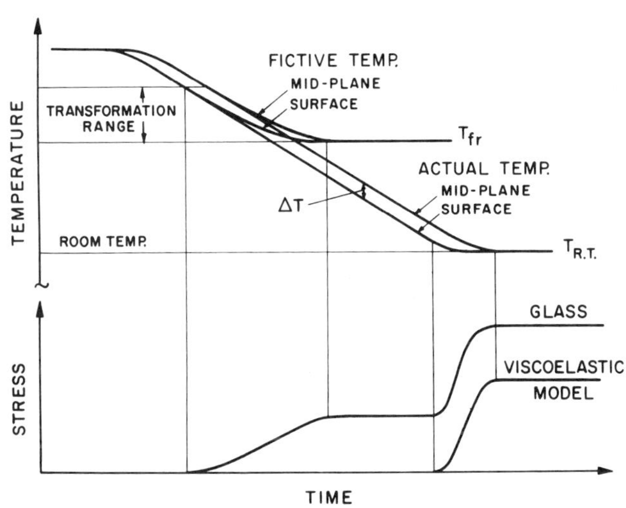

Figure 13-36. Temperature–time relationship for the generation of permanent stresses in glass during uniform cooling. (After Gardon and Narayanaswamy [29]. Reproduced with permission of The American Ceramic Society.)

the glass cools sufficiently. It may also be argued that, whereas much of the physical temperature gradient will ultimately yield stresses, many of the stresses due to the T_f gradient may themselves be released viscoelastically. (Generally, about a third survives.)

The key idea in both cases is the recognition that *the various layers of glass pass through the transition range at different instants of time.*

STRUCTURAL CONTRIBUTION FROM PERMANENT STRUCTURAL HETEROGENEITY. Again, this mechanism depends upon the different fictive temperatures of the various layers, but now this variation is due to the nonuniform cooling of the outside relative to the inside. A glass that is being rapidly cooled is likely to cool at a faster rate on the outside than on the inside, simply because of the low thermal diffusivity of glass. Now, consider the effect of the cooling rate in the V–T diagram, Fig. 13-1. If the cooling starts from a temperature well above the transition range, the outside layers of glass would pass along higher T_f (hence, larger volume) curves than those on the inside, a structural heterogeneity that would exist if the various layers of glass were torn apart. Hence, if the glass is to stay unfractured at room temperature, the elastic compatibility criteria will force the outside to develop compression and the inside to develop tension.

The difference between the two primary mechanisms may also be realized by examining the nature of a typical V–T diagram. A glass with a zero expansion coefficient in the glassy state may be expected to develop stresses due to the structural heterogeneities, transient or permanent, as long as it has a nonzero expansion in the liquid state (which is usually the case). Glass with a nonzero thermal expansion coefficient in the glassy state, which has been cooled from just above the transition range, does not develop a significant level of permanent structural heterogeneity. Most of the stresses in such a case are due to the viscoelastic contribution and the transient structural heterogeneity mechanism. Because the temperature gradients are small, the overall stresses that are generated are also small. For a common soda-lime silica flat glass cooled in this manner, the transient structural heterogeneity mechanism accounts for $\sim 40\%$ of the total stress developed. On the other hand, when this glass is quenched from well above the transition range, large temperature gradients, both physical and fictive, may develop. The magnitude of stresses developing from all three classes is very high. However, the viscoelastic contribution alone may account for 80–85% of the total.

Theoretical calculations of permanent stresses in glass may be carried out using the analysis described by Narayanaswamy [28]. For a soda-lime silica glass plate being cooled from both sides, he assumed a two-dimensional state of stress described by (Fig. 13-37)

$$\sigma(x, t) = \sigma_y(x) = \sigma_z(x), \qquad \sigma_x(x) = 0,$$

and

$$e(x) = e_y(x) = e_z(x), \qquad e_x(x) \neq 0.$$

After computing the physical temperature history $T(x, t)$ using the well-established solutions to heat-transfer problems, the fictive temperature history $T_f(x, t)$ is computed using Eq. (13.82) and the known strain point of glass as the reference temperature for reduced time calculations. The local, incompatible thermal strain is given by

$$e_{th}(x, t) = \alpha_g[T(x, t) - T_0] + \alpha_s[T_f(x, t) - T_0], \tag{13.92}$$

where α_g and α_S are the linear thermal expansion coefficients for the glass and the structure respectively, as indicated for a general property, α_p, in 13.6.5. The compatibility criteria dictate that the strain be independent of x, i.e.,

$$e(x, t) = e(t).$$

Hence, the difference $[e_{th}(x, t) - e(t)]$ generates stress in the individual layers of glass according to

$$\sigma_g(x, t) = [E/(1 - \nu)][e_{th}(x, t) - e(t)]. \tag{13.93}$$

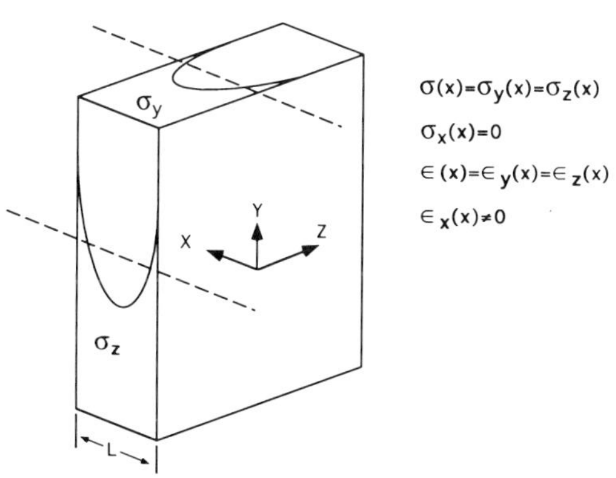

Figure 13-37. Thermal stresses in a semi-infinite glass plate.

However, these stresses also relax according to the viscoelastic law (Lee *et al.* (20)):

$$\sigma(x, t) = \int_0^t R[\xi(t) - \xi(t')] \frac{d\sigma_g(x, t)}{dt'} dt', \tag{13.94}$$

where R is a time-dependent auxiliary elastic modulus for two dimensions derived from $G(t)$. The stresses are subject to the condition that

$$\int_0^{L/2} \sigma(x, t)dx = 0.$$

A numerical solution of the preceding proceeds step by step from the starting temperature where both the initial strain and stress are zero. Note that in a purely viscoelastic model of calculating stresses, the right-hand term in Eq. (13.81) is not present, and the activation energy H for the temperature dependence of viscosity is not split into H_g (viscoelastic portion) and H_S (structural portion) in Eq. (13.82). Figure 13-38 shows a comparison of the measured stress profiles at various times with those calculated using the "structural model" having both the viscoelastic and structural contributions, and the "viscoelastic model" having only the viscoelastic contribution. In the latter calculation, an effective thermal contraction coefficient has been assumed to match the final stress magnitudes. As may be noted, the measured *transient* stress profiles are not even in qualitative agreement with the "viscoelastic model" and agree only with the "structural model."

13.9. Concepts of Annealing and Tempering

In glass technology, the word "annealing" implies removal of internal stresses. Generally, the source of stresses is the nonuniform cooling through the glass transition range. "Tempering" (or "toughening") is used to imply strengthening of glass. In particular, thermal tempering means strengthening by buildup of surface compressive stresses upon rapid cooling. Unless the V–T diagram of a glass has an unusual form (like that of fused silica), one expects to have some degree of temper produced in the glass by ordinary cooling. For an initial parabolic temperature distribution, with ΔT being the difference between the outside and the inside temperatures, the outside compression is proportional to $2\Delta T/3$ and the inside tension is proportional to $\Delta T/3$ (see Section 10.6 and Fig. 13-39). The neutral line is about one-fifth of the glass thickness below the surface. The magnitude of ΔT increases with the rate of cooling as well as with the glass thickness.

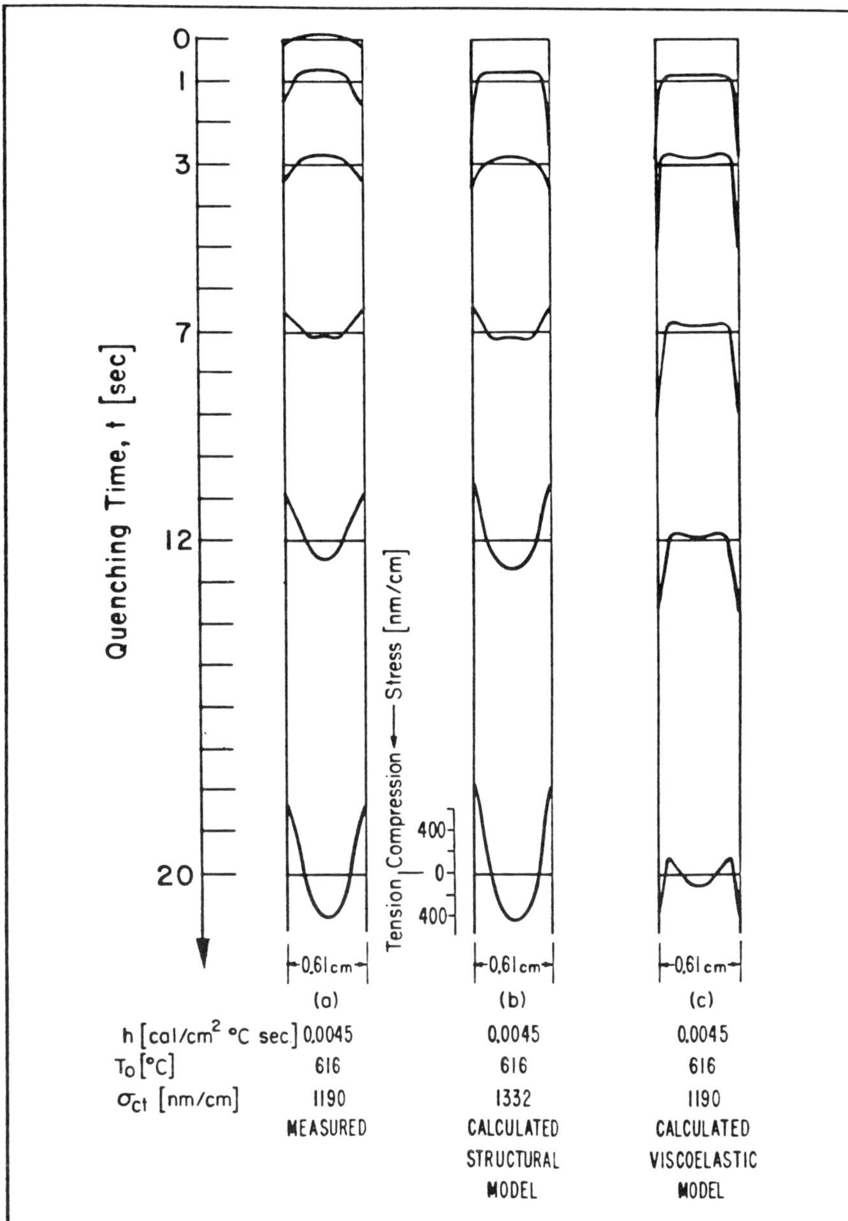

Figure 13-38. Comparison of measured transient stress distribution (a) with those predicted by (b) a model that has structural and viscoelastic mechanisms, and (c) a model with the viscoelastic mechanism alone. (After O. S. Narayanaswamy, *J. Am. Ceram. Soc.* **61**(3–4), 146–152 (1978). Reproduced with permission of The American Ceramic Society.)

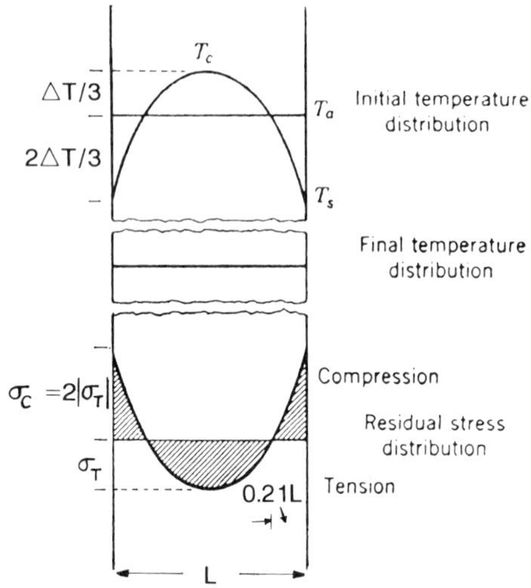

Figure 13-39. A parabolic temperature gradient produces surface compression which is twice the interior tension in magnitude.

Heat transfer calculations reveal that a surface compression of about 5 MPa develops in a soda-lime silicate glass plate 3 mm thick by normal cooling (rates 1–10°C/min) from slightly above the glass transition range (see Fig. 13-40). With increased cooling rate (often *assisted by forced-air cooling techniques*), it is possible to raise this compression to 100 MPa without too much difficulty (Fig. 13-41). Since the compression layer would be 0.6 mm thick from each side, there is a significant degree of protection against stressing events (such as wind pressure or impact of projectiles). One could conceivably try to increase the temper (e.g., for bulletproofing); however, glass fracture due to transient tension generated at the edges of the plate from preferential cooling becomes increasingly probable. Temperature profiles with specially designed conductive metal contact pads are significantly steeper than a parabola, and hence allow the generation of much higher surface compression before the interior tension reaches failure levels. A slow but more effective method of introducing large amounts of surface compression is by **ion exchange**. This will be studied in Chapters 14 and 18.

Because of the stored elastic energy in tempered glass, any fracture is often violent (shattering) and accompanied by the production of a large number of small pieces (dicing) that generally do not have any sharp edges. Much of the mechanical energy release goes into the surface energy of the pieces. Hence, the larger the degree of temper, the finer the dicing. It may be shown that even the highest rates of cooling employed fail to develop a large

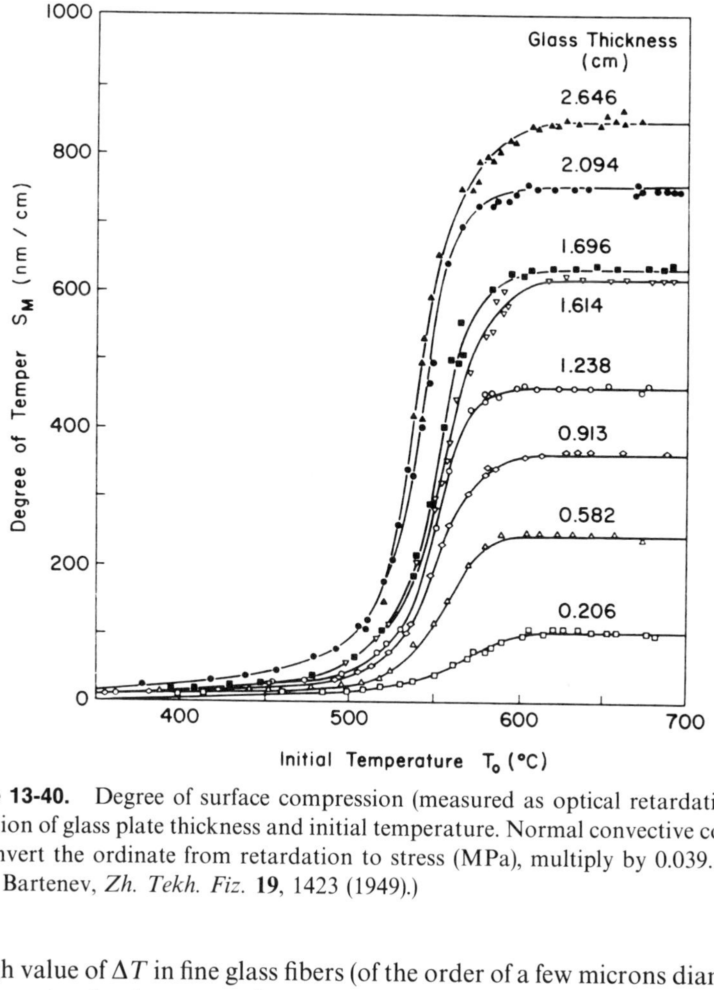

Figure 13-40. Degree of surface compression (measured as optical retardation) as a function of glass plate thickness and initial temperature. Normal convective cooling. To convert the ordinate from retardation to stress (MPa), multiply by 0.039. (After G. M. Bartenev, *Zh. Tekh. Fiz.* **19**, 1423 (1949).)

enough value of ΔT in fine glass fibers (of the order of a few microns diameter) to cause the development of any meaningful amount of compression. To introduce temper in glass fibers, one must employ the ion-exchange technique or cladding (Chapter 18).

Annealing of common glass products is carried out for two primary reasons: first, to reduce the risk of glass fracture from tensile stresses generated during manufacture due to nonuniform heating/cooling of the various parts, and during service where such stresses could be augmented by thermal and/or mechanical shocking; and second, to stabilize and homogenize the physical properties of glass such as the refractive index or the thermal expansion. Stabilization of refractive index with time clearly is of great importance for optical components. Likewise, instability and in-

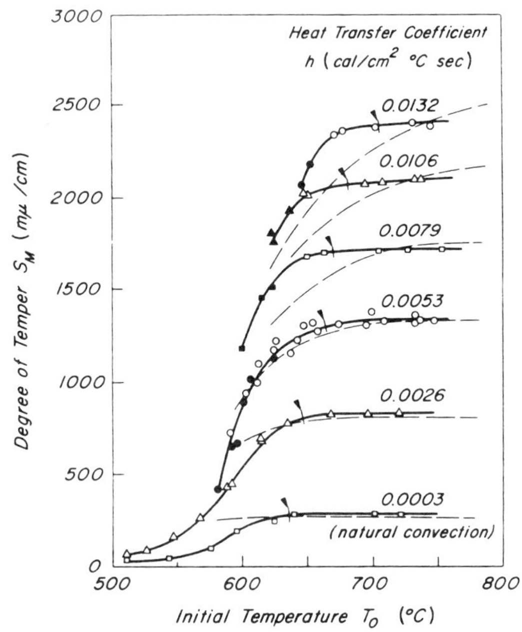

Figure 13-41. Degree of temper as a function of initial glass temperature and heat transfer coefficient. Forced convective cooling. Glass thickness 0.61 cm. To convert the ordinate from retardation to stress (MPa), multiply by 0.039. Experimental results (—○—); broken specimens (●); calculated by Bartenev's theory (----). (After Gardon, *Proc. 7th Int. Cong. Glass*, paper no. 79, Inst. Nat. du Verre, Charlroi, Belgium, 1965.)

homogeneity of expansion of the different parts of a mercury thermometer could cause the thermometer to deform.

One of the simpler procedures employed for annealing is to hold the glass at around the annealing temperature for 15 min, and then cool slowly (to avoid further buildup of stresses) to well below the strain point, after which the ware is cooled as rapidly as the thermal endurance of the glass will allow. If we assume a KWW relaxation law (Eq. (13.76)), an isothermal hold for 15 min at 10^{12} Pa·s viscosity (assumed shear modulus $G = 3 \times 10^{10}$ Pa) gives

$$S_{12}/S_{12}(0) = \exp[-\{900/(10^{12}/3 \times 10^{10})\}^{0.5}] = 0.0055.$$

In other words, essentially all the initial stress relaxes. At strain point (viscosity $= 10^{13.5}$ Pa·s), a similar release would require roughly eight hours. Because glasses in general do not all conform to a KWW relation, and because the value of G is not a universal constant, it is best to acknowledge the annealing point as the temperature where stress release occurs in a matter of minutes and the strain point as the temperature where the stress release

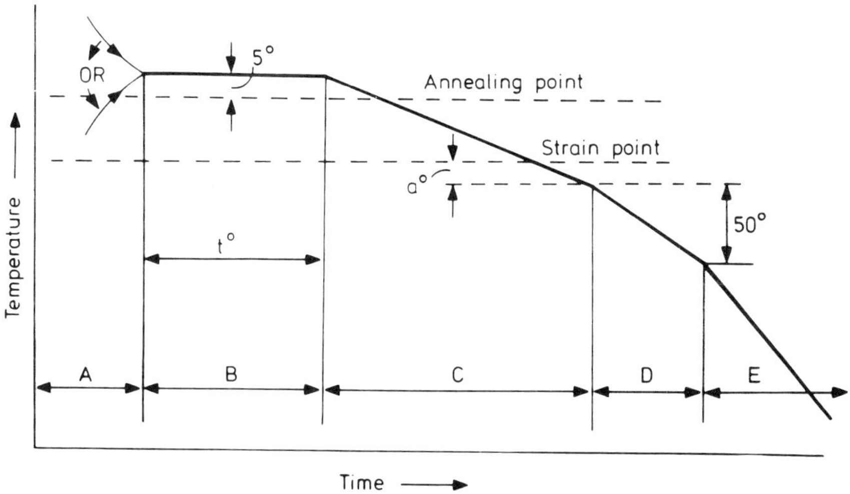

Figure 13-42. Suggested schedules for commercial annealing of soda-lime silica glassware. (Courtesy of Corning, Inc.)

takes a few hours.

A common practice to perform annealing is to follow a schedule shown in Fig. 13-42. The various regions of the schedule are given by:

$$
\begin{aligned}
A\,(\text{heating rate}) &= 839/[\alpha d^2]\,^\circ\text{C/min} \\
B\,(\text{holding time}) &= 97d \text{ min} \\
C\,(\text{slow cool}) &= 42.6/[\alpha d^2]\,^\circ\text{C/min} \\
X\,(^\circ\text{C below strain point}) &= 5^\circ\text{C for 0.3 cm thick plate} \\
&= 10^\circ\text{C for 0.6 cm thick plate} \\
&= 20^\circ\text{C for 1.3 cm thick plate} \\
D\,(\text{fast cool}) &= 2 \text{ times the slow cool} \\
E\,(\text{final cool}) &= \text{no greater than 10 times the slow cool} \qquad (13.95)
\end{aligned}
$$

where α is the thermal expansion coefficient of glass in $10^{-7}/^\circ$C units, and d is the thickness of the glass (cm), when cooled from one side only, and the half-thickness (cm), when the glass is cooled from both sides. The recommended values of A–E for three different values of α and three different plate thicknesses are listed in Table 13-4.

Table 13-4. Suggested Annealing Schedules for Glass Products[a,b]

Expansion coefficient of glass, per °C	Thickness of glass, mm	Cooling on one side: Heat rate, °C/min	Time t min	Temp, a,°C	Cool rate, °C/min			Cooling on two sides: Heat rate, °C/min	Time t, min	Temp. a, °C	Cool rate, °C/min		
		A	*B*	*C*	*C*	*D*	*E*	*A*	*B*	*C*	*C*	*D*	*E*
33×10^{-7}	3	130	5	5	12	24	130	400	5	5	39	78	400
	6	30	15	10	3	6	30	130	15	10	12	24	130
	13	8	30	20	0.8	1.6	8	30	30	20	3	6	30
50×10^{-7}	3	85	5	5	8	16	85	260	5	5	26	52	260
	6	21	15	10	2	4	21	85	15	10	8	16	85
	13	5	30	20	0.5	1.0	5	21	30	20	2	4	21
90×10^{-7}	3	50	5	5	4	8	50	140	5	5	14	28	140
	6	11	15	10	1	2	11	50	15	10	4	8	50
	13	3	30	20	0.3	0.6	3	11	30	20	1	2	11

[a] After G. M. McLellan and E. B. Shand, *Glass Engineering Handbook*. McGraw-Hill, New York, 1984.

[b] Annealing period: *A*, heating to 5°C above annealing point. *B*, hold temperature for time *t*. *C*, initial coding to a° below strain point. *D*, cooling, next 50°C. *E*, final cooling.

13.10. Summary of Glass Transition Range Behavior

Upon cooling a liquid through the supercooled region, the viscosity rapidly rises. Glass transition occurs when the viscosity is of the order of 10^{12} Pa·s. The intersection of the extrapolated liquid line and the glassy state line on the V–T diagram is called the "fictive temperature T_f" of the glass. In approximate language it may be stated that the structure of the glass corresponds to the structure of the equilibrium liquid at T_f at constant pressure.

Usually, the rapid rise in viscosity implies a rapid reduction in free volume. However, there are exceptions, particularly vitreous silica, where the volume of a corresponding crystal is higher than that of the glass. In such cases, the meaning of the free volume is not entirely clear. A more appealing description is the Adam–Gibbs concept, where the increase in viscosity with falling temperatures is ascribed to rapid loss of configurational entropy.

In the transition range, both the action of a mechanical impulse and a temperature impulse cause the system to be displaced from equilibrium. As a result, glass undergoes structural rearrangement. The physical properties, therefore, also vary. In the case of a mechanical impulse at constant temperature, the change in physical properties due to structural rearrangement is small when the applied deformations are small (i.e., for small displacement out of the equilibrium). For such a case, the assumption of linear viscoelasticity is generally applicable. Glass may behave as a combination of Hookean springs and Newtonian dashpots, often represented by a Maxwell element, a Voigt–Kelvin element, or a Burger element. The former two are based upon a single exponential relaxation/retardation law. The latter two simulate the delayed elasticity observed commonly in glass.

The change of temperature, on the other hand, brings about a large change in viscosity, causing the approach to equilibrium to become nonlinear. However, such nonlinearity, being due to the change in viscosity in part due to the changing temperature and due to the changing fictive temperature (structure), can still be converted to linearity by compressing or expanding the time scale to that at a reference temperature (again, as long as the deviations from the equilibrium are small) using the assumption of the thermorheological simplicity. At any point during the cooling of a glass, the structural change through a small step of physical temperature change can be found to be equivalent to a small change in the fictive temperature on the equilibrium liquid line. Because of nonlinearity, however, upon cooling through the glass transition range, glass retains a particular fictive history (or a memory) characteristic of the cooling path. The fictive history determines the value of the physical properties such as volume, specific heat, and thermal expansion coefficient that is the value of the property upon complete relaxation less the portion that is yet to be relaxed. Calculations show that the relaxation function is described as a sum of at least two Maxwellian functions with two different relaxation times. Properties of specimens where all the relaxation processes have not proceeded to near-completion thus display path sensitivity.

Thermodynamics suggest that glass is an unstable state with respect to the equilibrium liquid. However, the kinetics of the approach to the equilibrium are so sluggish that we may still apply the laws of thermodynamics to many, but not all, of the attributes of glass. In particular, glass not being an equilibrium state allows us to ignore the fact that glass has a residual positive (configurational) entropy at 0 K, otherwise required to be zero for equilibrium states by the Third Law. Cooling the supercooled liquid at infinitely slow rates appears to lead to negative entropy at 0 K. However, this entropy crisis is presumably avoided by speculating that at the Kauzmann temperature T_K, the number of possible arrangements for the atoms reduces to one, hence, the configurational entropy vanishes altogether, and the transition to the glassy state is imminent. This transition at T_K is believed to be a true thermodynamic second order. For all practical cooling rates, the transition is best called a "freezing" transition at which the structure freezes with one or more order parameters. For a thermodynamic description of the glassy state, one therefore needs specification of any $(n - 1)$ of the n thermodynamic variables P, V, T, and the various internal order parameters Z_i (describing the structure). Fictive temperature T_f or fictive pressure P_f, for instance, may be treated as internal order parameters. The thermodynamics requires that the specific heat and the isothermal compressibility increase in going from the glassy state to the supercooled liquid state. However, the change in the thermal expansion coefficient may have any sign. The Second Law, in particular, requires that there be a hysteresis of properties upon reheating relative to those upon the initial cooling. In other words, upon reheating at any practical rate, glass never retraces the initial cooling path.

Because of the changing volume of the individual layers as a function of the local temperature as well as the local rate of cooling, permanent stresses develop in glass products upon cooling through the glass transition range, which can potentially cause the glass to fracture. These stresses may be released, i.e., the product may be annealed by a suitable isothermal hold at around the annealing point, followed by a slow cool to below the strain point in order to avoid further buildup of stresses. Strengthening of the glass product can be achieved by a thermal tempering process, which is a rapid and uniform cooling of the glass surface to introduce beneficial surface compression.

Exercises

1. Sketch the expected strain vs. time behavior for a stabilized viscoelastic glass fiber exhibiting delayed elasticity (i) when the temperature is suddenly raised slightly (while maintaining the load), and (ii) when, after equilibrium is achieved, the load is suddenly halved.

2. Show that for a common soda lime silicate glass, the viscosity change per °C is equivalent to a change of about 12 MPa in pressure. (Hint: use Eq. (13.52).)

   [Answer: Substitution of $\Delta C_p \cong 0.25$ J/g, $\Delta\beta \cong 600 \times 10^{-7}$ K^{-1}, $V = 0.4$ cm^3/g, and $T = 800$ K yields $dP/dT = 12$ MPa/K.]

3. Show that a Ph.D. thesis project describing the measurement of viscosity in the 10^{20} Pa·s range is likely to be laughed at.

4. Calculate the increase in T_f of a soda-lime silicate glass when the cooling rate is one order of magnitude higher. (Hint: Use Eq. (13.91); $\Delta H \cong$ 100 kcal/mol.)

5. Since the glass is thermodynamically unstable with respect to the equilibrium liquid, the glass relaxes toward the liquid line at all times. This relaxation results in an enthalpy change which, in turn, may require heat exchange with the surroundings. Imagine a glass in the transition range having exactly the same temperature as the universe.
   (a) Does the relaxation of glass violate the Zeroth Law? Discuss.
   (b) What would the relaxation paths on the H–T curve be if the glass were to the left or to the right of the equilibrium curve, and if the heat exchange with the universe were allowed or disallowed?

6. The thermodynamic state of glass (at P, T_1) is described by (P, T_1, T_f) where T_f is the fictive temperature ($T_1 < T_f$). After a long time, the glass relaxes to the state (P, T_1, T_1). Show that the irreversible production of entropy in this process is $(1/2)\Delta C_p[(T_1 - T_f)/T_1]^2$, which on the H–T curve is the area enclosed by the glass line, the liquid line and the relaxation path, divided by T_1^2.

7. Sketch the variation of the physical temperature and the expected fictive temperature variation with location in a glass plate being cooled. Also sketch the density variation in an unbroken plate, and a fragmented one with preserved coordinates of the fragments. (Hint: See O. S. Narayanaswamy and R. Gardon, *J. Am. Ceram. Soc.* **61**(3–4), 145 (1978).)

8. Compute the variation of the room-temperature refractive index with real time for the glass specimens corresponding to crossover "A" after the crossover event. Show that the minimum in refractive index should be at around 0.72 hours.

9. Develop the fictive temperature versus reduced time relation for crossover "B."

References

1. R. W. Douglas, P. J. Duke, and O. V. Mazurin, *Phys. Chem. Glasses* **9**(6), 169–178 (1968).
2. O. V. Mazurin, in *Glass Sci. & Tech.*, Vol. 3 (D. R. Uhlmann and N. J. Kreidl, eds.), pp. 119–179. Academic Press, New York (1986).
3. D. Turnbull and M. H. Cohen, *J. Chem. Phys.* **52**(6), 3038–3041 (1970).
4. M. H. Cohen and G. S. Grest, *Phys. Rev.* **B20**(3), 1077 (1979).
5. P. K. Gupta, *Rev. Solid State Sci.* **3**(3&4), 221–257 (1989).
6. F. Simon, *Ergebn. Exakt. Naturwis.* **9**, 222 (1930).
7. W. Kauzmann, *Chem. Rev.* **43**, 219 (1948).
8. J. H. Gibbs and E. A. DiMarzio, *J. Chem. Phys.* **28**(3), 373–383 (1958).
9. P. J. Flory, *Proc. Roy. Soc.* **A234**, 60 (1956).
10. F. H. Stillinger, *J. Chem. Phys.* **88**(12), 7818 (1988).
11. J. Jäckle, *Phil. Mag. B* **44**(5), 533–545 (1981).
12. J. Jäckle, *Rep. Prog. Phys.* **49**, 171–231 (1986).
13. C. Angell, D. R. MacFarlane, and M. Oguni, *Ann. NY Acad. Sci.* **484**, 241 (1986).
14. G. Adam and J. H. Gibbs, *J. Chem. Phys.* **43**, 139 (1965).
15. P. K. Gupta and C. T. Moynihan, *J. Chem. Phys.* **65**(10), 4136 (1976).
16. A. Q. Tool, *J. Am. Ceram. Soc.* **29**, 240 (1946).
17. H. N. Ritland, *J. Am. Ceram. Soc.* **39**, 403 (1956).
18. S. Spinner and A. Napolitano, *J. Res. Natl. Bur. Standards* **70A**(2), 147 (1966).
19. O. S. Narayanaswamy, *J. Am. Ceram. Soc.* **54**(10), 498 (1971).
20. E. H. Lee, T. G. Rogers, and T. C. Woo, *J. Am. Ceram. Soc.* **48**(9), 480–487 (1965).
21. P. K. Gupta and J. Huang, *Proc. Phys. Non-cryst. Sol.* (L. D. Pye, W. C. LaCourse, and H. J. Stevens, eds.), pp. 321–326. Taylor & Francis, London (1992).

22. C. R. Kurkjian, *Phys. Chem. Glasses* **4**, 128 (1963).
23a. C. T. Moynihan, A. J. Easteal, J. Wilder and J. Tucker, *J. Phys. Chem.*, **78**, 2673 (1974).
b. C. T. Moynihan, A. J. Easteal, M. A. DeBolt, and J. Tucker, *J. Am. Ceram. Soc.*, **59**, 12 (1976).
24. J. S. Haggerty and A. R. Cooper, in *Phys. Non-cryst. Sol.* (J. A. Prins, ed.), p. 436. North-Holland Publishing Co., Amsterdam (1965).
25. R. W. Douglas, in *Phys. of Non-cryst. Sol.* (J. A. Prins, ed.), p. 397, 1965.
26. J. T. Littleton and W. L. Wetmore, *J. Amer. Ceram. Soc.* **19**, 243 (1936).
27. M. Hara and S. Suetoshi, *Rep. Res. Lab. Asahi Glass Co.* **5**, 126–135 (1955).
28. O. S. Narayanaswamy, *J. Am. Ceram. Soc.* **61**(3–4), 146–152 (1978).
29. R. Gardon and O. S. Narayanaswamy, *J. Am. Ceram. Soc.* **53**(7), 380 (1970).

Chapter 14

Permeation, Diffusion, and Ionic Conduction in Glass

14.1. Introduction

An atom is always in a state of agitation about its equilibrium site. Occasionally, it will have sufficient energy to jump over a potential barrier to the next site. Even in the absence of external electric, magnetic, or gravitational force fields, such jumps over a period of time will lead to a net displacement of the atom relative to the origin. We say that the diffusion of the particular atomic species has occurred. If r is the average jump distance, then it can be shown that the root-mean-square displacement $\sqrt{\overline{R^2}}$ after α jumps is given by

$$\sqrt{\overline{R^2}} = r\sqrt{\alpha}. \tag{14.1}$$

Thus, a Mexican jumping bean would be a certain average distance away from the starting point (known quite accurately on a statistical basis!) after a given amount of time. This "random walk" phenomenon is also the origin of the familiar "Brownian motion" when, for instance, a tea bag is let stand in hot water and the tea color gradually spreads over time.

To quantify the atomic diffusion in one dimension, consider two imaginary planes "1" and "2" a distance r apart along the x axis. Let there be m_1 diffusing atoms per unit area in plane 1 and m_2 in plane 2. If ν is the average frequency of successful jumps in one dimension, then the number of atoms diffusing out of plane 1 in time Δt is $m_1 v \Delta t$, half of which will be reaching plane 2. Likewise, the number arriving at plane 1 from plane 2 is $m_2 v \Delta t$. The net diffusive flux J per unit area per unit time out of plane 1 in the direction of plane 2 is given by

$$J = \text{net number of atoms out/(area} \times \text{time)} = (\tfrac{1}{2})(m_1 - m_2)v. \quad (14.2)$$

Since m_i can be related to the concentration n_i (number of atoms per unit volume) by $m_1/r = n_i$, we may rewrite (14.2) as

$$J = (\tfrac{1}{2})(n_1 - n_2)rv.$$

Assuming that concentration is a smooth, monotonic function of x,

$$n_1 - n_2 = -r\,dn/dx,$$

and so,

$$J = -(\tfrac{1}{2})r^2 v\,dn/dx = -(\tfrac{1}{2})r^2 v\,dc/dx, \quad (14.3)$$

where c is the concentration in moles/volume and J is now measured in units of moles per unit area per unit time. (Note: c = moles/volume = (number/volume)/(number/mole) = n/N, where N = Avogadro's number.) Equation (14.3) is of the form

$$J = -D\,dc/dx, \quad (14.4)$$

which is **Fick's first law of diffusion (in one dimension)**, where D is called the **diffusion coefficient**. The negative sign indicates that mass flows in the direction of decreasing concentration. Comparison of (14.3) and (14.4) yields

$$D = (\tfrac{1}{2})r^2 v. \quad (14.5)$$

For jumps in three dimensions, the numerical factor in this equation is $\frac{1}{6}$ instead of $\frac{1}{2}$. Appropriate units for the diffusion coefficient are cm^2/s (cgs units) and $m^2{\cdot}s^{-1}$ (SI units).

Fick's second law of diffusion is simply the mass conservation principle: the time derivative of the concentration should equal to the negative of the spatial derivative of the flux:

$$dc/dt = -dJ/dx. \quad (14.6)$$

Hence,

$$dc/dt = d(D\,dc/dx)/dx. \quad (14.7)$$

In particular, if D were a constant,

$$dc/dt = D\,d^2c/dx^2. \quad (14.8)$$

It may readily be realized that the absence of a concentration gradient does not prevent atoms from diffusing to the neighborhood sites. It only means that there is no net flux in any direction. Diffusion in the absence of any concentration gradient (to be more precise, in the absence of a chemical potential gradient) is called **self diffusion**. Diffusion due to a concentration gradient is called **chemical diffusion** (often, just "diffusion").

Since bonding in glasses is ionic, covalent, and metallic, diffusion of both inert and ionic species is of interest. Inert atoms and molecules diffuse through the glass network by hopping or squeezing through the interstitial spaces (free volume) as if the network were a sieve. Other than an elastic interaction during the "squeezing" through the channels and gateways, there really is no interaction with the ions that constitute the network. Thus, inert gases such as He, Ne, and Kr, and noble metals such as Au and Pt, diffuse through the network noninteractively. Diffusion of molecular gases such as N_2, CO_2, H_2O, and some ionizable atoms such as Cu may, for the most part, also fall under this category, except for an occasional possibility of interaction with ionically bonded network species. The passage of inert atoms through glass is more frequently called **permeation**, controlled by the property coefficient **permeability** K. Permeability is related to diffusivity by the relation

$$K = DS, \tag{14.9}$$

where S is the solubility defined by the ratio c_i/c_g, where c_i is the concentration of inert atoms dissolved in the glass, and c_g is the concentration in the gas phase. Thus, S is unitless, and K carries the same units as D.

The motion of ions such as Na^+, K^+, Ca^{2+}, Si^{4+}, O^{2-}, and F^- in oxide glasses is affected by the network species. The diffusing ions exchange sites with each other in such a way that there is no net accumulation of electrical charge. Thus, when one ionic species diffuses into the glass, an equivalent number of charges must move out of the glass. If the entering and the exiting species are the same, the diffusion is considered a case of self-diffusion. On the other hand, when the two species are different, a binary exchange is said to have occurred. When three or more species are involved, the diffusion is classified as a **multicomponent diffusion**.

The **electrical resistance** R (units: ohms or Ω) of a body is related to the applied electric potential across it through the familiar **Ohm's law**:

$$R = V/I, \tag{14.10}$$

where V is the applied potential (units: volts or V) and I the current (units: amperes or A). The resistance for a linear conductor may itself be written as

$$R = \rho L/A, \tag{14.11}$$

where L is the length, A the cross-sectional area, and ρ the resistivity (units: ohms·cm in cgs and ohms·m in SI;). **Electrical conductivity** σ is the reciprocal of electrical resistivity (units for conductivity: $\text{ohms}^{-1}\cdot\text{cm}^{-1}$ or mho/cm in cgs, and $\text{S}\cdot\text{m}^{-1}$ in SI. S = siemen $\equiv \text{ohm}^{-1}$).

If a potential is applied across a specimen of most oxide glasses, it can be shown that the electrical charge is carried essentially by the ions in the glass. An ionic mobility u may be defined as the average ionic drift velocity v per unit force $\mathfrak{F}$, i.e.,

$$u = v/\mathfrak{F}. \tag{14.12}$$

Now, $\mathfrak{F} = -d\mu/dx$, where μ is the chemical potential per ion, and $\mu \cong \mu^\circ + k_B T \ln c$, where μ° is the chemical potential in the standard state, and k_B is the Boltzmann constant. Hence,

$$\mathfrak{F} = -(k_B T/c)[dc/dx]. \tag{14.13}$$

Application of $cv = J$ and (14.4) yields

$$v = -(D/c)(dc/dx). \tag{14.14}$$

Substituting (14.13) and (14.14) in (14.12) gives

$$u = D/k_B T. \tag{14.15}$$

Stokes suggested that a particle of radius r moving with a velocity v in a viscous medium of viscosity η may experience a frictional force approximately equal to $6\pi r\eta v$. Assuming that the same viscosity applies on the atomic scale, equating the frictional force to $\mathfrak{F}$ in (14.12) and (14.15) yields

$$D = k_B T/6\pi r\eta, \tag{14.16}$$

which is called the **Stokes–Einstein relation**.

In the case of electrical conduction, the charge mobility u_e is defined by the drift velocity v per unit strength of the electrical field E, i.e,

$$u_e = v/E. \tag{14.17}$$

Since the electrical force per unit ion is zeE, where z is the valence of the ion and e the electronic charge,

$$(u_e/ze) = v/(zeE). \tag{14.18}$$

Comparison of (14.12) and (14.18) yields

$$u = u_e/ze,$$

which is an "obvious" conclusion that the charge drift velocity equals the ionic drift velocity times the charge per ion. This also allows (14.15) to be written as

$$u_e/ze = D/k_B T, \tag{14.19}$$

which is known as the **Einstein relation**. Electrical conductivity

$$\sigma = n(ze)u_e. \tag{14.20}$$

Substituting (14.19) in (14.20), we get

$$\sigma = n(ze)^2 D/k_B T = z^2 F^2 Dc/RT, \tag{14.21}$$

where F is the Faraday unit of charge ($= Ne$), and $c = n/N$. Equation (14.21) is called the **Nernst–Einstein relation**. Modifications to this relationship are discussed later.

Conduction of electrical charge by ions in glass implies that, upon the application of a dc voltage, the specimen will be gradually **electrolyzed**. Cations will be moving towards the cathode region, and anions, nonbridging oxygens, to the anode region. The electrode reactions in an alkali silicate glass can be stated as follows:

$$\text{Cathode region: } 2Na^+ + 2e \rightarrow 2Na^\circ$$

$$\text{Anode region: } O^{2-} \rightarrow 2e + \tfrac{1}{2}O_2\uparrow$$

The readjustment of the glass structure may be written as

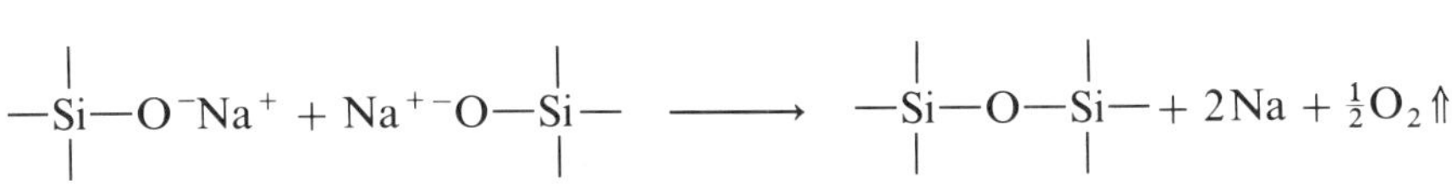

As a result of the electrolysis, two nonbridging oxygens collapse to form a bridge. The oxygen is liberated as gas at the anode, and the neutral alkali atoms are deposited on the cathode.

Because the ions are so much heavier than the electron, ionic mobilities are generally orders of magnitude smaller in comparison to electronic conductivities. One thus speaks of most oxide glasses as electrical insulators relative to most metals. On the other hand, *in glasses containing large amounts of transition metal oxides and chalcogenides, electronic conduction is well established to be the dominant mechanism of charge conduction.* As we shall see later, the way in which electrons conduct in these glasses is quite distinct from that in the metals. The electronic conduction in glasses is, therefore, often referred to as **semiconduction**.

We can readily imagine that mass and charge transport phenomena in glass are of great importance to many commercial processes. During the melting of glass, the last stages of homogenization are entirely controlled by the diffusion of various ionic species through an inhomogeneous mass. Removal of gas bubbles (otherwise called "seeds") depends critically upon the permeability and solubility of the various constituent gases. Chemical durability of glass (Chapter 17) is also at least a binary exchange phenomenon. Likewise, later in Chapter 18 we will show that the strength of glass is determined by stress corrosion, which in turn is related to mass transport. The insulating and dielectric properties of glass (discussed in Chapter 15) originating from the ionic nature of the charge transport are clearly of great technological interest—from the household incandescent lamp to microelectronic packaging.

14.2. More on the Atomic Theory of Diffusion and Electrical Conduction

14.2.1. Mechanisms of Diffusion

The important mechanisms of diffusion in glasses may be classified as follows.

An **exchange mechanism** is said to occur when two ions (or atoms) exchange sites with each other directly (Fig. 14-1a). In closely packed crystals, it may be difficult for two atoms to squeeze past each other. However, in open structures such as glasses, this mechanism may exist to a significant extent. In the **ring mechanism** (Fig. 14-1b), three or more atoms within a ring move to occupy each other's position. In the **vacancy mechanism**, the atom jumps to occupy a vacant site (Fig. 14-1c). An equivalent statement would be that the vacancy moves from site to site. The **direct interstitial mechanism** constitutes the jump of an interstitial specie from one interstitial site directly to another interstitial site (Fig. 14-1d). If, on the other hand, the interstitial atom moves into a normal lattice site, and the atom occupying the lattice site moves into an interstitial site (Fig. 14-1e), then this mechanism is called the **indirect interstitial** or the **interstitialcy mechanism**.

14.2.2. Correlation Effects

In discussing random walk, it was assumed that the jump probability of an atom is independent of the preceding jump. In most real situations, however, there is some correlation between the two jumps. For instance, if the atom jumped by a vacancy diffusion mechanism, then, in a situation where the

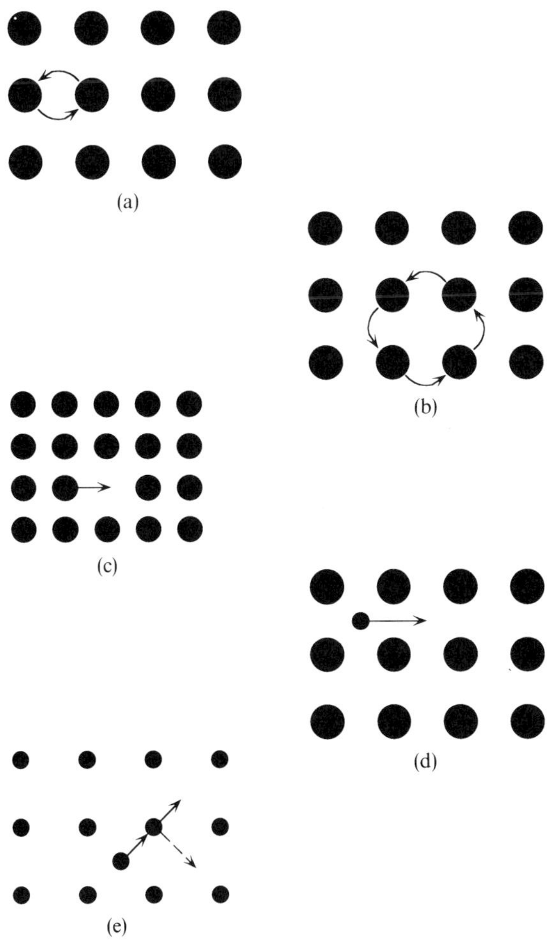

Figure 14-1. Mechanisms of diffusion in crystals. (a) Exchange mechanism where two atoms exchange sites with each other. (b) Ring mechanism where 3 or 4 atoms within a ring move to occupy each other's position. (c) Vacancy mechanism where the atom jumps to occupy a vacant site. (d) Interstitial mechanism where the atom on an interstitial site jumps directly to another interstitial site. (e) Interstitialcy mechanism with collinear (solid arrow) or noncollinear (dashed arrow) motion for the lattice atom. Here the atom from an interstitial site jumps to a normal site knocking its occupant to move to another interstitial site.

concentration of the vacancies is quite low, the likelihood of the atom jumping into the previous site is quite high. One must introduce a correlation factor f to relate the measured diffusion coefficient to a completely random walk situation, i.e.,

$$D_{\text{measured}} = fD_{\text{random}}.$$

Values of f for various diffusion mechanisms and various crystalline packing arrangements have been discussed by Manning (1). As an example, for the vacancy diffusion mechanism,

$$f = [1 + \langle\cos\theta\rangle]/[1 - \langle\cos\theta\rangle], \tag{14.22}$$

where $\langle\cos\theta\rangle$ represents the average of the cosines of angles between two successive jumps. Since $\langle\cos\theta\rangle$ is negative for the vacancy mechanism, it follows that $f < 1$ for vacancy diffusion. Also, for vacancy diffusion, Eq. (14.2) can be approximated as

$$f \cong 1 - (2/Z), \tag{14.23}$$

where Z is the number of nearest neighbors in the sublattice where the atom moves. It may be shown that, in 3-D face-centered cubic or hexagonal close-packed lattices, $f = 0.78$. (The reader should note that $f = 1$ means $D_{\text{measured}} = D_{\text{random}}$, i.e., no correlation, and $f = 0$ means highly correlated diffusion!)

For the interstitialcy mechanism,

$$f \cong 1 - (1/Z'), \tag{14.24}$$

where Z' is the number of neighboring interstitial sites to which an atom can jump.

14.2.3. Haven Ratio

If one examines the various mechanisms of diffusion carefully, one may recognize that correlation effects occur only in the movement of atoms by defects, and not in the movement of defects themselves. Thus, where the electrical conductivity may be solely described in terms of defect motion only, the Nernst–Einstein relation between diffusivity and conductivity should be rewritten as

$$\sigma = n(ze)^2 D/H_R k_B T, \tag{14.25}$$

where H_R, called the Haven ratio, equals the correlation factor in case the diffusion occurs by a vacancy mechanism. For the indirect interstial (the interstitialcy) mechanism,

$$H_R = 2(d/d')^2\ [1 - (1/Z')]. \tag{14.26}$$

The additional factor $2(d/d')^2$ accounts for the fact that the distance d' between one interstitial site and another interstitial site is different from d, which is the distance from a lattice site to the interstial site. Distance d' counts for ionic conduction, whereas d counts for diffusion. For a collinear interstitialcy, $d' = 2d$.

One may note that if the electrical conductivity is experimentally measured, then the application of the simple Nernst–Einstein relation (without the H_R) yields a diffusivity D_σ ($= k_B T\sigma/nz^2e^2$) that is related to the experimentally measured tracer diffusion coefficient D_{measured} by

$$D_{\text{measured}} = H_R D_\sigma. \tag{14.27}$$

Hence, comparison of the two measurements is often useful in determining the mechanisms of diffusion.

14.2.4. Isotope Effect

According to classical rate theory, the vibrational frequency of an atom is inversely proportional to the square root of its mass. Hence, two different mass isotopes m_a and m_b of a given element will have different diffusion coefficients (2) D_a and D_b, such that

$$(\{D_b/D_a\} - 1) = \Delta D/D = f(m_a/m_b)^{1/2} - 1, \tag{14.28}$$

where ΔD is the difference between their self-diffusion coefficients. Often, it may be necessary to use the "reduced mass" of each isotope, which takes into account the effect of the neighbors.

14.2.5. Temperature Dependence of Diffusion and Electrical Conduction

Suggested reading: D. Ravaine, *Kreidl Symp. J. Non-cryst. Sol.*, **73**, 295 (1985).

The temperature dependence of ionic transport in glass may be discussed in a manner similar to that described for viscosity. For a one-dimensional potential profile, the frequency ν of successful jumps may be written as

$$\nu = \nu_0 \exp[-\Delta H_D/k_B T], \tag{14.29}$$

where ΔH_D is the height of the potential barrier between two neighboring sites and ν_0 is the attempt frequency of the ion ($\cong k_B T/h$, h being Planck's constant). Substitution of (14.28) in (14.5) yields

$$D = (\tfrac{1}{2})a^2\nu_0 \exp[-\Delta H_D/k_B T],$$

which can be more generally written as

$$D = D_0 \exp[-\Delta H_D/k_B T], \tag{14.30}$$

where the pre-exponential factor $D_0 = \frac{1}{6}r^2\nu_0$ for jumps in three dimensions. Equation (14.30) is the Arrhenius relation for the ionic diffusivity.

When the potential is biased by an external electric field E, the frequency of jumping to the right becomes

$$\nu_+ = \nu_0 \exp[-\{\Delta H_D - (rze|E|/2)\}/k_B T],$$

where, as before, r is the jump distance and ze is the particle charge. The frequency of jumping to the left is given by

$$\nu_- = \nu_0 \exp[-\{\Delta H_D + (rze|E|/2)\}/k_B T].$$

Since the drift velocity $v = r(\nu_+ - \nu_+)$, and $rze|E|/2k_B T << 1$, it follows that

$$v = [r^2 ze|E|/2k_B T]\nu_0 \exp[-\Delta H_D/k_B T]. \tag{14.31}$$

Hence, from (14.20) and extending to three dimensions,

$$\begin{aligned} \sigma &= [nr^2 z^2 e^2/6k_B T]\nu_0 \exp[-\Delta H_D/k_B T] \\ &= \sigma_0 \exp[-\Delta H_D/RT], \end{aligned} \tag{14.32}$$

where ΔH_D, the **activation energy** for ionic transport, has now been written on a mole basis, R is the gas constant (not the resistance), and σ_0 is the **pre-exponential factor**. Equation (14.32), which, again, is an Arrhenius relation, describes the temperature dependence of the ionic conductivity. Note that σ_0 contains T. However, because of the temperature dependence of ν_0, this may be ignored in presenting the data by plotting $\log \sigma$ vs. $1/T$. Such a plot would generally exhibit a straight line with a slope $= -\Delta H_D/R$. The temperature dependence of resistivity written in the form

$$\log \rho = A + (B/T), \tag{14.33}$$

where A and B are constants, is called the **Rasch–Hinriksen relationship**. Comparison of (14.30) and (14.32) suggests that the activation energies for the ionic diffusivity and the ionic conductivity are expected to be identical.

To calculate activation energy from a structural consideration, Anderson and Stuart [3] suggested that ΔH_D *comprises two parts: an electrostatic binding energy to detach the ion from the present site, and a strain energy to push the ions through the interstitial "doorways."* An estimation of the binding energy can be obtained using the relation

$$E_b = [z_i z_O e^2/\varepsilon_\infty][\{1/(r_i + r_O)\} - \{1/r_e\}], \tag{14.34}$$

where z_i and r_i are the charge and the ionic radius of ion i, subscript O stands for oxygen, r_e is half the distance between equilibrium sites, and ε_∞ is the high frequency dielectric constant (Section 15.1). The strain energy E_s

required to dilate a spherical cavity of initial "doorway" radius r_d to radius r_i may be estimated by

$$E_s = 8\pi G r_d (r_i - r_d)^2 \tag{14.35}$$

where G is the shear modulus of the glass.

14.3. Measurement of Gas Permeability

Permeability of a glass to gases may be determined using an equipment described by Shelby [4]. To measure permeability, one must prepare the glass specimen in the form of a thin membrane either by blowing, or by grinding/polishing to a thickness of a fraction of a millimeter. The specimen is sealed between two concentric O-rings. One side of the specimen is connected to a gas supply line, and the other side to a mass spectrometer. Gas leakage through the O-rings is prevented by maintaining a vacuum of less than 10^{-2} torr around the specimen/O-ring holder. In one method, the permeating gas is continuously removed from the spectrometer side by pumping and maintaining a pressure differential P across the specimen. The flow rate is measured by the mass spectrometer. For such an experiment, permeability K is given by

$$K = Bd/AP, \tag{14.36}$$

where B is the flow rate and d and A are the specimen thickness and the area, respectively.

14.4. Measurement of Diffusion Coefficients

The measurement of diffusion coefficients in glass involves starting with a known concentration distribution of the desired species, allowing it to diffuse for a certain period of time, measuring the ending distribution of the diffused species, and fitting the two distributions to known solutions of the Fick's second law (Eqs. (14.7) or (14.8)) appropriate for the conditions of the experiment. Common experiments are discussed next.

14.4.1. Self-Diffusion Coefficients

Because the self-diffusion D_i involves no chemical potential gradient, the only way to measure D_i is by following the diffusion of a tracer isotope. The tracer may be radioactive or an enriched stable isotope. After a diffusion treatment for a time t, the specimen may be sectioned or etched layer by layer. Radioactivity counting or mass spectrometry may be used to determine the distribution of the tracer. As an example, in an "instantaneous source" experiment, a drop of aqueous solution of radioactive ^{22}NaCl may be deposited on a flat polished surface of glass. The initial condition is described by $c(x, t) = c_0\delta(x)$ for $t = 0$, where $\delta(x)$ is the Dirac's delta ($=1$ at $x = 0$; $= 0$ at $x \neq 0$). The boundary condition is given by

$$c(x, t) = 0 \qquad \text{for } x \to \infty \text{ and } t \geqslant 0$$

The solution of (14.8) for these initial and boundary conditions is the Gaussian distribution function

$$c = [c_0/\sqrt{(\pi Dt)}] \exp(-x^2/4Dt). \tag{14.37}$$

A plot of ln(^{22}Na radioactivity) on the ordinate scale and x^2 on the abscissa has a slope of $-1/[4Dt]$.

One other way to carry out the same experiment would be to measure the residual activity of the specimen after a given layer has been removed. (See Frischat [5].)

In a "constant source" experiment, the interfacial concentration is maintained at c_0. The initial and boundary conditions are described by $c = c_0$ at $x = 0$ and $t \geqslant 0$; $c = 0$ at $x > 0$ and $t = 0$. Also, $c = 0$ for $x \to \infty$ and $t \geqslant 0$. The solution of (14.8) for these conditions is given in the form of a "complementary error function" and is plotted in Fig. 14-2:

$$c = c_0 \operatorname{erfc}[x/2\sqrt{(Dt)}]; \qquad x \geqslant 0, \tag{14.38}$$

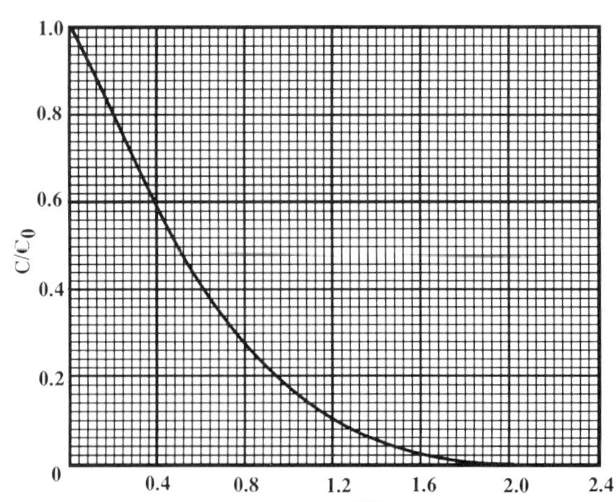

Figure 14-2. Concentration–distance curve for a semi-infinite diffusion couple ($x < 0$ and $x > 0$) with a step change (at $x = 0$) in initial concentration.

where $\mathrm{erfc}(z) = 1 - \mathrm{erf}(z)$, and

$$\mathrm{erf}(z) = [2/\sqrt{\pi}] \int_0^z \exp(-\beta^2)d\beta. \tag{14.39}$$

The error function has the property

$$\mathrm{erf}(0) = 0, \qquad \mathrm{erf}(-z) = -\mathrm{erf}(z), \text{ and } \mathrm{erf}(\infty) = 1.$$

Values of the error function, its complementary function, and its integral are tabulated by Crank (6). A numerical approximation is given by Hastings [7]. One may note that $\mathrm{erf}(3) \cong 1$. Hence, $c \to 0$ as $x \to 6\sqrt{(Dt)}$. For this reason, $6\sqrt{(Dt)}$ is often called the "diffusion distance."

Self-diffusion experiments are quite tedious to carry out. Foremost of the difficulties is in handling radioactivity, which requires elaborate safety procedures for processing and disposal. For the study of silicate glasses, it happens that the ^{31}Si isotope has a half-life of only about 2.6 h (barely allowing sufficient time to carry out a diffusion treatment), and oxygen diffusion experiments almost invariably have to be done using either the stable ^{18}O isotope or the $^{18}O[p, \alpha]^{15}N$ reaction [8]. Statistical errors of radioactivity counting, interference from other isotopes (particularly when the isotope is a beta particle emitter), layer thickness measurement, and misalignment of sectioning [9] are other major sources of errors.

14.4.2. Measurement of Chemical Diffusion Coefficients

The simplest of the experimental conditions to use for measuring a chemical diffusion coefficient is to place glasses with slightly different compositions together, and measure the concentration distribution after a suitable heat treatment. For the initial conditions of an extended source given by

$$c = c_0, \qquad x < 0; \; c = 0, \; x > 0 \text{ at } t = 0,$$

and boundary conditions

$$c = c_0, \qquad x \to -\infty; \; c = 0, \; x \to +\infty \text{ at } t \to \infty,$$

the solution of (14.8) is given by

$$c = [c_0/2] \, \mathrm{erfc}[x/2\sqrt{(Dt)}]. \tag{14.40}$$

Another situation of interest is the "sandwich" experiment. The reader may find it useful to refer to the book by Crank for solutions of Fick's second law for this and some other diffusion situations of interest.

The best way to measure the concentration distribution in such experiments is to use an electron probe analyzer (or an energy-dispersive x-ray spectrometer). The specimen after heat treatment may be sliced to expose a plane containing the diffusion direction. The surface is polished, and a conductive coating of carbon, about 300 Å thick, is vacuum-deposited. The analysis may be carried out using a micron-sized spot probe or a line probe of electrons with the specimen, which is stepped gradually to cover the entire diffusion zone. Common problems in the electron probe analysis are migration of the alkali under the influence of the electronic charge field, and overheating of the specimen due to the energy of the bombarding electrons. In addition, the determination of light elements (lower atomic weights than F) requires a specialized setup on the electron probe.

In many situations, the diffusion coefficient varies with the concentration. This means that Eq. (14.7) can not be solved unless the functional form of the concentration dependence is known. When this is not the case, the concentration dependence of D can be extracted from the diffusion data by performing a Boltzmann–Matano analysis [11]. Numerical methods of solving the diffusion equations are generally based upon finite difference algorithms, some of which are described by Crank [6].

14.5. Measurement of Electrical Conductivity

In the **blocking electrode method**, a small constant dc voltage is applied across a glass specimen and the current is monitored using an electrometer. The specimen is generally a polished disk (about 1 to 2 cm diameter × 1 to 2 mm). An Au–Pd film about 150 nm thick to function as electrodes is vacuum-evaporated on the specimen in a guard ring configuration (one side is full-coated; the other side has an inner circular coating surrounded by, but separated from, an outer ring called the "guard ring"). Electrical attachments (or mere point contacts) are made to the electrodes on the two sides as shown in Fig. 14-3 (after Tomozawa [12]). Parallel connection to the outer ring reduces any errors due to surface currents. One may expect that, because of the transfer of conducting alkali ions from the anode side of the specimen to the cathode side, polarization of the specimen occurs readily. Polarization leads to *essentially immediate decay of current with time* of current flow until no current flows. For this reason, dc conductivity measurements made using this technique should be cited with caution. (Use of voltages as little as 5–10 V to minimize polarization can help reduce the errors.)

A more useful dc measurement is the **nonblocking electrode method**, where the conducting alkali ions at the anode are continuously replenished from an essentially infinite source. The most convenient source is a bath of molten

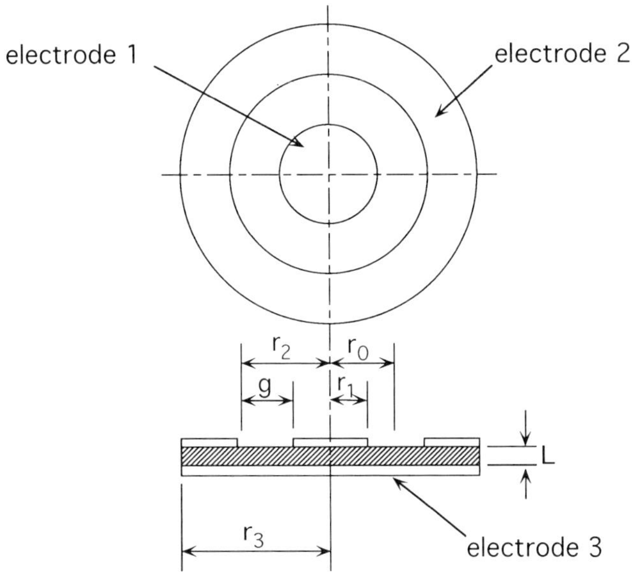

Figure 14-3. Guard-ring ("blocking electrodes") geometry to measure electrical conductivity.

alkali salt. An example is shown in Fig. 14-4, where the test specimen disk separates arms of a U-tube filled with molten salt. Platinum electrode wires dipped in the salt in each arm act to supply the current. The experiment is that of electrolysis of ions through the glass. One must therefore be careful here too in that the ions supplied should be exactly the ions being electrolyzed [13].

In the presence of electrode polarization effects, the recommended technique to measure the dc bulk conductivity is to perform ac measurements of the equivalent resistance R and capacitance C as a function of the ac frequency (generally from about 20 Hz to 1 MHz) using a capacitance bridge (see next chapter for details). A complex impedance Z^* (assuming an equivalent parallel circuit) may be written in the form

$$Z^* = Z' - jZ'' = R/[1 + j\omega RC] = R(1 - j\omega RC)/[1 + (\omega RC)^2], \quad (14.41)$$

where $j = \sqrt{(-1)}$. The real part $Z' = R/[1 + (\omega RC)^2]$, and the imaginary part $Z'' = \omega R^2C/[1 + (\omega RC)^2]$. One then plots Z'' against Z' for each value of ω. Semicircular arcs develop where the bulk conductivity is represented by the high-frequency arc, and the electrode polarization effects, if any, are given by the low-frequency arc. The intersection Z'_0 of the high-frequency arc with the Z'-axis (excluding the origin) (see Fig. 14-5) is then used to

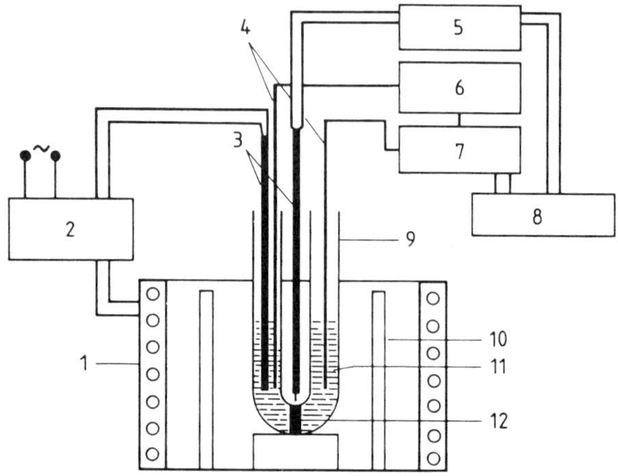

Figure 14-4. Nonblocking electrodes setup to measure dc conductivity. 1 = furnace; 2 = temperature controller; 3 = thermocouples; 4 = platinum electrodes; 5 = digital thermometer; 6 = dc power supply; 7 = ammeter; 8 = recorder; 9 = glass U-tube; 10 = steel shield; 11 = fused salt (nitrate or chloride); 12 = glass specimen. (After V. Jain, A. K. Varshneya, and P. P. Bihuniak, *Glastech. Ber.* **61**(11), 321 (1988). Reproduced with permission of the publisher.)

determine the true dc resistivity R_0 of the specimen. The true dc bulk conductivity is then given by the application of Eq. (14.11). (Under some experimental conditions, the arcs may only be portions of a circle or sums of circles that look more or less elliptical. The reader is advised to refer to Ravaine and Souquet [14], and to Kleitz and Kennedy [15].)

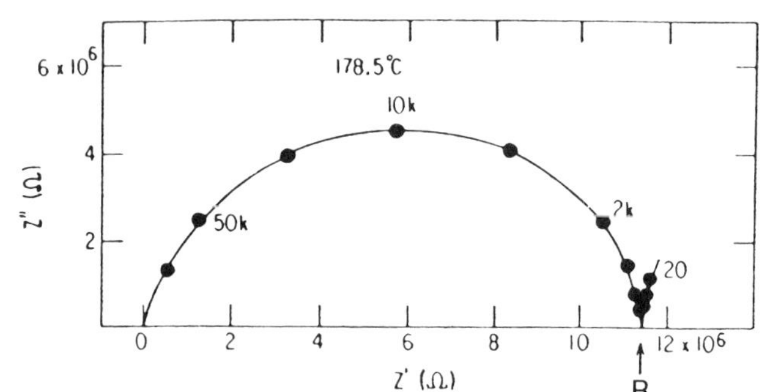

Figure 14-5. A plot of the imaginary vs. real parts of complex impedance for a $Li_2O{\cdot}B_2O_3$ glass at 178.5°C. Numbers near the data points are frequency in Hz. (After H. L. Downing, N. L. Peterson, and H. Jain, *J. Non-cryst. Sol.* **50**, 203 (1982). Reproduced with permission of Elsevier Science Publishers.)

In many of the microelectronic applications, a measurement of the sheet resistivity (sometimes referred to as the surface resistivity) is required. The sheet resistance R_{sh} (ohms per square or $\Omega/\square$) is related to the film resistivity ρ and thickness d by

$$R_{sh} = \rho/d. \tag{14.42}$$

Four point probes spaced on a square of side p are allowed to make contact with the film. A current I is allowed to flow through two adjacent probes, and the potential difference V across the other two probes is measured. If the contact points are exactly in a square pattern, then the sheet resistance is given by (see Meissel and Glang [16])

$$R_{sh} = [2\pi/\ln 2]\{V/I\}. \tag{14.43}$$

The reader may also find it beneficial to review a chapter on experimental techniques by Blumenthal and Seitz [17].

14.6. Data on Permeability and Molecular Diffusion of Gases in Glass

The permeation rates for helium as a function of the temperature in various glasses is shown in Fig. 14-6 (after Norton [18]). As would be expected, helium permeates through fused silica and Vycor most readily, since they are the most open-structured glasses. With the introduction of modifying cations, the permeation rate decreases. Lead borates, where the interstitial spaces may be blocked by the presence of the large lead ions, show very low permeation rates to He. In alkali silicate glasses, the permeability of He at 200°C decreases with the addition of the alkali. In Li silicates, the sharp decrease at 15–17 mol % Li_2O results from phase separation. Figure 14-6 also shows that permeability obeys an Arrhenius-type relationship with an activation energy Q. Again, as one might expect, Q should increase with the size of the permeating gas. Indeed, when $\sqrt{Q}$ is plotted against the molecular diameter of the permeating gas, one gets a good straight line having a positive slope.

It is reasonable to suggest, then, that permeability data reflect the characteristics of the distribution of interstitial spaces in glass. Care is needed, however, when the permeating species reacts with the glass network. An example [4] of this is the diffusion of H_2 in vitreous silica at 800°C, shown in Fig. 14.7. H_2 reacts with the silica network to form hydroxyls and silane groups:

$$\text{Si—O—Si} + H_2 \longrightarrow \text{SiOH} + \text{SiH}.$$

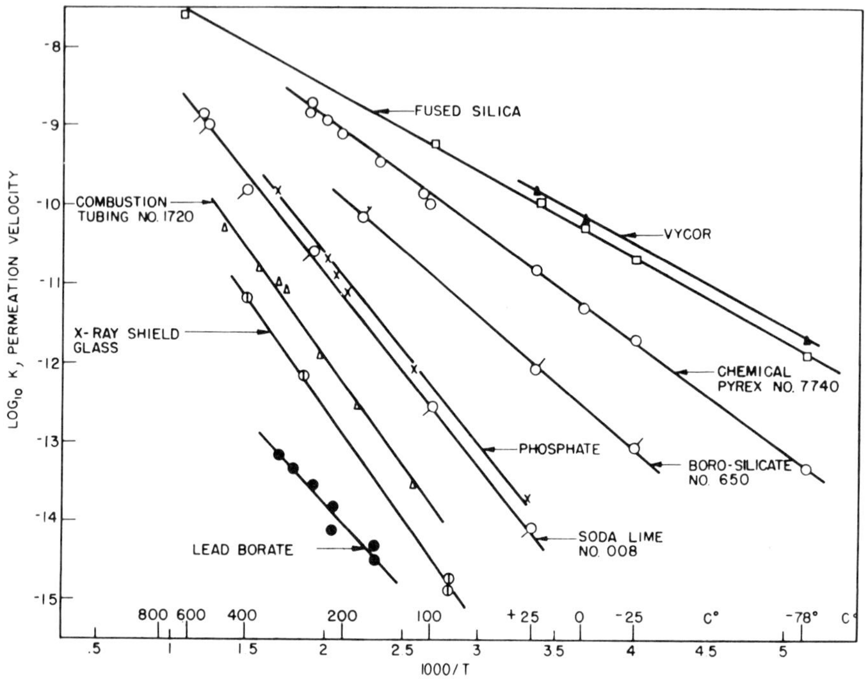

Figure 14-6. Permeation rate for helium in various glasses as a function of temperature, measured by Norton [18]. Permeation units are cm^3 of gas at STP per second per cm^2 of area per mm thickness per mm Hg pressure. (Reproduced with permission of The American Ceramic Society.)

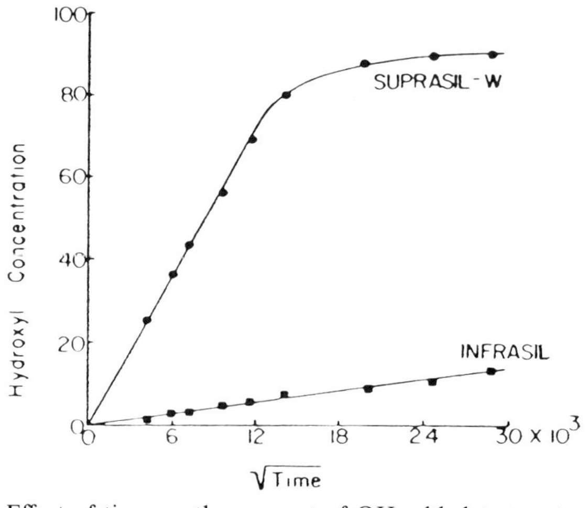

Figure 14-7. Effect of time on the amount of OH added to two types of vitreous silica by reaction with hydrogen gas at 800°C and 0.53 atm. The ordinate is normalized to a maximum OH concentration = 0.03%. (After J. E. Shelby, *Treat. on Mat. Sci. & Tech.*, Vol. 17, *Glass I*, p. 1. Academic Press, New York, 1979.)

Apparently, the rate of the reaction and the ratio of SiOH to SiH produced depend greatly on the mode of formation of the silica. As much as 0.03 wt. % SiOH without detectable amounts of SiH is formed in the synthetic silica, whereas electrically fused material shows SiOH and SiH, both only at parts-per-million levels.

Doremus has suggested that water diffuses in silica glass as molecules reacting with the network to form immobile hydroxyl ions. In silicate glasses, the diffusion coefficient of water remains relatively constant at low water concentrations, but increases exponentially at high water content [19].

14.7. Data on Ionic Diffusion and Electrical Conduction in Glass

Suggested reading: G. H. Frischat, *Ionic Diffusion in Oxide Glasses.* Trans Tech Publications, Aedermannsdorf, Switzerland, 1975.
M. D. Ingram, *Phys. Chem. Glasses* **28**(6), 215–234 (1987).
N. P. Bansal and R. H. Doremus, *Handbook of Glass Properties.* Academic Press, New York, 1986.

Diffusion of sodium ion as an impurity in fused-silica glass has been measured as a function of temperature by Frischat [20]. A plot of log D vs. $1/T$ over 200–1,000°C showed some departure from straight-line behavior (Fig. 14-8). The activation energy for diffusion was found to vary between 90 and 125 kJ·mol^{-1}(= 22–30 kcal/mol). Using a nonblocking electrode technique to measure dc conductivity in doped fused silicas, Jain *et al.* [21] also observed non-Arrhenius behavior. This was attributed to the relative contributions for alkali ion jumps via two different types of impurity sites present in fused silica, namely, the SiO^- groups and the $[AlO_4]^-$ groups. There is clear evidence [22] that the alkali diffusion in silica glasses depends significantly upon the technique of glass manufacture.

The effect of Al^{3+} ions on the diffusion of sodium ions in alkali aluminosilicate glasses containing large amounts of sodium and aluminum is shown in Fig. 5-25. The behavior of the activation energy with respect to the $[Al_2O_3/Na_2O]$ mole ratio has already been explained in Section 5.9.

The self-diffusion of alkali in binary alkali silicate and ternary soda–lime–silicate glasses has been measured by Johnson *et al.* [23]. The activation energy of sodium in the glassy state range is generally around 60–80 kJ·mol^{-1}, and that of potassium ion about 140–180 kJ·mol^{-1}. The activation energy shows a sudden rise in the glass transition range, followed by a decrease back to the 90–110 kJ·mol^{-1} range for sodium in the supercooled liquid range (Fig. 14-9). As expected, larger alkali ions have lower diffusivity and ionic conductivity. Likewise, higher-valency ions have

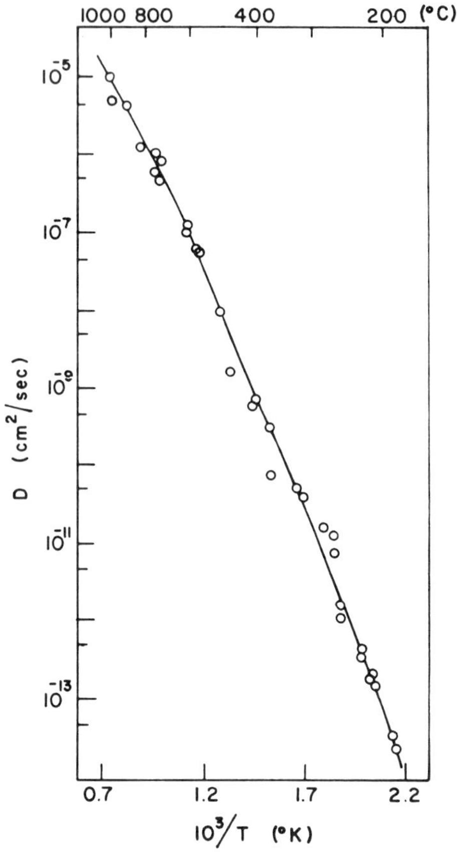

Figure 14-8. Diffusion coefficient of sodium in vitreous silica as a function of temperature. (After Frischat [20]. Reproduced with permission of The American Ceramic Society.)

lower diffusivity and conductivity. The activation energy for diffusion and conduction decreases with increasing alkali content in the silicate glasses, presumably because the continuing additions of alkali create nonbridging oxygens, causing the network to become more open for ionic motion. Conductivities are higher in chilled glasses compared to the well-annealed because of a larger interstitial free volume in the chilled specimens. The surface resistivity of alkali-containing glasses greatly depends upon humidity levels. A common soda-lime silicate glass may have as high as 10^{14} ohms/□ surface resistivity at room temperature and zero humidity, dropping to 10^8 ohms/□ at 60% relative humidity. This is ascribed to the extraction of

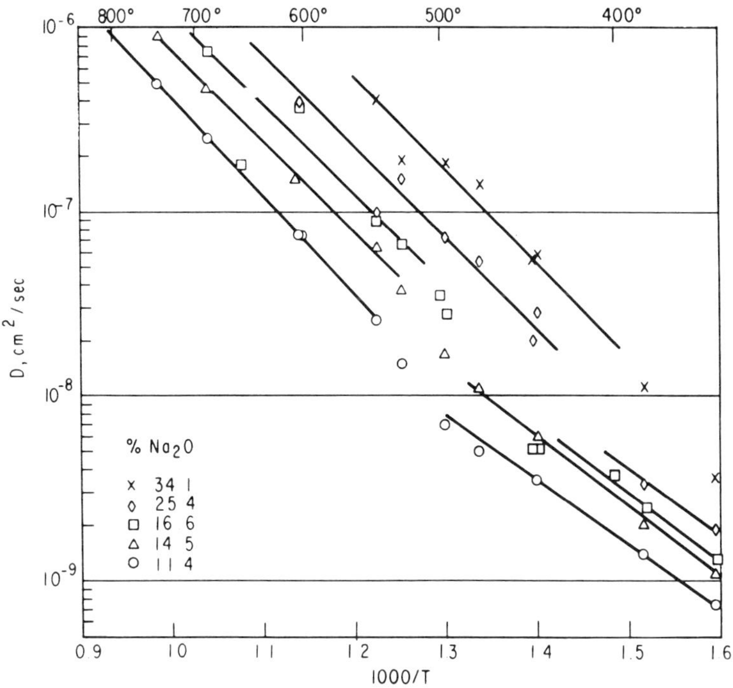

Figure 14-9. Diffusion coefficient of sodium in binary sodium silicate glasses. (Data of J. R. Johnson, R. H. Bristow, and H. H. Blau, *J. Am. Ceram. Soc.* **34**, 135 (1951). Replotted by R. H. Doremus, *Glass Science*, J. Wiley & Sons, Fig. 5, p. 161. J. Wiley & Sons, New York, 1973. Reproduced with permission of the publishers.)

Na^+ ions by water on the surface (a consequence of the chemical durability of glass, to be studied in Chapter 17).

Correlation factors have been determined by several workers to be as little as about 0.4 to as much as 1 in soda-lime silicate and Pyrex glasses. Thus, the exact atomic mechanisms of diffusion are not well established.

By far the most interesting feature in the diffusion and electrical conduction of alkali ions in glasses is **the mixed-alkali effect**. (It may perhaps be the most thought-provoking topic in all of glass science.) There are two aspects of mixed-alkali behavior:

1. Self-diffusivity of an impurity alkali ion is almost always lower than that of the majority alkali ion in a mixed-alkali glass regardless of the size relationship. Thus, diffusivities cross when plotted as a function of the mixed alkali composition (generally not at the 50–50 composition). This is shown in Fig. 14-10.

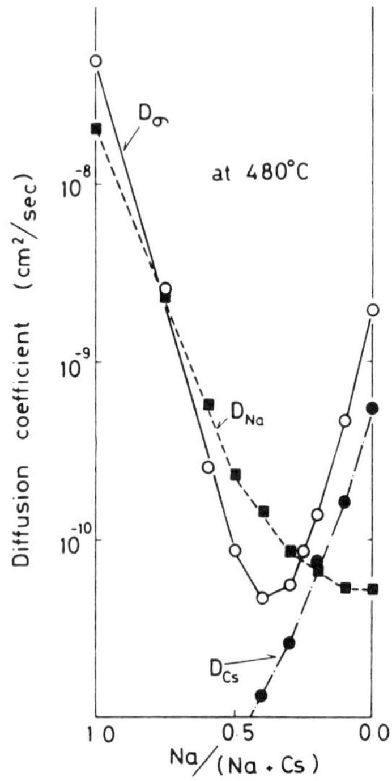

Figure 14-10. Diffusivity and conductivity data for Na^+ in the Na_2O–Cs_2O–SiO_2 system at 480°C. D_σ is the diffusivity calculated from the measured conductivity using the Nernst–Einstein relation. (After R. Terai, *J. Non-cryst. Sol.* **6**(2), 121 (1971). Reproduced with permission of Elsevier Science Publishers.)

2. For dc conduction in mixed-alkali glasses, one would expect that a linear extension of the Nernst–Einstein relationship would apply, i.e.,

$$\sigma_{\text{total}} = \sum_i [n_i e^2 D_i / f_i k_B T]. \tag{14.44}$$

If the mechanism of diffusion is not substantially different, then f_i may be assumed to be independent of the alkali composition. Figure 14-10 (after Terai) shows that this is not true. *The measured conductivity is a deeper minimum than that computed from Eq. (14.44)*. In addition, the composition at which the conductivity minimum occurs is distinctly shifted from the diffusivity crossover.

One is almost made speechless by the observation that *conductivity departs by as much as a factor of 10^3 to 10^6 from an additivity assumption between the conductivities of the end compositions* (see Fig. 14-11 as an additional

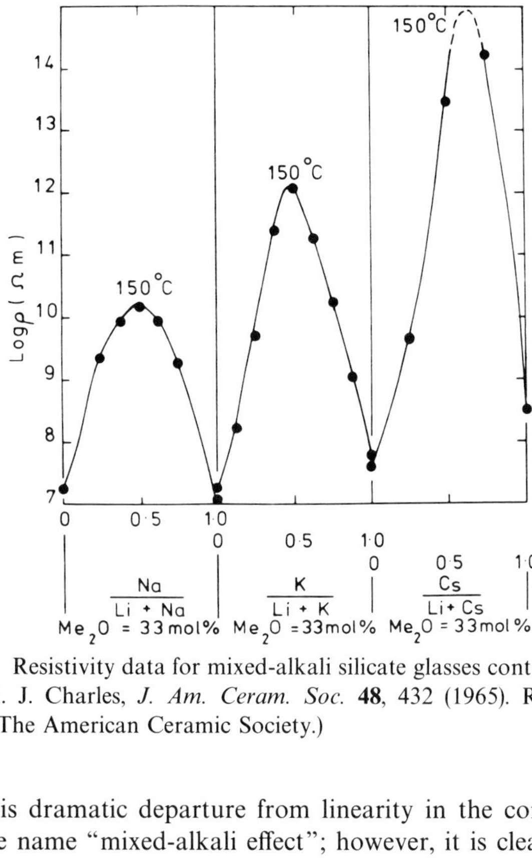

Figure 14-11. Resistivity data for mixed-alkali silicate glasses containing 30 mol % R_2O. (After R. J. Charles, *J. Am. Ceram. Soc.* **48**, 432 (1965). Reproduced with permission of The American Ceramic Society.)

example). This dramatic departure from linearity in the conductivity plot was given the name "mixed-alkali effect"; however, it is clear that, besides conductivity, other transport properties of mixed-alkali glasses, such as self-diffusivity, viscosity, dielectric loss, and internal friction, display significant departures from additivity. In most cases, it is noted that the non-linearity decreases with increasing temperature. Any proposed mechanism must somehow involve strong interaction between the different alkalis such that a large number of the majority alkali ions are immobilized (or their mobility is reduced).

Varshneya [24] suggested that mixed-alkali ions tend to pair, and prefer a direct exchange of sites. Hence, such pairs would contribute to "conductionless diffusion." The net dc conductivity would, therefore, be obtained by subtracting the probable number of such pairs from each term in Eq. (14.44). The agreement with measured data was claimed to be satisfactory. Varshneya, however, provided no basic explanation for the variation of the self-diffusivities.

Hendriksen and Bray [25] suggested that the positively charged alkali ion and the negatively charged site in which the ion sits will form an oscillating electric dipole. There exists a time-averaged interaction energy $\bar{W}$ between two oscillating dipoles (electrodynamic interaction) that would cause the energy levels of the oscillators to split by $\pm\bar{W}$. The ratio of ions n_+/n_- occupying the two states is given by the Boltzmann probability $\exp[-2|\bar{W}|/k_B T]$. For a typical mixed-alkali interaction energy $\bar{W} = 0.5$ eV, this population ratio can be as little as about 10^{-9}, i.e., most of the ions populate the lower energy level. The barrier height for diffusion (activation energy) ΔH_D is therefore increased, leading to decreased mobility of both the ions involved. The total conductivity is given by

$$\sigma = n_1 u_{e1} e + n_2 u_{e2} e + (\alpha - n_1 - n_2) u_{ea} e, \tag{14.45}$$

where α is the total number of alkali ions, n_1 and n_2 are the number of nonassociated ions "1" and "2" considering multiple interactions over an effective interaction range, and u_{ea} is the charge mobility of the associated ions. Hendriksen and Bray used empirical values of n_1, n_2, and u_{ea} and showed good agreement with measured conductivity and diffusivity data. Apparently, u_{ea}/u_{e1} varied between 0.0004 and 0.01.

The interaction energy would depend upon the oscillation characteristics, and hence on the masses of the alkali ions. However, the experiments suggest that the decrease in conductivity is more related to the difference in the ionic sizes rather than to the difference in masses [26].

Tomandl and Schaeffer [27] extended the electrodynamic interaction model by suggesting that the transport of alkali ion occurs via type-specific paths. In other words, a Na^+ ion prefers to move through paths of its type formed during the melting of glass. Likewise, K^+ would move through its preferred path, also formed during the same melting. The mobility would be greatly reduced when traveling ions interacted electrodynamically. Again, an interaction energy $\bar{W}$ would develop if the ions had different masses. Presumably, the type-specific migration in some ways accounts for the ion size effect.

LaCourse [28] suggested that alkali ions had type-specific sites formed during the preparation of glass. The "equilibrium" site consisted of a particular size (for stress considerations) and oxygen coordination depending upon the ionic size. A "defect" was formed whenever one alkali moved from its equilibrium site into the "other" alkali-type site. This would also mean that the "other" alkali ion has moved into the first ion's site. Defects themselves are more mobile, hence contributing to lowered viscosity. However, because of the energy requirement in forming defect sites, the alkalis were discouraged from jumping to sites suitable for the other type. This means that the mobility of each ion is effectively reduced by the "blocking" effect. A subsequent molecular dynamic study by Huang and Cormack [29] supported this notion by suggesting that, at least in the vicinity of 1:1 ratio of the mixed alkalis, the ions find themselves in deeper potential wells in the vicinity of the mixed alkali, causing them to have reduced mobility.

Moynihan and Lesikar [30] suggested that glass acts as a weak electrolyte. Alkali ion motion occurs via an interstitialcy mechanism or an interstitial pair mechanism where the concentration of such mobile species is governed by an equilibrium constant between the ion on an interstitial site (or a pair of ions on a pair of interstitial sites) and an ion on the regular site. As a result, even in a single alkali composition, only a small fraction (about 10%) of the alkali ions is actually mobile. Addition of a foreign ion further suppresses, by mass action, the concentration of the mobile species. In physical terms this could be achieved by preferentially occupying accessible interstitial sites or by forming more immobile mixed-alkali pairs.

Tomozawa and McGahay [31] noted that the mixed alkali glasses have nearly the same high-frequency conductivity as the single alkali glasses, even though their dc conductivity is much smaller. Since the conductivities are related to the dielectric constants (see next chapter), the dielectric relaxation strength ($= \varepsilon_s - \varepsilon_\infty$; ε_s = static dielectric constant, ε_∞ = high-frequency dielectric constant, see next chapter) of the mixed-alkali systems must be much larger than that of the single-alkali glasses. The origin of the large dielectric relaxation strength must be a large negative enthalpy of mixing in the mixed-alkali systems. The negative enthalpy presumably results from a local generation of hydrostatic stress due to the difference in ionic size. The stress also impedes ionic motion. The bottom line of the argument is that the difference in the size of the ion and the site it is jumping to creates the mixed alkali effect. With increasing temperature, stresses relax quickly and the mixed-alkali effect vanishes. The negative enthalpy may even change to positive, leading to possible phase immiscibility.

The reader may note that a widely accepted explanation for the "mixed alkali effect" has yet to be put forward.

14.8. Some Aspects of Diffusion and Conduction Phenomena

14.8.1. Ion Exchange

In ionic glasses, the diffusion due to the concentration gradient of one species necessarily involves counterdiffusion of a second species to maintain charge neutrality. The flux of ion A may be written as

$$-J_A = D_A\left[\frac{dC_A}{dx} + C_A\frac{F}{RT}\frac{\partial\phi}{\partial x}\right], \tag{14.46}$$

where ϕ is the electrostatic potential. The right-hand term is the ionic flux due to an electrostatic potential ϕ set up by the migration of the ions. Similarly, the flux of B may be written as

$$-J_B = D_B\left[\frac{dC_B}{dx} + C_B\frac{F}{RT}\frac{\partial\phi}{\partial x}\right]. \tag{14.47}$$

Application of the requirements for electrical neutrality, $J_A + J_B = 0$, and preservation of stoichiometry $C_A + C_B = 1$, and solving for J_A gives

$$-J_A = \frac{D_A D_B}{[D_A C_A + D_B C_B]}\frac{dC_A}{dx}. \tag{14.48}$$

This implies that the effective diffusion coefficient $\bar{D}$ is

$$\bar{D} = \frac{D_A D_B}{[D_A C_A + D_B C_B]}. \tag{14.49}$$

Equation (14.49) is known as the **Nernst–Planck equation**. Note that as $C_A \to 1$, $\bar{D} \to D_B$, which is the diffusion coefficient of the impurity. Occasionally, it may be preferable to have a thermodynamic activity factor n in the numerator of Eq. (14.49).

Ion-exchange phenomena in glass are important technologically in glass pH electrodes and ion-exchange strengthening. When a glass containing ions M^+ is in contact with a solution, different surface equilibria potentials develop at different pH as a result of the electrochemical reactions between functional groups at the glass surface and the solution. The kinetics of the potential development are controlled by the exchange current densities between protons and alkali ions [32]. In **ion-exchange strengthening** (also called **ion-exchange stuffing** or **chemical tempering**), the glass is immersed in a molten salt bath having alkali ions larger than those in the host glass. These large alkali ions from the bath diffuse into the glass, exchanging sites with the smaller host ions. If the temperature of the exchange is lower than the glass transition temperature, then, in effect, the large ions are squeezed into smaller spaces, generating a surface compression. Clearly such a squeeze on the surface of the glass will create surface compression. It is shown in Chapter 18 that this surface compression leads to glass strengthening. For a Na^+-containing glass, baths containing KNO_3 are generally used.

The applicability of Eq. (14.49) to describe the diffusion profile of the invading ion has been demonstrated convincingly by several workers [33, 34]. Within the diffusion zone, one may imagine that the glass resembles a stack of layers of mixed-alkali glasses with varying mixed-alkali ratio. Thus, the D_A and D_B are not to be assumed to be constant diffusivities in the parent glass, but are the composition-dependent self-diffusion coefficients of the two alkalis in the various glasses, as shown in Fig. 14-10. The reader may note, however, that the concentration dependence of the diffusivity is often ignored, and a solution of the form given by Eq. (14.40) is utilized.

The development of stresses during ion exchange requires a consideration of two competing processes: the stress production due to the molecular stuffing, and the relaxation of these stresses due to glass fluidity and other mechanisms.

The production of stress due to molecular stuffing may be calculated [35] by analogy to thermal stresses. Consider a semi-infinite slab that has been immersed in a salt bath. Diffusion essentially occurs only in the $\pm x$-direction (thickness), if we ignore the ends. The dimensional changes upon ion exchange will be as shown in Fig. 14-12. If all the x-elements were free to expand in all directions, the resultant shape of the body would be that shown in Fig. 14-12b. Because of the free expansion, there are no stresses; the equilibrium elongation of each x-element in the y-direction is given by

$$(e_y)_x = (BC)_x, \tag{14.50}$$

where B is the linear network dilatation coefficient, i.e., the amount of linear expansion/exchange of a concentration unit. Dilatation occurs because the exchanging ions have different partial molar volumes in glass. The relationship of B to C is analogous to that of the linear thermal expansion coefficient to the temperature T.

If the exchange is carried out at temperatures well below the glass transition temperature, then the body is not free to expand: the uneven dilatations in the y and z directions are suppressed because of compatibility criteria (Appendix I). The body swells freely in the x-direction, but expands uniformly in the y- and z-directions, so that the net elongations, e, are given by the average expansion of the x-element over the thickness $2L$:

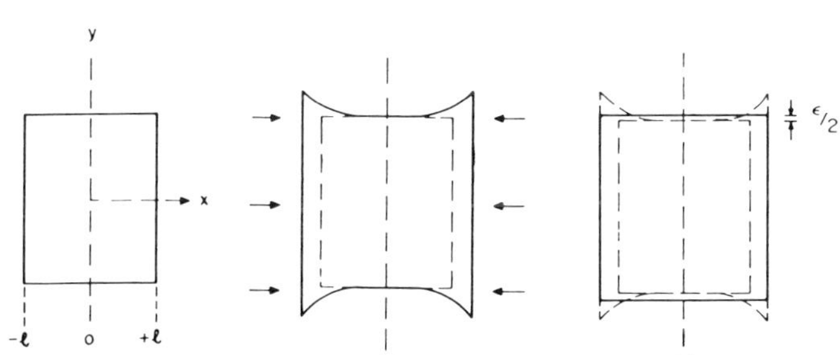

Figure 14-12. Dimension changes after ion exchange. (a) No ion exchange. (b) Ion exchanged and unrestrained. (c) Ion exchanged and restrained.

$$(e_y)_x = (e_z)_x = [1/(2L)] \int_{-L}^{+L} (BC)_x \, dx. \tag{14.51}$$

The true strain of an x-element, however, is the difference between the equilibrium and the net expansions, i.e.,

$$(\varepsilon_y)_x = (\varepsilon_z)_x = (BC)_x - [1/(2L)] \int_{-L}^{+L} (BC)_x \, dx. \tag{14.52}$$

The stresses σ_{ii} are hence given by

$(\sigma_{zz})_x = 0,$ and

$$(\sigma_{yy})_x = (\sigma_{zz})_x = -[(BC)_x E/(1-v)] + [E/\{2(1-v)L\}] \int_{-L}^{+L} (BC)_x \, dx. \tag{14.53}$$

where E is Young's modulus and v is the Poisson ratio. If E, v, and B do not vary with composition, then the expected stress profile follows the concentration distribution C_x. The average concentration given by the integral on the RHS is generally quite small for a 2–3 mm thick glass plate. Thus, in the interior as $C \to 0$, $(\sigma_{yy})_x \to$ small tension, corresponding to the right-hand of Eq. (14.53). On the surface, however, $(\sigma_{yy})_x \to$ large compression because of the large value of $(C)_x = \pm L$. Note that

$$\int_{-L}^{+L} (\sigma_{yy})x \, dx = 0.$$

A stress profile measured by ion-exchanging a $20Na_2O{\cdot}15B_2O_3{\cdot}65SiO_2$ (mol %, $T_g = 600°C$) in molten KNO_3 for 12 h at 400°C is shown in Fig. 14-13. As Varshneya [36] has shown, the measured stress profiles yield values of the network dilatation coefficient B that are generally smaller by a factor of 3 to 4 compared with those calculated from the molar volume changes. Because the exchange temperature was significantly below the T_g, such a large discrepancy may not be explained on the basis of normal viscous relaxation phenomena. Apparently, there must additionally be low-temperature relaxation mechanisms such as the initial accommodation of the differing ionic size, possibly related to the volume viscosity (Section 9.7). Low-temperature relaxation phenomena in glasses are not well understood. Other possible sources of disagreement could be (i) differential thermal contraction of the surface from the exchange temperature to the room temperature because of a changed composition; (ii) effect of geometry and slicing on the stress profile; (iii) the measuring instrument, which measures the optical retardation due to the difference of the principle stresses, and not due to individual components; and (iv) the use of physical properties such as the thermal expansion coefficient, elastic moduli, and the stress–optical coefficient, which correspond to those of the as-melted glass structures, and

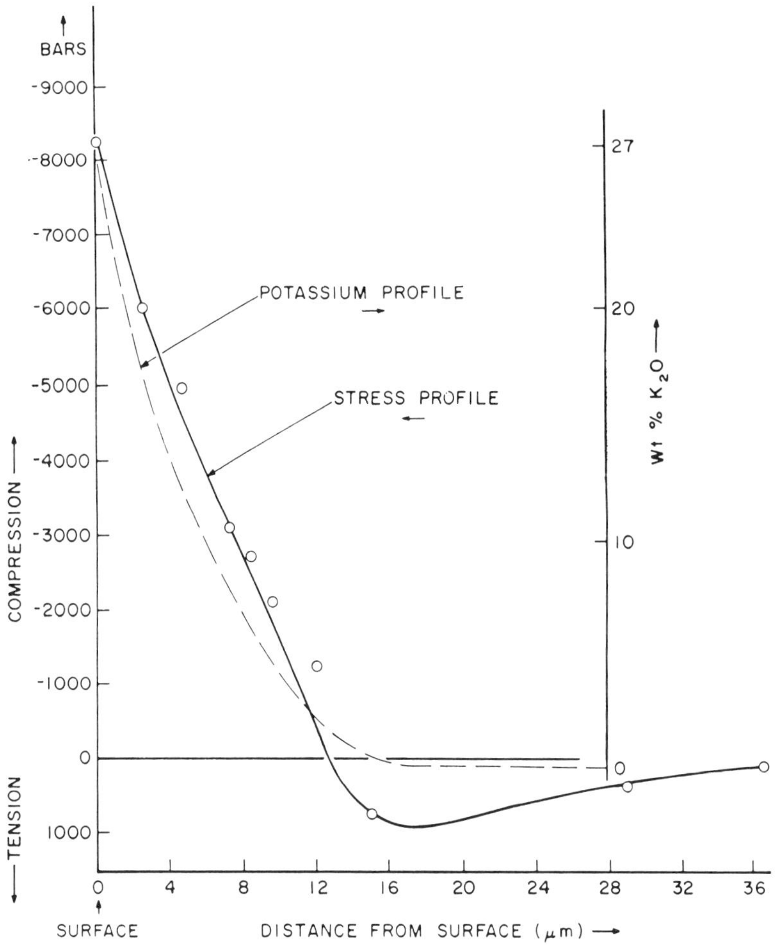

Figure 14-13. Stress profile in a sodium borosilicate glass sample after ion exchange at 400°C for 12 h. (After A. K. Varshneya, *J. Non-cryst. Sol.* **19**, 355 (1975). Reproduced with permission of Elsevier Science Publishers.)

not to the stuffed structures. (See Section 18.12 for practical techniques of ion-exchange strengthening.)

Despite the lack of understanding of details, the ion-exchange process can be used to strengthen a variety of glasses (invariably alkali-containing) to develop as much as 300–400 MPa surface compression readily. Typical experiments require 4 to 24 hours of exchange at 50–200°C below the glass transition temperature. Typical diffusion profiles are 40–300 μm deep. Because of the low magnitude of interior tension, chemically tempered glasses rarely dice, in contrast to thermally tempered glasses. Also, glass wafers and fibers as little as 100 μm thick can be strengthened using the ion-exchange process. Often it is possible to employ an electric field to accelerate the diffusion kinetics. In this **field-assisted ion exchange**, the alkali ions in the glass rapidly move towards the cathode region and are replaced by the bath alkali ions from the anode region. As expected, the concentration profiles of the two ions depend upon their mobility ratio. When the mobility of the glass ions is larger than that of the entrant ions, the latter move into the glass in the form of a "wall." On the other hand, when the entrant ions have larger mobility, the concentration profiles are more complex. The analysis of field-assisted ion exchange has been carried out by Abou-el-leil and Cooper [37], and by Shaisha and Cooper [38].

It has also been shown that the stresses generated by the ion exchange themselves affect the kinetics of diffusion. Tension, in effect, makes the average site larger; hence, the diffusion coefficient is larger. The opposite is the case when compression exists [39, 40].

14.8.2. Multicomponent Diffusion

When more than two species must take part in diffusion concurrently, the phenomenology of diffusion is described as multicomponent diffusion, following Onsager's linear extension of Fick's first law:

$$-J_i = \sum D_{ij}\, dc_j/dx \qquad (j = 1 \text{ to } n - m), \tag{14.54}$$

where n is the number of independent components and m is the number of restraints in the system. The D_{ij} are known as the **interdiffusion coefficients**. D_{ii} is the coefficient of i in its own concentration gradient, and D_{ij} $(i \neq j)$ is the coefficient of i in the gradient of j. Application of conservation of species leads to

$$\frac{dc_i}{dt} = \frac{d}{dx}\left[D_{ii}\frac{dc_j}{dx}\right] \qquad (j = 1 \text{ to } n - m). \tag{14.55}$$

For a simple case where semi-infinite slabs of two glasses are mated to each other with planar interface (diffusion couples), it can be shown [41] that the concentration profile of each species after a diffusion time t is described as a sum of error functions of the type (14.39), the number of error functions in the sum being equal to $(n - m)$. Figure 14-14a shows the distribution of the various species for a diffusion couple having the initial glass compositions H ($22.8K_2O{\cdot}15.6SrO{\cdot}61.3SiO_2$, analyzed wt. %) and L ($18.1K_2O{\cdot}18.0SrO{\cdot}64.1SiO_2$) following a diffusion anneal. The **uphill nature of the diffusion**

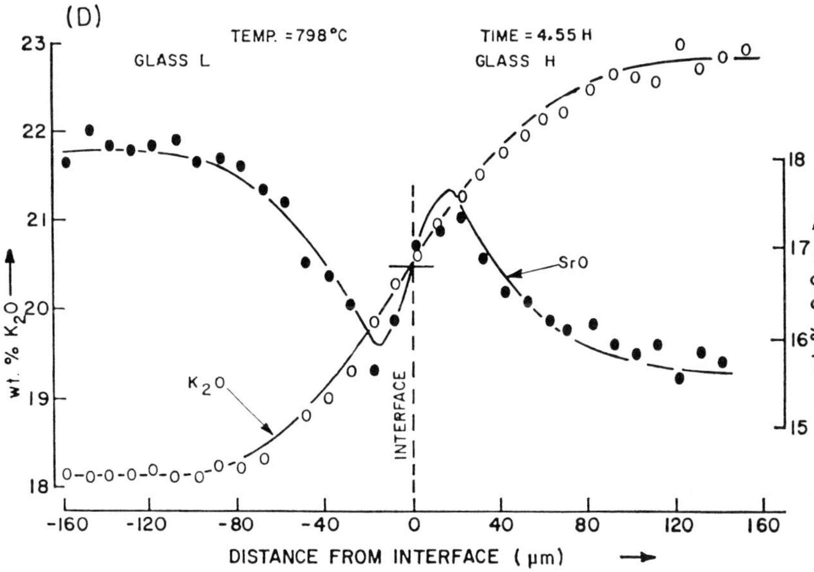

(a)

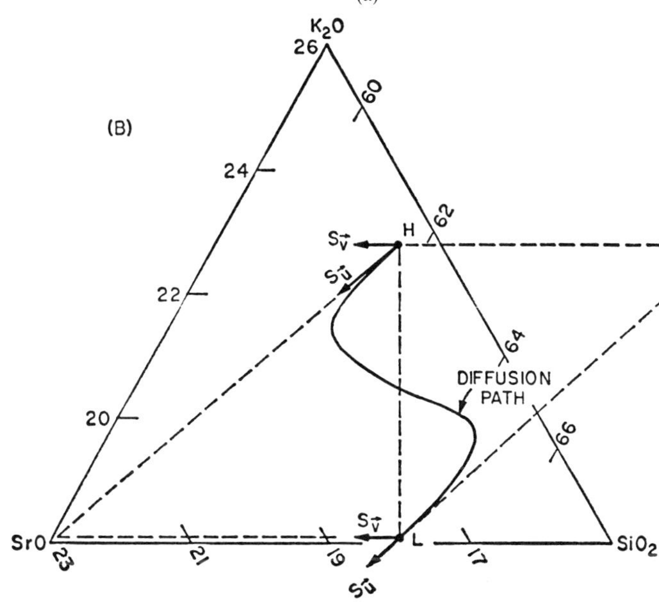

(b)

Figure 14-14. (a) Electron microprobe analysis of concentration with distance in the K_2O–SrO–SiO_2 system whose termini compositions are given by points H and L at 798°C. (A. K. Varshneya and A. R. Cooper, *J. Am. Ceram. Soc.* **55**(6), 312 (1972). (b) Diffusion paths for the H–L couple. S_u and S_v are "fast" and "slow" eigenvectors of the diffusion matrix. (After Varshneya and Cooper. Reproduced with permission of The American Ceramic Society.)

(diffusion against the gradient) near the interface is particularly striking. On a ternary diagram, the diffusion path (the locus of all the compositions) in such a case is an "S"-shaped curve (Fig. 14-14b). The phenomenology of multicomponent diffusion has been treated extensively by Gupta and Cooper [42].

14.8.3. Fast Ion Conduction

Suggested reading: K. S. Sidhu, S. Singh, S. S. Sekhon, S. Chandra, and A. Kumar, *Phys. Chem. Glasses* **32**(6), 255 (1991).
M. D. Ingram, *Phys. Chem. Glasses* **28**(6), 215–234 (1987).

It may be realized that the magnitudes of ionic conductivity in common glasses are generally quite low. The highest at room temperature are about 10^{-5} $Ohm^{-1}\cdot cm^{-1}$. In recent years, there has been considerable interest in the use of solid electrolytes for rechargeable battery or microbattery operations such as those needed for electric automobiles and heart pacemakers. This has led to an intense research effort into developing **fast ion conductors** or **superionic conductors**. Examples of glass families that have been found to have high conductivities are

(i) $AgX–Ag_2Ch–M_mCh_n$, where X = halogen, Ch = O, S, Se, and M = B, Si, Ge, P, V, Cr, Mo, W, Ta.
(ii) $CuI–Cu_2O–M_mO_n$.
(iii) $LiI–Li_2Ch–M_mS_n$.
(iv) $Li_2O–Al_2O_3–SiO_2–P_2O_5$.
(v) Alkali niobates, tantalates, aluminates, gallates, bismuthates, tungstates, and molybdates.

In the first family, mixtures of AgI with Ag_3AsO_4, Ag_2SeO_4, $Ag_2Cr_2O_7$, or $Ag_2Mo_2O_7$, generally in the molar ratio 4:1, have been made [43] by quenching extruded droplets in liquid N_2. Conductivity of these glasses follows an Arrhenius equation and approaches 0.1 $S\cdot m^{-1}$ at room temperature with an activation energy of only 18 to 28 $kJ\cdot mol^{-1}$. (To compare, copper conductivity at room temperature is 6×10^7 $S\cdot m^{-1}$.) The charge carriers are thought to be Ag^+ ions moving through channels rich in the halide ions.

The copper family of glasses have been reported [44] to have conductivities comparable to those of the Ag family.

The lithium family has been made [45] using the gel route to glass formation.

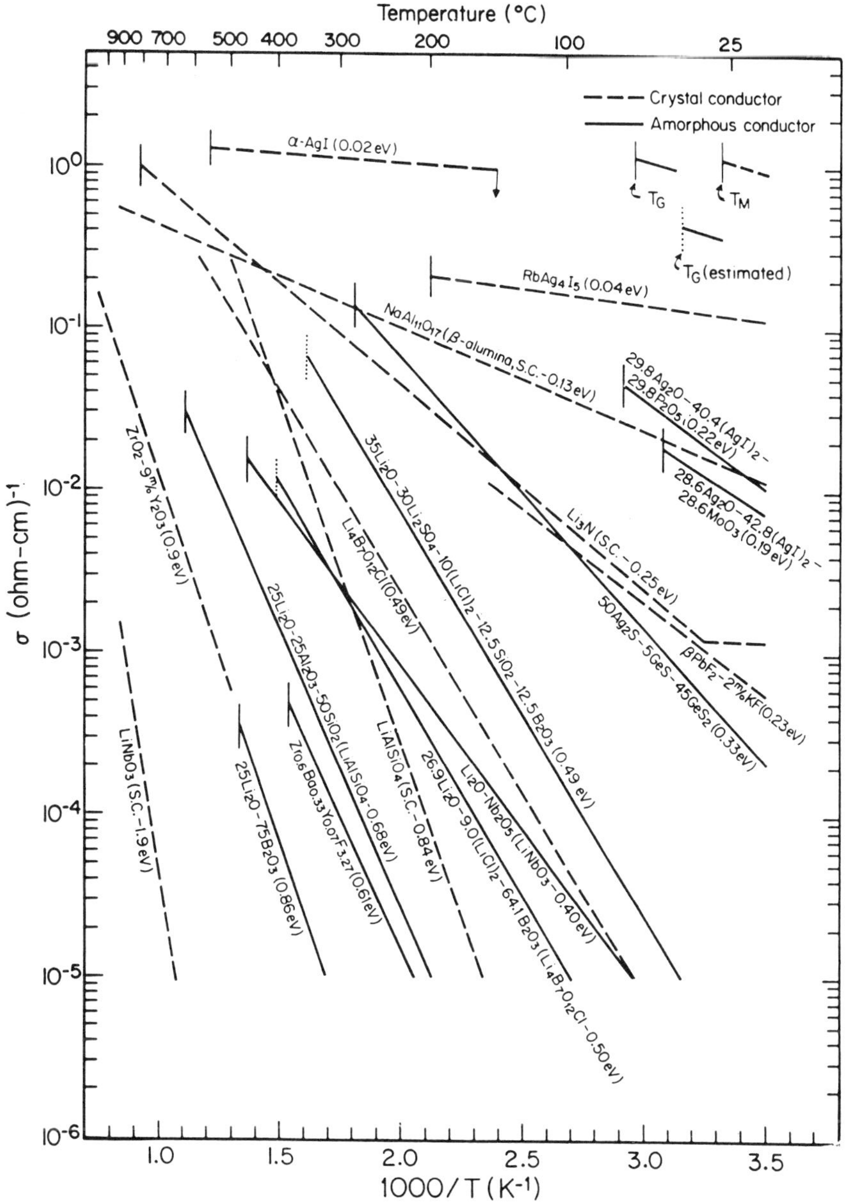

Figure 14-15. Ionic conductivities as a function of reciprocal temperature for a number of fast Ag, Li, and F conducting glasses compared to those of crystalline fast ion conductors. (After H. Tuller and M. W. Barsoum, *J. Non-cryst. Sol.* **73**, 331 (1985). Reproduced with permission of Elsevier Science Publishers.)

The alkali niobates, tantalates, etc., are made by quenching at cooling rates approaching 10^7 K·s^{-1} using twin rollers. Their conductivities are generally of the order of 10^{-4} S·m^{-1} at room temperature. Apparently, the glasses are more conducting than their crystal counterparts. Conductivities follow an Arrhenius-type relationship with temperature. An example is shown in Fig. 14-15 (after Tuller and Barsoum, *J. Non-cryst Sol.* **73**, 331 (1985).) It has been argued [46] that this fast conduction is due to large interstitial channels between corner-connected octahedra such as NbO_6.

14.8.4. Anionic Conduction

Cations are not the only charge-carrying species in glasses. Research into fast ion conducting glasses led to the discovery of anionic conduction in glass. It has been shown [47], for instance, that in $(65 - y)PbO \cdot yPbX_2 \cdot 35SiO_2$ glasses, where X = F, Cl, Br, I, the charge may be carried by both the halide ion and by Pb^{2+}. These glasses can be melted within 5 to 10 minutes by heating a suitable batch to 800°C in a vitreous silica crucible. Log (conductivity) at 100°C for these glasses as a function of y is shown in Fig. 14-16. The conductivity increases monotonically with PbF_2 addition,

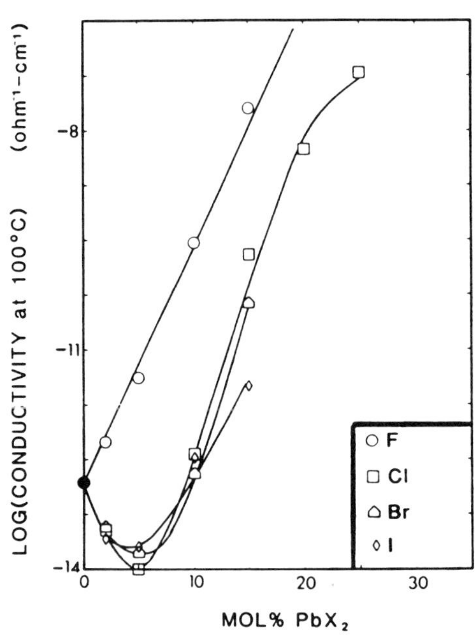

Figure 14-16. Effect of composition on the conductivity of $(65 - y)PbO \cdot yPbX_2 \cdot 35SiO_2$ glasses. (After J. Coon and J. E. Shelby, *J. Am. Ceram. Soc.* **73**(2), 379 (1990). Reproduced with permission of The American Ceramic Society.)

but goes through a minimum for the other halide systems. Apparently, in the fluoride system, the nonbridging F^- ion is the dominant charge carrier, being roughly the same size as the bivalent Pb^{2+}. However, for the other three halides, initial additions of the halide are believed to form bridging halogen bonds, leading to a decrease in conductivity. In this region, Pb^{2+} ions may be more mobile than the Cl^-, Br^-, or I^-. At higher values of y, the halide is believed to form nonbridging links to Pb, leading to an opening of the network, and hence to an increased conductivity.

In addition to the fluorosilicates, fluoroborates and fluorophosphates have also been shown to display fluoride ion conductivity. Reported [48] conductivities approach about 10^{-4} ohm^{-1} cm^{-1} at 200°C. High halide ion conductors have potential use as halide sensors.

Exercises

1. A disk, 0.5 mm thick, of $15Na_2O{\cdot}10CaO{\cdot}75SiO_2$ (wt. %) glass has been sandwiched between two semi-infinite thickness glass disks of identical composition. The middle disk contains trace amounts of radioactive Na. The sandwich is heat-treated at 727°C. Assume that $D_0 = 8 \times 10^{-6}$ cm^2/s and $\Delta H_D = 16$ kcal/mol in Eq. (14.30). Calculate the concentration distribution after (i) 1 hour, (ii) 400 hours of diffusion anneal.

2. Continuation of Chapter 7, Question 4: A 3-mm thick semi-infinite plate of $15Na_2O{\cdot}10CaO{\cdot}75SiO_2$ (mol %) glass is ion exchanged from both sides in KNO_3 bath for 16 hours such that the surface concentration is $2Na_2O{\cdot}13K_2O{\cdot}10CaO{\cdot}75SiO_2$ (mol %) glass, and the K^+ penetration profile can be approximated by a constant diffusion coefficient $D = 10^{-10}$ cm^2/s. Calculate the expected stress profile. (Measured stresses are generally a factor of 3 to 5 lower than those calculated; see text.)

3. A semi-infinite thick slab of $Na_2O{\cdot}3SiO_2$ glass is ion-exchanged from two sides in KNO_3 bath at 350°C for 100 hours. It is observed that the penetration profile of K^+ is approximately described by a constant diffusion coefficient $D = 10^{-10}$ cm^2/s. Assume that the density of the glass remains unchanged. Calculate the weight gain per unit area normal to the diffusion direction for the slab.

4. A tube of fused-silica glass, with one end closed, containing molten RNO_3 (R = alkali) is immersed in a molten salt bath containing $NaNO_3$. Pt wire electrodes are introduced in the outer bath (B) as well as inside the tube (A). The silica contains traces of Na and K impurities. Sketch the variation of current with time after a DC voltage is applied, if:

(i) R = Na; A = anode; B = cathode; long-time experiment at 500°C. (ii) R = K; A = anode; B = cathode; long-time experiment at 500°C. (iii) After (ii), the polarity is switched, temperature is maintained. (iv) After (ii), the temperature is raised to 550°C and the experiment is continued for a long time. (v) After (iv), the temperature is lowered back to 500°C and the experiment is continued for another long time. Compare the results of (v) with (ii).

5. What glass and metal seal combinations would you recommend for the following lamp applications:
(i) A household incandescent lamp, where the maximum service temperature requirement is a sustained electrical insulation between the current leads at about 150°C, and there is little concern for thermal shock.

(Answer: Although the lamp envelope is made from the most inexpensive soda lime silica glass, the mount structure is actually made using a potash soda lead glass that has a longer working range and a much higher electrical resistivity than the soda lime silica glass. The seal is made over "Dumet" wire, which is copper-clad 42Ni–Fe alloy. Because it is a composite wire, Dumet has an anisotropically thermal expansion coefficient. The expansion of the glass is "matched" well with the radial expansion of the alloy.)

(ii) A photoflash lamp, where the envelope experience a sudden temperature increase, but the outer wall temperature does not exceed ~200°C. Because of the temperature pulse, there is some need for superior thermal shock resistance relative to that of the soft glasses.

(Answer: These lamps are made from sodium borosilicate glasses having an α of $\sim 53 \times 10^{-7}/°C$. The seal is made using kovar alloy.)

(iii) A glass–halogen lamp where the service temperature requirement is a sustained 600°C (hence, the glass must also have high electrical resistivity at 600°C).

(Answer: These lamps are made using alkaline earth–aluminosilicate glasses. Because of the absence of alkalis, they have very high resistivities. The seal is made over Mo metal.)

References

1. J. R. Manning, *Diffusion Kinetics for Atoms in Crystals.* Van Nostrand, Princeton, New Jersey, 1968.
2. A. Schoen, *Phys. Rev. Lett.* **1**, 138 (1958).
3. O. L. Anderson and D. A. Stuart, *J. Am. Ceram. Soc.* **37**(12), 573–580 (1954).

4. J. E. Shelby, in *Treat. on Mat. Sci. & Tech.*, Vol. 17, eds. M. Tomozawa and R. H. Doremus. p. 1. Academic Press, New York, 1979.
5. G. H. Frischat, *Ionic Diffusion in Oxide Glasses*. Trans Tech Publishers, Aedermannsdorf, Switzerland, 1975.
6. J. Crank, *The Mathematics of Diffusion*. Oxford at the Clarendon Press, 1956.
7. C. Hastings, in *Handbook of Mathematical Functions* (M. Abramovitz and I. A. Stegun, eds.), p. 299. U.S. Dept. of Commerce, 1964.
8. R. Robin, A. R. Cooper, and A. H. Heuer, *J. Appl. Phys.* **44**, 3770 (1973).
9. A. K. Varshneya and A. R. Cooper, *J. Am. Ceram. Soc.* **55**(4), 220–223 (1972).
10. A. K. Varshneya and A. R. Cooper, Jr., *J. Am. Ceram. Soc.* **55**(6), 312–317 (1972).
11. P. G. Shewmon, *Diffusion in Solids*, p. 29. McGraw-Hill, New York, 1963.
12. M. Tomozawa, in *Treatise on Mat. Sci. & Tech.*, Vol. 12, *Glass I* (M. Tomozawa and R. H. Doremus, eds.), p. 296. Academic Press, New York, 1977.
13. V. Jain, A. K. Varshneya, and P. P. Bihuniak, *J. Am. Ceram. Soc.* **72**(5), 843–845 (1989); *ibid.*, **73**(2), 409–414 (1990).
14. D. Ravaine and J. L. Souquet, *Phys. Chem. Glasses* **18**(2), 27–31 (1977).
15. M. Kleitz and J. H. Kennedy, in *Fast Ion Transport in Solids* (P. Vashishta, J. N. Mundy, and G. K. Shenoy, eds.), pp. 185–188. Elsevier, Amsterdam, 1979.
16. L. I. Meissel, in *Handbook of Thin Film Technology*, pp. 13-1 to 13-13. McGraw-Hill, eds. L. Meissel and R. Glang. New York, 1970.
17. R. N. Blumenthal and M. A. Seitz, in *Electrical Conductivity in Ceramics and Glass*, Part A, (N. M. Tallan, ed.), pp. 35–178. Marcel Dekker, New York, 1974.
18. F. J. Norton, *J. Am. Ceram. Soc.* **36**, 90 (1953).
19. R. F. Bartholomew and H. L. Hoover, *Glastech. Ber.* **56K**, 671 (1983), quoted by M. Tomozawa, *J. Non-cryst. Sol.* **73**(1), 199 (1985).
20. G. H. Frischat, *J. Am. Ceram. Soc.* **51**, 528 (1968).
21. V. Jain, A. K. Varshneya, and P. P. Bihuniak, *J. Am. Ceram. Soc.* **73**(2), 409–414 (1990).
22. V. Jain, "On the Nature of Ionic Conductivity in Fused Silica Glasses," PhD. thesis, submitted to Alfred University (1987).
23. J. R. Johnson, R. H. Bristow, and H. H. Blau, *J. Am. Ceram. Soc.* **34**, 135 (1951).
24. A. K. Varshneya, *J. Am. Ceram. Soc.* **57**(1), 37–39 (1974).
25. J. R. Hendriksen and P. J. Bray, *Phys. Chem. Glasses* **13**(4), 43 and 107 (1972).
26. D. E. Day, *J. Non-cryst. Sol.* **21**(3), 343–372 (1976).

27. G. Tomandl and H. Schaeffer, *J. Non-cryst. Sol.* **73**, 179–196 (1985).
28. W. C. LaCourse, *J. Non-cryst. Sol.* **95&96**, 905–912 (1987).
29. C. Huang and A. N. Cormack, *Proc. VII Phys. of Non-cryst. Sol.*, pp. 31–35. Taylor & Francis, London, 1992; *J. Mat. Chem.* **2**, 281–287 (1992).
30. C. T. Moynihan and A. V. Lesikar, *J. Am. Ceram. Soc.* **64**(1), 40–46 (1981).
31. M. Tomozawa and V. McGahay, *J. Non-cryst. Sol.* **128**, 48–56 (1991).
32. F. G. K. Bauke, *J. Non-cryst. Sol.* **73**, 215–231 (1985).
33. A. K. Varshneya and M. E. Milberg, *J. Am. Ceram. Soc.* **57**(4), 165–169 (1974).
34. J. P. Lacharme, *C. R. Acad. Sci. Ser. C*, **270**(16), 1350–1353 (1970).
35. A. R. Cooper and D. A. Krohn, *J. Am. Ceram. Soc.* **52**(12), 665–669 (1969).
36. A. K. Varshneya, *J. Am. Ceram. Soc.* **58**(3–4), 106–109 (1975).
37. M. Abou-el-leil and A. R. Cooper, *J. Am. Ceram. Soc.* **62**(7–8), 390–395 (1979).
38. E. E. Shaisha and A. R. Cooper, *J. Am. Ceram. Soc.* **64**(1), 34–36 (1981).
39. G. Dumais and A. K. Varshneya, *J. Am. Ceram. Soc.* **68**(7), C165–166 (1985).
40. M. Tomozawa, *J. Non-cryst. Sol.* **73**, 197–204 (1985).
41. A. K. Varshneya and A. R. Cooper, *J. Am. Ceram. Soc.* **55**(6), 312–317 (1972).
42. P. K. Gupta and A. R. Cooper, *Physica* **54**(1), 39 (1971).
43. R. J. Grant, M. D. Ingram, L. D. S. Turner, and C. A. Vincent, *J. Phys. Chem.* **82**(26), 2838–2844 (1978).
44. C. Liu and A. Angell, *Sol. St. Ionics* **13**, 105 (1984).
45. J. P. Boilot, Ph. Colomban, and N. Blanchard, *Sol. St. Ionics* **9/10**, 639 (1983).
46. A. M. Glass, K. Nassau, and D. H. Olson, in *Fast Ion Transport in Solids* eds. P. Vashishta, J. N. Mundy and G. K. Shenoy, pp. 707–712. Elsevier, Amsterdam, 1979.
47. J. Coon and J. E. Shelby, *J. Am. Ceram. Soc.* **73**(2), 379–382 (1990).
48. C. Gressler and J. E. Shelby, *J. Appl. Phys.* **66**(3), 1127–1131 (1989).

Chapter 15

Dielectric Properties

15.1. Introduction

Suggested reading: T. G. Barnes, *Foundations of Electricity and Magnetism*. D. C. Heath & Co., Boston, 1965.

When a dc voltage V is applied to a large parallel-plate capacitor having a narrow vacuum gap d between the plates of area A each, the plates are charged with $+q$ and $-q$ charges per unit area immediately. The electrical flux density or the **electric displacement vector $\boldsymbol{D}$** is defined by the integral form of Maxwell's third field equation (Gauss's law) stating that the surface integral of the normal component of $\boldsymbol{D}$ over an arbitrary enclosed surface is equal to the net charge enclosed by the surface:

$$\int \boldsymbol{D} \cdot \mathbf{n}\, dS = Aq, \tag{15.1}$$

where $\mathbf{n}$ is a unit vector normal to a surface element dS. In a parallel-plate capacitor, the flux lines originate from the charge on the positively charged plate and end at the negative charge on the opposite plate. Thus, the $\boldsymbol{D}$ lines are parallel to $\mathbf{n}$, with the result that Eq. (15.1) reduces to

$$D_0 = q, \tag{15.2}$$

where the subscript 0 has been used to denote vacuum. The **electrical field strength vector E** between the plates is defined by

$$\boldsymbol{D}_0 = \varepsilon_0 \mathbf{E}, \tag{15.3}$$

where ε_0 is the **absolute permittivity (dielectric constant)** of free space, which in SI units has the value $8.855 \times 10^{-2}\,\mathrm{C}^{-2}\mathrm{N}^{-1}\mathrm{m}^{-2}$ ($\equiv$farad m^{-1}). (The reader should note the use of the SI system. Older books may have $\boldsymbol{D}_0 = 4\pi q$, and $\boldsymbol{D} = E$.) E is related to the potential through

$$V = \int_0^d \mathbf{E} \cdot \mathbf{j}\, dy = Ed, \tag{15.4}$$

where $\mathbf{j}$ is a unit vector in the direction of the plate separation (assumed to be along the y-axis). The capacitance C_0 of the parallel plates is given by

$$\begin{aligned} C_0 &= Aq/V \\ &= A\varepsilon_0/d, \end{aligned} \tag{15.5}$$

by substituting for q and V. Let us now replace the vacuum between the plates by a **dielectric** that is *not a good conductor of electricity*. The action of a dielectric between the plates is to reduce the applied voltage, or if the voltage were to be maintained constant, then the charge on the $+q$ plate should increase to $+(q + P)$ and on the $-q$ plate to $-(q + P)$. The increase in charge P is called the **polarization** and can be considered to result from the appearance of $-P$ "bound" charges on the dielectric surface adjacent to the $+q$ plate. In vectorial representation,

$$P = \mathbf{n} \cdot \mathbf{P}.$$

The flux lines originating from the $+P$ "bound" charges on the $+(q + P)$ plate end on the $-P$ "bound" charges on the adjacent dielectric surface (see Fig. 15-1). The capacitance C of the new system is now increased to $A(q + P)/d$, and the new electric flux density is $(q + P)$. The ratio C/C_0 (which is greater than 1) is called the **static relative permittivity** ε_s (or the **relative dielectric constant**). Since the net electric displacement D may be written as

$$D = D_0 + P, \tag{15.6}$$

we have, using the basic definitions,

$$\varepsilon_s = C/C_0 = (q + P)/q = D/D_0 = D/\varepsilon_0 E. \tag{15.7}$$

From (15.6) and (15.7), we get several other useful relations:

$$\varepsilon_s - 1 = P/q = P/\varepsilon_0 E = \chi, \tag{15.8}$$

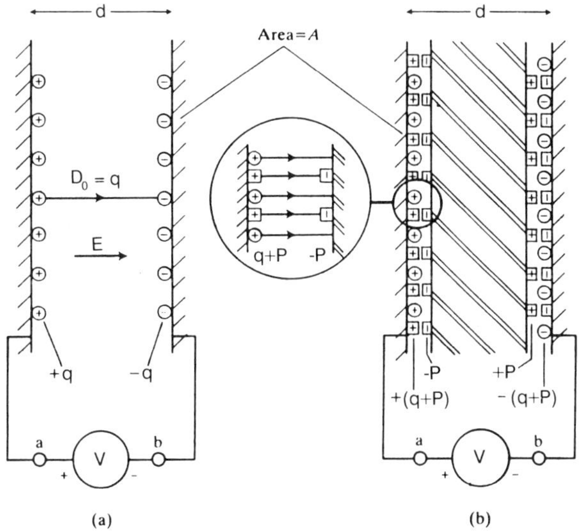

Figure 15-1. (a) A parallel-plate capacitor connected to a potential difference V. With vacuum between the plates, the electric displacement is D_0 (=charge q). The flux lines start from the + charges on the positive plate and end on the − charges on the negative plate. (b) Upon introducing a dielectric between the plates, polarization charges $\pm P$ appear on the dielectric facing the plates. This causes the charges on the plates to become $(q + P)$ and $-(q + P)$. Inset shows exaggeration of the dielectric face region nearest the plate. (Adapted after B. K. P. Scaife, *Principles of Dielectrics*, Fig. 1.2, p. 3. Oxford Science Publ., Clarendon Press, Oxford, 1989. Reproduced with permission of Oxford University Press.)

where χ is called the **dielectric susceptibility**, and

$$D = \varepsilon_0 E + P = \varepsilon_s \varepsilon_0 E = \varepsilon E, \tag{15.9}$$

where ε is called the **electric permittivity** of the medium (similar to ε_0 of vacuum). Had we carried the vectorial representation, we would have arrived at

$$\boldsymbol{D} = \boldsymbol{D}_0 + \mathbf{P} = \varepsilon_0 \mathbf{E} + \mathbf{P} = \varepsilon_s \varepsilon_0 \mathbf{E} = \varepsilon \mathbf{E}. \tag{15.10}$$

In general, the electric displacement vector $\boldsymbol{D}$ need not be parallel to the electric field vector $\mathbf{E}$; hence, ε is a second-rank tensor. However, in isotropic bodies such as glass, the vectors $\boldsymbol{D}$ and $\mathbf{E}$ are parallel; hence, ε may be treated as a scalar quantity. Also, from (15.10),

$$\mathbf{P} = (\varepsilon_s - 1)\varepsilon_0 \mathbf{E} = (\varepsilon - \varepsilon_0)\mathbf{E} = \alpha \mathbf{E}, \tag{15.11}$$

which defines the **polarizability** α of the medium.

Again, the polarization vector is assumed to be parallel to the field vector **E** for glass such that the polarizability α is a scalar quantity. Henceforth, the use of bold letters in this chapter will be dropped, in effect treating all quantities as scalars.

It may be recognized that dielectric polarization involves the appearance of induced charges as **dipoles** (equal charges of opposite sign separated by distance d). These dipoles create a depolarization field whose direction is opposite to the field that generates the polarization. With time, some of the induced charges tend to migrate towards each other, and hence the depolarization field will decay. A capacitor with a real dielectric medium filling the space may thus be equated to various combinations of resistance and capacitance. (The presence of vacuum between the capacitor plates implies the existence of a pure capacitor.) For a series combination of capacitance C_s and resistance R_s, it may be shown [1] that for the charging of a capacitor with an applied step voltage, V_0, the charge Q has a time dependence:

$$Q = C_s V_0[1 - \exp(-t/\tau)], \tag{15.12}$$

where $\tau = R_s C_s$ is the time constant. The buildup of the full charge is thus not instantaneous. For discharging of the capacitor, the charge Q decays according to

$$Q = C_s V_0 \exp(-t/\tau). \tag{15.13}$$

The preceding discussion also implies that the sudden application of a constant field produces an instant polarization P_∞ that approaches a static value P_s over a period of time. The two polarization values may be defined in terms of Eq. (15.11) as

$$P_\infty = (\varepsilon_\infty - 1)\varepsilon_0 E, \qquad \text{and } P_s = (\varepsilon_s - 1)\varepsilon_0 E. \tag{15.14}$$

Here, ε_∞ may be called the **instantaneous dielectric constant**.

If an alternating field represented by

$$V^* = V_0 \exp(j\omega t), \tag{15.15}$$

where $j \equiv \sqrt{-1}$ and ω is the angular frequency, is applied to the capacitor plates, then an alternating field develops with the field strength E^* given by

$$E^* = V^*/d = (V_0/d)\exp(j\omega t) = E_0 \exp(j\omega t) \tag{15.16}$$

The complex displacement flux D^* for a real dielectric is, in general, not in phase with E^* and is given by

$$D^* = D_0 \exp j(\omega t - \delta), \tag{15.17}$$

where δ is the phase difference. For the alternating field, one can rewrite Eq. (15.10) in terms of D^* and E^* as

$$D^* = \varepsilon^* \varepsilon_0 E^*, \tag{15.18}$$

where ε^* is the **complex dielectric constant** of the form

$$\varepsilon^* = \varepsilon' - j\varepsilon''. \tag{15.19}$$

The real part, ε', is the **dielectric constant**, and the imaginary part, ε'', is the **loss factor**. ε^* approaches ε_s at low frequencies, and ε_∞ at high frequencies. These terms may now be renamed as the **zero-frequency** and the **high-frequency dielectric constants**, respectively. The **complex capacitance** C^* may be defined from

$$C^* = \varepsilon^* C_0. \tag{15.20}$$

It follows that

$$\varepsilon^* \varepsilon_0 = (D_0/E_0)\exp(-j\delta) = (D_0/E_0)(\cos\delta - j\sin\delta) \tag{15.21}$$

and hence, the ratio

$$\varepsilon''/\varepsilon' = \tan\delta, \tag{15.22}$$

where δ is the phase angle defined by Eq. (15.17). The term $\tan\delta$ is called the **loss angle**. (The reader may note the similarities between the treatment of properties as complex numbers included in Section 13.2.1, here, and later on in Chapter 19.)

The frequency dependence of ε' and ε'' is generally described in terms of the Debye relations; these are

$$\varepsilon' = \varepsilon_\infty + \frac{\varepsilon_s - \varepsilon_\infty}{1 + \omega^2\tau^2}, \tag{15.23}$$

$$\varepsilon'' = (\varepsilon_s - \varepsilon_\infty)\frac{\omega\tau}{1 + \omega^2\tau^2}, \tag{15.24}$$

where τ is called the **relaxation time** for the process. The variation of ε' and ε'' as a function of frequency is shown in Fig. 15-2. These curves are referred to as dielectric dispersion and absorption (or loss), respectively, and are symmetric about $\omega\tau = 1$. Note that the loss curve displays a single relaxation maximum at this location. Most real dielectrics show broader distributions than those predicted by Eqs. (15.23) and (15.24), presumably because of a spectrum of relaxation times.

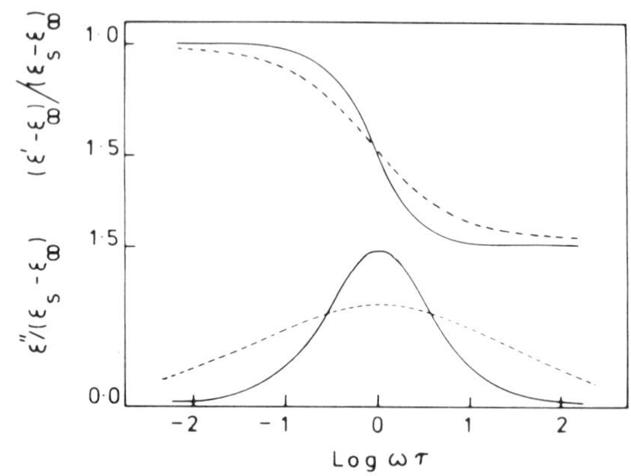

Figure 15-2. The dependence of the dielectric constant and dielectric loss on frequency.

Recall that the current I is the time derivative of the charge on the plate, i.e., $I = A\dot{D}$. In an alternating field, this leads to

$$I^* = Aj\omega\varepsilon^*\varepsilon_0 E^* = [j\omega\varepsilon_0\varepsilon' + \omega\varepsilon_0\varepsilon''](V^*A/d). \tag{15.25}$$

The expression within the square brackets is of the form $(\cos\phi - j\sin\phi)$, which implies that I leads V by a phase difference ϕ, causing a power loss. Since $\tan\phi = \varepsilon''/\varepsilon' = \cot\delta = \tan[(\pi/2) - \delta]$, it follows that $\phi = (\pi/2) - \delta$. In a loss-free dielectric, the current leads the voltage by $\pi/2$. Also, $I^* = j\omega C^* V^*$. If we write Ohm's law as $V^* = Z^* I^*$, where Z^* is the **complex impedance**, then

$$Z^* = 1/j\omega C^*. \tag{15.26}$$

The reciprocal of the complex impedance is the **complex admittance** Y^*.

As a material, dielectrics are basically insulators to semiconductors of electricity. When an alternating electric field is applied to a relatively more conducting dielectric slab, the equivalent circuit may be described by a resistance R_p ($=1/G_p$, G_p being called the conductance) and a capacitance C_p in parallel (Fig. 15-3). The real parts of the **complex electrical conductivity** (defined below), σ', and the dielectric constant, ε', are given by

$$\sigma'(\omega) = G_p(\omega)(d/A), \tag{15.27}$$

and

$$\varepsilon'(\omega) = [C_p(\omega)/\varepsilon_0](d/A), \tag{15.28}$$

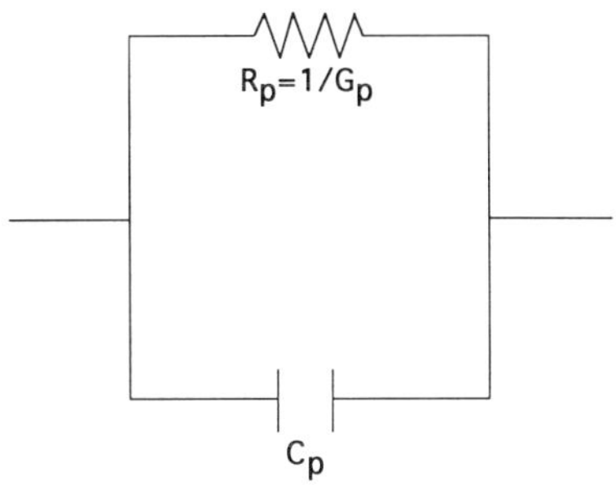

Figure 15-3. Equivalent circuit of a real dielectric.

where both the resistance and the capacitance are, in general, frequency-dependent. For the parallel circuit, the admittances, $G_p(\omega)$ (from Eq. (15.27)) and $j\omega C_p(\omega)$ (from Eq. (15.28)) of the two components are additive. The response of the circuit may be expressed in one of the following four mechanisms [2,3]:

1. Complex admittance $Y^*(\omega)$ or conductivity $\sigma^*(\omega)$

$$Y^*(\omega) = G_p(\omega) + j\omega C_p(\omega) = Y' + Y'', \tag{15.29}$$

$$\sigma^*(\omega) = Y^*(\omega)(d/A) = \sigma' + j\omega\varepsilon_0\varepsilon' = \sigma' + \sigma''. \tag{15.30}$$

2. Complex capacitance C^* or permittivity ε^*:

$$C^*(\omega) = Y^*(\omega)/j\omega = C_p(\omega) - jG_p(\omega)/\omega = C' - jC, \tag{15.31}$$

$$\varepsilon^*(\omega) = [C^*(\omega)/\varepsilon_0](d/A) = \varepsilon' - j\sigma'/\omega\varepsilon_0 = \varepsilon' - j\varepsilon''. \tag{15.32}$$

3. Complex impedance Z^* or resistivity ρ^*:

$$\begin{aligned} Z^* &= 1/Y^*(\omega) = \{G_p(\omega)/[G_p^2(\omega) + \omega^2 C_p^2(\omega)]\} \\ &\quad - \{j\omega C_p(\omega)/[G_p^2(\omega) + \omega^2 C_p^2(\omega)]\} \\ &= Z' - jZ'', \end{aligned} \tag{15.33}$$

$$\rho^* = Z^*(A/d) = \{\sigma'/[\sigma'^2 + \omega^2\varepsilon_0^2\varepsilon'^2]\} - \{j\omega\varepsilon_0\varepsilon'/[\sigma'^2 + \omega^2\varepsilon_0^2\varepsilon'^2]\} = \rho' - j\rho''. \tag{15.34}$$

4. Complex electric modulus M^*:

$$\begin{aligned} M^* &= 1/\varepsilon^* = \varepsilon_0(A/d)/C^*(\omega) \\ &= \varepsilon_0(A/d)j\omega Z^* = \varepsilon_0 j\omega\rho^* = \varepsilon_0\omega\rho'' + j\varepsilon_0\omega\rho' = M' + jM''. \end{aligned} \tag{15.35}$$

To study dielectric relaxation in such conducting dielectrics, the loss factor, $\sigma/\omega\varepsilon_0$, corresponding to the dc conductivity, σ, is subtracted out of the observed total ε'' (see Eq. (15.32)). The residual complex dielectric constant ε_p^* is then written as

$$\varepsilon_p^* = \varepsilon' - j(\varepsilon'' - \sigma/\omega\varepsilon_0) = \varepsilon' - j\varepsilon_p''. \tag{15.36}$$

In ionically conducting glasses, this treatment presents a problem because the frequency corresponding to the peak of the dielectric loss ε_p'' itself arises from σ. This may be demonstrated by assuming an extreme case of an "ideal solid electrolyte," where G_p and C_p are frequency-independent. For such a case, $\varepsilon' = \varepsilon_s$, which is also frequency-independent, and from (15.32),

$$\varepsilon^* = \varepsilon_s - j\sigma/\omega\varepsilon_0. \tag{15.37}$$

For the equivalent circuit, the current through the resistor $I_R^* = V^*/R_p$, and that through the capacitor $I_C^* = C_p\dot{V}^*$. The total current I^* is given by

$$I^* = I_R^* + I_C^* = [V^*/R_p] + [C_p\dot{V}^*]. \tag{15.38}$$

For the applied ac voltage $V^* = V_0 \exp(j\omega t)$, Eq. (15.38) can be solved to give

$$Z^* = V^*/I^* = R_p[1/\{1 + j\omega\tau\}], \tag{15.39}$$

where $\tau = R_pC_p$ is the **conductivity relaxation time** characterizing the decay of the electric field due to the dc conduction process under the constraint of a constant displacement vector D. In other words,

$$E = E_0 \exp(-t/\tau). \tag{15.40}$$

Since $\varepsilon_s = [C_p/\varepsilon_0](d/A)$, and $R_p = d/\sigma A$, we get

$$\tau = R_pC_p = \varepsilon_s\varepsilon_0/\sigma. \tag{15.41}$$

Hence, it follows from (15.37) and (15.35) that

$$\varepsilon^* = \varepsilon_s - j\varepsilon_s/\omega\tau = \varepsilon_s[1 + j\omega\tau]/j\omega\tau \tag{15.42}$$

and

$$M^* = M_s j\omega\tau/[1 + j\omega\tau], \qquad \text{where } M_s = 1/\varepsilon_s. \tag{15.43}$$

Plots of the real and imaginary parts of the various quantities as a function of the angular frequency ω, and in the complex plane (by plotting the imaginary part against the real part), are shown in Fig. 15-4. For convenience,

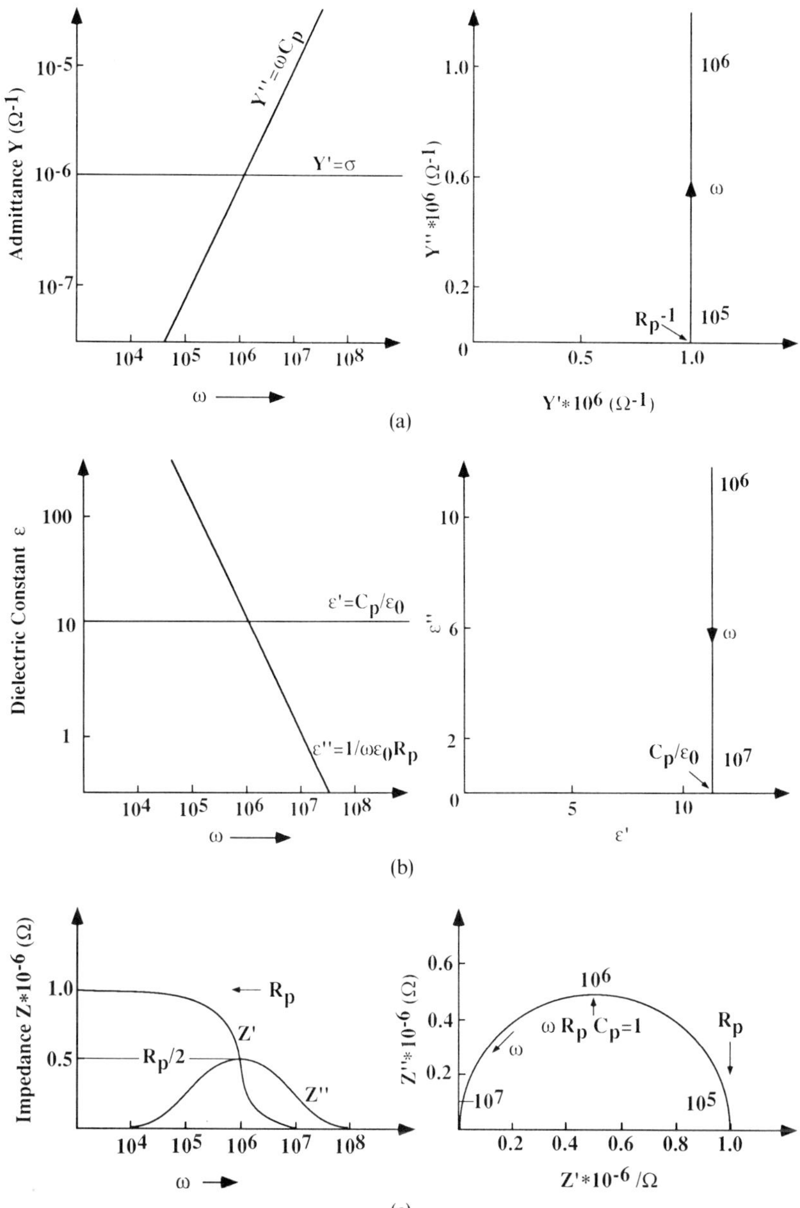

Figure 15-4. Complex plane plots of the various quantities associated with the measurement of dielectric properties. The dielectric is assumed to have an equivalent circuit as shown in Fig. 15-3. Assumed $R_p = 10^6\ \Omega$, and $C_p = 10^{-12}$ F. (a) Admittance *Y*. (b) Dielectric constant ε. (c) Impedance *Z*.

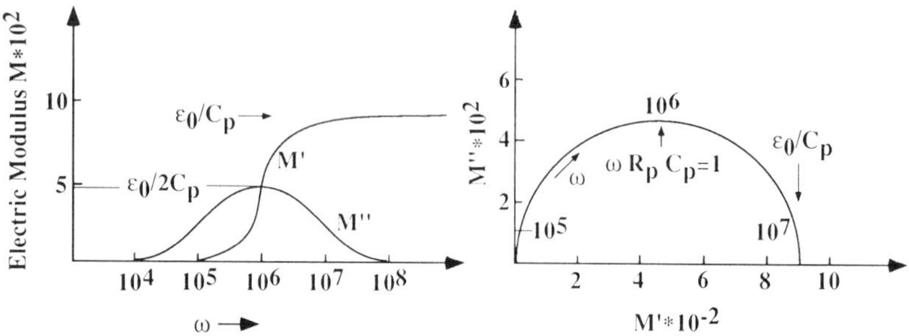

(d)

Figure 15-4. (d) Electric modulus M. (After Hodge *et al.* [2]. Reproduced with permission of Elsevier S. A., Lausanne.)

R_p, C_p, and (d/A) have been set equal to $10^6\ \Omega$, 10^{-12} F, and 1, respectively. Since $d/A = 1$, $Y^* = \sigma^*$, and $Z^* = \rho^*$. It may be noted that the plots for $\log(Y^*)$ and $\log(\varepsilon^*)$ are straight lines (e.g., $Y' = G_P$, which is constant with ω, and $Y'' = \omega C_p$), and ε' and σ' do not show any dispersion. On the other hand, Z', Z'', M', and M'' vs. $\log(\omega)$ display dispersion and absorption behavior of a single Debye relaxation process, the peak in Z'' and M'' being at $\omega_{max} = 1/\tau$. Complex plane plots are semicircles with the peak being, again, at $\omega_{max} = 1/\tau$. The behavior of M'' or Z'' represents **conduction loss**. Hence, many authors prefer not to use the ε^* formalism, but use M^* and Z^* (or ρ^*). Further, $M' \to 0$ as $\omega \to 0$ implies a lack of restoring force for the flow of electrical charge in a conducting dielectric in a constant electrical field. This is analogous to the shear modulus at low frequencies at constant shear stress. For this reason, M is called the **electric modulus**.

All dielectrics suffer what is known as a **breakdown** when the applied voltage exceeds a critical value, called the **dielectric strength**. The breakdown results in a changeover from insulating to conducting behavior, usually in a matter of microseconds. Dielectric breakdown in glass generally has two causes: (i) intrinsic, in an extreme case of polarization, where under the action of high electric fields the electrons are accelerated to the point of causing ionization of atoms; and (ii) thermal, where a material absorbs large amounts of energy by virtue of a high dielectric loss factor, causing it to heat up, further lowering its electrical resistivity, increasing current flow, increasing heating, and so on.

Dielectric constants are important for many applications. High-magnitude capacitors obviously require high dielectric constants. On the other hand, in integrated circuit applications, the substrates are required to have low dielectric constants for higher interconnect signal speeds between chips. As shown in Fig. 15-5 (after Ref. 4), the delay time per length of

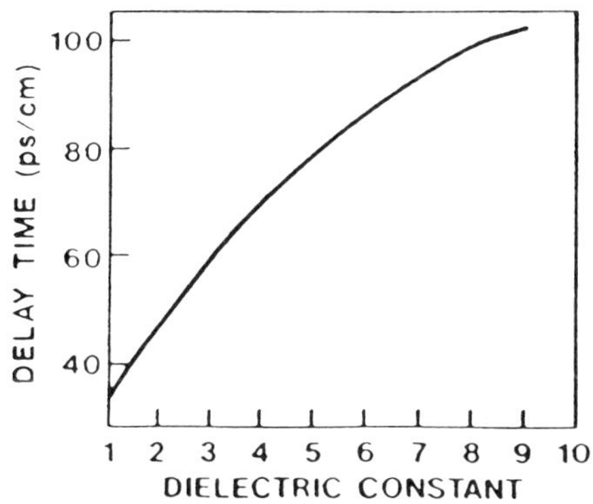

Figure 15-5. Propagation delay time vs. the dielectric constant. (After McDowell and Beall [4]. Reproduced with permission of The American Ceramic Society.)

the interconnect can be decreased significantly by lowering the dielectric constant.

15.2. Measurement of Dielectric Properties

Dielectric properties are measured as a function of the frequency of the alternating voltage using commercially available ac bridges similar to a Wheatstone bridge. An example is the bridge shown schematically in Fig. 15-6. The variable capacitors C_B and C_N are adjusted till the detector shows

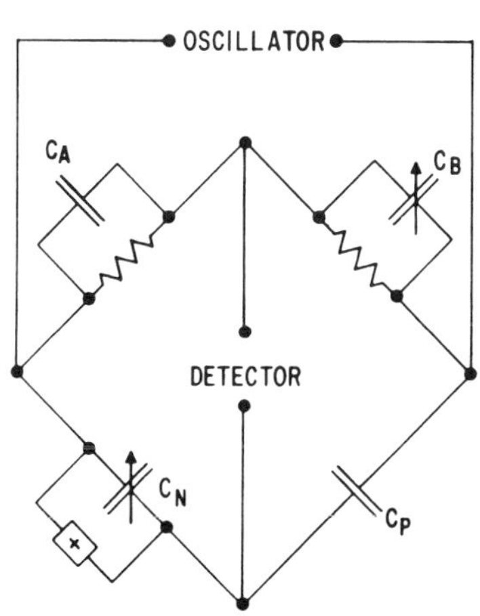

Figure 15-6. Schematic diagram of a bridge circuit to measure dielectric properties. (After N. P. Bansal, and R. H. Doremus, *Handbook of Glass Properties*, Fig. 14.1, p. 451. Academic Press, New York, 1986.)

minimum current, first with the unknown specimen X attached in parallel to C_B, and next without it. The real part $C'(\omega)$ is given by the difference of C_B values, whereas tan δ is proportional to $C_B - C_N$.

15.3. Data on Dielectric Properties

The dielectric constant of glasses decreases with frequency less steeply than predicted by the Debye function, supporting the notion established earlier, in the chapter on the glass transition range behavior, that the single relaxation-time concept is not valid for glasses. Dielectric constants increase with increasing temperature.

The dielectric constant of glasses is generally in the range of 4–11 at 1 MHz and 20°C. Fused silica is about the lowest, having $\varepsilon' = 3.8$. The dielectric constant decreases slightly by additions of B_2O_3, is almost unaffected by addition of TiO_2, and increases considerably by addition of alkalis, alkaline earths, and other network-modifying oxides. Typical values of ε' for soda lime glasses are 7–10, whereas those for sodium borosilicates are around 4.5–8. Some internally nucleated glass-ceramics in the B_2O_3–P_2O_5–SiO_5 system have been shown [4] to have dielectric constants in the 3.8–4.5 range, suggesting potential application as substrates for microelectronic packages substituting for Al_2O_3 where ε' is about 9. Dielectric constants may be increased sharply by additions of heavy metal oxides such as BaO, CdO, PbO, and Bi_2O_3.

The frequency variation of the loss tangent tan δ often displays peaks whose positions depend upon the glass composition. A typical spectrum for an alkali-containing glass is shown in Fig. 15-7 (after Stevels [5]). The magnitude of the tan δ may typically be as little as 0.002 in the low-loss region, compared to as much as 0.03 in the high-loss region. According to Stevels, the total loss may be split into four components:

(i) Conduction losses due to the mobility of the network modifiers. With increasing applied field frequency, the ions are increasingly unable to respond to the field, and hence this component decreases with frequency. Likewise, since the mobility increases with temperature, the conduction loss component increases with temperature.

(ii) Relaxation losses that occur because of the relaxation, again, of the network modifier ions during jumps over potential barriers. These losses are also observed mostly at low frequencies. The relaxation time of these losses is essentially the same as that in the mechanical relaxation spectrum.

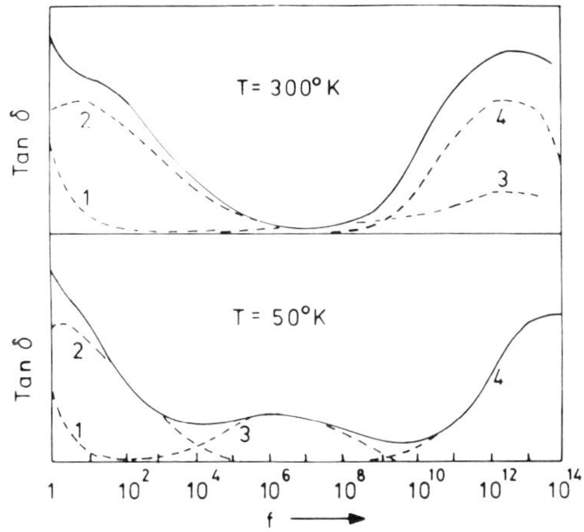

Figure 15-7. Schematic diagram of the dependence of the dielectric loss on frequency and temperature. (After Stevels [5]. Reproduced with permission of the National Academy Press.)

(iii) Resonance losses occur when the applied frequency is the same as the characteristic vibration frequency of the atoms.
(iv) Deformation losses occur when the atomic substructures begin to vibrate, resulting in local deformations.

It may be concluded that tan δ might be reduced by having low-mobility network modifiers. Indeed, glasses of the type (BaO,CaO)–Al_2O_3–SiO_2 and PbO–B_2O_3–SiO_2 have [6, 7] values of tan δ as low as about 0.0012 at 25°C. A glass of the composition 29.4CdO·68.6Bi_2O_3·2.0SiO_3 (wt. %) is claimed [8] to have tan $\delta = 0.0003$ at 1 MHz and 25°C.

Dielectric strengths are generally of the order of 10^7 V/cm. Because of ionic mobility, lower strengths are obtained at elevated temperatures.

References

1. See, for example, V. V. Daniel, *Dielectric Relaxation*, p. 2. Academic Press, New York, 1967.
2. I. M. Hodge, M. D. Ingram, and A. R. West, *J. Electroanalyt. Chem.* **74**, 125 (1976).
3. P. B. Macedo, C. T. Moynihan, and R. Bose, *Phys. Chem. Glasses* **13**(6), 171–179 (1972).

4. J. F. McDowell and G. H. Beall, *Ceram. Trans.* (K. M. Nair, R. Pohanka, and R. C. Buchanan, eds.), Vol. 15, pp. 259–278. American Ceramic Society, Columbus, OH, 1990.
5. J. M. Stevels, in *Non-cryst. Sol.* (V. D. Frechette, ed.), p. 412. J. Wiley, New York, 1960.
6. D. Bahat, *J. Mater. Sci.* **4**, 855 (1969).
7. F. Bischoff, *Glastech. Ber.* **28**, 98 (1955).
8. B. V. J. Rao, *J. Am. Ceram. Soc.* **45**, 555 (1962).

Chapter 16

Electronic Conduction

16.1. Introduction

The exploitation of the electronic conductivity property of many glasses is arguably the high-tech development of the last two decades. This is the only area of glass science so far where a Nobel Prize has been awarded. Electronic conduction is responsible for the photoconductivity property in Se and As–Se glasses that have been the heart of the photocopying process. Some electronically conducting glasses display switching behavior between insulator and semiconductor states. These are utilized in computer memory devices. Electronic conduction in amorphous hydrogenated silicon (written as *a*-Si:H) and its alloys is responsible for its photovoltaic behavior, which is the basis of solar cell technology. The need to reduce our dependence on fossil fuels as the source of energy is likely to enhance our efforts in this direction.

Electronic conductivity implies conduction of electrical charge by either electrons (negatively charged) or electron holes (positively charged). In comparison to the usually high electronic conductivity in metals, the electronic conductivity in glasses is orders of magnitude smaller. For this reason, electronic conduction in amorphous solids is labelled **semiconduction**. The subject is enormously complex; the physics in particular is far beyond the scope of this book. The reader may find it useful to review the following:

Richard Zallen, in *Physics of Amorphous Solids*, John Wiley & Sons, New York, 1983, Chapters 5 and 6.
S. R. Elliott, in *Physics of Amorphous Materials*, Longman Scientific and Technical, Chapters 5 and 6, Harlow, Essex, 1990.
C. Kittel, in *Solid State Physics*, 4th ed. 1971, John Wiley & Sons.
David Adler, in *Electrical Conductivity in Ceramics & Glass*, Part A (N. M. Tallan, ed.), pp. 1–34. Marcel Dekker, New York, 1974.
J. D. Mackenzie, in *Electrical Conductivity in Ceramics & Glass*, Part B (N. M. Tallan, ed.), pp. 559–618. Marcel Dekker, New York, 1974.
Amorphous Semiconductors, *Topics in Applied Physics*, Vol. 36 (M. H. Brodsky, ed.), Springer-Verlag, New York, 1979.
Noncrystalline Semiconductors, Vols. I–III (M. Pollock, ed.). CRC Press, Boca Raton, Florida, 1987.
J. Non-crystalline Solids, Elsevier Science Publishers, Amsterdam, **77–78** (1985).

16.2. Concepts of Electronic Conduction in Amorphous Solids

Two glass families are known to show significant electronic conduction: oxide glasses containing large amounts of transition metal oxides, where the bonding is mixed ionic and covalent, and chalcogenides and tetrahedral glasses, which are essentially covalently bonded solids. Both of these are significantly different from metals, where electronic conductivity is the dominant mode of charge transfer.

Conduction in most metals is generally understood in terms of the free electron theory. The weakly bound electrons from the valence shell of the constituent atoms are considered to move about essentially freely. Any interaction with the core atoms is considered negligible. Under the influence of an externally applied electric field, these free electrons move in a direction opposite to the field, and hence conduct electricity. The electrical resistance at low temperatures arises primarily because of scattering of the electron waves with lattice imperfections and impurities. At higher temperatures, the resistivity increases because of interactions with phonons (lattice vibrations). Thus, the resistivity of very pure copper at 4 K is of the order of 10^{-11} ohm·cm, increasing to about 1.7×10^{-6} ohm·cm at room temperature.

In ionic crystals, the free electron model is not adequate because of strong interaction of the electrons with the periodic potential of the ions. Electronic conductivity in ionic crystals is best explained in terms of the band theory of solids. Under the one-electron, independent particle assumption, band theory suggests that there are a set of stationary states available to any one electron. The electrons are distributed among these states according to the

Fermi–Dirac distribution, which gives the probability $f(E)$ that a given state of energy E will be occupied in an ideal electron gas in thermal equilibrium:

$$f(E) = 1/[1 + \exp\{(E - \mu)/k_B T\}], \tag{16.1}$$

where μ is the chemical potential and k_B is the Boltzmann constant. At 0 K, $\mu =$ the Fermi energy E_F, defined as the energy of the topmost filled level.

In a periodic potential, the Hamiltonian operator H in the Schrödinger wave equation includes, in addition to the electron kinetic energy term, a crystal potential term to account for the interaction of the electron with all of the other ions and electrons in the crystal. The eigenfunctions of the Schrödinger wave equation for a periodic potential are **Bloch functions**:

$$\Psi(\mathbf{k}, \mathbf{r}) = e^{i\mathbf{k}\cdot\mathbf{r}} u(\mathbf{k}, \mathbf{r}), \tag{16.2}$$

where **k** is the wave vector, **r** is a distance vector, and the function $u_{\mathbf{k}}(\mathbf{r})$ remains invariant with a lattice translation.

Under the influence of an externally applied electric field, an electron has to be able to move from one energy state to another as yet unoccupied. It moves relative to the lattice with an effective mass, m^*, which may be greater or less (even negative) than the rest mass. When the electron energy is plotted against the wave vector k (see Fig. 16-1, in one dimension for convenience),

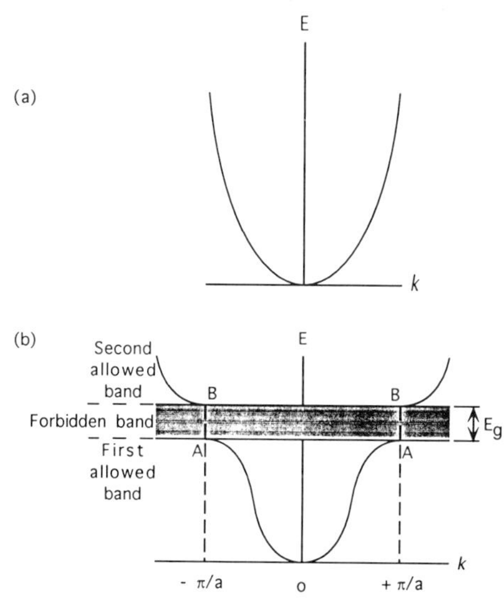

Figure 16-1. Plot of energy vs. wavevector k for (a) a free electron, (b) an electron in a monatomic linear lattice of lattice constant a. The energy gap E_g is associated with the first Bragg reflection at $k = \pm\pi/a$. Other gaps are found at $k = \pm n\pi/a$.

the allowed energies occur in the form of bands separated by **band gaps, or ranges of forbidden (inaccessible) energies**. The allowed electron states within the bands are **extended states**. *Each electron is delocalized, i.e., it has equal probability of being found in the vicinity of any equivalent ion core within the solid.* The energy discontinuities occur at $k = \pm n\pi/a$, corresponding to the Bragg reflections, where a is the lattice constant. The region $-\pi/a$ to $+\pi/a$ is called the first **Brillouin zone**; $-2\pi/a < k < -\pi/a$ and $2\pi/a > k > \pi/a$ is the second Brillouin zone, and so on. The total number of available states in any given band is $2N$, where N is the number density of equivalent atoms. If the band is completely filled, i.e., all the available states are occupied, and the Fermi energy lies within the gap, then the material would behave as an **insulator** at absolute zero temperature (Fig. 16-2a). A considerable amount of thermal energy is required to excite the electrons from any of the extended states in the lower **valence band** to jump across the band gap into the **conduction band**. Examples of insulators are diamond, Si, and Ge at 0 K, which have eight valence electrons per primitive cell, completely filling the s and p bands with no overlap from the higher band. In most **metals**, the Fermi level lies within the topmost occupied band, such that there are still many unoccupied states for electrons (Fig. 16-2b or c). Hence, this band is the conduction band, and the metal is a **good conductor**. Examples of substances having such behavior would be alkali and noble metals where there is exactly one electron per atom, whereas two electrons are required to completely fill the s band. It is also possible that, because of band overlap, electrons may populate the second band before completely filling the first (Fig. 16-2b). Examples in this category are the alkaline earth elements.

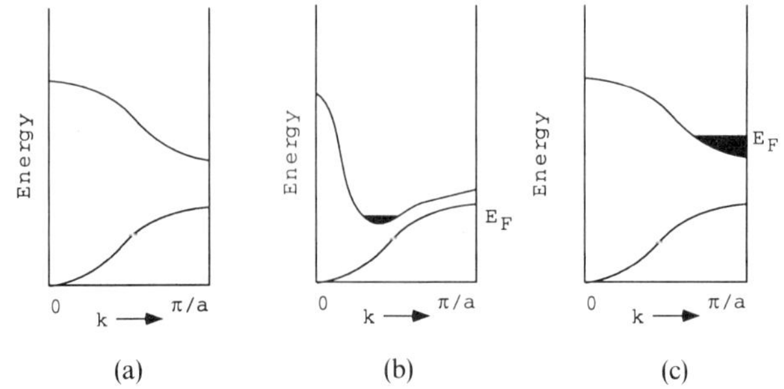

Figure 16-2. Occupied states and band structures giving (a) an insulator because of the energy gap, (b) a metal or a semimetal because of the band overlap, and (c) a metal because of the electron concentration. In (b), the overlap may be small with a few electron states involved. (After C. Kittel, *Introduction to Solid State Physics*, 4th edition, Fig. 10, p. 316. J. Wiley & Sons, New York, 1971. Reproduced with permission of the publishers.)

Although each atom has two electrons sufficient to fill a band, this does not happen. The band structure is such that some of the electrons near the Fermi level occupy the higher band. Because of the low carrier density in the conduction band, these may be called **semimetals**. In semimetals, one band is almost filled and the other is nearly empty at 0 K. By changing pressure or temperature, it may be possible to change the band gap to cause a metal $\Leftrightarrow$ insulator transition (Yb is an example of this type of behavior). Note that the electron states in either case, metal or insulator, remain extended. In **semiconductors**, the valence band is filled at 0 K and there are no electrons in the conduction band. As the temperature is raised, the energy gap separating the valence and the conduction bands is small enough that some electrons are able to jump across from the valence band into the conduction band. The electrons in the conduction band and the **holes** created by the vacancy of the electrons in the valence band both may contribute to electrical conduction (**intrinsic semiconduction**). At low enough temperatures, it is possible to have semiconduction due to defects (**extrinsic semiconduction**). If the impurity atom donates an electron, for instance pentavalent P impurity in Si crystal, then the semiconductor is called ***n*-type** (see Fig. 16-3a). A trivalent impurity such as Al or B in Si, on the other hand, picks up an electron, creating a positive hole (Fig. 16-3b). Such a semiconductor is called ***p*-type**. Whether electrons or holes are the charge carriers can be determined by carrying out a **Hall effect** experiment. *The sign of the Hall constant is the same as the sign of the charge carriers*. Another test would be to determine the sign of the thermoelectric potential. If the two ends of a specimen are held at two different temperatures, the charge carriers move to the colder end. An important observation to make for semiconduction is that not only is the conduction thermally activated, but the concentration of charge carriers also increases with temperature as a result of more carriers excited across the band gaps. Hence, *unlike the metals, the electrical conductivity in a semiconductor increases with the temperature.*

Because of the lack of periodicity in amorphous structures, the band theory breaks down. All that really means is that there is no reciprocal space, and one cannot write the electron states in the form of bands of the form $E(\mathbf{k})$. An appropriate description, however, would be to use the density of states $N(E)$, defined such that $N(E)dE$ is the number of electron states per unit volume having energies between E and $E + dE$. Suitable units of $N(E)$, for instance, are $\mathrm{cm}^{-3}\,\mathrm{eV}^{-1}$. The integral of the density of states function, along with the occupancy factor $f(E)$ between energy band levels, gives the total number, n, of carriers within that band; for instance, for the conduction band,

$$n = \int_{E_c}^{+\infty} N(E)f(E)dE. \tag{16.3}$$

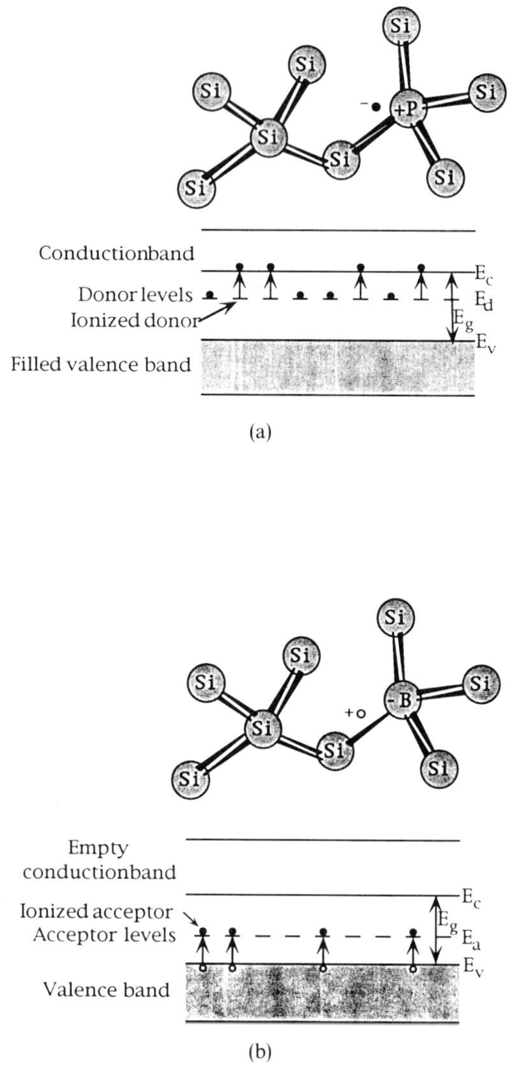

Figure 16-3. Schematic diagram of an *n*-type semiconductor. (a) The crystal or glass structure where a Si atom is replaced by a P impurity atom, and (b) the energy levels of the donor impurity atoms, with some ionized donors that have donated electrons to the conduction band.

The total conductivity σ may be written as

$$\sigma = -e\int_{0}^{+\infty} N(E)\beta(E)k_{\mathrm{B}}T\frac{\partial f(E)}{\partial E}\,dE, \tag{16.4}$$

where e = the electronic charge, and $\beta(E)$ = the mobility of the charge carrier having an energy E.

It has been shown that if short-range interactions between electrons are dominant, then the electronic density of states is primarily determined by the short-range order. Thus, the sharp features in the density of states spectrum of crystals—the **Van Hove singularities**—get smeared out; however, the overall spectrum remains roughly the same. Anderson [1] showed that, in addition, *the introduction of disorder causes the electron states to become localized.* The state is localized in the sense that an electron placed in such a region will not diffuse to other regions at absolute zero with potential fluctuations. Thus, localized electrons have zero conductivity (see Fig. 16-4). Extended states or delocalized states, on the other hand, have a finite conductivity (even at $T \to 0$) that has a power law dependence in the $E > E_c$ region (see Fig. 16-4). This also means that *conduction within the localized states may occur only by phonon-assisted hopping (quantum-mechanical tunneling from one state to the other*). Because all the states within the $E_c - E_v$ gap are necessarily localized, $E_c - E_v$ is called the **mobility gap**. The distinction between extended and localized electron states is shown in Fig. 16-5. Whereas in the extended state, the electron wavefunction is extended over the entire crystal, in the localized mode, the wavefunction is essentially confined to a few atoms. Away from the core atom, the wave function amplitude Ψ decreases exponentially as $\exp[-\alpha r]$, where α is called the

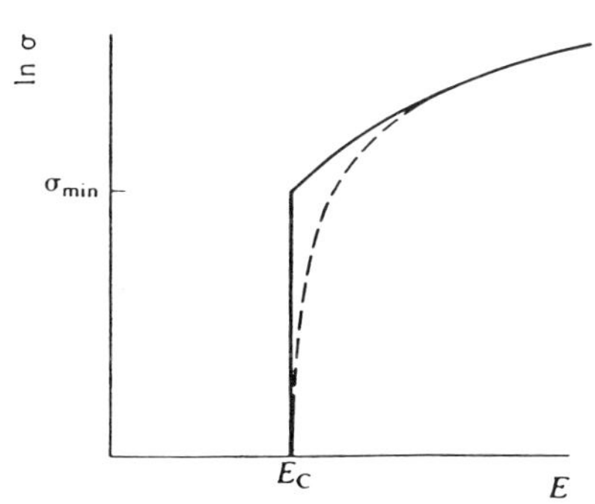

Figure 16-4. Behavior of dc conductivity near a mobility edge as suggested by Mott and scaling theories (after A. Abrahams, P. W. Anderson, D. C. Licciardello, and T. V. Ramakrishnan, *Phys. Rev. Lett.* **42**, 673 (1979).

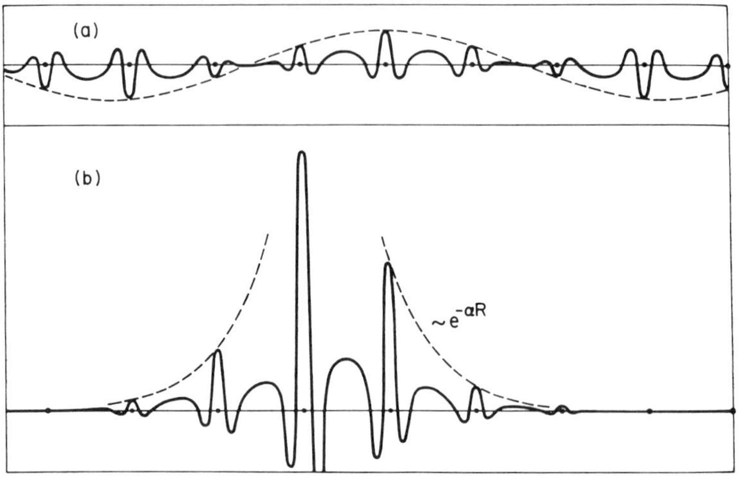

Figure 16.5. (a) A Bloch-type extended electron state. (b) An Anderson-localized electron state. (After R. Zallen, *The Physics of Amorphous Solids*, Fig. 5.10, p. 226. J. Wiley & Sons, New York, 1983. Reproduced with permission of the publishers.)

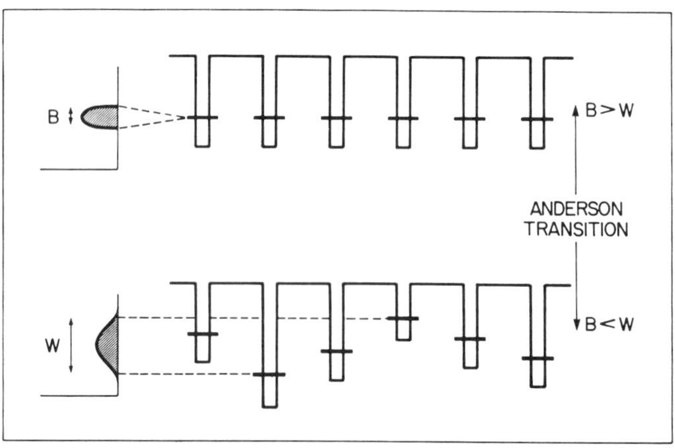

Figure 16-6. One-electron tight-binding picture for the Anderson transition. When the width W of the disorder exceeds the overlap width B, transition to disorder-induced localization occurs. (After R. Zallen, *The Physics of Amorphous Solids*, Fig. 5.13, p. 232. J. Wiley & Sons, New York, 1983. Reproduced with permission of the publishers.)

inverse localization length. The transition from extended states on the metal side to localized states on the insulator side is called the **Anderson transition** and is shown schematically in Fig. 16-6. In the absence of any disorder, the electron energy bandwidth B relative to the ion cores is to be considered against the magnitude of the random potential W when the atomic potentials have a distribution of values due to the disorder. When $B > W$, the substance remains mostly metallic. However, when $B < W$, the electron states become localized and the material behaves like an insulator. In crystals, the localization occurs only as a result of impurities or defects; hence, the contribution to overall conductivity is extrinsic. However, localization is an intrinsic characteristic of the amorphous solids.

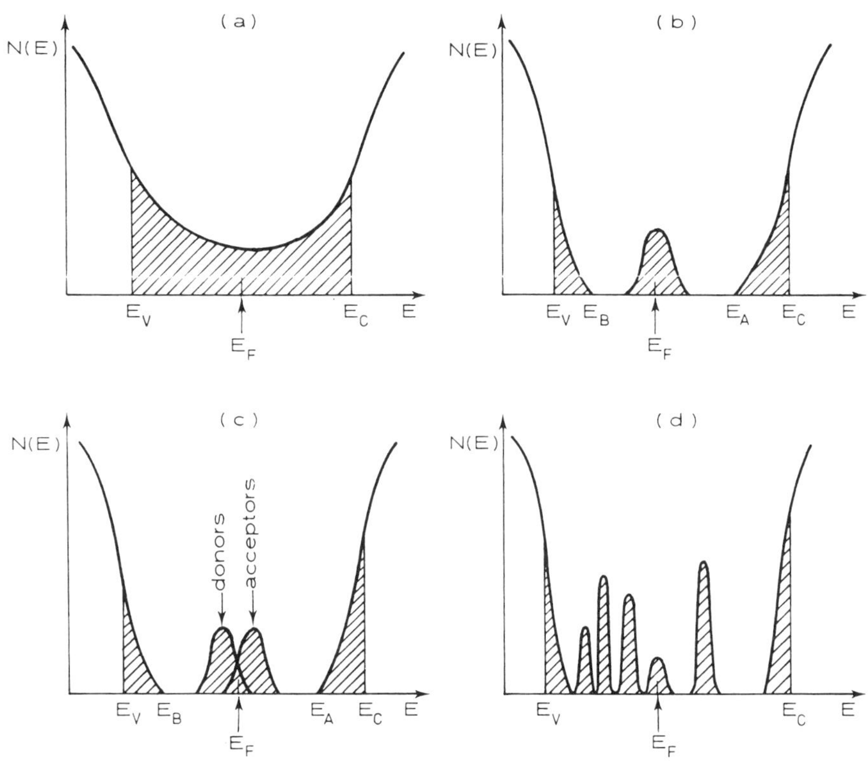

Figure 16-7. Density of states for amorphous semiconductors. (a) The CFO model, showing tailing of states causing overlap. (b) The Davis–Mott model, showing a band of compensated levels near the middle of the gap. (c) The Marshall–Owen model. (d) A "real" glass with defect states. (After P. Nagels, in *Amorphous Semiconductors* (M. H. Brodsky, ed.), Fig. 5.1, p. 117. Springer-Verlag, Berlin, 1979. Reproduced with permission of the publisher.)

There are essentially two different models suggested for the density of states in amorphous solids (see Fig. 16-7). The fundamental premise of both the models is the *tailing of states into the gap at both the valence and the conduction band edges* (E_v and E_c, also called **mobility edges**) because of structural fluctuations such as bond angle distortions. In the extreme case of the **Cohen–Fritzsche–Ovshinsky (CFO) model** [2], the tailing is so pronounced that they actually overlap in the mid-gap (Fig. 16-7a). It may be shown that the electron states lying within the mobility gap are the ones that are localized. The Fermi level is essentially pinned in the midgap. (By "pinning," one means that the zero-temperature position of the Fermi level does not shift much with a few donors and acceptors.) The primary objection to the CFO model is the high transparency observed in the chalcogenide glasses below a well-defined absorption edge, implying the presence of an actual gap within the tailing.

The second model is due to **Davis and Mott** [3]. They suggest that the tailing of the localized states is confined to a fraction of an electron volt within the gap, along with a band of compensated levels originating from defects such as dangling bonds and vacancies near the midgap (Fig. 16-7b). By a dangling bond, we mean a broken or unsatisfied bond. The true gap in the Davis–Mott model is $E_A - E_B$. In a modification by Marshall and Owen [4], the midgap band was split into donor and acceptor levels with some overlap (Fig. 16-7c). The Fermi level was again assumed to be pinned in the midgap.

On the basis of the Davis–Mott model, there are three mechanisms of electronic conductivity in amorphous solids, schematically shown in Fig. 16-8. At high temperatures, an electron that has been excited to lie in extended states such as C behaves as a conduction electron, and the **band transport mechanism** applies. The electron motion is occasionally interrupted by self-trapping and release events because of localized states just below it. Thus, its overall mobility is reduced. However, the maximum effect on conductivity comes from the density of charge carriers, which itself is proportional to $\exp[-(E_c - E_F)/k_B T]$, where E_c is the energy of the mobility edge on the conduction band side. Since $(E_c - E_F) \cong E_g/2$, where E_g is the gap energy, σ is given by

$$\sigma \cong \sigma_0 \exp(-E_g/2k_B T), \tag{16.5}$$

where σ_0 is of the order of 10^2–10^4 $\text{ohm}^{-1}\ \text{cm}^{-1}$. Thus, the activation energy for conduction is half the gap energy (calculated from the measured optical absorption edge, Section 19.2.4).

At intermediate temperatures, conduction occurs by **hopping of carriers**

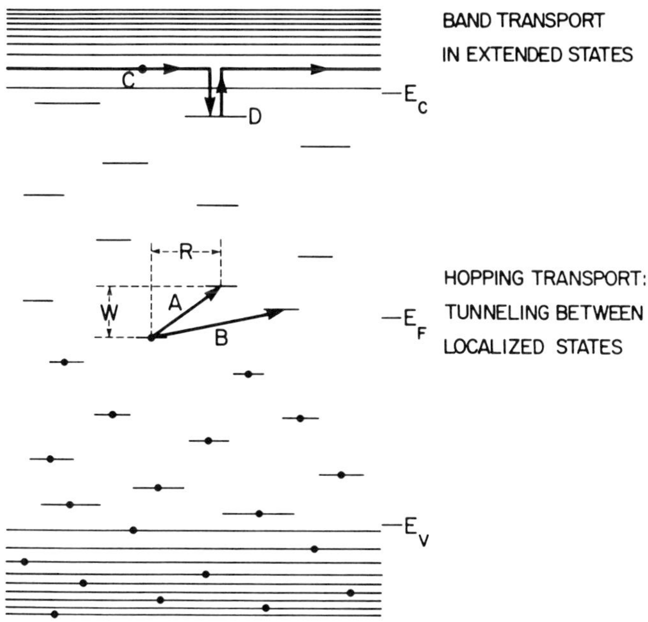

Figure 16-8. Electron conduction mechanisms in amorphous semiconductors. Vertical axis represents increasing energy, horizontal axis represents distance. (After R. Zallen, *The Physics of Amorphous Solids*, Fig. 6.12, p. 276. J. Wiley & Sons, New York, 1983. Reproduced with permission of the publishers.)

excited from the valence band edge into the localized tail states within the gap:

$$\sigma \cong \sigma_0 \exp[-(E_b - E_F + \Delta W_H)/k_B T], \tag{16.6}$$

where E_b is the energy of the valence band tail, and ΔW_H is the activation energy for hopping.

At lower temperatures, **variable-range hopping conduction** in a band of localized states at the Fermi energy may be observed, as advanced by Mott (5). The term variable-range is intended to mean that the range R through which an electron hops varies with temperature. Mott showed that

$$R \cong [\alpha k_B T N(E_F)]^{1/4}, \tag{16.7}$$

where $N(E_F)$ is the density of states near E_F, and α is the inverse localization length. Hence,

$$\sigma \sim \exp(-A/T^{1/4}). \tag{16.8}$$

A somewhat different picture of conductivity in amorphous solids is given in terms of the **small polaron** concept. A localized charge carrier such as an electron is likely to distort its surroundings. Such distortion may be experienced over a few atoms physically. The distortion tends to lower the energy of the carrier; hence, the resulting attraction will cause the carrier to be self-trapped. We may imagine the motion of a localized electron to be the motion of the electron plus its distorted surroundings. The name given to this entity is the small polaron (see Fig. 16-9). It has a larger effective mass than the effective mass of an electron in extended states discussed earlier. When one considers the interaction between electrons and phonons where a large number of interionic distances may be involved, the spatial extent of the distortion may be even larger. In such cases one speaks of **large polarons**. Both the polaron sizes may form in either of the polar (e.g. ionic) or the non-polar (e.g. covalently bonded) solids. In the former they are called the **dielectric polaron** (Fig. 16-9a), and in the latter, the **molecular polaron** (Fig. 16-9b). Mott and Davis [6] suggested that for the small polaron process the

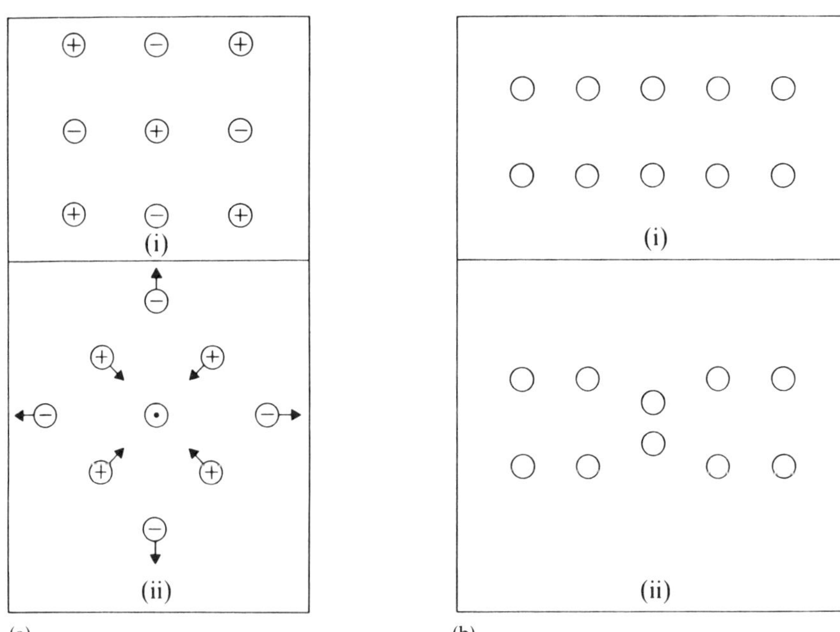

Figure 16-9. Schematic diagram of polaron formation. (a) "Dielectric" polaron, where an excess electron is added in (i) at the location of the central + charge, causing ionic readjustment (ii). (b) "Molecular" polaron, in which an excess hole is added (i) to an array of rare gas atoms, causing atomic readjustment (ii).

conductivity is given by

$$\sigma = \nu c(1-c)\frac{e^2}{Rk_B T}\exp[-2\alpha R]\exp[-W/k_B T], \tag{16.9}$$

where c and $(1-c)$ are the concentrations of the two valence states of the ion, ν is a phonon frequency, α is the inverse localization distance, R is the mean separation between the two valence state ions, and W is an activation energy given by

$$W = W_H + \tfrac{1}{2}W_D, \tag{16.10}$$

where W_D is the distortion energy (called the Miller–Abrahams energy), and W_H is the hopping energy (equal to half the binding energy of the small polaron).

It is now understood that the density of states in a real amorphous solid looks much like Fig. 16-7d, with well-defined localized peaks distributed within the gap.

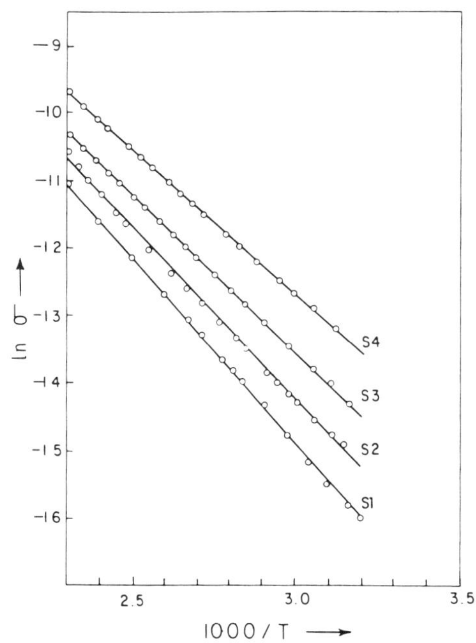

Figure 16-10. Variation of conductivity of $xV_2O_5\cdot(100-x)TeO_2$ glasses with the reciprocal of temperature. ($x = 10$, 20, 30 and 40 for compositions S1, S2, S3 and S4, respectively.) (After N. Chopra, A. Mansingh, and G. K. Chadha, *J. Non-cryst. Sol.* **126**, 194 (1990). Reproduced with permission of Elsevier Science Publishers.)

The high-temperature band-type transport mechanism is believed to apply to many chalcogenide glasses such as As_2Se_3. In oxide glasses containing a high percentage of a transition metal oxide, the electronic conductivity apparently does not follow band-type behavior. This is shown [7], for instance, for V_2O_5–TeO_2 in Fig. 16-10. Conductivity is intrinsic and not defect-controlled: it increases with the amount of the transition metal oxide. The conductivity also follows an Arrhenius-type behavior; however, the activation energies for conduction are of the order of 0.4–0.5 eV (E_g predicted to be 0.8 to 1 eV). The optical data, on the other hand, reveal the band gap to be about 2.3 eV. Apparently, blown films have roughly the same dc conductivity as bulk glasses. It has been found [8] in the V_2O_5–P_2O_5 system that the conductivity increases as the ratio of the two valence ions approaches unity.

Among the Fe_2O_3-containing glasses, the intrinsic electronic conductivity of Fe_2O_3–TeO_2 glasses is known to be several orders of magnitude higher than that in Fe_2O_3-containing silicates, borates, and phosphates. Again, a band-type transport mechanism cannot apply because too low a value of E_g is predicted. The conductivity for $10Fe_2O_3{\cdot}(90 - x)TeO_2{\cdot}xPO_{2.5}$ glasses is

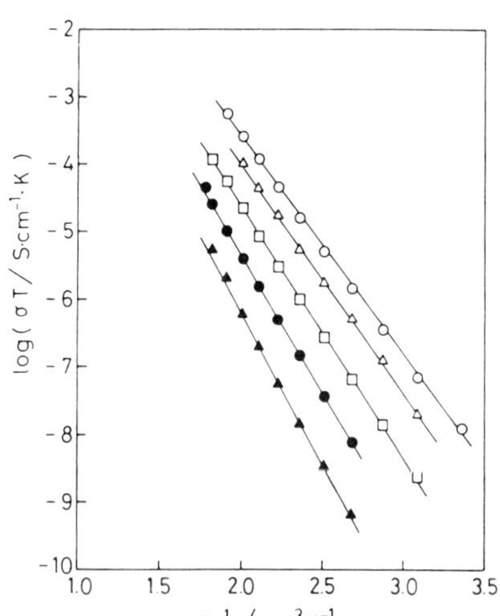

Figure 16-11. Variation of log (σT) vs. $1/T$ for $10Fe_2O_3{\cdot}(90 - x)TeO_2{\cdot}xPO_{2.5}$ glasses. ($x = 20$ mol % (○); $= 30\%$ (△); $= 40\%$ (□); $= 50\%$ (●); $= 60\%$ (▲)]. (After K. Tanaka, T. Yoko, M. Nakano, M. Nakamura, and K. Kamiya, *J. Non-cryst. Sol.* **125**, 264 (1990). Reproduced with permission of Elsevier Science Publishers.)

shown [9] in Fig. 16-11. For the Fe_2O_3–TeO_2–P_2O_5 glasses, the activation energy W is measured to be in the range of 60–80 kJ·mol^{-1}.

There has been a suggestion that electron transport at low temperatures in all glasses can be best explained in terms of small polaron motion rather than invoking variable-range hopping conduction. In V_2O_5–P_2O_5 and FeO–P_2O_5 glasses where V^{4+}/V^{5+} and Fe^{2+}/Fe^{3+} are involved, the excess electrons may form small polarons, and the polaron hopping then looks like

$$V^{4+}\text{—O—}V^{5+} \longrightarrow V^{5+}\text{—O—}V^{4+}$$

and

$$Fe^{2+}\text{—O—}Fe^{3+} \longrightarrow Fe^{3+}\text{—O—}Fe^{2+}.$$

As an example, in the vanadium system, the small polaron constitutes not only the V^{4+}, but also involves the oxygen ion at the intermediate location. The polaron then hops over to the adjacent V^{5+}, converting it to V^{4+}, and so on.

16.3. Defects in Amorphous Solids and Their Management

Perhaps the most dramatic consequence of Mott's and Anderson's concepts of electronic conductivity in amorphous solids has been the notion that *even glasses may have frozen-in intrinsic structural defects and may be extrinsically doped to have n-type or p-type defects—characteristics thought to be exclusively associated with periodic potentials.* For instance, a continuous random network (CRN) is allowed to have odd-membered rings, not present normally in a periodic lattice. If a chalcogenide glass has two types of atoms, then having an odd-membered ring requires bonding between like atoms. Such **"wrong" bonds** may be considered defects if their number is much smaller than that of the bonds between unlike atoms. The most recognized and the simplest form of intrinsic defect, however, is the **dangling bond**, shown schematically for tetrahedral amorphous solids in Fig. 16-12. An atom in a chalcogenide glass may be considered as having dangling bonds if its normal covalent coordination given by Mott's "$8 - N$" rule (N = number of valence electrons) is not obeyed. A single neutral dangling bond, commonly denoted D^0, contains one unpaired electron; it can donate as well as accept an electron. By acquiring an electron it becomes negatively charged; by giving up an electron it becomes positively charged. The two defects are commonly denoted D^- and D^+, respectively. The three electronic defects, namely, D^+, D^0, and D^-, are unoccupied, singly occupied, and doubly occupied sites, respectively. These are sketched for a-Se in Fig. 16-13. The lattice distortion

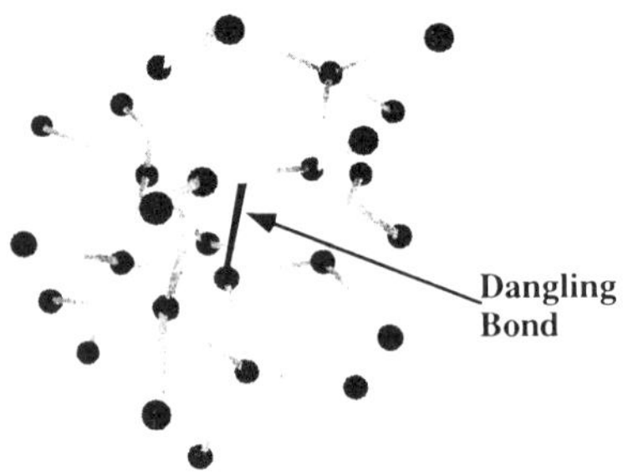

Figure 16-12. Isolated dangling bond (blackened stick) in a tetrahedrally coordinated continuous random network simulating *a*-Si or *a*-Ge.

around the D^- site is negligible, quite large around the D^+ site, and intermediate around the D^0 site. It is suggested that the reaction

$$2D^0 \longrightarrow D^+ + D^- \tag{16.11}$$

is exothermic, as a result of the strong electron–lattice (phonon) interaction. Thus, the variation of their energies versus a configuration coordinate, e.g., a bond displacement, looks like that shown in Fig. 16-14. Transfer of an electron between two D^0 sites without a configurational change would increase the energy by an amount U, called the **Hubbard correlation energy**, equal to the coulombic repulsion energy of bringing two electrons in close proximity. However, because of the distortion energies associated with the defect sites, an energetically more favorable reaction occurs by a configuration change of 2 D^0 to D^- and D^+. The net energy of occupied and unoccupied paired sites is an amount $-U_{\text{eff}}$ relative to the two single-electron unpaired sites. Anderson suggested that the energies of two-electron

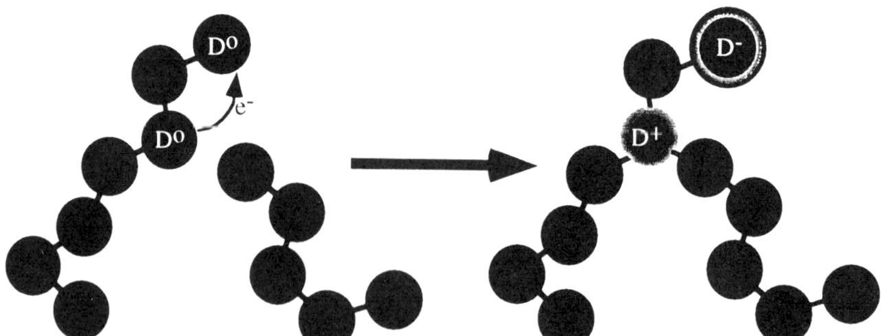

Figure 16-13. Formation of charged defects in chalcogenide glasses. (a) An electron is transferred from one of the D_0 sites, causing (b) the formation of a 3-coordinated Se atom with a + charge (D^+ defect), and a singly coordinated D^- defect.

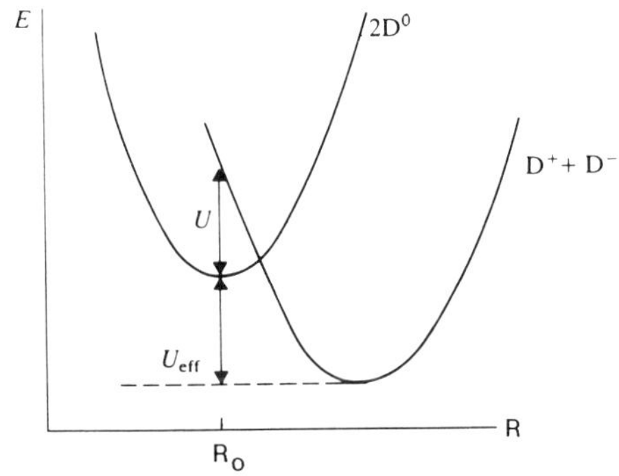

Figure 16-14. Configuration coordinate diagram for the formation of a D^+–D^- defect pair. If the configuration coordinate R_0 is to remain unchanged, then an energy U (Hubbard energy) would be needed to convert the $2D^0$ centers to a pair of D^+–D^- defects. However, subsequent relaxation of R actually results in overall lowering of energy by U_{eff}.

states are continuously distributed as **gap states**. The lower midgap band in Fig. 16-15 corresponds to a neutral dangling bond containing one electron, and hence acts like a donor. The upper band corresponds to a charged version of the same state with an electron added to it. The upper band is neutral when unoccupied, i.e., when the electron has not been added to it. In the unoccupied form it acts like an acceptor. The centers of the two bands are separated by U. In effect, the two-electron density of states distribution causes the gap to be only a pseudo-gap. (Note that the Cohen–Fritzsche–Ovshinsky model suggested an overlapping density of states distribution for one-electron states.)

In normally prepared, evaporated *a*-Si, the primary defect is the single dangling bond. The density of states of such defects is generally of the order of 10^{20} cm^{-3} eV^{-1}. Annealing *a*-Si to obtain a more equilibrated glass does not give any significant reduction in defects [10]. One approach to reduce the number of dangling bond defects is to hydrogenate the *a*-Si. Glow discharge decomposition of silane on a glass substrate at about 200–400°C, or rf sputtering of a Si target in Ar/H_2 plasma, readily yields *a*-Si:H with about 1–10 wt. % hydrogen, which is about 10 times more than that needed to satisfy all the dangling bonds. Hydrogenation thus removes the dangling bond gap states, creating Si–H bonding states deep in the valence band, thereby passivating *a*-Si. The density of the dangling bond states drops [11] to about 10^{16} cm^{-3} eV^{-1}. The optical gap increases by about 0.4 eV. It is then possible to carry out a systematic and controlled doping to yield point defects of the *n*-type where electrical conduction occurs by electrons in

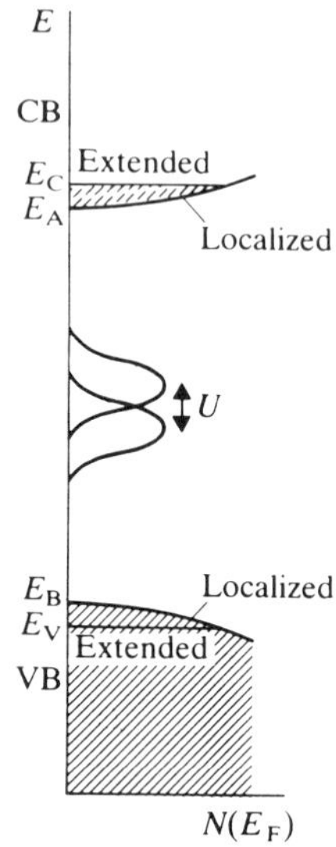

Figure 16-15. Density of states showing localized band-tail states, a dangling bond (lower midgap band), and double-occupancy bands (upper midgap band). The dangling bond and the double-occupancy bands differ by the Hubbard energy, U (assumed positive). (After S. Elliott, "Physics of Amorphous Materials", 2nd ed., Fig. 6.3b, p. 343, Longman Scientific and Technical Essex UK, 1990. Reproduced with permission of the publisher.)

extended states above E_c, and p-type where the conduction occurs by holes in extended states below E_v. In the normal a-Si, the density of occupied gap states is so high that they cannot be doped much by extrinsic means. In fact, it may be said that because of the high density of gap states in a-Si, earlier studies found it difficult to resolve other features in the gap such as band tailing and interpret the changes in the activation energy for conduction with temperature, etc.

Mixing about 1% of gases such as PH_3 or B_2H_6 with SiH_4 and carrying out plasma deposition of amorphous film produces a-Si:H:P, where the group V element P can donate an (n-type), or a-Si:H:B, where the group III element B may accept an electron (p-type). Figure 16-16 shows [12] that the electronic conductivity of a-Si:H increases by six orders of magnitude upon gradual doping by PH_3. On the other hand, its conductivity decreases by 10^3 at first upon introduction of B_2H_6 and increases thereafter to a net gain of 10^6, suggesting that the a-Si:H produced normally is slightly n-type. These well-controlled, specific-type amorphous semiconductors have found wide application in the solar cell operation discussed later.

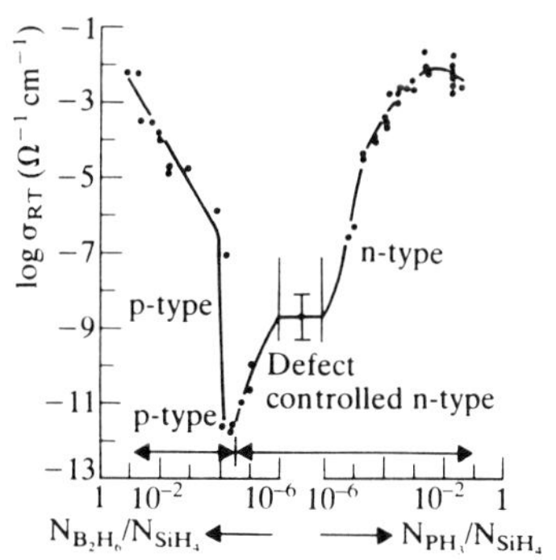

Figure 16-16. Room-temperature conductivity of *n*- and *p*-type *a*-Si specimens plotted as a function of the gaseous impurity ratio. The center refers to undoped specimens. (After Spear and LeComber [12]. Reproduced with permission of Taylor & Francis, Inc.)

16.4. Photoconductivity, Photoluminescence, and Xerography

The phenomenon of excess electronic conduction following exposure to photons is called **photoconductivity**. It is also possible that the excited carrier returns to its ground state via a radiative emission of a wavelength longer than that of the impinging photon. The phenomenon of radiative emission is called **photoluminescence**, and the emitted wavelength is supposed to have undergone a **Stokes shift**.

Mott suggested that the charged dangling bonds can explain photoconductivity in chalcogenides. The D^+ and D^- act as effective traps for electrons and holes generated by photon excitation. (Note: an electron–hole pair in a bound state is called an **exciton**.) The capture of electrons or holes converts these to excess D^0 centers that are subsequently released to the valence or conduction band, and hence give rise to excess conductivity $\Delta\sigma$:

$$\Delta\sigma = e(\Delta n\mu_n + \Delta p\mu_p) \tag{16.12}$$

where Δn and Δp are the increases in the number of the n and p charge carriers compared to those at thermal equilibrium, and the μ's are the corresponding mobilities.

The rate of generation of additional charge carriers is determined by the quantum yield η, which is the number of electron–hole pairs generated per photon. Once they are generated, the excess carriers disappear by recombination through three processes: direct recombination of a free electron and a free hole, capture of an electron by a center having a localized hole, and capture of a hole by a center with a bound electron.

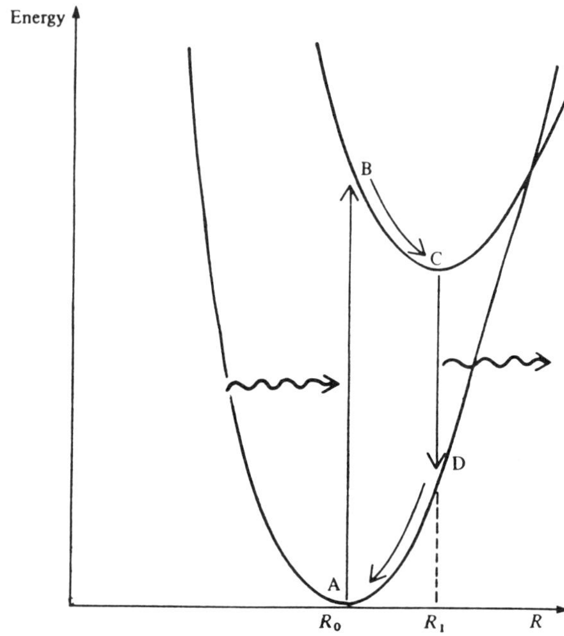

Figure 16-17. Schematics of photoluminescence. A photon absorption by the impurity ion at A kicks it to B. The surrounding ions then relax the configuration from the initial R_0 to R_1, causing the ion to move to C and lose the energy by a phonon emission. The impurity ion then makes a transition to D, emitting a photon, and finally to the original location A by configuration relaxation and phonon emission.

If a specimen across which a voltage has been applied is illuminated by a short burst of photons of energy higher than the gap through a semi-transparent front electrode, then a current of carriers, all of the same type, drifts towards the back electrode. The oppositely charged carriers are collected by the front electrode. Because of hopping from site to site among the localized states where the intersite distance varies, and trapping and release mechanisms having a distribution of residence time (called **multiple trapping**), the drift mobility of the carriers shows strong dispersive behavior, far greater than that achieved by an ordinary diffusion process. Apparently, there is no marker on the current–time curve to signify a specific transit time. This phenomenon is called **dispersive transport** and is basic to carrier motion through localized states in amorphous solids.

Photoluminescence in chalcogenides may also be explained by excitation of an electron from the D^- center by a photon of energy $h\nu_E$ to the conduction band (or the hole from a D^+ center to the valence band); the subsequent recombination with D^0 yields the charged defect D^-, again accompanied by a radiative emission of energy $h\nu_{PL}$ Stokes-shifted in energy by an amount $2W_d$, where W_d is the distortion energy. The overall process may be written as

$$D^- + h\nu_E \rightarrow D^0 + e \rightarrow \; + h\nu_{PL}. \tag{16.13}$$

The luminescence may be explained using a configuration coordinate diagram (Fig. 16.17). The optical excitation brings the ground state A (e.g., site D^-) to B (site D^0) having the same configuration coordinate R_0. It is drawn vertically because the optical excitation/de-excitation is a relatively fast process compared to thermal excitation/relaxation (Franck–Condon principle). Interaction with lattice then results in the release of a phonon of energy W_d, and the state moves to the point C having a configuration coordinate R_1. At this point, recombination with an electron causes a photon of energy

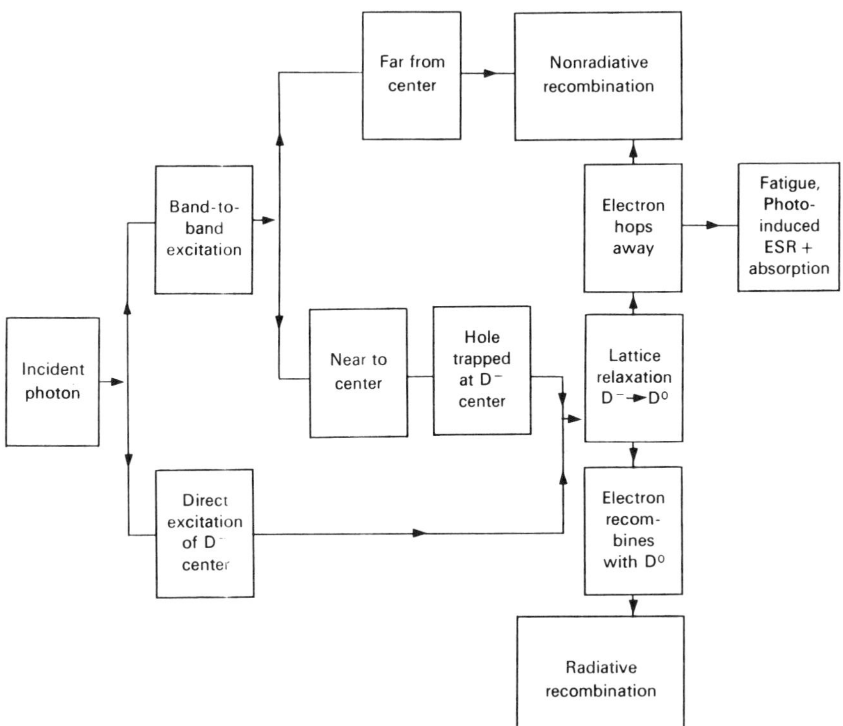

Figure 16-18. Block diagram for photogeneration and recombination mechanisms for chalcogenides. (The D^- could be exchanged for D^+ if the roles of the electron and holes are reversed.) (From Davis [13]. Reproduced with permission of Springer-Verlag.)

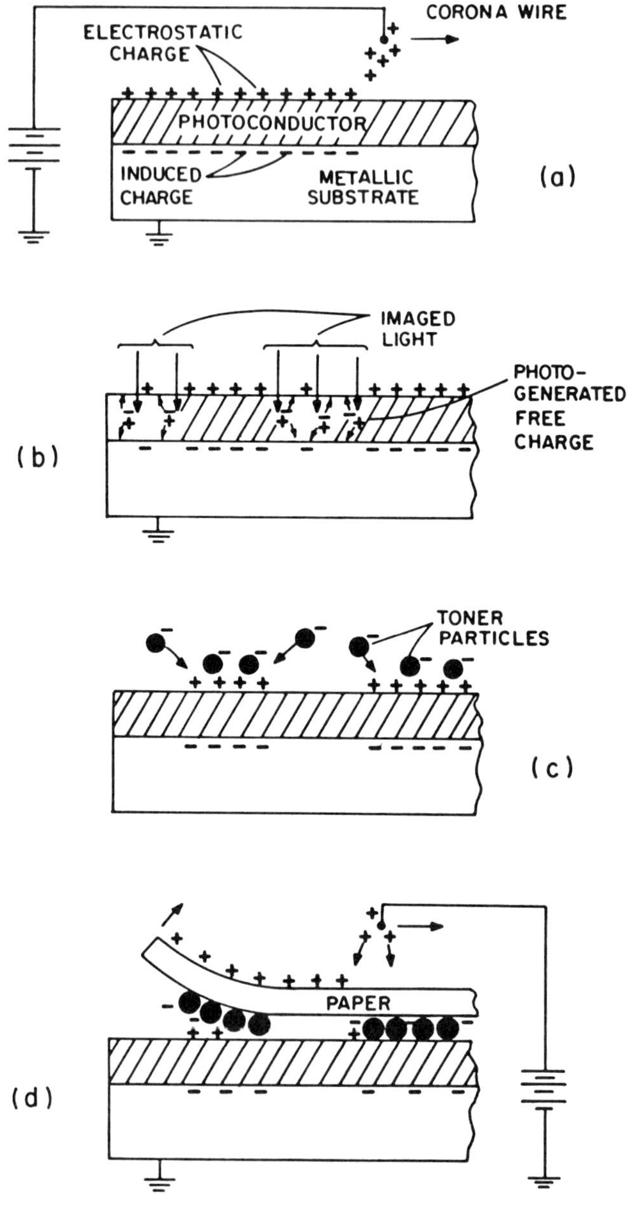

Figure 16-19. Principle of xerography: (a) Charging, (b) exposure, (c) development, and (d) transfer. (After R. Zallen, *The Physics of Amorphous Solids*, Fig. 1.12, p. 29. J. Wiley & Sons, New York, 1983. Reproduced with permission of the publishers.)

CD to be emitted, bringing the defect back to D^- still having coordinate R_1. Relaxation of the lattice with phonon emission (energy difference between D and A) restores the atomic configurations. A block diagram of the various steps in photogeneration and recombination mechanisms for chalcogenides is shown [13] in Fig. 16-18.

The principle of photocopying or xerography, schematically shown in Fig. 16-19, is based essentially upon the photoconductivity property of chalcogenide glasses such as Se and As_2Se_3 (having band gaps of about 2 eV). (For more details, see Zallen, pp. 28, 29.) Briefly, a thin film about 50 μm thick is deposited on a large-area metal substrate. Positive charges are uniformly deposited on the top surface of the film; balancing negative charges develop on the glass–metal substrate interface. When light strikes the photoconductor film, photons are absorbed and electron–hole pairs are created by excitation over the band gap. The electrons move to the top, neutralizing the positive charges, and the holes move towards the glass–metal interface, neutralizing the negative charges there. Next, the toner particles are applied to the top. Toner particles are negatively charged, and hence they attach themselves to the non-annihilated positive charges on the top surface of the glass film. The "image" is then transferred to a positively charged paper and fused to it by gentle heat. Reset status is obtained by cleaning off the toner and erasing the previous image by flooding with light.

16.5. Photovoltaics and Solar Cells

The solar cell application is based upon the photovoltaic phenomenon. Electron–hole pairs are created by excitation due to incident photons having greater-than-gap energies and are separated from each other by a rectification action. The rectification can be achieved by a ***p–n* junction** or ***p–i–n* structure**. (The "*i*" stands for intrinsic or undoped.) In initially separated *p*- and *n*-type materials, the Fermi levels are different relative to the valence and conduction bands (Fig. 16-20a). When brought in contact, the diffusion of charge across the junction causes excess negative charge on the *p* side and excess positive charge on the *n* side. This double layer of charge (Fig. 16-20c) creates an internal electric field set up in a narrow region either side of the junction directed from the *n* to the *p* side. The field itself opposes further diffusion. Excitation by photons at the junction creates an electron–hole pair. The electric field of the double layer drives the electron toward the *n* region and the hole toward the *p* region such that, in equilibrium, the Fermi level is constant across the junction, as shown in Fig. 16-20b, and there is no net flux by diffusion and drift of either the electrons or the holes across the junction, creating a depletion region (Fig. 16-20d). The internal electric field

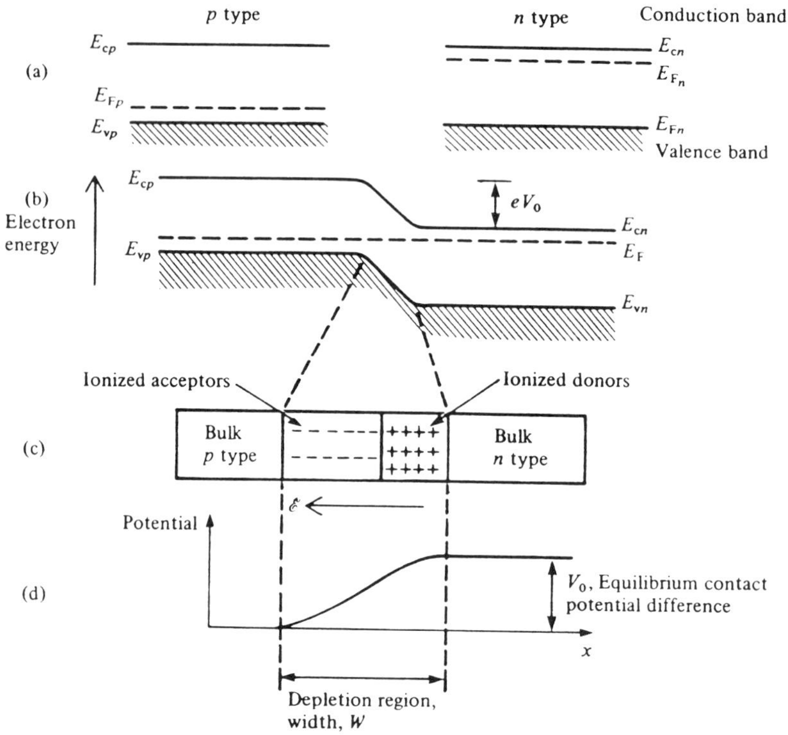

Figure 16-20. Schematic representation of a *p–n* junction: (a) initially separated *p*- and *n*-type materials, (b) the energy band distribution after the junction is formed, (c) the space-charge layers of ionized impurity atoms within the depletion region, *W*, and (d) the potential distribution at the junction. (After J. Wilson and J. F. B. Hawkes, *Optoelectronics*, 2nd Ed., Fig. 2.20, p. 62. Prentice-Hall, 1989. Reproduced with permission of the publisher.)

thus sets up a contact potential difference V_0, the maximum value of which is the gap voltage. When the ends are connected, a current flows from the *n* side to the *p* side. By having an intrinsic layer between the *p–n* junction, the long wavelength absorption is increased.

The advantages of *a*-Si:H instead of *c*-Si are twofold. Firstly, the absorption coefficient of *a*-Si:H for daylight spectrum ranges from 10^4 to $10^{5.5}$ cm^{-1}, which is roughly 10–50 times that of *c*-Si. Thus, a film of *a*-Si as thin as 1 μm is sufficient to absorb most of the solar spectrum, whereas a film of *c*-Si 50 μm thick would be needed for equivalent performance. Secondly, the cost of producing large-area thin films of *a*-Si is far less than that of producing *c*-Si.

A typical solar cell is constructed by depositing on a thin substrate soda-lime glass first, a conducting ITO (indium-doped tin oxide) film followed by about 200 Å thick *p*-type *a*-Si:H, next a 5,000 Å to 1 μm thick *i*-type *a*-Si:H followed by about 400 Å thick *n*-type *a*-Si:H layer, and finally, an aluminum electrode metal layer. The optical gap is estimated to be about 1.55 eV. A short-circuit current density of 14 mA/cm^2 and an open circuit voltage of 0.9 V for a 1.5 mm thick film solar cell is about normal (14). The efficiency of a *p–i–n* structure for AM1 (notation for solar illumination intensity = 100 mW/cm^2 received at sea level on a sunny day with the sun at its highest) is generally of the order of 8–10%, which is not economical enough as yet. Nonetheless, the procedures for making *a*-Si:H and its alloys have been developed to the point that pocket calculators needing no batteries (solar cells that work "even in the dark") cost about $4.

16.6. Switching and Computer Memory Devices

According to Fritzche [15], negative resistance, switching, and memory phenomena may be classified into five different *I–V* characteristics, shown schematically in Fig. 16-21, and described as follows.

(a) A **negative resistance device** may display a positive dI/dV up to some characteristic voltage V_{th}, at which point the slope rapidly reverses. In the presence of a small resistor load, the negative resistance can be switched from point A (load line tangent to the *I–V* curve) to point B (the intersection of the loadline with the negative resistance arm). The complete path is retraceable except for a small hysteresis when I is switched rapidly.

(b) A **current-controlled negative resistance device with memory** has two stable states. In the first state it behaves like (a) up to a current I_m. If the current is increased, the second state becomes active, and the *I–V* curve now follows the conductive arm even if the voltage is dropped to zero or reversed. The device has a memory in the sense that it remains in either state. The first state can be brought back, i.e., the memory of the second state can be erased, only if the current is increased above I_m and the device is switched off rapidly. In principle, all points on the *I–V* curve can be made stable.

(c) A **switching device** remains in a low-conductivity state until V_{th}, the threshold voltage, at which point the operating characteristics switch to a high-conductivity arm rapidly traversing along the load line. There is no stable operating point on the load line (shown dashed). The low conductivity state is called the *off* state, high-conductivity state is called

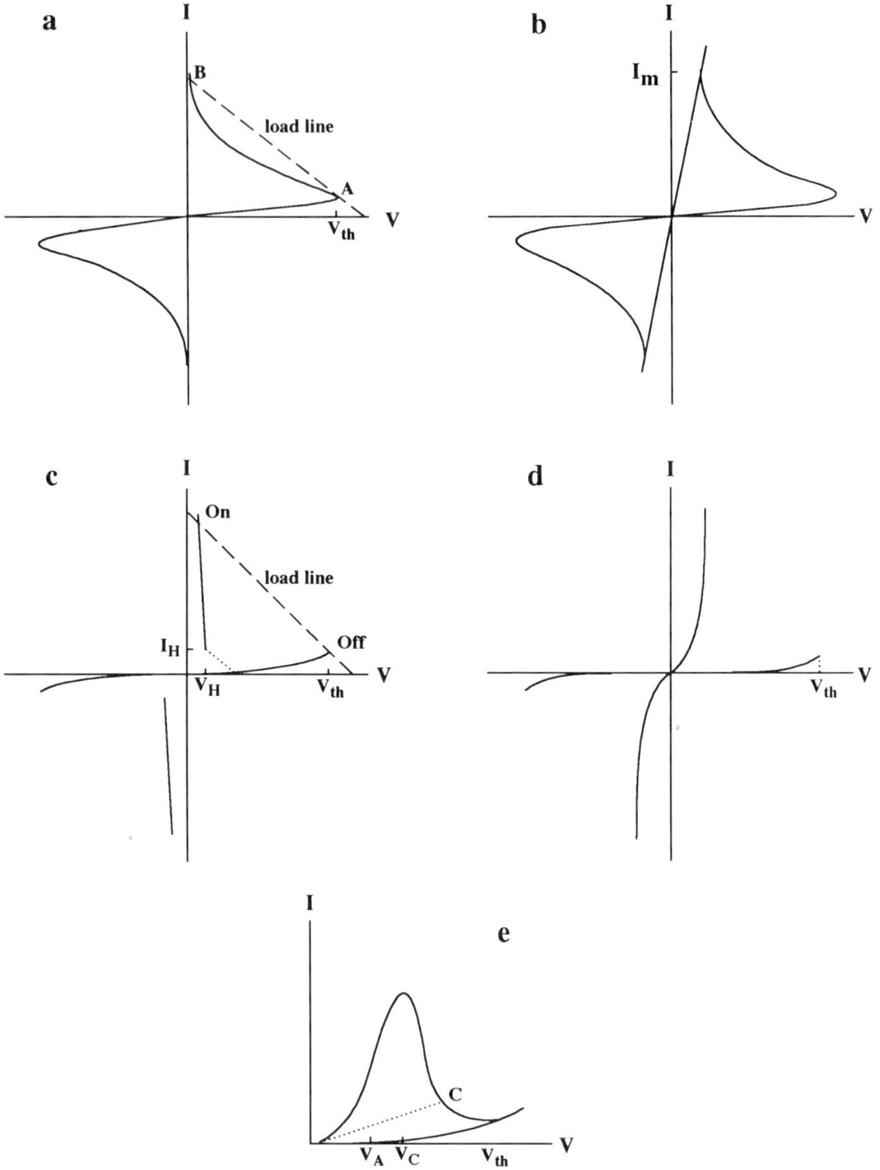

Figure 16-21. Classification of switching and memory characteristics. (a) Negative resistance device. (b) Negative resistance device with memory. (c) Switching device. (d) Switching device with memory. (e) Voltage-controlled negative resistance device with memory. (After Fritzsche [15]. Reproduced with permission of Plenum Press, Inc.)

the *on* state. When the current is reduced to the level I_H, called the minimum holding current (typically 0.1–0.5 mA), the state switches to the *off* curve. The V–I behavior mirrors itself when the polarities are reversed. In the *on* state, the resistivity is typically 0.08 ohm·cm. Apparently, the V_{th} is lower for a subsequent switch. Also, the switching time at V_{th} varies statistically. This switching delay time can be reduced by applying higher overvoltage (Fritzsche [15], p. 324).

(d) A **switching device with memory** has no I_H. Once switched to the *on* state, it will remain in this state even if turned off until reset by short current pulses.

(e) A **voltage-controlled negative resistance device** starts out having low resistance and turns to negative resistance above an applied voltage V_C and subsequently to a high resistance beyond V_{th}. The initial low resistance can be brought back by applying pulses of voltage exceeding V_A. The device may stay in varying states of memory, depending upon the location of the point C from which the voltage is reduced rapidly.

A large number of electronically conducting oxide glasses and almost all the chalcogenides display one or other of these types of behavior. The *on* and *off* states of the switching devices with (type (d)) and without (type (c)) memory are particularly useful for binary information storage in a computer. Following the announcement by Ovshinsky [16] in 1968 of the discovery of switching in a 0.5 μm thick evaporated film of $Te_{48}As_{30}Ge_{10}Si_{12}$ chalcogenide glass between two carbon electrodes of contact area about 10^{-4} cm^2, the interest in amorphous switches leaped overnight. Amorphous switches clearly offered a technological advantage over their crystalline counterpart switches because the latter had to be extremely pure. In hindsight, this is what brought on the computer revolution of the 1970s and 1980s, which brought us from cranking adding machines to 100 Mb desktop personal computers.

A significant amount of research that followed the discovery suggested that the switching was a transition between a low-conductivity glassy state and a high-conductivity crystalline state brought on locally by joule heating. There is now sufficient evidence that, whilst such is true for some systems, the *on* state in other systems experiences little joule heating and has its current decaying abruptly at the V_H. There are no thermal phenomena to explain a discontinuous, abrupt ending. The former, where joule heating brings about a controlled and reproducible local crystallization, are the type (d) switches and are called **memory switches**. The latter, where no joule heating associated memory is evident, are the type (c), called **threshold switches**. Three-dimensionally crosslinked chalcogenide glasses have generally a high enough glass transition that they act only as threshold switches.

It is believed that the *on* state in both the switches is associated with a high-carrier-density (about 10^{18}–10^{19} cm^{-3}) electron–hole plasma having a mobility $\cong 10\ cm^2/V\ s$. There is no space-charge buildup because both the electrons and holes are the carriers. The losses of charge carriers due to recombination and transfer at the electrodes are continuously balanced by an injection of both the electrons and holes tunneling into the glass from the electrode region.

The memory action is brought about by thermally driven controlled crystallization of an *on*-state filament. A reset pulse revitrifies the crystals, and subsequent rapid cooling brings the switch to an *off* state. Different crystallization mechanisms are known to operate in different materials. For instance, in $As_{55}Te_{35}Ge_{10}$ one observes nearly single-phase growth of As_2Te_3 from the anode because of electromigration of Te to the anode; none are observed at the cathode (Fritzsche [15], p. 352). In compositions such as $Te_{81}Ge_{15}Sb_4$ based upon the $Ge_{17}Te_{83}$ eutectic, Te crystals separate first, followed by the crystallization of the Ge-enriched phase at a higher temperature.

References

1. P. W. Anderson, *Phys. Rev.* **109**, 1492 (1958).
2. M. H. Cohen, H. Fritzsche, and S. R. Ovshinsky, *Phys. Rev. Lett.* **22**, 1065 (1969).
3. E. A. Davis and N. F. Mott, *Philos. Mag.* **22**, 903 (1970).
4. J. M. Marshall and A. E. Owen, *Philos. Mag.* **24**, 1281 (1971).
5. N. F. Mott, *Philos. Mag.* **19**, 835 (1969).
6. N. F. Mott and E. A. Davis, *Electronic Processes in Non-crystalline Materials*. Clarendon Press, Oxford, 1979.
7. N. Chopra, A. Mansingh, and G. K. Chadha, *J. Non-cryst. Sol.* **126**, 194–201 (1990).
8. J. D. Mackenzie, in *Electrical Conductivity of Ceramics & Glasses*, Part B (N. M. Tallan, ed.). Marcel Dekker, New York, 1974.
9. K. Tanaka, T. Yoko, M. Nakano, M. Nakamura, and K. Kamiya, *J. Non-cryst. Sol.* **125**, 264–271 (1990).
10. M. H. Brodsky, in *Amorphous Semiconductors* (M. H. Brodsky, ed.), p. 1. Springer-Verlag, Berlin, 1979.
11. S. Elliott, *Physics of Amorphous Materials*. Longman Scientific and Technical, 2nd ed., Harlow, Essex (1990).
12. W. E. Spear and P. G. LeComber, *Philos. Mag.* **33**, 935–949 (1976).

13. E. A. Davis, in *Amorphous Solids and the Liquid State* (N. H. March, R. A. Street and M. Tosi, eds.), p. 521. Plenum Press, New York, 1985.
14. D. E. Carlson and C. R. Wronski, in *Amorphous Semiconductors* (M. H. Brodsky, ed.), pp. 287–329. Springer-Verlag, Berlin, 1979.
15. H. Fritzsche, in *Amorphous and Liquid Semiconductors* (J. Tauc, ed.), pp. 313–359. Plenum Press, New York, 1974.
16. S. R. Ovshinsky, *Phys. Rev. Lett.* **21**, 1450 (1968).

Chapter 17

Chemical Durability

17.1. Introduction

The term "chemical durability" is used to refer to the resistance of the glass surface to chemical agents, in particular, to fluids commonly encountered during service. Ordinarily, one thinks of glass as a durable material. However, the observations of white deposits on windows and household glassware after repeated dishwashing are common proofs of the lack of such durability in terms of the reactions that occur between glass surfaces and apparently "harmless" fluids such as water. The resistance to attack by beverages for human consumption determines the use of glass as a container. One may have to worry about the extraction of Pb from PbO-containing "crystal" glass containers due to storage of wines over long periods. The use of glass as an optical component is determined, to a great extent, by its ability to withstand long-term exposure to atmospheric humidity. Likewise, the function of many electronic devices where glass has been used in a variety of packaging applications such as thick films is adversely affected by the corrosion of glass by humid atmosphere. On the positive side, it is in fact the excellent chemical durability of glasses to common elements of nature such as water, humidity, oxygen, weak acids, and weak bases that allows the applications of glass to a large variety of products. No matter what they say, beer still does not taste as good out of a metal can as out of a glass bottle!

Because of its good durability relative to most metals, glass is the leading candidate for "fixation" or "immobilization" of radioactive waste. It is hoped that if radioactive waste is melted into a glass matrix, the radioactive elements would remain contained or immobilized over centuries of possible attack by flowing water, and such a container would not "rust away."

The subject of chemical durability contains two primary areas: durability against fluids of various pH including water, and weathering or weatherability against elements of the atmosphere — in particular, humidity. In this chapter, we will understand the various reactions involved in durability and weatherability and study how glass composition influences such behavior.

17.2. Mechanisms of Durability and Weatherability

In both durability and weatherability, the reaction begins at the surface and spreads inwards as a result of either the diffusion of the invading elements into the glass or the breakup of the glass network bonds by the attacking species. Consequently, there are three mechanisms of attack:

1. leaching of mobile ions out of glass;
2. uniform dissolution of the glass network; and
3. preferential dissolution of glass.

During leaching, invading ions from the fluid medium enter the glass structure through the interstices and exchange sites with host ions in the glass on a no-net-transfer-of-electrical-charge basis. In turn, the host ions of the glass work their way gradually to the glass surface, which is what is meant by **leaching**. Because alkali ions (R) are the most mobile ions in the glass, leaching of alkali ions is the dominant durability problem and may be expressed by the following surface reaction:

$$\equiv\text{SiOR} + (\text{H}^+ + \text{OH}^-) \longrightarrow \equiv\text{Si—OH} + \text{ROH}. \quad (17.1)$$

In the next step, the proton on the Si–OH group ("silanol" group) exchanges sites with an inner R^+. In other words, the proton diffuses inwards and the R^+ moves outwards. The Si–OH ultimately combines with an Si–OR to form a bridging bond $\equiv$Si—O—Si$\equiv$, or the "siloxane" group:

$$\equiv\text{Si—OR} + \text{HO—Si}\equiv \longrightarrow \equiv\text{Si—O—Si}\equiv + \text{ROH}. \quad (17.2)$$

Since *the leaching process is interdiffusion-controlled*, the concentration profile of R in glass surface would be solutions of Fick's second law of diffusion with appropriate boundary conditions (see Section 14.4), and the *amount of*

R in the leachate would increase as a function of $(\text{time})^{1/2}$. Note also that in a weathering experiment on a soda lime silicate glass, the NaOH so extracted would react with CO_2 of the atmosphere, forming $NaHCO_3$, which is the white deposit formed on windows over a period of time.

A completely different reaction occurs where OH^- group in a basic solution of ROH breaks the network:

$$\equiv\text{Si}-\text{O}-\text{Si}\equiv + (\text{R}^+ + \text{OH}^-) \longrightarrow \equiv\text{Si}-\text{OH} + \text{RO}-\text{Si}\equiv. \tag{17.3}$$

The bridge is broken to form a more open "gel-like" structure on the surface. With time, more of the R^+OH^- molecules are able to break up more of the bridging oxygen bonds, resulting in a gradual dissolution of the network. The surface appears corroded or etched. *The rate of such a corrosion reaction or etching is generally linear with time.* When the dissolution is uniform, the ratio of the various species in the attacking medium is the same as that in the glass.

Preferential dissolution occurs because of the phase microstructure of glass. Some regions of glass contain compositions that are more readily attacked by the medium than others. In such a case, the ratio of the dissolved species in the medium would be different than that of the parent glass.

In all cases, the rate of attack is a thermally activated process such that a plot of $\ln (Q)$ vs. $1/T$, where Q equals the amount of material extracted at temperature T, is a straight line with a negative slope proportional to the activation energy E_{act}. The values of E_{act} are about the activation energies for the diffusion of alkali ions in the glass (about 100–120 $kJ{\cdot}mol^{-1}$). For most soda lime silicate glasses, Q increases by a factor of 1.5 to 2 per 10°C rise in temperature [1].

17.3. Measurement of Chemical Durability

Chemical durability may be readily measured by conducting simple experiments: immersing a specimen of glass in a given quantity of fluid contained in, for instance, a covered, nonreacting polymer beaker at a controlled temperature in a shaker bath. The quantity of material leached is obtained either by measuring the weight change or by flame photometric/atomic absorption/induction-coupled plasma (ICP) analyses of the leachate. *The experiments can be conducted on either a bulk specimen or a pulverized specimen.* The latter experiments require shorter periods of time because of the much larger exposed surface area. The choice here depends largely upon the intended application. Glass surface composition is generally quite different from that of the bulk as a result of volatilization of particular species during glass-forming processes. Hence, where durability of the as-formed surface is the issue, the experiments are best conducted on specimens having as-formed surfaces. Also, since the glass constituents are likely to affect the attacking medium characteristics, *the experiments can be performed in closed (fixed quantity) medium, or in open (flowing) medium* where the fluid is continually replenished. ASTM-recommended methods to conduct durability measurements are described in C225-85.

In the closed-medium experiments, the pH increases with time because of the extraction of the alkali. As we will note later, the increased pH encourages network dissolution. Closed-medium experiments also suffer from the additional possibility that reaction products are deposited on the glass surface, acting as barriers for further reaction. Thus, the relationship of the amount of material extracted with time is rarely constant. The use of a gentle shake (provided by a shaker bath) to reduce the deposit problem is often recommended.

Weathering experiments generally involve the measurement of the weight gain of a specimen of known surface area kept in, say, a 90°C–90% RH chamber for a specific period of time.

The extent and the mode of chemical attack may also be analyzed using surface characterization instruments such as infrared reflection spectroscopy (IRRS), Auger electron spectroscopy (AES), secondary ion mass spectroscopy (SIMS), and energy dispersive x-ray analysis (EDAX). Because all charged particle beam analysis instruments affect the local composition during the analysis, one must be careful in the collection of meaningful data.

17.4. Data on Chemical Durability

The concentration profile of hydrogen and sodium in the glass surface of a soda lime silicate glass after immersion in water at 90°C for 560 hours, shown in Fig. 17-1 (after Lanford [2]), confirms the basic interdiffusion between mobile alkali ions and exchangeable species from water. However, Lanford found that three hydrogens exchanged for each sodium. In addition, it has been observed that in potassium-containing glasses, the interdiffusional mobilities of the exchanging species are about the same. Both these pieces of evidence suggest that, at least at low temperatures, the exchangeable species in water are not protons but H_3O^+ ("hydronium") ions, which have ionic radii that are roughly the same as, or not much smaller than, that of K^+. In such a case, the leaching reactions might be better

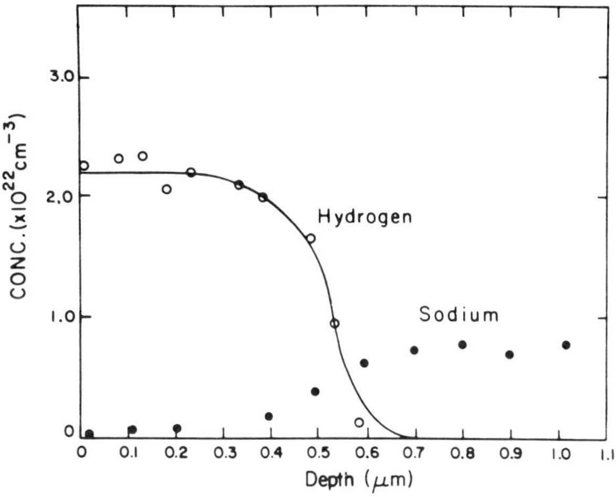

Figure 17-1. Profile of hydrogen and sodium in a soda lime silica glass held for 560 h in water at 90°C. (After Lanford [2].)

expressed as

$$\equiv Si—O^-R^+ + 2(H^+ + OH^-) \longrightarrow \equiv Si—O^-H_3O^+ + ROH, \tag{17.4}$$

$$\equiv Si—O^-H_3O^+ + R^+O^-—Si\equiv \longrightarrow \equiv Si—O—Si\equiv + ROH + H_2O. \tag{17.5}$$

The H_2O molecule presumably stays within the interstices, whereas ROH is able to gradually move outwards.

Chemical durability of glass in water during the leaching stage is a mass transport property, and hence its relationship to the kinetics of ionic diffusion is direct. One expects that vitreous silica with nearly 100% bridging oxygens and only trace mobile cations in the network should be one of the most durable glasses upon immersion in water. Addition of alkalis not only opens up the network, but also introduces mobile cations within the interstices; hence, the durability should decrease greatly. The addition of alkaline earths to an alkali silicate glass introduces blocking species within the network; hence, the durability of the glass would improve. Likewise, the addition of small amounts of Al_2O_3 (2–3%) increases the connectivity (by soaking up NBOs); this should result in durability improvement. Glass container compositions for beverage applications generally utilize this beneficial property of the small alumina addition to bring about marked improvement in durability over compositions to which alumina is not added.

Table 17-1 Ranking of the Chemical Durability of Several Commercial Glasses in pH 2 to 8 Media

Relative durability	Glasses	Manufacturer's designation	
		Corning	Owens-Corning
1	Vitreous silica		
2	Pyrex borosilicate	7740	KG-33
	Aluminosilicate, low alkali	1710	EZ-1
	Fiber glass "E"		
	Borosilicate, mixed alkali	3320	EN-2
	Borosilicate, low	7059	EM-1
	alkali	7800	N-51A
3	Commercial soda-lime with 1–3% Al_2O_3	0080	R-6
	Potash–soda–lead	0010	KG-1
		0120	KG-12
	High lead	7570	EG-4
		8871	
	Laboratory soda-lime + 5% Al_2O_3		
	Laboratory soda-lime + 2% Cs_2O		
	Lithium–cesium lime silicate		
Have a transformed surface layer			
4	Soda-lime		
	Soda–potash–lime		
5	Binary lithium silicates, 10–33% Li_2O		
6	Binary sodium silicates, 15–25 mol% Na_2O		
7	Binary sodium silicates, 25–40 mol% Na_2O		
	Binary potassium silicates, 15–25 mol% K_2O		

The chemical durability of several commercial compositions in aqueous solutions of pH = 2 to 8, ranked 1 to 7 ("1" = most durable, "7" = least durable), is shown in Table 17-1. As expected, silica is the most durable glass. This is followed by aluminosilicates and borosilicates, where the high durability is indeed due to a high degree of network connectivity.

In an acidic medium, pH < 7, the durability reactions generally remain of the leaching type at early times. Thus, in a durability experiment on a soda lime silicate glass, the leachate contains mostly Na_2O and little, if any, CaO and SiO_2. However, the observation of surface cracking in modified E-glass having 5% Na_2O or K_2O upon immersion in HCl (pH 2) suggests [3] that the diffusing species in acidic media are H^+ and not H_3O^+, as in the reaction with water. The suggested reaction is similar to (17.1):

$$\equiv\text{Si—ONa} + \text{H}^+ \longrightarrow \equiv\text{Si—OH} + \text{Na}^+. \qquad (17.6)$$

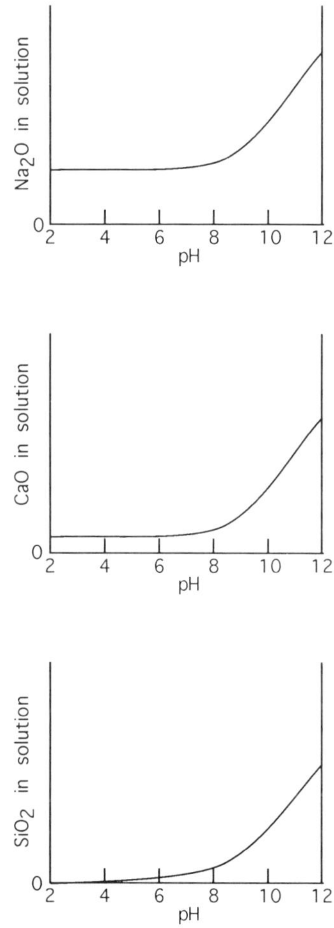

Figure 17-2. Effect of pH on the extraction of the various species from a soda lime silica glass.

Apparently, the exchange of Na^+ in the glass with the smaller-sized proton creates tensile stress in the surface, which leads to cracking (see Sections 14.8.1 and 18.10).

In basic media, pH > 7, the durability reaction turns to network dissolution. More of the CaO and SiO_2 is extracted into the leachant. The behavior is schematically shown in Fig. 17-2, and contrasted for a soda lime silicate "E-glass" as dissolution velocity in Table 17-2.

Within the binary alkali silicates, the durability decreases in the order Li → Na → K. The formation of the silica-rich film and glass corrosion in water at 100°C as a function of time for 10–46% R_2O–SiO_2 glasses are shown schematically in Fig. 17-3. It may be noted that the silica-rich film forms early in K_2O–SiO_2 glass, followed by the highest corrosion rate compared to others. It is possible, as suggested earlier, that K^+ is slightly bigger than the H_3O^+ ions, so that a tensile stress develops in the surface after ion exchange during leaching (see Section 14.8.1 on ion exchange). Such a tensile stress would cause cracking in the surface, which, in turn, would accelerate the glass corrosion.

An example (4) of infrared reflection spectroscopy on a $Li_2O{\cdot}2SiO_2$ glass is shown in Fig. 17-4. Curve 1 is the reflection off freshly polished surface and shows the presence of the Si–NBO peak at about 900 cm^{-1} (marked "NSX"). Upon exposure to 85% RH at 40°C, the NSX peak is reduced a small amount. The peak is reduced considerably when the glass is immersed at 50°C in static demineralized water having an initial pH of 6.5. In addition, a peak marked "S" characteristic of the stretching vibration in Si–O–Si bridging oxygens in vitreous silica appears at 1,100 cm^{-1}. Clearly, the water immersion causes leaching of alkali, leaving a silica-rich surface

Table 17-2 Dissolution Velocity of a Soda Line Silicate "E" Glass[a]

pH	Rate (cm s^{-1}) × 10^{13}
1.3	5.3
2	2.2
3–5	1.5
6	3.3
8	10.5
9	34.0
10	121
11	192

[a] Z. Boksay and G. Bouquet, *Phys. Chem. Glasses*, **21**, 110 (1980).

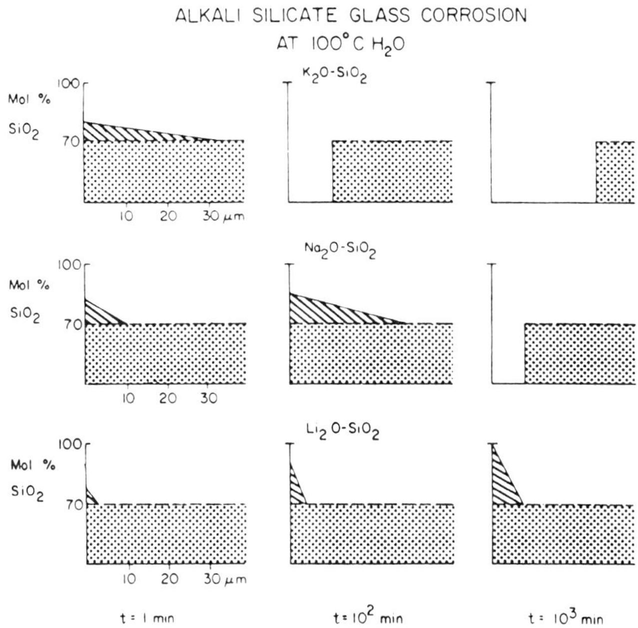

Figure 17.3. Formation of silica-rich films on the surface of silicate glasses containing 30 mol % alkali resulting from static aqueous corrosion at 100°C. (After Hench [4]. Reproduced with permission of Elsevier Science Publishers.)

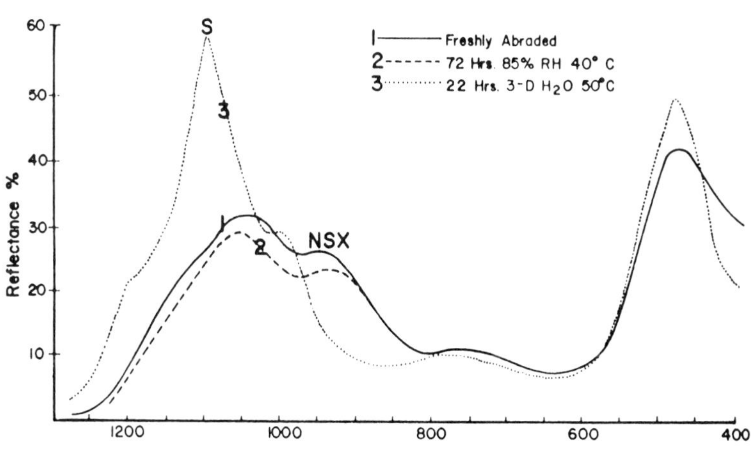

Figure 17-4. Infrared reflection spectra of $30Li_2O–70SiO_2$ after various corrosion reactions. (After Hench [4]. Reproduced with permission of Elsevier Science Publishers.)

as per reactions (17.1) and (17.2). On the other hand, exposure to 85% RH causes buildup of Li ions on the surface, which rapidly brings about surface corrosion, leaving Si–OH gel still containing Li ions on the surface as per reaction (17.3), which does not resemble vitreous silica.

Appropriate mathematical models of the extraction process involve a linear summation of contributions due to diffusion-controlled ($Q = k_1 t^{1/2}$) and corrosion-controlled reactions ($Q = k_2 t$). One such model assuming a concentration-independent diffusion coefficient is described by Doremus [5].

All silicate glasses are readily dissolved by HF acid. Apparently, HF attacks the Si–O bond, converting it to silicofluoride complexes. Glasses of the composition $75P_2O_5{\cdot}20Al_2O_3{\cdot}5RO$ (wt. %, R = Pb, Zn, Be, etc.) are resistant to HF attack. Phosphoric acid also attacks most silicate glasses at temperatures $>200°C$.

For glass-reinforced-cement applications, one needs to have alkali-resistant glass fibers. Such glasses are of the general composition $50\text{–}75SiO_2{\cdot}5\text{–}30ZrO_2{\cdot}0\text{–}20MO{\cdot}0\text{–}25R_2O{\cdot}<20Al_2O_3$ (wt. %). For a high-temperature alkali vapor resistance application, namely, in the sodium vapor lamp seal, the glass composition is approximately $45Al_2O_3{\cdot}36CaO{\cdot}14BaO{\cdot}5MgO$ (wt. %). In some variations, the substitution of some of the CaO and MgO by SrO, Y_2O_3, and small amounts of B_2O_3 may be helpful.

For immobilization high-level radioactive waste [6], the basic requirements are (i) low melting, preferably below, 1,400°C; (ii) no phase separation; (iii) resistance to crystallization during slow cooling; (iv) high durability; and (v) high level of predictability of properties in repeated melts. A simulated borosilicate glass composition designed in the U.S. has the composition

$$43.44SiO_2{\cdot}6.55Al_2O_3{\cdot}10.11B_2O_3{\cdot}3.74K_2O{\cdot}11.44\ Na_2O{\cdot}3.16Li_2O{\cdot}\\ 0.9MgO{\cdot}1.0MnO{\cdot}0.35P_2O_5{\cdot}3.61ThO_2{\cdot}0.91TiO_2{\cdot}\\ 0.5\ CaO{\cdot}12.15Fe_2O_3{\cdot}0.58\ UO_2\ (\text{wt. }\%),$$

plus the balance 1.56% that comprises a host of oxides such as BaO, SrO, and Cs_2O, each less than 0.3%. In actual practice, the batch consists of spent nuclear fuel and oxide additions to make up the full composition. After melting, approximately 1,700 kg glass is stored in a 61 cm OD × 59–60 cm ID × 300 cm height stainless steel cylinder, screw-capped, and transferred to a storage site. Because all of the materials handling during batching, melting, and pouring must be performed remotely, the need for a full control over properties, particularly the viscosity, is imperative.

One commercial exploitation of preferential dissolution behavior is the manufacture of Vycor, which has already been discussed in Section 4.4. Sodium borosilicate glasses lying within the immiscibility dome (Fig. 4-12a) are phase-separated into interconnecting soluble and insoluble matrices [7].

skeleton is dried and sintered. The drying phase is perhaps the most critical of all the steps. Residue deposited along the channels, which are generally about 4 nm in diameter, making up about 25–40% void volume, may generate enormous capillary stresses (Section 20.3.3), which have the potential to fracture the body.

In contrast to the leaching behavior of Vycor-type glasses, the high chemical resistance of the laboratory Pyrex-type glasses is achieved by making sure that the sodium borosilicate composition is near the outside (silica-rich side) of the immiscibility dome (Fig. 4-12a), and by rapid cooling of the formed products. (The droplet phase is kept below 2–5 nm size.)

17.5. Methods of Improving Chemical Durability

Techniques to improve chemical durability of glass are based upon either reducing the concentration of alkali ions in the surface, or forming barrier coatings to impede the approach of the attacking fluid to the glass surface.

The easiest way to reduce alkali concentration in the surface is to fire-polish a product during forming. Fire-polishing of an Na_2O-containing glass causes Na_2O to volatilize, forcing two nonbridging oxygens to form a bridging bond. The surface thus becomes silica-rich, and consequently its durability increases. Another method often used is treatment with wet SO_2 mixed with O_2:

$$\equiv\text{Si—ONa} + \text{NaO—Si}\equiv + SO_2 + \tfrac{1}{2}O_2 + 3H_2O \longrightarrow \equiv\text{Si—O}^-H_3O^+ + H_3O^+\ {}^-\text{O—Si}\equiv + Na_2SO_4. \quad (17.7)$$

The sulfate layer produces a bloom on the surface that is washed away by rinsing with hot water. The resultant glass surface has larger hydronium ions blocking further diffusion. Instead of wet SO_2 and O_2, one can also utilize wet-SO_3 with the same effective result. Inhalation of SO_2 can cause serious injury to the lungs, and the release of SO_2 into the atmosphere is the primary cause of acid rain. Hence, current OSHA and EPA regulations require that SO_2 treatment be performed in closed chambers with scrubbers to retrieve unspent gas.

Similar results are obtained with wet Cl_2 and F_2 treatments, although the glass surface reactions are not well understood. Again, because of the toxic nature of these gases, such treatments are rarely employed.

Coating of metal oxides and polymers such as lacquer, Teflon, Lexan (a GE product), and polyimides can reduce the activity of water on the glass surface. Thus, these act as barriers to further attack by the medium.

Exercise

1. A glass of the composition $4Na_2O{\cdot}CaO{\cdot}5P_2O_5$ is immersed in 1,000 mL of deionized water for 150 min at room temperature. Assume that any ions coming out of the glass dissolve in the water. Analysis of the solution for Na_2O and CaO shows the presence of 38 ppm Na_2O and 9.5 ppm CaO. A specimen weight loss of 0.1539 g was observed. Show that these data are consistent with a uniform dissolution mechanism.

References

1. P. B. Adams and H. V. Walters, in *Glass Engineering Handbook* (G. W. McLellan and E. B. Shand, eds.), 3rd Ed., pp. 3–4. McGraw-Hill, New York, 1984.
2. W. A. Lanford, *Nucl. Instr. Methods* **149**, 1 (1978).
3. A. G. Metcalfe, M. E. Gulden, and G. K. Schmitz, *Glass Technol.* **12**(1), 15–23 (1971).
4. L. L. Hench, *J. Non-cryst. Sol.* **19**, 27–39 (1975).
5. R. H. Doremus, in *Treatise on Mat. Sci. & Tech.* (M. Tomozawa and R. H. Doremus, eds.), Vol. 17, *Glass II*, pp. 41–69. Academic Press, New York, 1979.
6. Suggested reading: (a) G. G. Wicks, in *Treatise on Mat. Sci. & Tech.*, Vol. 26, *Glass IV* (M. Tomozawa and R. H. Doremus, eds.), pp. 57–118, Academic Press, New York, 1985. (b) C. M. Jantzen, in *Adv. In Fusion & Processing of Glass* (D. F. Bickford, E. N. Buolos, F. Olix, W. E. Horsfall, J. N. Lingscheit, W. C. LaCourse, F. E. Woolley, F. Harding and L. D. Pye, eds.), Paper no. 24. American Ceramic Society, Columbus, Ohio, 1988.
7. Suggested reading: W. Vogel, *Chemistry of Glass*, p. 113, Am. Ceram. Soc., Columbus, Ohio, 1985.

Chapter 18

Strength and Toughness

18.1. Introduction

Applied stress at the point of failure is termed the strength. A measurement of strength generally consists of applying an increasing magnitude of stress in a defined-geometry specimen until failure occurs. Units of strength are the same as those of stress, e.g., dyn/cm^2 in cgs, MPa in SI, and psi in English systems.

It is a common experience that glass products break readily when subjected to mechanical or thermal shock. On the other hand, bulletproof glass and glass fibers that are "stronger than steel" are not unheard-of, either. We could summarize our experience as follows:

Strength of	common glass products	about 14–70 MPa.
	freshly drawn glass rods	about 70–140 MPa.
	abraded glass rods	about 14–35 MPa.
	wet, scored glass rods	about 3–7 MPa.
	armored glass	about 350–500 MPa.
	handled glass fibers	about 350–700 MPa.
	freshly drawn glass fibers	about 0.7–2.1 GPa.

In contrast, most metals including steels fail around 140–350 MPa. Two other common observations that contrast failure in metals with that in glass are

(i) Most metals neck or yield plastically before actual failure occurs, whereas glasses do not show any ductility prior to the failure. We say that the metals show ductile failure and the glasses show brittle failure.
(ii) Whereas a given group of metal specimens prepared identically shows a narrow distribution of yield stress and ultimate fracture stress, glass specimens produced apparently identically are likely to display a large variation of strength.

All this is despite the fact that the elastic properties and the density of glasses are generally of the same order as that of most metals!

The ductile behavior of metals is easy to understand. Crystalline structures such as those of the metals contain a large number of dislocation defects, which can move readily, and new dislocations created under the influence of applied stresses. An irreversible movement of dislocations appears as ductility or plastic deformation. In glasses there are no dislocations in the structure, and hence there is no apparent ductility upon the application of stress. Some plastic deformation, under a microhardness indentation, and irreversible compaction, under a high magnitude of hydrostatic stress, do occur in glass (see Chapter 8). However, the amount of permanent strain involved is small compared to that due to dislocation motion in most metals.

We know now that the strength of brittle solids is also compromised by structural flaws. Flaws in solids act as stress intensifiers, making it possible for the solid to fracture at stress magnitudes lower than that theoretically needed to pull atoms apart. *Whereas in metals the dislocations move without breaking atomic bonds, in glasses the flaws grow, creating, in principle, two new surfaces. Because flaws grow, there is fatigue over the duration of stress application, not predicted by earlier theories of the strength of brittle solids.*

In this chapter, our objective is to understand the marked influence of specimen size, surface condition, and environment on the measured strength of glass. After a brief mention of the experimental techniques of strength measurement for glasses, we will consider the theoretical strength of a "flawless" and a flawed brittle solid, and explore the description of strength in terms of flaw growth kinetics and fracture statistics.

18.2 Theoretical Strength of a Flawless Brittle Solid

The amount of work done in pulling two atoms apart such that two new surfaces are created is the theoretical strength of a flawless solid. Consider the pulling of the atom located at a_0 away from an atom located at the origin (Fig. 18-1). Because of the nature of interatomic forces, the stress reaches a maximum σ_m, after which a decrease is experienced asymptotically until the separation becomes much too large for the atoms to be considered bonded, and two new surfaces having fracture surface energy γ_f each are created. The instantaneous stress, σ, may be approximated by a sinusoidal form

$$\sigma = \sigma_m \sin(2\pi x/\lambda), \quad (18.1)$$

where x is the increase in atomic separation. At small separation, $\sin(2\pi x/\lambda) = (2\pi x/\lambda)$, and hence

$$\sigma = \sigma_m(2\pi x/\lambda). \quad (18.2)$$

According to Hooke's law,

$$\sigma = Ex/a_0, \quad (18.3)$$

where E is the Young's modulus and x/a_0 is the strain. Combining (18.2) and (18.3), we get

$$\sigma_m = \lambda E/2\pi a_0. \quad (18.4)$$

The energy of two freshly fractured surfaces should equal the work of separation, which is the area under the stress–distance curve approximated as a half sine wave. Thus,

$$2\gamma_f = \frac{1}{2}\int_0^{\lambda/2} \sigma_m \sin(2\pi x/\lambda)dx = \lambda\sigma_m/2\pi. \quad (18.5)$$

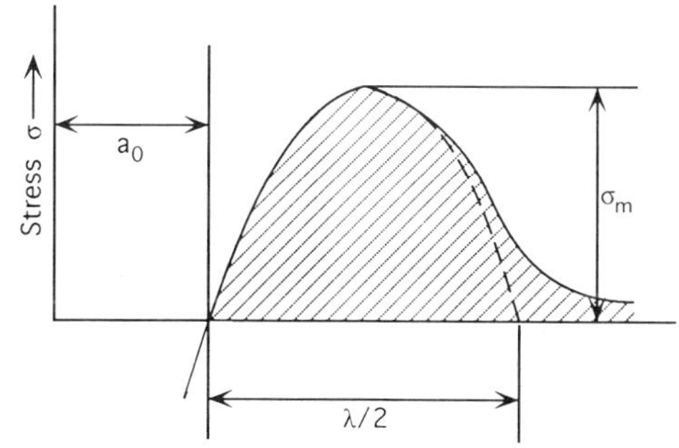

Figure 18-1. Variation of stress σ with increasing interatomic spacing.

Combining (18.4) and (18.5), we get

$$\sigma_m = [\gamma_f E/a_0]^{1/2}. \tag{18.6}$$

Equation (18.6) is the Orowan [1] expression for the theoretical strength of "perfect" brittle solids.

If we use $E = 70$ GPa, $\gamma_f = 3.5$ J/m^2, and $a_0 = 0.2$ nm, we get $\sigma_m = 35$ GPa $= E/2$. (Recall: 1 Pa = 1 N·m^{-2}, and 1 N·m = 1 J. The reader is to be cautioned not to mix up γ_f with the interfacial surface energy γ, commonly quoted for glasses at 0.35 J/m^2. The normally measured interfacial energy is that of a relaxed surface having adsorbed species such as water. In a fresh fracture, the atomic bonds are separated, with no adsorbed species such as water. There may also be some plastic deformation energy requirement for the bond-stretching, in addition to irreversible effects such as acoustic and photon emission, and conversion of some energy to heat.) Strengths approaching this figure within a factor of 5 have now been achieved for many freshly drawn fine glass fibers under very careful drawing conditions.

18.3. Strength of a Flawed Brittle Solid: Griffith's Analysis

It was Inglis (2) who pointed out that flaws in solids act as stress concentrators. Upon the application of an applied stress σ_a, the stress σ_{yy} at the tip of an assumed elliptical hole or flaw (Fig. 18-2) with minor axis $2b$

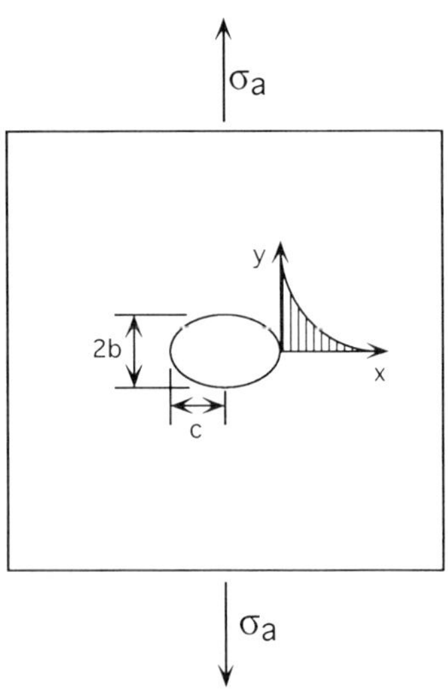

Figure 18-2. Elliptical flaw in a glass plate under tension.

and major axis $2c$ in a plate (z-plane) is given by

$$\sigma_{yy} = \sigma_a[1 + (2c/b)]. \tag{18.7}$$

This means that when σ_a is such that $\sigma_{yy} = \sigma_m$, the specimen will fracture. Since the radius of curvature at the tip of the ellipse $\rho = b^2/c$, Eq. (18.7) yields

$$\sigma_{yy} = \sigma_a[1 + 2(c/\rho)^{1/2}].$$

Now, since $c >> \rho$,

$$\sigma_{yy} = 2\sigma_a(c/\rho)^{1/2}. \tag{18.8}$$

The curvature ρ at the crack-tip in a brittle solid approximates atomic dimensions a_0,

$$\sigma_f = [\sigma_a]_{failure} = (\tfrac{1}{2})[a_0/c]^{1/2}[\gamma_f E/a_0]^{1/2} = (\tfrac{1}{2})[\gamma_f E/c]^{1/2}. \tag{18.9}$$

Equation (18.9) implies that a flaw having c as small as about 50 μm can cause fracture at as little as 35 MPa.

Griffith (3) suggested that having an applied stress such that the stress at the tip of the flaw exceeds the theoretical strength cannot be the sole criterion for strength. The flaw must exceed a critical length before it would propagate into a fracture. Griffith argued that the change $\Delta\Psi$ in the potential energy of a plate with an elliptical flaw of length $2c$ is given by

$$\Delta\Psi = -\frac{\pi c^2 \sigma_a^2}{2E} + 4c\gamma_f. \tag{18.10}$$

The first term represents the loss in strain energy, and the second the approximate gain in surface energy. The linear surface energy term causes $\Delta\Psi$ to increase with c initially until c becomes c^*, at which point $\Delta\Psi$ begins to decrease due to the dominance of the parabolic term. The c^* is found by differentiating with respect to c and setting the expression equal to zero. This gives

$$\sigma_f = [\sigma_a]_{failure} = [2\gamma_f E/\pi c^*]^{1/2}. \tag{18.11}$$

(Note: This equation is valid for the plane-stress condition. Replace E by $E/(1 - \nu^2)$, where ν = Poisson ratio, for a plane-strain condition.)

Equation (18.9) or (18.11) hints at the effect of the environment through its effect on the surface energy. Thus, since contact with water lowers γ_f, $[\sigma_a]_{water} < [\sigma_a]_{air}$. Also apparent is the statistical nature of the failure in brittle solids. *Flaws generally occur on a free surface and have a wide distribution of lengths*, which gives them a wide distribution of severity in terms of their $[c/\rho]$ ratio. We may argue that *fracture would occur at the site of the most severe flaw, much like the case of the weakest link in a chain.*

These flaws are called **Griffith flaws**. Although γ_f and E depend on glass composition somewhat (varying about 20%), the variation in c can be of several orders of magnitude. It is not a surprise, therefore, that *strength of glass is not intrinsic to its composition, but may depend greatly on methods of processing.*

Even though the length of a typical Griffith flaw may be as little as 5 to 50 μm, its width is of atomic dimensions. For this reason, Griffith flaws are not easily observable, even with the help of the highest-power optical microscope. Some efforts have been made to visualize Griffith flaws by decorating the flaw sites using exposure to sodium vapor above 350°C [4] and using ion exchange with a eutectic mixture of $LiNO_3$–KNO_3 salt droplets [5]. In both these cases, the decoration in the form of extended cracks probably originated at the flaw sites, but clearly did not highlight the flaw geometry itself. Use of high-resolution electron microscopy on thin silica films has shown some evidence [6] of an elliptical crack having about 1.5 nm tip curvature.

The most important drawback in Griffith's analysis is the lack of time effect. Since it is known that glass fatigues over long duration of stress application, any consideration of σ_f must include a time effect. The reconciliation is made through the application of fracture mechanics.

18.4. Elementary Fracture Mechanics Concepts

It may be realized from (18.11) that $\sigma_f[\pi c]^{1/2} = [2\gamma_f E]^{1/2}$ = constant, depending upon the material and its environment. E is determined by the material, whereas γ_f is determined both by the material and the environment. If $\sigma_a[\pi c]^{1/2} = K$, the **stress intensity factor**, and the quantity $\sigma_f[\pi c^*]^{1/2} = K_c$, the **critical stress intensity factor**, then fracture would occur when $K \rightarrow K_c$

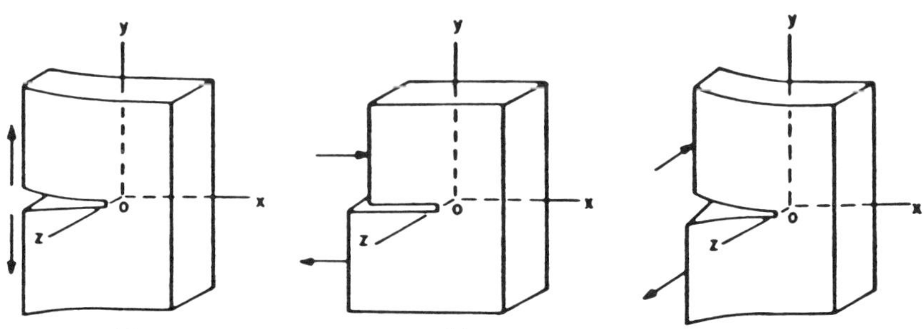

(a) (b) (c)

Figure 18-3. The different modes of crack extension. (a) Opening. (b) Sliding. (c) Tearing.

because of changes in σ_a or c. Suitable units for K are MPa·m$^{1/2}$ and N·mm$^{-3/2}$ in SI units, and ksi·in.$^{1/2}$ in English units. Note 1 N·mm$^{-3/2}$ = 0.0316 MPa·m$^{1/2}$ and 1 ksi·in.$^{1/2}$ $\cong$ 1.1 MPa·m$^{1/2}$.

According to Irwin [7], there are three possible modes of independent kinematic movements that cracked faces could have relative to each other (Fig. 18-3). In **mode I (opening mode)**, the fracture plane (the y-plane) and the direction of propagation (x-direction) are perpendicular to the direction of the applied stress (y-direction). **Mode II (sliding mode)** is a failure due to shearing where the applied stress (σ_{xy}) acts to slide fracture planes (y-plane) over each other along the direction of propagation (x). **Mode III (tearing mode)** is also a shear failure, except that the direction of fracture propagation (x) is perpendicular to that of the stress (σ_{zy}) on the y-plane. Stress fields at a distance r and polar angle θ from the tip of a sharp slit crack can then be defined in the form of a generalized equation:

$$\sigma_{ij} = K(2\pi r)^{-1/2} f_{ij}(\theta), \tag{18.12}$$

where K represents three stress intensity factors K_{I} (pronounced K-one), K_{II}, and K_{III} corresponding to the three modes of failure. For a single crack of length $2c$ extending in an infinite body subjected to a uniform remotely applied stress field $[\sigma_{ij}]_\infty$, the K's are given by [8]

$$\begin{aligned} K_{\mathrm{I}} &= [\sigma_{yy}]_\infty [\pi c]^{1/2}, \\ K_{\mathrm{II}} &= [\sigma_{xy}]_\infty [\pi c]^{1/2}, \\ K_{\mathrm{III}} &= [\sigma_{zy}]_\infty [\pi c]^{1/2}. \end{aligned} \tag{18.13}$$

There are thus three critical values of K's: K_{Ic} (pronounced K-one-cee), K_{IIc}, and K_{IIIc}. The fracture mechanics criteria for failure, then, are the approach of either of the stress intensity factors to their critical values. *Of the three critical values, we are generally concerned with K_{Ic} for failure in brittle solids that represents both failure plane as well as propagation perpendicular to the principal tension.*

Irwin also defined the term strain energy release rate **G** equal to K^2/E for the plane-stress condition, and equal to $K^2(1 - \nu^2)/E$ for the plane-strain condition. **G** is the change in the strain energy of the body when a unit change occurs in crack length. (Note, as before, that plane-strain calculations for mode I and mode II failure can be obtained from the plane-stress formulae by replacing E with $E/(1 - \nu^2)$. Henceforth, we will assume plane-stress condition.) For an elastic crack in an infinitely wide plate,

$$\mathbf{G}_{\mathrm{Ic}} = K_{\mathrm{Ic}}^2/E = \sigma_{\mathrm{f}}^2 \pi c^*/E.$$

Hence,

$$\sigma_{\mathrm{f}} = [E\mathbf{G}_{\mathrm{Ic}}/\pi c^*]^{1/2}, \tag{18.14}$$

which gives the basic definition that $\mathbf{G}_{\mathrm{Ic}} = 2\gamma_{\mathrm{f}}$.

Likewise, $\mathbf{G}_{\mathrm{IIc}} = K_{\mathrm{IIc}}^2/E$, and $\mathbf{G}_{\mathrm{IIIc}} = K_{\mathrm{IIIc}}^2/2\mu$ (μ = shear modulus) under conditions of plane-stress.

We describe **toughness** of a solid as its ability to absorb energy prior to failure. Mathematically, it is the area under the true stress–true strain curve. In brittle solids, such a plot has little meaning on a macro scale. We must talk about the energy required to extend the flaw from some initial size to the critical size. Thus, $\mathbf{G}_{\mathrm{Ic}}$ would be a measure of the **fracture toughness**. By convention, however, the critical stress intensity factor K_{Ic} $[=(E\mathbf{G}_{\mathrm{Ic}})^{1/2}]$ itself is called the fracture toughness of the material. (Many of the classic papers, such as those by Griffith and Irwin on strength and fracture mechanics, are included in Ref. 9.)

The inclusion of the dynamic changes in σ_a and c all lumped into K makes it possible to include the time element into proper perspective while considering fracture in brittle solids.

18.5. Glass Fatigue

Griffith's theory was challenged by two observations in particular:

(i) Under constant load, the time-to-failure varies inversely with the load applied.
(ii) Under increasing load (constant strain rate), higher stress to failure is needed for a higher straining (or stressing) rate.

Both these observations of **delayed failure** are manifestations of **fatigue**: the first is termed **static fatigue**, and the second, **dynamic fatigue**. The most cited static fatigue experiments are those of Mould and Southwick [10]. These authors first obtained the strength σ_N of pristine glass rods immersed in liquid N_2 (called the **inert strength**). They then plotted the strengths of rods of the same glass abraded in different manners using different grits as a fraction of the inert strength against log(t = time-to-failure). All the curves were linear for the most portion, were approximately parallel to each other, and had negative slopes (Fig. 18-4). By replotting the data of σ/σ_N vs. $\log(t/t_{0.5})$, where $\log t_{0.5}$ is the time corresponding to $\sigma/\sigma_N = 0.5$, it could be shown that all the data fell on a master curve (Fig. 18-5). In the linear portion, one can write

$$\sigma/\sigma_N = -A\log(t/t_{0.5}) + B. \tag{18.15}$$

The experiments of Mould and Southwick showed that flaws apparently

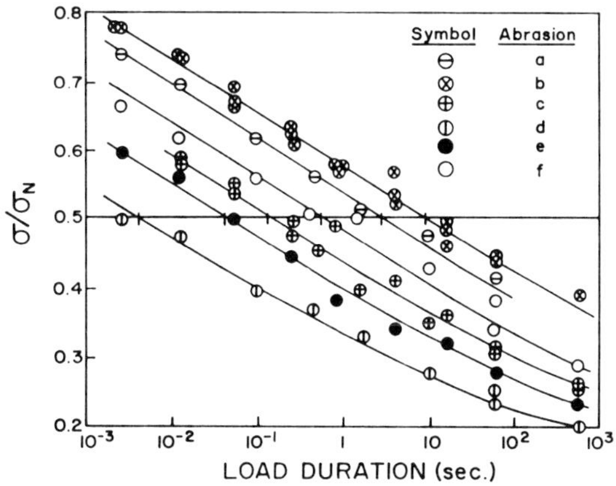

Figure 18-4. Plot of the time to fracture (x-axis) as a function of the normalized stress for various conditions of surface abrasion. (Abrasion code: a = severe grit blast; b = mild grit blast, emery cloth to stress; c = 150 grit; d = 600 grit; e = 320 grit; f = 150 grit, parallel to stress.) (After Mould and Southwick [10]. Reproduced with permission of The American Ceramic Society.)

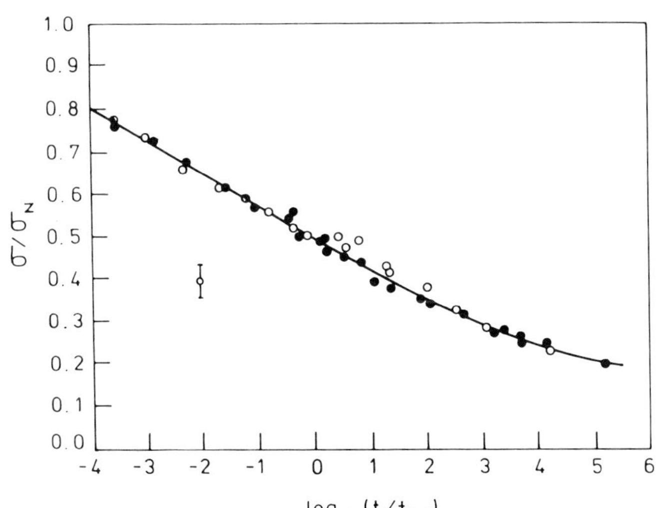

Figure 18-5. Universal static fatigue curve of Mould and Southwick.

grow with time during the application of stress; *the growth velocity must be lower at lower stresses* than at higher stresses, thus allowing a longer delayed failure.

The dynamic fatigue experiments are conducted by applying stress at varying rates and plotting σ_f against log(stressing rate). Typical results (Fig. 18-6) show that higher stressing rates allow the specimens to sustain higher stresses. Apparently, higher stressing rates allow lesser time for flaws to grow, making it possible for the specimen to sustain higher stresses.

Both the static and dynamic fatigue experiments are now reconciled

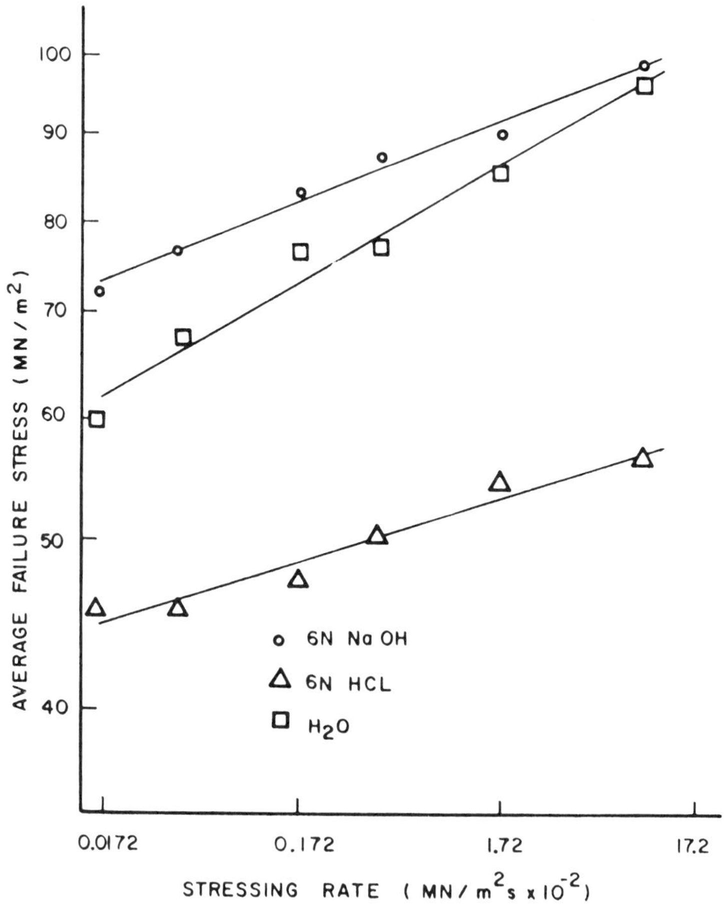

Figure 18-6. Dynamic fatigue of glass. (After J. E. Ritter, Jr., and R. P. LaPorte, *J. Am. Ceram. Soc.* **58**(7–8), 265 267 (1975). Reproduced with permission of The American Ceramic Society.)

through direct measurements of what is called the **subcritical crack velocity** as a function of the applied stress or the stress intensity factor by Wiederhorn and his associates [11]. Wiederhorn used a microscope/cathetometer to measure the growth velocity of cracks of length *a* as a function of the applied load *P* on double cantilever specimens of microscope slides of soda-lime glass. The measurements were made with the specimens immersed in liquid N_2, gaseous N_2 at 25°C of different RH, and water. The measurements of load and crack length could be converted to stress intensity factor K_I using the formula

$$K_I = [Pa/w^{1/2}b^{1/2}d^{3/2}][3.467 + 2.315\, d/a], \tag{18.16}$$

where *b* is the "web" thickness, and the meaning of the other terms is as in Fig. 18-7.

A plot of the log(crack velocity) vs. K_I is shown in Fig. 18-8. In general, *three regions of subcritical crack growth behavior* are observed, after which the crack has grown to the critical size required to fulfil the Griffith energy release criterion, and a spontaneous failure is imminent:

(i) Region I, where the crack velocity increases exponentially with applied load as well as humidity.
(ii) Region II, where the velocity is independent of the applied stress, but still depends upon the humidity level.
(iii) Region III, where the velocity is independent of the humidity level, but depends, once again, exponentially on the applied load. The slope in region III is much steeper than in region I. At the end of region III, $K_I \rightarrow K_{Ic}$, and "spontaneous" failure follows. The spontaneity of the failure is agreed, generally, to be a velocity exceeding about 0.1 m/s. Thus, as a convention, velocities less than about 0.1 m/s are the "subcritical" crack velocities that cause the growth of a flaw up to its critical size.

Wiederhorn and Bolz [11] showed that the crack velocities depend upon glass composition (Fig. 18-9). There apparently existed a fourth region at very low velocities where the V–K_I plot tended to become vertical. Thus, there appeared a **static fatigue limit** of the applied stress below which crack did not grow (to any measurable extent). In normal alcohols, the crack growth is essentially similar to that in nitrogen gas and depends upon the partial pressure (RH) of the dissolved water [12]. In addition, Wiederhorn *et al.* showed that the crack velocity in region I depended not only upon composition, but also on temperature (Fig. 18-10) and the environment—for instance, the pH of the medium—suggesting the possibility that subcritical crack growth may be thermally activated stress corrosion process. Wieder-

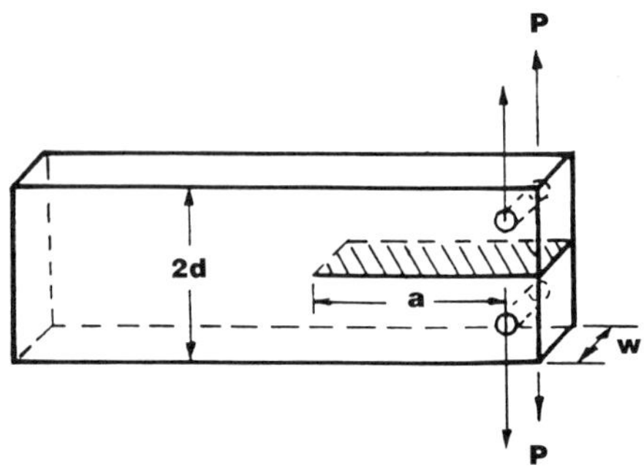

Figure 18-7. Double cantilever beam specimen configuration to measure crack velocity. Hatched areas represents the uncracked portion of the specimen ("web") between the guiding notches. (After Wiederhorn [11a]. Reproduced with permission of The American Ceramic Society.)

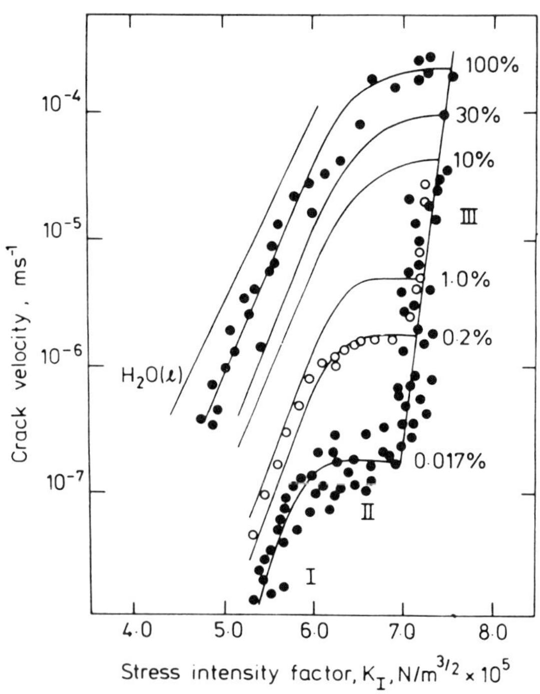

Figure 18-8. Crack velocity vs. applied intensity factor for soda lime silica glass in N_2 gas of varying levels of humidity. (After Wiederhorn [11a]. Reproduced with permission of The American Ceramic Society.)

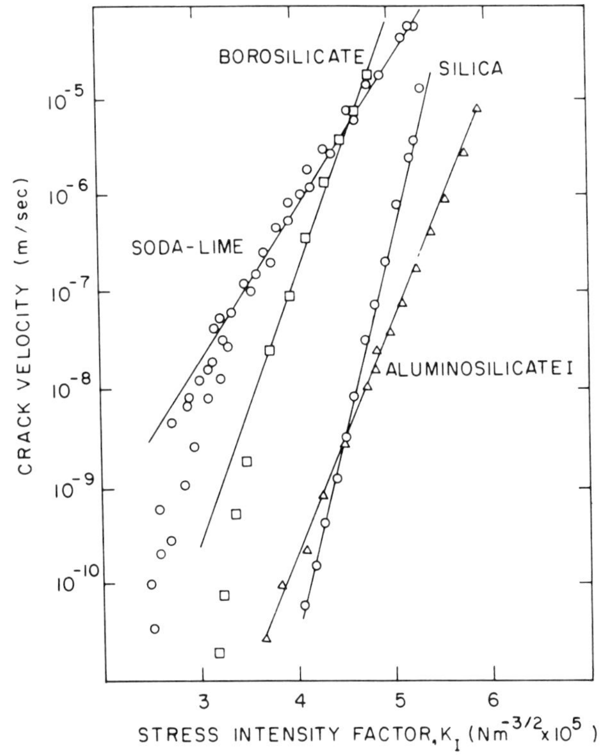

Figure 18-9. Effect of glass composition on crack velocity–K_I behavior of the various glasses tested in water at room temperature. (After Wiederhorn and Bolz [11b]. Reproduced with permission of The American Ceramic Society.)

horn suggested that the behavior in region II could be explained by assuming that the crack velocity in region II was limited by the rate of water transport to the crack tip. As indicated by Quakenbush and Frechette [13], the existence of region II is crack-size-dependent. If the crack is sufficiently small, then water migrates to the tip as rapidly as the crack growth, and there is no region II. Crack propagation studies in vacuum indicate that region III does not occur in all glasses. Fused silica, Vycor, and Pyrex show only region I prior to catastrophic failure.

The dependence of the crack velocity V upon K_I is more popularly expressed in terms of a power law (instead of exponential functions) as

$$V = AK_I^n, \tag{18.17}$$

where A and n are constants. n is called the **stress-corrosion susceptibility coefficient** or the **crack-growth exponent** and ranges between 12 and 35 for

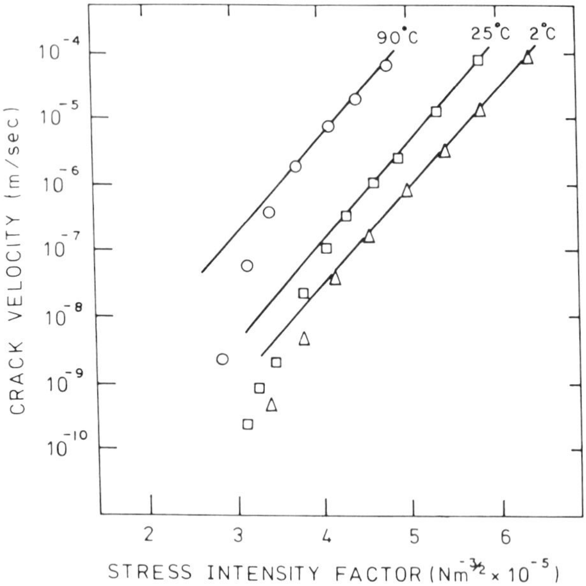

Figure 18-10. Effect of temperature on the crack velocity–K_I curves for a soda lime glass in water. (After Wiederhorn and Bolz [11b]. Reproduced with permission of The American Ceramic Society.)

most glasses: higher values are observed for immersion in strong HCl, and for silica glass.

For a generalized case of a flaw of size a,

$$K_I = \sigma_a Y c^{1/2}, \tag{18.18}$$

where Y is a geometry parameter ($= \pi^{1/2}$, for instance, for a slit crack in an infinite plate as in Eq. (18.13)). Now, since $V = dc/dt$, Eqs. (18.17) and (18.18) yield $dc = 2K_I\, dK_I / Y^2 \sigma_a^2$, which when integrated between the limits of time $t = 0$ and $t = t_f$ gives

$$t_f = 2[K_{Ii}^{2-n} - K_{Ic}^{2-n}]/A Y^2 \sigma_a^2 (n-2).$$

Since $K_{Ii}^{2-n} >> K_{Ic}^{2-n}$ for large values of n, the preceding expression reduces to

$$t_f = 2[K_{Ii}^{2-n}]/A Y^2 \sigma_a^2 (n-2). \tag{18.19}$$

Now,

$$K_{Ii} = (\sigma_a/\sigma_{Ic}) K_{Ic}, \tag{18.20}$$

where σ_{Ic} is the critical fracture stress in an inert environment. It follows that

$$t_f = 2[\sigma_a/\sigma_{Ic}]^{-n}[K_{Ic}^{2-n}/\sigma_{Ic}^2]/[AY^2(n-2)]. \tag{18.21}$$

The time-to-failure $t_{0.5}$ at half the inert strength (as discussed in the Mould and Southwick experiments) is given by

$$t_{0.5} = 2[0.5^{-n}][K_{Ic}^{2-n}/\sigma_{Ic}^2]/[AY^2(n-2)]. \tag{18.22}$$

Dividing (18.21) by (18.22), we get

$$t_f/t_{0.5} = [0.5^n][\sigma_a/\sigma_{Ic}]^{-n}. \tag{18.23}$$

After rearranging and taking logarithms of both sides, we get

$$\log[\sigma_a/\sigma_{Ic}] = [-1/n]\log[t_f/t_{0.5}] + \log(0.5). \tag{18.24}$$

Except for the logarithmic function on the left side, Eq. (18.24) would yield a "universal fatigue curve" similar to the experimental observations of Mould and Southwick (Eq. (18.15)) as long as the value of n remains unchanged in a given group of specimens with a given type of flaws.

Now, $d\sigma/dc = [d\sigma/dt]/V$. Hence, substituting (18.17) and (18.18),

$$\sigma_a^n\, d\sigma = \{[d\sigma_a/dt]/AY^n c^{n/2}]dc.$$

Integrating both sides over the extent of the crack growth, and assuming that the stressing rate $[d\sigma/dt]$ is constant,

$$\int_0^{\sigma_f} \sigma_a^n\, d\sigma = \{[d\sigma_a/dt]/AY^n\} \int_{c_i}^{c^*} c^{-n/2}\, dc.$$

This yields

$$\sigma_f^{n+1}/(n+1) = \{2[d\sigma_a/dt]/AY^2(2-n)\}\{c^{*(2-n)/2} - c_i^{(2-n)/2}].$$

Since $c^{*(2-n)/2} << c_i^{(2-n)/2}$, and $c_i^{1/2} = K_{Ic}/\sigma_{Ic}$, we get

$$\sigma_f^{n+1} = 2(n+1)[d\sigma_a/dt][\sigma_{Ic}^{n-2}]/[n-2)AY^2K_{Ic}^{n-2}] = B[d\sigma_a/dt],$$

where B is a constant. This shows that a plot of $\log(\sigma_f)$ versus $\log(d\sigma_a/dt)$ has a slope of $1/(n+1)$. The gentle upward slope is as observed in dynamic fatigue experiments.

18.6. Mechanism of Strength Based upon Slow Crack Growth

The experimental observations of static and dynamic fatigue under different environments support the notion that the strength of glass is determined by a thermally activated, stress-assisted crack growth. The most successful explanation using this argument has been put forward by Hillig and Charles [14]. These authors suggested that flaws in glass grow by stress corrosion. The resulting subcritical crack velocity V is given by

$$V = V_0 \exp[\{-\Delta E + (\sigma \Delta v/3) - v_m \gamma_f/\rho\}/RT], \tag{18.25}$$

where ΔE is the activation energy of the chemical reaction in the stress-free state, Δv is the activation volume, and v_m is the molar volume of the glass. The third term within the brackets accounts for changes in the surface curvature. The stress varies along the sides of the flaw according to the Inglis criterion (Eq. (18.8)), being the highest at the tip, and the lowest along the wall. Corrosion of the tip causes c to increase, whereas corrosion of the sides causes ρ to increase. At high values of applied stress, the differential corrosion causes the c/ρ of the flaw to increase, i.e., the flaw sharpens (Fig. 18-11b). This, in turn, causes increased stress at the flaw tip, which leads to more rapid propagation of the crack and, ultimately, to failure. On the other hand, in the presence of low applied stress, the c/ρ does not increase much; it may even decrease, leading, ultimately, to flaw blunting (Fig. 18-11c).

Because of thermal activation, there is very little corrosion at very low temperatures, such as that of liquid N_2. One would tend to get essentially no subcritical crack growth, or, in other words, the "inert" strength would approach the theoretical strength. As the temperature rises, more and more sub-critical crack growth occurs, leading to lower observed strengths. The possibility of a fatigue limit is indicated by the situation when changes in $\sigma \Delta v/3$ are compensated by changes in $v_m \gamma_f/\rho$.

While much of the Hillig–Charles explanation is widely accepted, some problem areas do remain. The chemistry of stress-assisted corrosion has been

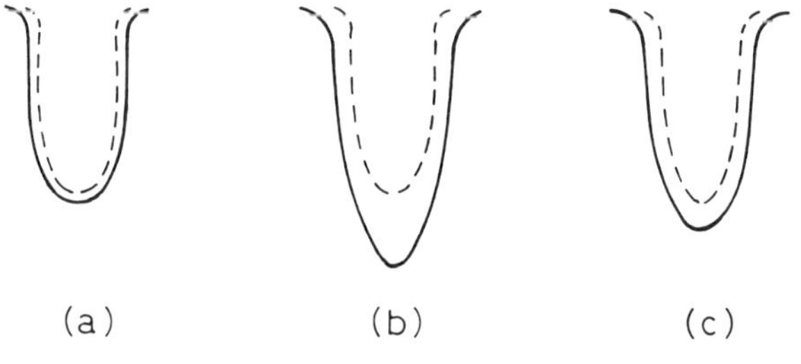

Figure 18-11. Effect of stress corrosion on the geometry of the crack tip. (After Hillig and Charles [14].)

assumed to be bridging oxygen-bond breaking by water adsorption (Eq. (17.3)). Glass components react with water at the crack tip and change its pH. Such a reaction in silica glass would result in acidic pH at the crack tip, whereas the ion exchange of mobile alkalis from soda lime silicate glasses would produce a more corrosive basic solution at the crack tip. However, the role of stress-assistance in increasing the corrosion rate is not clear. Also, there is some question as to whether crack healing would occur if the applied stress were removed. From Griffith's energy criterion, it is a natural expectation that if the crack did not reach its critical length, then the removal of the applied stress should lead to crack closure. However, irreversibility such as plastic deformation at the crack tip, if any, and acoustic emission would prevent crack healing. Crack healing in water or air is observed [15] experimentally in soda lime glasses; however, it is not readily obvious how stress corrosion chemistry would reverse its course at the tip of an advancing crack front. An atomistic model to explain corrosion and crack healing has been suggested by Michalske and Freiman [16]. One may also argue that, since there is no corrosion in vacuum, there is no mechanism provided for crack propagation in vacuum. As indicated earlier, crack growth of the region I type has been observed in 10^{-5} torr vacuum for fused silica, Vycor, and a Pyrex borosilicate glass. It is possible that structural or interstitial water released at the crack tip is responsible for the subcritical crack growth in these glasses.

18.7. Elementary Fractographic Analysis

A fracture in glass leaves its "fingerprints" behind, which can be recognized to gain useful information regarding the cause of the fracture. Fractographic analysis is basically a postmortem analysis where the confidence level is high when all the pieces are available. The best source for learning the details of fractographic analysis in brittle solids is a book by Frechette [17].

Examination of the fracture is easier with very careful lighting conditions, a low-power pocket magnifier, or a low-power stereomicroscope with a portable reflected-light accessory. A typical fracture in glass is shown in Fig. 18-12. Fracture itself is a kinetic phenomenon; it originates from a critical-sized flaw and propagates across the body. Surrounding the origin is a shiny region called the **mirror**, which ends in a fine radial fibrous or misty texture, called the **mist**. Surrounding the mist are wider and deeper radial ridges with slivers of glass lifted out, called the **hackle**, leading ultimately to crack branching. In the mirror region, the fracture front travels with increasing velocity. The mist begins to form as the fracture velocity approaches its terminal velocity, generally equal to about 0.6 times the velocity V_s of

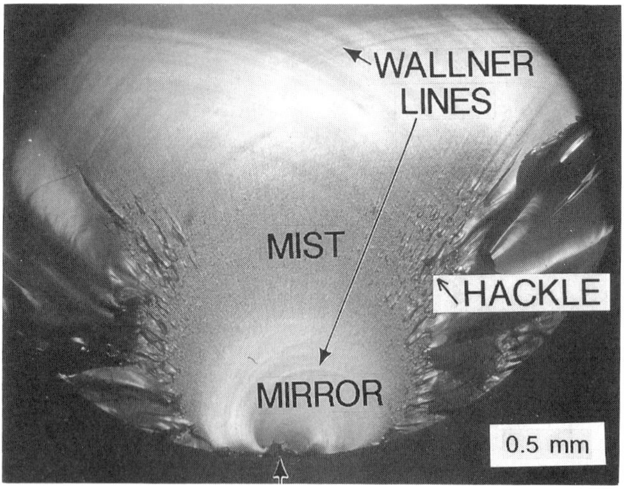

Figure 18-12. Fracture surface of a soda lime silica glass rod broken in bending at room temperature. (80 ×.)

transverse acoustic (shear) waves in the medium (typically $V_s \cong 3{,}000\ \mathrm{m\ s^{-1}}$). In rare cases of extraordinarily fast loading, the terminal velocity has been reported to be as much as 95% of the transverse acoustic wave velocity. The hackle region represents the motion of the fracture front at terminal velocity along parallel but slightly different coplanar steps. Much of the elastic energy is liberated in the hackle region; hence, the width of the hackle reflects the degree of violence associated with the fracture. After the hackle region, the fracture decelerates. In its onward journey, it may encounter a defect or a surface. Interaction of the fracture front with stress waves originating from the defect or the surface produces **Wallner lines** (Figs. 18-12 and 18-13). Fracture travels faster in the tensile region, and hence the Wallner lines are advanced in the direction of increasing tension. **Twist hackle** (Fig. 18-13), also called river patterns, is formed when the tension at the head of the crack front is turned. The crack front breaks up into a group of closely spaced, nearly parallel, but non-coplanar surfaces. The fracture surface is completed by the segments curving over each other. Long, needle-shaped slivers of glass are often lifted out at the curve, making the marks look like tributaries joining into a river. In an end view, the twist hackle would look like a series of steps. The twist hackle is observed often when the trailing edge of a fracture front extends into the compression region. When fracture occurs in the presence of a corrosive medium such as water, **scarps** may be produced. These scarps represent smooth boundaries between the liquid-filled and the dry regions of the fracture front. When crack branching occurs, the maximum **angle of forking** θ depends [18] upon the ratio of the principal stress

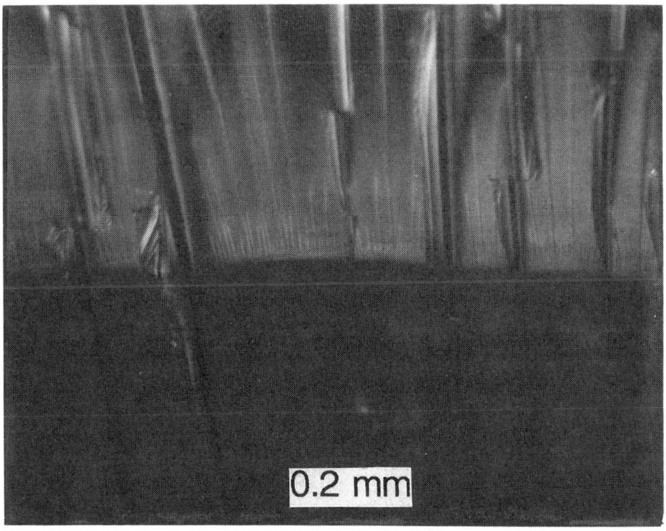

Figure 18-13. Differential interference contrast fractograph showing twist hackle (river pattern) and Wallner lines. The fracture propagated from lower left (the origin is not included) to right and upper right (edge). The primary Wallner line is the curved boundary of the dark region: the shade contrast on either side is caused by the tilt of the fracture planes. Faint secondary Wallner lines may be seen originating from the primary. (Courtesy of V. D. Frechette, Alfred University, Alfred, New York.)

magnitudes: $\theta = 180°C$ when $\sigma_y = \sigma_x$ (uniform biaxial stress); 90° when $\sigma_y = \sigma_x/2$ (cylindrical vessel bursting due to internal pressure); 45° when $\sigma_y = 0$ (simple tension); and 15° when $\sigma_y = -\sigma_x$ (simple twist).

If a hard spherical ball is dropped on a plate of glass, a thin plate may flex such that tension is created on the opposite side. Cracking, if any, would then originate on the tension side. For a given force of impact, the flexural stress decreases as d^{-2}, where d is the thickness of the plate. When the plate is thick and the impacting object is relatively smooth or blunt, the impacting face develops [19] **Hertzian** or bearing stresses that increase as $d^{3/2}$. Fracture may start as a circular crack perpendicular to the tensile radial stress around the circle of contact, moving vertically downwards at first, followed by flaring out to form a conoidal fracture reaching the back surface. This type of fracture is termed a **Hertzian cone** fracture (Fig. 18-14). Radial cracks may also form if the impact is violent.

Pointed projectiles produce more complicated fracture. Of particular interest, discussed below, are the cracks produced by a Vickers microhardness indenter, shown schematically in Fig. 18-15. A recent review on this subject is by Cook and Pharr (20). A model of cracking, suggested by Lawn *et al.* (21), is shown in Fig. 18-16. The penetration of the pointed object creates a zone of plastic deformation of radial dimension b around the

(a)

(b)

Figure 18-14. Hertzian cone caused by high-velocity impact from a small, hard body against the center. (a) Top view. (b) Cross-sectional. (Courtesy of J. R. Varner, Alfred University, Alfred New York.)

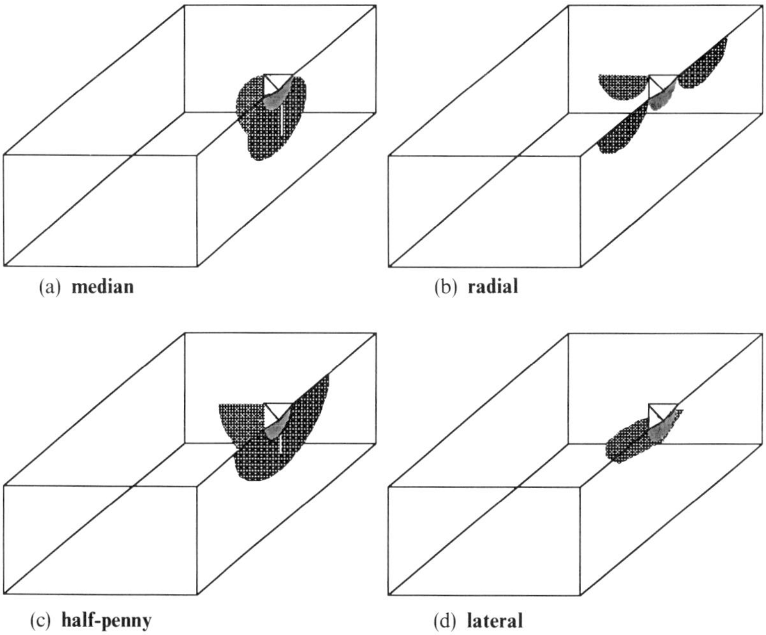

Figure 18-15. Idealized crack morphologies at the site of a Vickers indentation contact. (a) Median cracks originating at the edge of the plastic deformation zone and propagating deeper into the glass along median planes, formed during loading. The cracks are circular, or parts of a circle truncated by the deformation zone. (b) Radial (or Palmqvist) cracks emanating at the ends of the diagonal near the surface in the form of near-elliptical shape propagating radially outwards along median planes containing the direction of loading, formed during unloading. (c) Half-penny crack formed when the median and radial cracks join during unloading. (d) Lateral cracks upon unloading. These originate at the base of the deformation zone, propagating perpendicular to the direction of loading, and curving towards the surface.

immediate area of contact (characteristic dimension a). Depending upon the glass type, the plastic deformation zone may include some irreversible densification of the glass network. The elastic field outside the plastic deformation zone reaches its maximum intensity at full loading and may be considered as a superposition of an ideally elastic (reversible) field (Fig. 18-16b) and a residual (irreversible) field in the unloaded solid (Fig. 18-16c). The residual field arises from the mismatch tractions exerted on the surrounding matrix by the deformed material within $r \leq b$, reaching a maximum value upon full loading but remaining unchanged upon unloading. At the boundaries of the plastic deformation zone, shear-activated fault interfaces, similar to slip lines in crystalline materials, lead to nucleation of cracks at

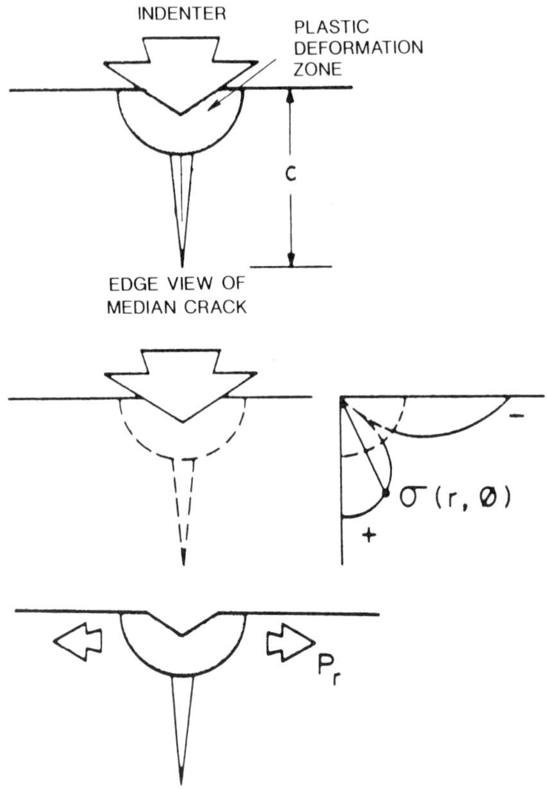

Figure 18-16. Idealized Vickers indentation. (a) During loading, a plastic deformation zone develops surrounded by (b) an elastic field that is tensile subsurface and compressive close to the surface, and (c) a "residual" hemispherical tensile field. Upon unloading, only the residual field remains. Median cracks develop because of the elastic field during loading. Radial and lateral cracks develop because of the residual field during unloading. (After Lawn *et al.* [21]. Reproduced with permission of The American Ceramic Society.)

their intersections. At low loads, nothing may happen during the loading cycle. However, during the unloading cycle the compressive elastic field in the surface region diminishes, leaving a tensile residual field. The tensile residual may cause the nucleation of **radial cracks** (also called **Palmqvist cracks**; see Shetty *et al.* [22]). These cracks propagate radially along the median planes close to the surface (Fig. 18-15b). At higher loads, growth of penny-shaped **median cracks** propagating downwards along the median planes occurs during loading because of the subsurface tensile elastic stress field surrounding the plastic deformation region (Figs. 18-15a, 18-16b, and

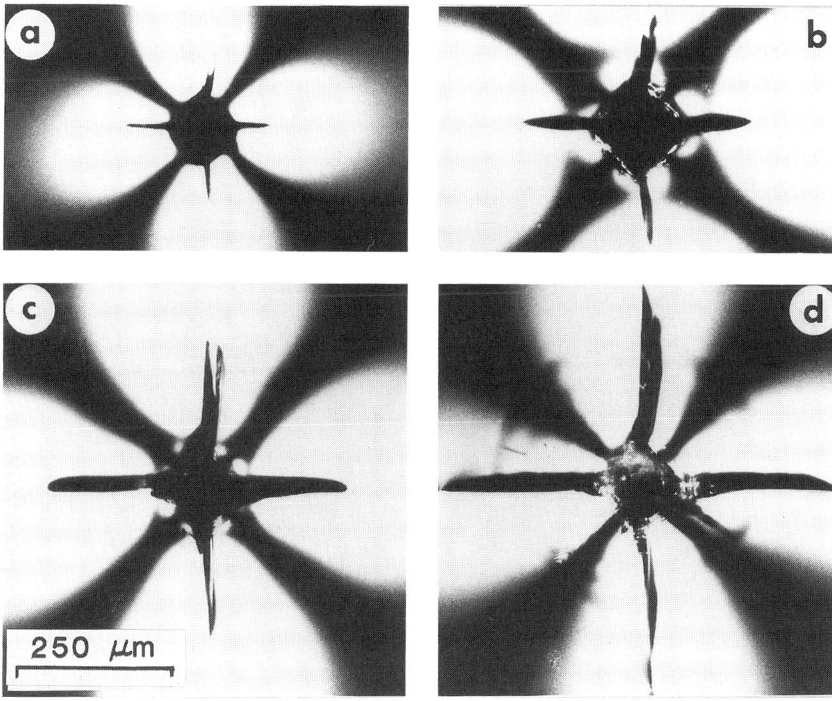

Figure 18-17. Top view of the cracking in Vickers indentation. (a) Partial loading, load = 47 N, (b) Median cracks at maximum load = 90 N. (c) Partial unloading, load = 30 N, (d) Upon full unloading. Circular lateral cracks may be seen as shadows in (d). Width of field = 720 μm. (After Marshall and Lawn (23). Reproduced with permission of Chapman & Hall. Photographs Courtesy of B. R. Lawn, National Institute of Standards and Technology, Gaithersburg, Maryland.)

18-17a and b). In addition to the median/radial crack system, the gradual retreat of the compression zone due to elastic recovery brings the tensile region closer to the shear fault lines. Thus, upon unloading after a high loading, circular **lateral** cracks develop [23] originating at the base of the plastic zone, in addition to the radial cracks closer to the surface (Figs. 18-15d, 18-17d). With continued unloading, all the cracks grow in the residual field. Median cracks grow mostly towards the surface so as to join with the radially growing radial cracks, making a **half-penny crack** (Figs. 18-15c). Occasionally, the circular fronts of the lateral cracks may grow up to the surface, in which case chipping occurs.

The basic morphological features just summarized are often sufficient to trace back the origin of the fracture, the sign of stresses at the surfaces (the origin invariably would be on the tensile surface). A judgment of the violence associated with the fracture, and counting the number of fragments produced, can often give at least a semiquantitative idea of the stresses that led to the fracture.

18.8. Fracture Statistics

Suggested reading: R. H. Doremus, *J. Appl. Phys.* **54**(1), 193–198 (1983).
J. R. Matthews, F. A. McClintock, and W. J. Shack, *J. Am. Ceram. Soc.* **59**(7–8), 304–308 (1976).
G. J. DeSalvo, Technical Report No. WANL-TME-2688, Westinghouse Electric Co., 1988.
C. A. Johnson, Technical Report No. 79CRD212, General Electric Corporate Research & Development, 1979.

Fracture statistics are based upon the weakest-link criterion, according to which the body will fail when the stress at any defect is sufficient to cause unstable crack propagation of that defect. Examples of weakest-link distributions are the Weibull distribution, flaw density distributions, and extreme-value distributions. Of these, the **Weibull distribution** is the most popular because of its mathematical simplicity and closer agreement with experimental data.

The Weibull theory states that the cumulative probability of failure P for a body is given by

$$P = 1 - \exp[-R], \tag{18.26}$$

where R is the risk of rupture given by

$$\begin{aligned} R &= \int_V [(\sigma - \sigma_u)/\sigma_0]^m \, dV \qquad (\text{for } \sigma > \sigma_u), \\ &= 0 \qquad\qquad\qquad\qquad\qquad\;\; (\text{for } \sigma < \sigma_u), \end{aligned} \tag{18.27}$$

where V is the volume, σ is the stress over some volume element dV, σ_u is the minimum stress level for expectation of failure (the fatigue limit), σ_0 is a normalizing stress parameter (unfortunately having the units of stress × $(\text{volume})^{1/m}$), and m is called the **Weibull modulus**. Often σ_u is set at zero without introducing much error.

The basic expression for R can often be integrated to the form

$$R = Y_v V[(\sigma - \sigma_u)/\sigma_0]^m = Y_v V[\sigma/\sigma_0]^m, \qquad \text{assuming } \sigma_u = 0,$$

where Y_v is called the loading factor for a body with volume flaws. $Y_v V$ may be called **effective volume**. For the case of uniform tension, the stress σ is constant, and hence, from Eq. (18.27),

$$R = V[\sigma/\sigma_0]^m,$$

i.e., $Y_v = 1$.

For the case of a rectangular beam subjected to pure bending, the stress varies linearly through the height. The tensile stress at a distance y from the neutral axis is given by

$$\sigma = \sigma_m[y/(h/2)],$$

where σ_m is the maximum stress (happens to be on the surface), and h is the height of the beam.

Since $dV = wL\,dy$, where w is the width and L the length of the beam, the risk of rupture R due to volume flaws may be written as

$$R = \int_0^{h/2} [2y\sigma_m/h\sigma_0]^m wL\,dy, \qquad \text{or } R = [V/2(m+1)][\sigma_m/\sigma_0]^m,$$

which shows that $Y_v = 1/[2(m+1)]$. Likewise, it can be shown that for three-point loading (see Section 18.10),

$$Y_v = 1/[2(m+1)^2],$$

and for four-point loading

$$Y_v = [m(L_i/L_o) + 1]/[2(m+1)^2],$$

where L_i is the inner span and L_o is the outer span. If the flaws are distributed on the surface only, then Eq. (18.27) may be modified to

$$R = \int_A [\sigma/\sigma_0]^m\,dA \tag{18.28}$$

$$= Y_s A[\sigma/\sigma_0]^m,$$

where Y_sA may be called the **effective surface area**. Again, by simple integration, it may be shown that

$$\begin{aligned}
Y_s &= 1 && \text{for uniform tension}\\
&= [(w/h) + \{1/(m+1)\}]/2[1 + (w/h)] && \text{for pure bending,}\\
&= [(w/h) + \{1/(m+1)\}]/2[\{1 + (w/h)\}\{m+1\}] && \text{for three-point loading,}\\
&= [m(L_i/L_o) + 1][(w/h) + \{1/(m+1)\}]/2[\{1 + (w/h)\}\{m+1\}] && \text{for four-point loading.}
\end{aligned}$$

It may be recognized that Y and V are material and loading constants that can be lumped back into a redefined σ_0 (with units of stress) such that a simplified form of the Weibull distribution is

$$P = 1 - \exp\{-[\sigma/\sigma_0]^m\}. \tag{18.29}$$

The derivative of the Weibull distribution $dP/d\sigma$ gives the probability density function p, which is the probability that the specimen will fail within σ and $\sigma + d\sigma$, i.e., the strength distribution

$$p = dP/d\sigma = [m/\sigma_0][\sigma/\sigma_0]^{m-1} \exp[-(\sigma/\sigma_0)^m]. \tag{18.30}$$

Plots of Eqs. (18.29) and (18.30) are shown in Fig. 18-18. Note the skewness of the strength distribution.

Taking logarithms of the survival probability $(1 - P)$ twice, we get from Eq. (18.29):

$$\ln \ln[1/(1 - P)] = m \ln \sigma + \ln[1/\sigma_0^m]. \tag{18.31}$$

This means that a plot of $\ln \ln[1/(1 - P)]$ vs. $\ln \sigma$ is a straight line whose slope is the Weibull modulus m (Fig. 18-19). *Typical values of m are between 5 and 15, with 10 as an average.* Occasionally, one may get more than one straight-line section, indicating the presence of a multimodal distribution of flaws.

One could rewrite the Weibull distribution to include the effects of time under stress (which represents static fatigue) and the specimen size (which reflects volume or surface effects as shown earlier) as:

$$P = 1 - \exp\{-[\sigma/\sigma_0]^m[t/t_0]^r[L/L_0]^q\}, \tag{18.32}$$

where r and q are Weibull exponents for time and dimension, respectively. Equations (18.29) and (18.32) are instructive in the sense that they suggest *relatively large increases in fractional failure when any of the processing stresses, time under stress, or dimensions over which stresses are distributed are increased even a small amount.* A typical illustration could be by showing that a glass plant operating at 2% glassware cracks due to processing thermal stresses would experience a 60% increase when the expansion coefficient of the glass rises only 5% from the usual $35 \times 10^{-7}/°\text{C}$. Note that conventional tests in an R & D laboratory on the glass (see Section 18.10) will not detect any statistically significant change in strength due to the slight expansion increase, but the number of scrap glass barrels will surely rise on the factory floor. Likewise, examining stressed glassware for breakage after a specific time of holding in the warehouse is quite useful for quality control.

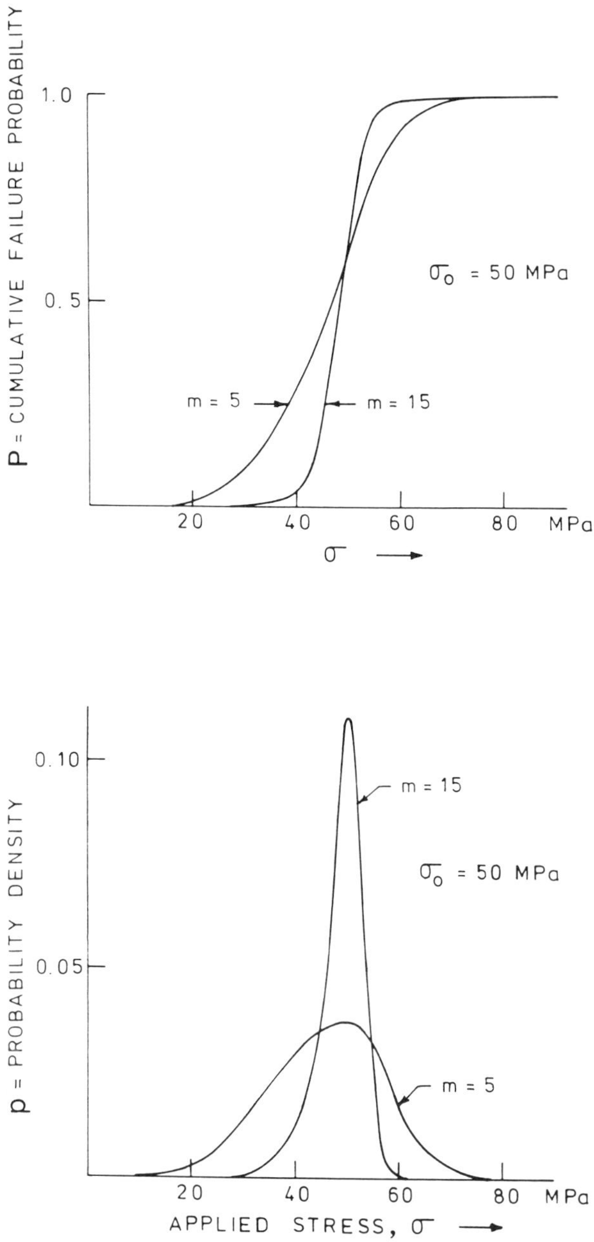

Figure 18-18. Weibull distribution. Top figure is the cumulative fracture probability, P, plotted against the applied stress σ for a normalizing stress $\sigma_0 = 50$ MPa. Lower figure is the probability density, p $(=dP/d\sigma)$, that a failure will occur between stress σ and $\sigma + d\sigma$. Strength distribution is narrower for a larger value of the Weibull modulus, m.

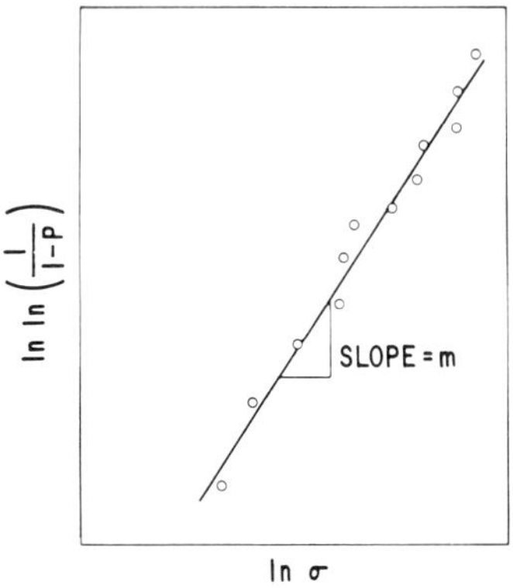

Figure 18-19. Weibull plot of the survival probability (1-P) vs. the applied stress σ.

Because of the varying values of Y, groups of specimens with identical flaw distribution are likely to yield different strength distributions when tested in varying modes.

18.9. Life Prediction

Suggested reading: A. G. Evans and S. M. Wiederhorn, *Int. J. Fracture* **10**(3), 379–392 (1974).

If a crack velocity vs. K_I curve is available, then life prediction at constant stress may be made using stepwise integration of the curve in small time steps Δt. The subcritical crack velocity V at some applied stress intensity factor causes the crack to grow Δc, where $\Delta c/\Delta t = V$. A new K_I corresponding to length $c + \Delta c$ is computed, which then determines a new V, and so on. The accumulated time when c equals specimen dimensions gives the life expectancy.

If no V–K_I is available but the time-to-failure is known at least at one other constant applied stress, then one may proceed as follows. Equation (18.24) may be rewritten as

$$\log t_f = -n \log \sigma_a + [n \log \sigma_{Ic} + \log t_{0.5} + n \log (0.5)].$$

For a given material, the quantity within the square brackets is a constant, and hence we may write

$$\log t_f = -n \log \sigma_a + C, \tag{18.33}$$

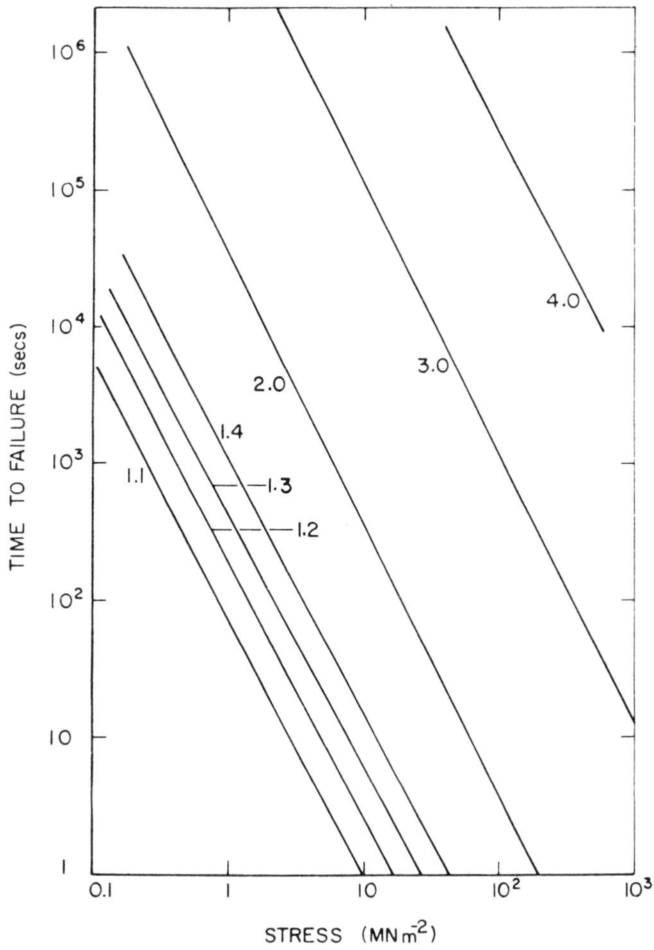

Figure 18-20. Proof test diagram for soda lime silica glass in water. The numbers below each line refer to the proof stress/applied stress ratio. (After Evans and Wiederhorn [24]. Reproduced with permission of Kluwer Academic Publishers.)

which may be used for life prediction at some other unknown constant applied stress.

One other method of estimating life at constant stress is to perform **proof testing**. Proof testing often involves subjecting a group of specimens to a stress higher than the known service stress. The survival of the specimens at this **proof stress** σ_p guarantees that K_I at the tip of the most serious flaw is less than K_{Ic}, i.e.,

$$K_{Ii}/\sigma_a = [K_I]_{proof}/\sigma_P < K_{Ic}/\sigma_{Ic}.$$

Hence,

$$K_{\mathrm{Ii}} < (\sigma_{\mathrm{a}}/\sigma_{\mathrm{P}})K_{\mathrm{Ic}}. \tag{18.34}$$

Substituting (18.34) in Eq. (18.19), we get a minimum time-to-failure

$$t_{\min} = 2[\sigma_{\mathrm{p}}/\sigma_{\mathrm{a}}]^{n-2}/[K_{\mathrm{Ic}}^{n-2}\sigma_{\mathrm{a}}^{2}AY^{2}(n-2)]. \tag{18.35}$$

In this equation, $t_{\min}$ is independently a function of the $[\sigma_{\mathrm{p}}/\sigma_{\mathrm{a}}]$ ratio and σ_{a}. A plot of $\log[t_{\min}]$ vs. $\log[\sigma_{\mathrm{a}}]$ yields (24) a family of shallow lines as shown in Fig. 18-20, which can be used for life prediction after the specimens have survived a proof stress. Clearly, the larger the ratio $[\sigma_{\mathrm{p}}/\sigma_{\mathrm{a}}]$, the longer the survival time at any given applied stress.

If Weibull probability data of fast fracture are available, then one does not need proof testing for life prediction. Instead, the procedure recommended by Evans and Wiederhorn may be followed to yield a $t_{\min}$ for a given fracture probability.

One question often asked is, "Would the specimens weaken after a proof test?" The question is based upon the argument that subcritical growth of flaws is likely to have occurred during proof-testing, which could lead to weakening. The counterargument is that the testing would have weeded out the low-strength specimens, and hence the average failure strength σ_{f}^{*} of the remainder is likely to be higher than the average failure strength σ_{fi} of the initial population. Evans and Wiederhorn have shown that $\sigma_{\mathrm{f}}^{*} > \sigma_{\mathrm{fi}}$, so long as the Weibull exponent $m < (n - 2)$, where n is as defined in Eq. (18.17).

18.10. Experimental Measurement of Glass Strength and Fracture Toughness

Conveniently shaped specimens (generally circular or rectangular cross-section rods/beams) are subjected to increasing loads using a universal mechanical tester, for instance, an Instron machine. Strength is given by the load divided by the area at the point of fracture.

This simple experiment, unfortunately, is loaded with problems. If a uniaxial pull is carried out, one must make sure, by using universal ball joint motion, that the pull is indeed along the length of the specimen and that there is no shear component of the pull acting particularly at the grips. Often it helps to use self-aligning stranded wire and a specimen that has a larger cross-section at the ends to be gripped than in the middle.

A more convenient way to test the strength is to perform centrally loaded three-point, or preferably, four-point beam bending. The measured strength in such testing is also called the **modulus of rupture (MOR)** (see ASTM C158-84). In a three-point loading scheme (Fig. 8-3, Chapter 8), a rod of glass is supported on two pegs (generally steel pins placed in a grooved steel plate) and loaded with a force F in the middle (again, using a steel pin). The stress σ on the mid-vertical plane at a point y away from the neutral axis is terminated by the well-known formula

$$\sigma = \frac{[My]}{I}, \tag{18.36}$$

where M = maximum bending moment = $(F/2)(L/2)$, L = length of the beam between the pegs, and I = geometric moment of the beam cross-section. $I = bh^3/12$ for a rectangular beam (b = width, h = height), and $I = \pi a^4/4$ for a circular rod of radius a. Maximum tension is on the convex surface. Thus, $y = h/2$ for a rectangular beam, or a for a circular rod.

In the four-point bending experiment (Fig. 8-4, Chapter 8), the maximum tensile stress is uniform along the outer convex surface between the inner pegs. The value is obtained by substituting $M = (F/2)[(L_1 - L_2)/2]$ in Eq. (18.36).

In the MOR tests, the biggest problems are the friction at the pegs and the applicability of the linear beam theory. Depending upon the beam length-to-height ratio, the center deflection should not exceed the beam height. For recommended specifications on fixtures, and the corrections needed when the center deflection appears large compared to the beam height, the reader is advised to refer to Baretta *et al.*, West, and Quinn, cited at the end of Chapter 8.

Plate specimens may be tested under a biaxial ring test [25]. In a burst-pressure test, containers are pressurized using fluids or gas until failure occurs. Strength measurements may also be performed by impact testing, generally by dropping a stainless steel ball from a known height. For instance, eyeglass lenses are expected to survive a 50-inch drop of a 5/8 in. diameter steel ball.

To ensure that the failure does not originate from an experimental artifact, such as a gound edge, it is strongly advised that strength tests be carried out, in general, in conjunction with a determination of the fracture origin. To keep the pieces of the specimen together after failure, one may use adhesive tape, at least on the nontensile surfaces.

Because failure generally originates at the surface from the site of a severe flaw, most as-received products show a wide variation in measured strengths because of a wide distribution of surface flaws. Strength measurements, therefore, involve breaking a number of specimens. *The recommended minimum number is generally 10 for the measurements to be within ±15% of the average.*

If the intention is to examine the effects of a certain processing step not related to the surface, then it is an accepted technique to abrade the specimen with a known distribution of flaws and obtain what is called the **abraded MOR**. Dry SiC grit (of a specific mesh size) may be dropped from a fixed height on the specimen surface to be tested (see ASTM C158-84) and exposed away from the edges using a mask.

Fracture toughness and the stress-corrosion susceptibility coefficient n of glass are measured by subjecting specimens of a defined initial crack geometry to a variety of loading techniques and measuring the time-to-failure under constant stress (static fatigue), failure stress as a function of stressing rate (dynamic fatigue), or crack velocity. Appropriate equations to calculate K_{Ic} and n were discussed earlier. Popular standard specimen configurations and loading conditions are the double cantilever, double torsion, and the notched bend methods outlined by Freiman [26].

Fracture parameters may be determined from fractographic features. This branch has been of particular interest in recent years. Figure 18-21 shows a redrawn schematic of Fig. 18-12. Each of the boundaries are characterized by their radii. The initial flaw of size a_i (semiminor axis on the fracture plane)

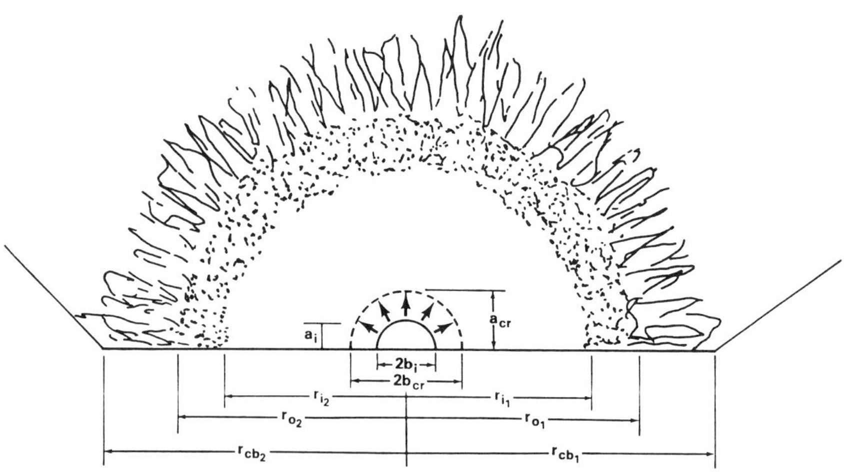

Figure 18-21. Features observed on a glass fracture surface: r_i, mist boundary; r_0, hackle boundary; r_{cb}, macroscopic crack branching; a_i, initial flaw size; a_{cr}, size at which flaw extends catastrophically; b_i, the initial semimajor axis of the flaw at the open surface. (The dimensions are often asymmetrical.) (After J. J. Mecholsky, A. C. Gonzalez, and S. W. Freiman, *J. Am. Ceram. Soc.* **62**(11–12), 577 (1979). Reproduced with permission of The American Ceramic Society.)

and semimajor axis b_i grow to critical size a_{cr} and b_{cr}, at which point K_{Ic} is reached and the flow propagates rapidly. (Unfortunately, there are no markings on the surface at a_{cr}.) Experimentally it is observed that

$$\sigma_f = (\sigma_a + \sigma_r) = M_j r_j^{-1/2}, \tag{18.37}$$

where $j = 0, 1, 2, 3$ refers to critical flaw, mirror–mist boundary, mist–hackle boundary, and crack branching, respectively; σ_f is the true stress at failure; σ_a is the applied stress; and σ_r is any residual stress, if present. r_j is the radial distance. The terms M_j are called mirror constants. The validity of Eq. (18.37) is demonstrated in Fig. 18-22, where the $\log(\sigma_f)$ has been plotted against $\log r_1$ for a variety of glasses with solid lines drawn at slope $= \frac{1}{2}$. Also,

$$(a_{cr} b_{cr})^{1/2} = c^* = \text{constant}. \tag{18.38}$$

In effect, c^* is the equivalent radius of a semicircular crack that would lead to spontaneous crack propagation under the applied stress [27].

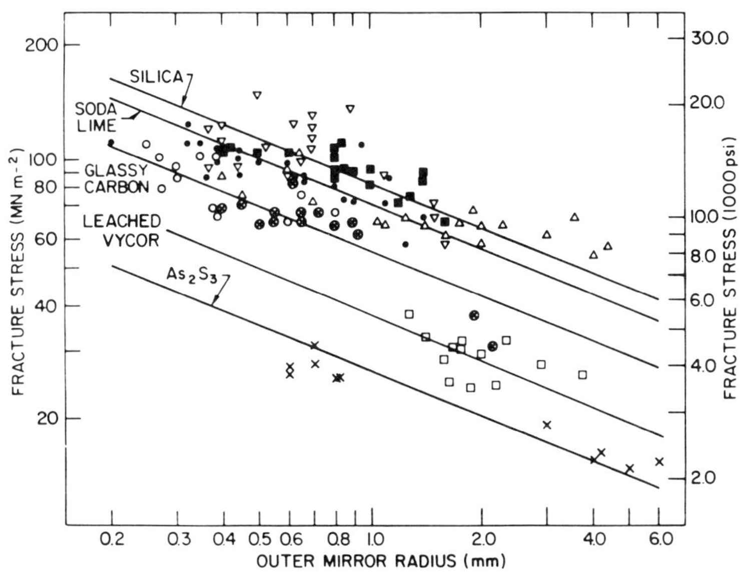

Figure 18-22. Plot of log(fracture stress) vs. outer mirror radius for various glasses. Solid lines are drawn with slope $-\frac{1}{2}$. As_2S_3 (×); vitreous carbon (○); borosilicate (■); soda lime silica (•); aluminosilicate (▽); lead silicate (⊕); leached Vycor (□). (After J. J. Mecholsky, R. W. Rice, and S. W. Freiman, *J. Am. Ceram. Soc.* **57**, 440 (1974). Reproduced with permission of The American Ceramic Society.)

It has been found that

(i) $r_1/c^* = 5$ to 7 in the absence of residual stress, and
(ii) $M_{\text{mirror-mist}} = 1.7\text{–}1.9\ \text{MPa·m}^{1/2}$ and $M_{\text{mist-hackle}} = 2.0\text{–}2.1\ \text{MPa·m}^{1/2}$.

Now, for the elliptical flaw with equivalent semicircular radius c^*,

$$K_{\text{Ic}} = \sigma_{\text{f}}[1.2\pi c^*]^{1/2}/\phi, \tag{18.39}$$

where ϕ is an elliptical integral of the second kind:

$$\phi = \int_0^{\pi/2} [1 - \{(b^2 - a^2)/b^2\} \sin^2 \theta]^{1/2}\, d\theta$$

$$= \pi/2 \qquad \text{for } b = a \text{ (half-penny shape),}$$

we get

$$M_j = \phi K_{\text{Ic}}/[1.2\pi c^*/r_j]^{1/2}. \tag{18.40}$$

When M_2 is plotted against K_{Ic}, data on most brittle solids fall in a band around a hackle/flaw size ratio of about 13.

Vickers indentation and Hertzian impact methods may be used to determine K_{Ic}. The Hertzian impact test is discussed by Abou-el-leil (28). For the Vickers indentation, Anstis *et al.* (29) have developed the following expression for K_{Ic} in $\text{MPa·m}^{1/2}$:

$$K_{\text{Ic}} = \Omega P[E/H]^{1/2}[c_0]^{-3/2}, \tag{18.41}$$

where $2c_0$ is the post-indentation equilibrium length of the radial crack on the surface from end to end (>3 times the indentation diagonal), P is the load in Newtons, E is the Young's modulus (MPa), H is the hardness (MPa), and Ω is a material-dependent constant for Vickers-produced cracks (=0.016 for glasses). Because of the problems associated with a complex crack system, the values of K_{Ic} so obtained are not very reliable and are at best within about 20% of those measured by more conventional techniques. It is often recommended that this measurement be called the **indentation toughness** K_{IS}. Some research [30] has also been performed to employ the indentation itself as a precracked specimen with a defined K_{I}, and allow these cracks to propagate under different conditions of loading.

18.11. Data on Strength and Fatigue Parameters

As indicated in Section 18.3, strength is not an intrinsic property to quote. Majority of research interest has been on the issue as to how close one could get to the theoretical strength predicted by Eq. (18.6). In all such studies,

Table 18-1 Fracture Surface Energies of Some Glasses[a]

Glass	Environment	γ_f (J/m^2)
Pyrex	Air, 20°C, 20% RH	4.7
	Air, 22°C, 40% RH	4.0
	N_2 (gas), 27°C, <0.1% RH	4.5–4.8
	$N_2(l)$, 77 K	4.7
	Water, 20°C	2.5
Soda-lime (or float glass)	Air, 20°C, 20% RH	3.9
	$N_2(l)$, 77 K	4.1
	$N_2(l)$, 77 K	4.5–4.6
	N_2 (gas), 27°C, <1% RH	3.8–3.9
	Air, 22°C, 40% RH	3.5
	Vacuum, 10^{-1} torr	5.0
Fused silica	N_2 (gas), 27°C, <1% RH	4.3–4.4
	$N_2(l)$, 77 K	4.6
	Air, 22°C, 40% RH	3.7
Aluminosilicate	N_2 (gas), 27°C, <1% RH	4.6–4.7
	$N_2(l)$ 77 K	5.2
	Air, 22°C, 40% RH	3.7

[a] After Ref. 26, Table IV. Reproduced with the permission of the Academic Press.

specimens have been glass fibers, the interest being motivated by the potential application of fibers in long-distance telecommunications (see Chapter 19). Some of the highest values reported are 14 GPa for fused silica, 8.5 GPa for S-glass, and 6.3 GPa for E-glass (31). The reported high strengths undoubtedly reflect processing in a corrosion-free and abrasion-free environment, as well as the low total surface area, which decreases the absolute number of surface defects.

Fracture toughness K_{Ic} and the stress-corrosion susceptibility coefficient n have been studied for several glasses. Typical values for fused silica, conventional soda lime silicate, Pyrex borosilicate, aluminosilicate ($57\%SiO_2 \cdot 15\%Al_2O_3 \cdot 5\%B_2O_3 \cdot 7\%MgO \cdot 10\%CaO \cdot 6\%BaO$), and a lead silicate ($35\%SiO_2 \cdot 4\%Na_2O \cdot 61\%PbO$) are 0.75, 0.7, 0.77, 0.85, and 0.63 MPa·m$^{1/2}$, respectively. Values of n generally range between 12 and 35, although occasionally values as high as 70 have been reported. The rather low values of K_{Ic} relative to those found typically for metals (15–50 MPa·m$^{1/2}$ for aluminum alloys, 50–200 MPa·m$^{1/2}$ for steels, and 75–125 for titanium alloys) show why metals may be preferred in applications requiring high strength under load over long periods. Even polycarbonates have K_{Ic} of the order of 3 MPa·m$^{1/2}$ despite their low fracture strengths.

Chalcogenide glasses in the Ge–Sb–As–Se–Te systems have been reported (32, 33) to have even lower K_{Ic} values. These range between 0.15 and 0.3 MPa·m$^{1/2}$.

Fracture surface energies γ_f for a number of glasses are shown [26] in Table 18-1. Variation of γ_f vs. Young's modulus and Al_2O_3 and B_2O_3 content in a compilation of roughly 300 separate records of various glasses is shown by Freiman *et al.* [34]. There are some interesting trends, perhaps with E, but many of the trends are inconclusive.

18.12. Strengthening and Toughening

Essentially everything, including metals, ceramics, glass, dust particles, and even human hands, coming into contact with glass has the potential of causing surface damage, resulting in a lowered practical strength. Strengthening and toughening of glass products, therefore, is of considerable importance to glassware manufacturers. At stake are issues involving product safety in the hands of consumers, including legal costs involved in personal injury and property damage; process yields during manufacture; and competition from alternate materials. The last item involves issues such as lightweighting, where glass containers should be made lighter for a higher strength/weight ratio to retain their market share.

By now the reader should understand the difference between strength and toughness. Strength σ_f is the stress, almost invariably tensile, at failure. Toughness of a brittle solid is the area under the stress-crack growth curve. Fracture toughness K_{Ic} is a measure of this. Strengthening would lead to toughening; the reverse is generally but not always true. The underlying concepts to strengthen glass, i.e., increase σ_f, are essentially five:

1. reduce the severity of the flaws (reduce K_I);
2. control the environment in the immediate vicinity of the crack tip (increase K_{Ic});
3. use polymeric coatings to reduce risk of surface damage (reduce c);
4. introduce compression in the surface (increase σ_a to reach σ_f level); and
5. crack pinning, deflection or crack-tip shielding (increase K_{Ic}).

Firepolishing glassware during glass-forming is the simplest way to reduce the severity (c/ρ) of the flaws. The flow of glass rounds out or even closes the flaws such that either c decreases or ρ increases or both. Likewise, **etching**, generally using 2–6% HF, can evenly dissolve glass such that flaws are reduced. (Note: higher-strength HF can be quite corrosive and is likely to have a weakening effect on the glass.) In many instances, autoclaving glassware under pressurized steam or water can also remove the flawed surface. Because subsequent glass handling by the consumer will always produce flaws on exterior surfaces, these methods are helpful only during glass processing or for strengthening unexposed glass surfaces.

Control of the environment in the immediate vicinity of the crack tip and reduced risk of surface damage can both be achieved by coating with nonpolar polymers such as lacquers, Teflon, Lexan, UV-curable polyacrylates, or silanes. The polymers act as barriers to reduce the thermodynamic activity of water and provide an inert environment around the crack tip such that not only is the chemical durability improved, but also the fracture surface energy is increased, thus causing increased strength. Additionally, the risk of surface damage to glass products is reduced by "hot end" and "cold end" coatings. The hot end coatings essentially consist of SnO_2 or TiO_2 deposited by passing a mixture of $SnCl_4/TiCl_4$ vapors and steam near the entrance to the annealing lehr. Both the oxides are believed to increase the abrasion resistance of the glass surface, thus preventing them from being damaged by contacts with metal surfaces (particularly the conveyor belt). They do not, however, protect glass surfaces from getting abraded by rubbing against each other. The cold end coatings are stearates (soaps and lubricants) applied by spraying at the exit end of the annealing lehr at a temperature not so high as to impart thermal shock, but high enough to dry them. The increased lubricity reduces risk of damage during packing in cartons.

Compressive stresses in the glass surface are helpful in strengthening because any applied tension must overcome the compression before subcritical crack growth can occur. Surface compression is readily introduced by thermal tempering and by chemical tempering, which includes ion-exchange stuffing, glazing, and surface crystallization. Glass cooled through the transition range naturally develops a beneficial compression in the surface and a balancing, but relatively harmless, tension in the interior (Section 13.8). In short, a steep temperature gradient across the body (cooler on the outside, warmer on the inside), and a cooling rate gradient (faster cooling on the outside), help achieve the desired stress profile. The processing of glass to develop −70 to −200 MPa (−10,000 to −30,000 psi) surface compression is called **thermal tempering**. The magnitude of surface compression increases with increased cooling rate, thermal expansion coefficient of the glass, and the difference between the expansion coefficient of the glass and the supercooled liquid. Thus, thermal tempering normally involves cooling or quenching a glass rapidly from about 50 to 100°C above its transition range using jets of cool air. Occasionally, contact with a liquid medium — for instance, molten tin, molten salts, or oils — or contact with solid pads may also be used to effect rapid quenching. Commercially available fully tempered flat glass is air jet-cooled and generally has -100 MPa surface compression. Because the stress profile is more or less parabolic (Section 13.8) for the case of air-tempering, higher surface compression would lead to the development of higher interior tension, which in turn poses increased risk of tension-driven glass fracture from an internal or edge flaw during the tempering process. Hence, thermal tempering is rarely carried out to develop surface compression higher than about -200 MPa. Thermal tempering is not very effective for a thin-walled glass product, such as a fiber, because it is not easy to establish a meaningful temperature gradient or a cooling rate gradient across the fiber diameter by normal means.

Because of a high level of stored elastic energy, thermally tempered products dice into a large number of pieces when a sharp object penetrates through the compressive layer. The pieces so produced are dull-edged and pose little risk of body injury in terms of cuts. The number of pieces and their size is determined by the degree of temper; the higher the temper, the larger the number and the smaller the size. The use of tempered glass instead of the normal product in places such as doors in building access areas, bathroom sliding doors, and automobile windows (except the front windshield) is, hence, government-regulated. (For suggested reading, see Ref. 35.)

The **ion-exchange stuffing process** involves immersing the glass in a bath of molten alkali salt for typically several hours at temperatures close to, but less than, the strain point of the glass (Section 14.8.1). For effective strengthening, two features must be considered: the magnitude of surface compression, and the depth of this compression layer below the surface, otherwise called the **case depth**. A high magnitude of surface compression (of the order of -500 to -700 MPa), but a shallow case depth (of the order of 30–40 μm) may be sufficient for ordinary reusable beverage containers. However, for aircraft windshield applications where bird hits are a major concern, the presence of a substantially increased case depth (>300 μm) is vital. For compression to develop on the outside surface, the size of the bath alkali ions should be greater than that in the host glass. Hence, a KNO_3 bath is typically used for a sodium-containing glass, and a $NaNO_3$ bath is employed to strengthen a lithium-containing glass. The stress builds up as a result of the dominance of stress production by ion stuffing over stress relaxation due to glass fluidity. The stress production is directly related to the concentration of alkali ions in the glass and their interdiffusion coefficient with the salt ions, which typically has an activation energy of the order of 120–160 kJ·mol^{-1}. Hence, for an overall successful chemical tempering, first, the glass composition should have a high concentration of mobile ions; second, since the relaxation is viscosity-controlled, which has a relatively high activation energy (300 600 kJ·mol^{-1}), the lower the exchange temperature the greater the net residual stress. At temperatures above the annealing point, stresses produced are essentially relaxed continuously. Products of fused silica or a Pyrex-type borosilicate glass having only about 4% alkali are not suitable for chemical tempering because they fail to meet the first requirement. Common soda lime silicate glass, unfortunately, is generally not very suitable because any efforts to increase the case depth lead to the relaxation of surface compression. Glasses that meet both the requirements reasonably for a commercial operation are mixed alkali glasses, in particular the Li_2O–Na_2O–Al_2O_3–SiO_2 system. Apparently, additions of ZnO and P_2O_5 in the glass composition are generally helpful.

Because a typical stress profile in the diffusion zone is quite steep (Fig. 14-13), the interior tension is usually small, and hence, glass products can be tempered to high levels of surface compression (−200 to −700 MPa) without the risk of tension-driven failure from an internal or edge flaw. For the same reason, dicing upon penetration of the compressive layer does not occur, and even fine glass fibers may be chemically tempered.

There are some interesting variations applied on the chemical tempering process. These are (i) combinations of thermal and chemical tempering; (ii) two-step chemical tempering, where the glass may be immersed in a bath containing smaller ions at above the strain point, and subsequently exchanged in a bath containing larger ions at below the strain point; (iii) a three-step process where ion size for exchange and air reheat steps are carefully separated; (iv) contact pastes; (v) vapor-phase exchange; (vi) electrical field assistance; (vii) microwave assistance; and finally (viii) sonic assistance. Some of these techniques are reviewed by Bartholomew and Garfinkel [36] and by Varshneya and LaCourse [37]. Smaller ions at high temperatures can penetrate deeply into the glass: a subsequent exchange with larger ions can result in a significantly deeper compressive layer relative to that obtained in a single-step process. Contact pastes made with exchange salt mixed with an inert carrier such as clay have found much use in stained glass products, particularly for Cu and Ag staining.

Because of the long processing time requirement, most large-scale commercial glass products are not very amenable to the ion-exchange strengthening technique. To increase the kinetics, electric-field assistance has been studied. The setup requires electrical separation of the molten salt by the wall of the product; electrodes are inserted on each side, and the current is passed for a certain length of time in a forward direction, and reversed for half the time. (See Abou-el-leil and Cooper [38] for the mathematics of the field-assisted exchange.) The electrical separation itself requires that at least some portion of the glass product stay above the molten salt line. Unfortunately, the portion above the salt line then develops high tensile stresses on the surface after exchange (by stress reversal) and acts as the Achilles' heel. For this reason, the field-assisted ion-exchange process is not a commercial process even for containers where the geometry is natural for salt separation.

Microwave-assisted ion exchange has been shown [39] to yield a greatly increased depth of ion exchange. However, beneficial strengthening has yet to be reported. It is possible that increased depth of penetration results from unrecorded increases in the ion-exchange temperature, which would also result in increased stress relaxation. Likewise, although sonic-assisted ion exchange appears to improve strength somewhat [40], large increases in strength have yet to be reported. Again, it is possible that local atomic relaxation is also assisted by the same sonic wave, resulting in no significant strengthening.

Glazing is based upon developing a surface layer whose thermal contraction coefficient is lower than the matrix. Upon cooling, the higher contracting matrix forces the surface layers to be in compression with the matrix developing the balancing tension. The compression is limited to the lower expansion layer, which can be as thin as desired. Glazing can be accomplished in many ways. The simplest one is to fuse a lower-softening layer on all outer surfaces (encasement) without significantly distorting the main body of the product. Glass pastes may be applied by a screenprinting process, or by flame/plasma-spraying. Away from the edges, the body acts like a sandwich or composite cylinder "bead" seal, for which the stress formulae are given by Varshneya (41). For an engineering computation of stresses, one needs to consider the total contraction of the glaze from its set point down to the room temperature relative to that of the substrate glass over the same range. The set point may be thought of as an effective temperature, below which the glass acts essentially as a solid and above which, essentially as a liquid.

Low-expansion surface layers may also be produced by dealkalization of the surface with steam, or SO_2 treatments used in improving chemical durability that produce silica-rich surface layers. In addition, if a small ion after exchange causes the outer layers of glass to have a lower thermal contraction coefficient, then just the one-step process of above-strain-point exchange with smaller ions can result in the development of surface compression. Likewise, vapor-phase treatments at above strain point with salts that have smaller ions, for instance, CuCl, may also produce surface compression.

Strengthening by **surface crystallization** is actually based upon two principles. Firstly, the relative contraction of the surface from the set point of the glass to room temperature after the onset of surface crystallization is less for the surface layers than for the glass substrate. Secondly, second-phase inclusions act to increase the effective K_{Ic}, which leads to strengthening for an assumed initial flaw distribution. Surface reduction of heavy metal ions such as Pb by H_2, which causes PbO to reduce to metallic Pb clusters (see Swiler *et al.* [42]) and inclusion of fine ductile metal whiskers in the glass also increase glass strength by increasing K_{Ic}. In fiber-reinforced plastics, the strength of the composite is actually lower than that of a pristine bundle of fibers, but fracture does not occur till much higher strains. In other words, the toughness is considerably higher than that of either of the components.

Effective increase of K_{Ic} of composites by management of second-phase materials such as ductile and brittle precipitates, whiskers, and continuous fibers is a lengthy and complex topic by itself. Advanced readers may find it useful to review Evans [43], and several books on fiber reinforcement. The key idea is that the fracture surface energy of the composite is not a weighted sum of the γ_f's of the individual components. There are many possible mechanisms for rather large increases in the composite γ_f. The detailed mechanisms depend upon the interaction of the advancing crack front with the second phase. Some of these are sketched in Fig. 18-23. A ductile metal precipitate or whisker encountered by the crack front absorbs energy by plastic deformation in the stress field caused by the crack tip. Thus, the crack is arrested or pinned. When the second-phase particle is brittle, the stress fields around it may cause the crack to deflect. Since this also leads to absorption of energy, the fracture toughness would increase. In crack shielding, when a crack encounters a second-phase particle, the local stress intensity factor at the crack-tip is reduced by an amount ΔK_I. The net increase in fracture toughness is therefore ΔK_{Ic}, which is the value of $-\Delta K_I$ at the critical condition. Crack shielding may also occur if whisker inclusions develop "microcracks" around them to absorb energy or to reduce the stresses ahead of the crack tip. The microcracks are formed in the stress field of the advancing crack, primarily because sufficient stresses due to a thermal expansion mismatch with the glass matrix pre-existed. Crack-bridging occurs when second-phase obstacles are bypassed by the advancing crack and are left behind intact as a ligament to interlock the matrix and offer resistance for further opening of the crack at a given applied stress. Subsequent fiber fracture, interfacial debonding (in friction-bonded composites), and fiber pullout result in expense of energy as acoustic waves and frictional dissipation. These mechanisms lead to strengthening as well as toughening. In C or SiC continuous aligned-fiber lithium aluminosilicate glass matrix composites (both brittle components), the tensile loading behavior is initially elastic until a crack passes through the matrix. The bypassed fibers carry the load and bridge the cracks in the matrix, until the fibers fail by gradual pullout. Ultimate strengths as high as 500 MPa in such composites are achievable.

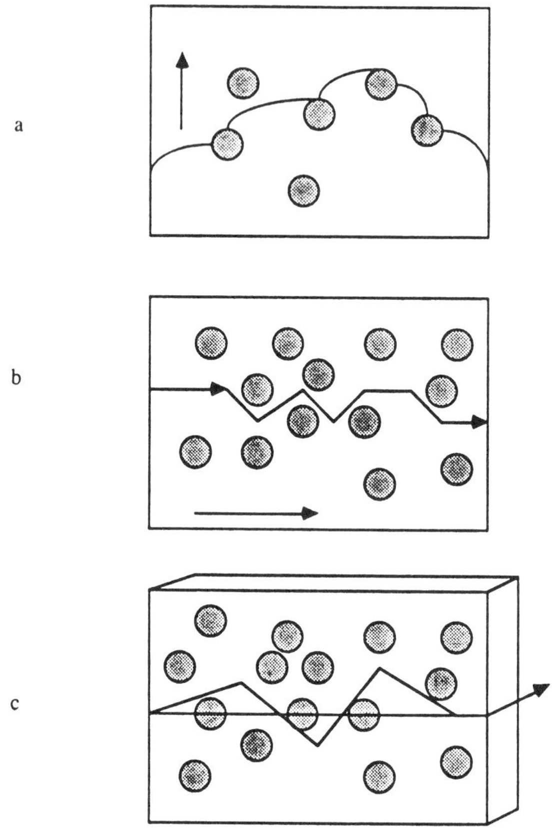

Figure 18.23. Toughening mechanisms due to second-phase precipitates. (a) Bowing around a set of pinning obstacles. (b) Deflecting around the particles. (c) Twisting in addition to deflection to avoid the particles. (After D. J. Green, R. H. J. Hannink, and M. V. Swain, *Transformation Toughening of Ceramics*, Fig. 24, p. 86. CRC Press, Boca Raton, Florida, 1989. Reproduced with permission of the publisher.)

Exercises

1. Groups of glass specimens having an identical cross-sectional area, square and circular, are being tested for strength under (i) uniaxial stress, (ii) three-point bend, and (iii) four-point bend modes. Will there be any difference in the observed strengths? Why?

2. Why does a concentric ring test give a more reliable measure of the strength of glass plates than a four-point mcthod on bcam-shapcd specimens of the same plate?

3. The difference between the breaking strengths of specimens before and after ion-exchange strengthening is usually significantly less than the surface compression. Why? What if the surface flaws from the glass were removed (by etching with dilute HF) prior to ion exchanging? (Hint: See D. H. Roach and A. R. Cooper, *J. Amer. Ceram. Soc.* **71**(4), C192 (1988).)

References

1. E. Orowan, *Z. Krist.*, **A89**, 327–343 (1934).
2. C. E. Inglis, *Trans. Inst. Nav. Archit.* **55**, 219–241 (1913).
3. A. A. Griffith, *Phil. Trans. Roy. Soc.* **A221**, 163–198 (1921).
4. E. N. Andrade and L. C. Tsien, *Proc. Roy. Soc.* **159A**, 346 (1937).
5. F. M. Ernsberger, *Proc. Roy. Soc.* **A257**, 213–223 (1960).
6. Y. Bando, S. Ito, and M. Tomozawa, *J. Am. Ceram. Soc.* **67**(3), C36–37 (1984).
7. G. R. Irwin, *J. Appl. Mech.* **24**, 361 (1957).
8. Suggested reading: J. R. Rice, in *Fracture*, Vol. 2 (H. Liebowitz, ed.), pp. 190–311. Academic Press, New York, 1968. Also for a summary, see Isa Bar-On, *Engineered Materials Handbook*, Vol. 4, *Ceramics and Glasses*, pp. 645–651. ASM, 1991.
9. J. M. Barson, *Fracture Mechanics Retrospective—Early Classic Papers (1913–1965)*. ASTM Publication, 1987.
10. R. E. Mould and R. D. Southwick, *J. Am. Ceram. Soc.* **42**, 542–547; 582–592 (1959).
11. See, for example: S. M. Wiederhorn, *J. Am. Ceram. Soc.* **50**, 407–414 (1967); S. M. Wiederhorn and C. H. Bolz, *J. Am. Ceram. Soc.* **53**, 545–548 (1970).
12. S. W. Freiman, *J. Am. Ceram. Soc.* **57**, 350–353 (1974).
13. C. L. W. Quakenbush and V. D. Frechette, *J. Am. Ceram. Soc.* **61**, 207–210 (1978).
14. W. B. Hillig and R. J. Charles, in *High Strength Materials* (V. F. Zackay, ed.), pp. 682–705. J. Wiley, New York, 1965.
15. D. G. Halloway and B. Stavrinidis, in *Strength of Inorganic Glasses* (C. R. Kurkjian, ed.), pp. 285–290, Plenum Press, New York, 1985.
16. T. A. Michalske and S. W. Freiman, *J. Am. Ceram. Soc.* **66**(4), 284–288 (1983).
17. V. D. Frechette, *Fracture Analysis of Brittle Materials*. Am. Ceram. Soc., 1990.
18. F. W. Preston, *J. Am. Ceram. Soc.* **18**(6), 175 (1935).
19. J. L. Glathart and F. W. Preston, *Glass Technol.* **9**(4), 89–100 (1968).

20. R. F. Cook and G. M. Pharr, *J. Am. Ceram. Soc.* **73**(4), 787–817 (1990).
21. B. R. Lawn, A. G. Evans, and D. B. Marshall, *J. Am. Ceram. Soc.* **63**(9–10), 574–581 (1980).
22. D. K. Shetty, I. G. Wright, P. N. Mincer, and A. H. Clauer, *J. Mat. Sci.* **20**, 1873–1882 (1985).
23. D. B. Marshall and B. R. Lawn, *J. Mat. Sci.* **14**(8), 2001–2012 (1979).
24. A. G. Evans and S. M. Wiederhorn, *Int. J. Fract.* **10**(3), 379–392 (1974).
25. See J. B. Wachtman, W. Capps, and J. Mandel, *J. Materials* **7**(2), 188–194 (1972).
26. S. W. Freiman, in *Glass Sci. & Tech.*, Vol. 5 (D. R. Uhlmann and N. J. Kreidl, eds.), pp. 21–78. Academic Press, New York. 1980.
27. G. K. Bansal, *J. Am. Ceram. Soc.* **59**(1–2), 87–88 (1976).
28. M. M. Abou-el-leil, *Fracture Mech. of Cer.*, Vol. 7 (R. C. Bradt, A. G. Evans, D. P. H. Hasselman, and F. F. Lange, eds.), pp. 197–209. Plenum Press, New York 1986.
29. G. R. Anstis, P. Chantikul, B. R. Lawn, and D. B. Marshall, *J. Am. Ceram. Soc.* **64**(9), 533–538 (1981).
30. P. K. Gupta and N. J. Jubb, *J. Am. Ceram. Soc.* 64(8), C112–C114 (1981).
31. See P. K. Gupta, in *Strength of Inorganic Glasses* (C. R. Kurkjian, ed.), pp. 351–362. Plenum Press, New York, 1985.
32. B.-D. Michaels and G. H. Frischat, *J. Am. Ceram. Soc.* **64**(11), C150–151 (1981).
33. A. N. Sreeram, A. K. Varshneya, and D. R. Swiler, *J. Non-cryst. Sol.* **130**, 225–235 (1991).
34. S. W. Freiman, T. L. Baker, and J. B. Wachtman, in *Strength of Inorganic Glasses*, (C. R. Kurkjian, ed), pp. 597–608. Plenum Press, New York, 1985.
35. R. Gardon, in *Glass Science & Tech.* (D. R. Uhlmann and N. J. Kreidl, eds.) Academic Press, New York, **5**, 145–216 (1980).
36. R. F. Bartholomew and H. M. Garfinkel, in *Glass Sci. & Tech.*, Vol. 5, (D. R. Uhlmann and N. J. Kreidl, eds.), pp. 217–270. Academic Press, New York, 1980.
37. A. K. Varshneya and W. C. LaCourse, in *Ceramic Transactions: Proceedings of the Third Int. Conf. on Adv. in the Fusion and Processing of Glass.* Ceramic Transactions, vol. 29, pp. 365–378. The American Ceramic Soc., 1993. In press.
38. M. Abou-el-leil and A. R. Cooper, *J. Am. Ceram. Soc.* **62**(7–8), 390–395 (1979).
39. Z. Fathi, I. Ahmed, A. Lodding J. H. Simmons, and D. E. Clark. To be published.

40. H. Wang, W. C. LaCourse, and P. F. Johnson, in *Proc. 1991 Dalian Conference on Glass*, pp. 226–230. China Science and Technology Press, Beijing, 1992.
41. A. K. Varshneya, in *Treatise Mat. Sci. & Tech.*, Vol. 22, *Glass III* (M. Tomozawa and R. H. Doremus, eds.), pp. 241–306. Academic Press, New York, 1982.
42. D. R. Swiler, A. K. Varshneya, and P. Carville, *J. Am. Ceram. Soc.* **70**(4), C75–77 (1987).
43. A. G. Evans, *J. Am. Ceram. Soc.* **73**(2), 187–206 (1990).

Chapter 19

Optical Properties

19.1. Introduction

If one were to list the major technological developments that have added to the comforts of living, glass lenses as an aid to failing vision, glass windows to bring daylight into housing structures while providing protection from the harsh elements of nature, glass in the light bulb to provide a similar function after dark, and glass fibers for enhanced communication capabilities would perhaps rank very high.

For the glass window and the glass bulb to function, the optical transmission property is important. The correction for the failing vision and glass fiber communication are based upon the refractive, optical dispersion and transmission properties of glass. Beside these, there are many more technological advances where the optical properties of glass play a pivotal role. These properties are the subjects of this chapter.

Before we begin, we should note that, in the visible (VIS) region, light rays are generally referred to by their wavelength λ stated in nanometers (SI units) or angstroms. In the infrared (IR) region, the wavenumber $(=1/\lambda)$ with units cm^{-1} is used. In the ultraviolet (UV) region, sometimes the energy E(eV) of the photons is used. For conversion, $1\ eV = 1239.8/\lambda$ (nm).

When light travels through a medium, a part of it is reflected from the front surface, a part is absorbed, and the balance is transmitted.

The **reflectance** R at normal incidence for light waves traveling from one dielectric medium to another is given by

$$R = [(n_1 - n_2)/(n_1 + n_2)]^2, \tag{19.1}$$

where the **refractive index**, n, is the ratio of the velocities of light in vacuum to that in the medium; it is unitless.

When light waves travel from one medium of refractive index n_1 into another medium of refractive index n_2, the incident and the refracted rays remain in the same plane and the angles of incidence i and refraction r (angles between the ray and the surface normal) are related through **Snell's law**:

$$\sin r/\sin i = n_1/n_2. \tag{19.2}$$

When $n_1 > n_2$, $r = 90°$ at the critical angle of incidence ϕ_C (called **Brewster's angle**). Upon $i > \phi_C$, the incident ray is total internally reflected. The bending of light rays as they pass from one medium to another is the key to the design of lenses in eyeglasses, microscopes, telescopes, and cameras.

The **absorption coefficient**, a, is defined by **Beer–Lambert's law** for intensity I transmitted through a distance z in a medium:

$$I = I_0 \exp[-az], \tag{19.3}$$

where I_0 is the incident intensity. The units of a are cm^{-1}. (Note: this unit has yet to be replaced by m^{-1}.) Occasionally, the Beer–Lambert law is written as

$$I = I_0\, 10^{-a'z}. \tag{19.4}$$

We may think of the absorption as due to a solution of the absorbing species in the medium, in which case it is useful to define a **molar extinction coefficient** δ as

$$a = \delta c, \tag{19.5}$$

where c is the concentration (mol L^{-1}) of the absorbing centers. δ is expressed as $L\cdot mol^{-1}\cdot cm^{-1}$. The **absorbance** A and the **optical density** D are given by

$$A = -\ln[I/I_0], \tag{19.6}$$

$$D = -\log[I/I_0]. \tag{19.7}$$

One other important criterion in the design of optical components to be considered is the correction for chromatic aberration resulting from the dispersion of light, i.e., the variation of n with wavelength λ. Away from an absorption peak, n may be expressed by **Cauchy's dispersion formula**:

$$n = A' + \frac{B'}{\lambda^2} + \frac{C'}{\lambda^4} + \cdots, \qquad (19.8)$$

where A', B', C', ... are constants. It may be noted that, according to this relation, the refractive index decreases with increasing wavelength. While this is true for the most part, the relation breaks down in the region where the material strongly absorbs light rays. This is explained in Section 19.2.1.

Useful parameters in the correction for chromatic aberration are the **Abbe number** ν_d and the **partial dispersion**. The Abbe number is defined by

$$\nu_d = (n_d - 1)/(n_F - n_C), \qquad (19.9)$$

where the F-line is the blue $\lambda = 486.1327$ nm line from the hydrogen spectrum, the C-line is the red $\lambda = 656.2725$ nm line, also from the hydrogen spectrum, and the d-line is the yellow $\lambda = 587.5618$ nm line from the He spectrum. The partial dispersion for any two wavelengths x and y is given by

$$P_{x,y} = (n_x - n_y)/(n_F - n_C). \qquad (19.10)$$

19.2. Some Theoretical Concepts

Because light waves are electromagnetic in nature, and glass is a dielectric, the interaction of light with glass is closely related to its dielectric properties. The reader is advised to skim through Chapter 15 before continuing this section.

19.2.1. Dispersion

Maxwell's relations may be used to derive the relationship between the refractive index, n, and the dielectric constant, ε:

$$n = [\varepsilon\mu/\varepsilon_0\mu_0]^{1/2} = [\varepsilon_s\mu_s]^{1/2}, \qquad (19.11)$$

where ε_0 is the absolute permittivity of the free space, ε is the electric permittivity of the medium, ε_s is the relative dielectric constant (see Chapter 15), and μ_0, μ, and μ_s represent the corresponding terms for magnetic permeability. It happens that for most nonferromagnetic materials, μ_s is nearly 1, so that we get the relation

$$n = [\varepsilon_s]^{1/2}. \qquad (19.12)$$

Apparently, this relation works only for some simple gases. The reason for the apparent non-obedience is the frequency dependence of both the ε_s and n. In Chapter 15, we described ε_s as a complex number ε^*:

$$\varepsilon^* = \varepsilon' - j\varepsilon'', \qquad j \equiv \sqrt{(-1)}. \tag{19.13}$$

In a similar manner, we may write the complex refractive index n^* as

$$n^* = n_R - jn_I. \tag{19.14}$$

The imaginary refractive index n_I is called the **extinction coefficient**. It causes attenuation: the amplitude of an electromagnetic wave traveling through the medium decreases by a factor $\exp[-2\pi n_I/n_R]$. The absorption coefficient in the medium as defined in Eq. (19.3) can be shown to be

$$a = 4\pi n_I/\lambda n_R = 4\pi n_I/\lambda_0, \tag{19.15}$$

where λ_0 and λ are the wavelengths of the incident radiation in vacuum and medium, respectively. For light waves traveling from one medium and falling at the surface of a dielectric at normal incidence, the reflectance R is given by

$$R = (n_1^* - n_2^*)^2/(n_1^* + n_2^*)^2, \tag{19.16}$$

where n_1 and n_2 are the complex refractive indices of the two media. For air, $n_2^* = 1$, such that Eq. (19.16) can be approximated to

$$R = [(n_R - 1)^2 + n_I^2]/[(n_R + 1)^2 + n_I^2], \tag{19.17}$$

which reduces to (19.1) for the case of highly transparent media such as glasses. For most metals, $n_R \to 0$, and hence, $R \to 1$.

From Maxwell's equations, the relationship between the complex dielectric constant and the complex refractive index may be written as

$$\varepsilon^* = n^{*2} = [n_R - jn_I]^2 = (n_R^2 - n_I^2) - 2jn_R n_I. \tag{19.18}$$

Thus,

$$\varepsilon' = (n_R^2 - n_I^2), \qquad \text{and } \varepsilon'' = 2n_R n_I. \tag{19.19}$$

If in Eq. (15.10), we assume that the total polarization $\mathbf{P}$ is due to N_L dipoles per unit volume, each contributing an average polarizability α', then

$$\mathbf{P} = N_L\alpha'\mathbf{E}', \tag{19.20}$$

where $\mathbf{E}'$ is the local electric field on the average dipole. In dense dielectrics such as a glass, $\mathbf{E}'$ is the sum of the macroscopic field $\mathbf{E}$ and a Lorentz field contribution equal to $\mathbf{P}/3\varepsilon_0$, i.e.,

$$\mathbf{E}' = \mathbf{E} + \mathbf{P}/3\varepsilon_0. \tag{19.21}$$

Equation (15.8) can be rewritten in the generalized form as

$$\mathbf{P} = [\varepsilon^* - 1]\varepsilon_0\mathbf{E}. \tag{19.22}$$

Equations (19.20), (19.21), and (19.22) yield

$$[\varepsilon^* - 1]/\{N_L[\varepsilon^* + 2]\} = \alpha'/3\varepsilon_0, \tag{19.23}$$

which is called the **Clausius–Mosotti relation**. N_L is the number of dipoles per unit volume and is identified as the Loschmidt number (= the number per unit volume in an ideal gas at 0°C and 760 mm Hg = $2.687 \times 10^{25}\ \text{m}^{-3}$). Avogadro's number N is related to N_L by

$$N = N_L V_m,$$

where V_m is the molar volume. Hence, from (19.18) and (19.23),

$$N\alpha'/3\varepsilon_0 = V_m(n^{*2} - 1)/(n^{*2} + 2) = R_m, \tag{19.24}$$

which is called the **Lorentz–Lorenz relation**. R_m is called the **molar refractivity**. According to these relations, we could determine α' in the gas phase by measuring the refractive index, and then calculate the dielectric constant in the condensed phases at much higher density. Equation (19.24) may also be written as

$$n^{*2} = (1 + 2Y)/(1 - Y) \qquad \text{where } Y = N\alpha'/3V_m\varepsilon_0, \tag{19.25}$$

which shows that a low value of V_m and high α' result in a high n^*.

In an alternating electric field, forced oscillations of the electron remove a certain amount of energy from the electromagnetic radiation that causes the polarization. Part of the energy difference is radiated back by the electron, and part is converted to heat through interaction with the neighboring atoms. Thus, an electron oscillating in the electric field of an electromagnetic radiation is a damped, anharmonic oscillator with a frictional force proportional to the velocity of oscillations. For such a model, it can be shown that the frequency-dependent polarization for a single oscillator is given by

$$\mathbf{P} = \frac{(e^2/m)\mathbf{E}'}{(\omega_0^2 - \omega^2) + j\gamma\omega}, \tag{19.26}$$

where m is the mass of the electron, $\mathbf{E}'$ as before is the local field, γ is the frictional constant (a damping term that has the dimensions of time^{-1}), ω is the angular frequency, and ω_0 is the resonance angular frequency of a single oscillator given by

$$\omega_0^2 = \frac{K}{m} - \frac{N_L e^2}{3m\varepsilon_0}, \tag{19.27}$$

where K is the force constant. Again, using the expression for the local field that includes the Lorentz field correction for a dense dielectric, and rearranging terms, we get

$$\varepsilon^* = 1 + \frac{(N_L e^2/m\varepsilon_0)}{(\omega_0^2 - \omega^2) + j\gamma\omega - (N_L e^2/3m\varepsilon_0)}. \tag{19.28}$$

Writing $(N_L e^2/m\varepsilon_0)^{1/2} = \omega_p$ = angular plasma frequency,

$$\varepsilon^* = n^{*2} = (n_R - jn_I)^2 = 1 + \frac{(\omega_p^2)}{[\omega_0^2 - (\omega_p^2/3)] - \omega^2 + j\gamma\omega}. \tag{19.29}$$

Multiplying both the numerator and denominator in the fraction on the right-hand side by the conjugate $[\{\omega_0^2 - (\omega_p^2/3)\} - \omega^2 - j\gamma\omega]$, and separating the real and imaginary parts, we get

$$n_R^2 - n_I^2 = 1 + \frac{(\omega_p^2)(\bar{\omega}^2 - \omega^2)}{(\bar{\omega}^2 - \omega^2)^2 + \gamma^2\omega^2} \tag{19.30}$$

and

$$2n_R n_I = \frac{(\omega_p^2)\gamma\omega}{(\bar{\omega}^2 - \omega^2)^2 + \gamma^2\omega^2)}, \tag{19.31}$$

where $\bar{\omega}^2 = [\omega_0^2 - (\omega_p^2/3)]$, which can be solved simultaneously for n_R and n_I.

A plot of n_R and n_I as a function of angular frequency is shown in Fig. 19-1 with corresponding behavior of the ε' and ε'' (for small γ). In contrast to the behavior in gases where the resonance occurs at ω_0 (because no Lorentz field correction is needed), the resonance behavior in dense medium occurs at a lower frequency $\bar{\omega}$. Except for the resonance region, n_R essentially increases with ω (decreasing wavelength). This is termed **normal dispersion**. One expects light waves to be transmitted in this wavelength region. The region where n_R decreases with ω is called the region of **anomalous dispersion**. *In the anomalous region where n_I is large, light waves will be absorbed.* At slightly higher frequencies where ε' is negative, the refractive index is mostly imaginary. A medium with an imaginary refractive index cannot propagate electromagnetic waves. As a result, the dielectric totally reflects all light incident upon it as if it were metallic; see Eq. (19.17). If light is successively reflected from the same material in the narrow region where ε' is negative, then the reflected light will be intensified in this wavelength range. These

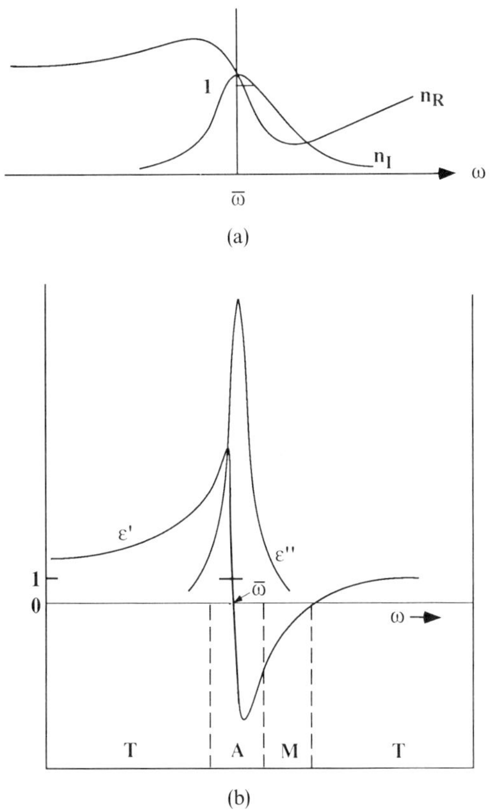

Figure 19-1. Angular frequency dependence of (a) the refractive index (n_R = real part; n_I = imaginary part) and (b) the dielectric constant (ε' = real part; ε'' = imaginary part) for a dielectric having a small γ (see text). Regions of transmission, absorption and metallic reflection are marked *T*, *A*, and *M* respectively.

residual rays are called **reststrahlen**. The region of reststrahlen is important [1] in producing selective filters in the far IR. For real dielectrics, there may be several types of oscillators, each having their resonance frequencies $\bar{\omega}_1$, $\bar{\omega}_2, \ldots$. The refractive index spectrum would then show several regions of anomalous dispersion, each corresponding to its characteristic frequency.

19.2.2. Scattering

For randomly distributed particles, whose size r is much smaller than the wavelength of light ($r < \lambda/4\pi$), the scattering of light is essentially a diffraction phenomenon. Each electron in the medium becomes a source of secondary light whose frequency is unchanged to a large extent (about 99.9%) from that of the incident radiation. This form of elastic scattering is called **Rayleigh scattering**. We assume that the concentration of the scattering centers is low, and hence that they are independent scatterers. For a natural, unpolarized incident radiation of intensity I_0, the intensity I_s of the scattered light is given by

$$I_s/I_0 = (9\pi^2 NV^2/2d^2\lambda^4)\ [(n^2 - 1)/(n^2 + 1)]^2\ [1 + \cos^2\theta], \qquad (19.32)$$

where N is the number of scattering centers per unit volume, V is the volume of the scattering center, n is the ratio of the refractive index of the scatterer to that of the surrounding medium, θ is the angle of observation, and d is the distance of the observation detector from the scattering center. For isotropic scattering centers, the scattered light at 90° is fully polarized in the plane normal to the direction of the incident light. The λ^{-4} dependence of Rayleigh scattering is what gives rise to the blue color of the sky during daytime, and the red color of the sky at dawn and dusk for an observer on the earth. The absorption coefficient, a, in a glass of refractive index n with density fluctuations acting as the source of Rayleigh scattering can be shown to be given by [2]

$$a = [8\pi^3/3\lambda^4](n^2 - 1)^2 \beta k_B T_f, \qquad (19.33)$$

where β is the isothermal compressibility, k_B is the Boltzmann constant, and T_f the fictive temperature of the glass. Thus, a low refractive index, a low compressibility, and a low fictive temperature could give a low absorption coefficient.

The 0.1% of the intensity of incident light is inelastically scattered and is called **Raman scattering**. The photon exchanges a quantum of rotational or vibrational energy with the scattering atom. The perturbation of the light wave of angular frequency Ω puts the atom into mixed quantized angular momentum states with oscillation frequencies ω_1, ω_2, etc. The re-emission of energy thus occurs at frequencies $\Omega \pm \omega_1$, $\Omega \pm \omega_2$, etc. The emissions at the lower energies, $\Omega - \omega_1$, $\Omega - \omega_2$, etc., are called **Stokes lines**, and those at higher energies $\Omega + \omega_1$, $\Omega + \omega_2$, etc. (which require an extra source of energy such as thermal energy) are called **anti-Stokes lines**.

When the particle size is comparable to λ, the scattered intensity is phase-shifted and is given by **Mie scattering theory**. The absorption coefficient a of N spheres of dielectric constant ε^* $(=\varepsilon' - j\varepsilon'')$ embedded in a medium of refractive index n_0 is [3]

$$a = 18\pi NV n_0^3 \varepsilon''/[\lambda\{(\varepsilon' + 2n_0^2) + \varepsilon''^2\}]. \qquad (19.34)$$

For glasses having phase separation, the Mie theory is generally replaced by the **Rayleigh–Gans approximation** theory, where it is assumed that the individual volume elements of the scattering centers are independent scatterers; however, the phase difference between light waves scattered from different volume elements on the particle may give rise to interference effects. It is assumed that the particles have a large refractive index difference from the surrounding medium. For simplicity, it is further assumed that both the particles and the surrounding medium have an average refractive index n_0; however, the particles have a composite refractive index profile with a nucleus having n_1 and a covering layer having n_2 such that the difference of either n_1 or n_2 from n_0 is large (of the order of 0.05). Under these conditions, the scattered intensity at a distance d and angle θ away is given by [4]

$$I_s/I_0 = [(4\pi^2 V^2/d^2\lambda^4)](n_0 - 1)^2[(1 + \cos^2\theta)/2]|\Phi(\theta, \lambda)|^2, \quad (19.35)$$

where

$$\Phi(\theta, \lambda) = \frac{1}{V}\int \exp(j\delta)dV$$

is the scattering function, where δ = phase difference between two rays scattered from different points in the particle. $\Phi(\theta, \lambda)$ can be obtained analytically for simple geometric shapes. At $\theta = 0^\circ$, $\Phi(\theta, \lambda) = 1$. In other directions, $|\Phi(\theta, \lambda)| < 1$.

19.2.3. Absorption in the Visible Region (Colors in Glass)

Selective absorption of various wavelengths in the visible region gives rise to the appearance of colors in glass. The source of absorption is of three types: (i) electron transitions within the unfilled orbits of the transition elements; (ii) plasma resonance, defined by Eq. (19.29); and (iii) electron transitions across the band gap, described in Chapter 16.

There are four possible transition element series in the periodic table. Of these, the first series from Sc to Ni (including the divalent Cu ion) produces strong colors because of absorption of selected wavelengths when they are present in glass even in amounts less than 1%. The absorption spectra of the transition elements in glass are described by Bamford [5] and generally explained in terms of either **ligand field theory** (which evolved from crystal field theory) or molecular orbital theory. The reader is referred to Bates [6] and Paul [7] for a detailed treatment of the two theories. The more popular of these, the ligand field theory, is outlined below.

The transition elements are characterized by an unfilled inner electron shell. The first transition series from Sc to Ni (including divalent Cu), has an electronic shell structure $1s^2;2s^22p^6;3s^23p^63d^n;4s^2$, where one to nine electrons out of a total of ten $3d$ electrons may be unfilled. The second transition series from Y to Pd has a (1–$4p$ filled)$4d^n;5s^{1-2}$ structure. Likewise, the third, the rare-earths from Ce to Lu, has (1–$4d$ filled)$5s^25p^6;6s^2;4f^n;5d^{0-1}$ structure. The 15 elements of the actinide series beginning with Th are considered the fourth transition series and are homologous to the third series. The transition ion, denoted M^{m+}, is surrounded by negative ions or oriented dipoles called ligands in glass. Together, they form a "complex." The electric field of the ligands causes the degenerate energy levels of the d-electron in M^{m+} to split. Although Laporte's selection rule forbids d^n–d^n transitions, because it requires that the change in the angular momentum $\Delta L = \pm 1$, the rule is relaxed somewhat because of some mixing between the $3d$-transition metal orbital and the metal $4p$- or ligand $3p$- orbitals. As a consequence, an allowed electronic transition from a ground state d^n shell to an excited state d^n shell absorbs the energy difference from the incident radiation, resulting in a "weak" (molar extinction coefficient 0.01–200 L mol^{-1} cm^{-1}) absorption band in the 6,000–50,000 cm^{-1} range, which spans the near UV through the near IR. The absorption bands are broadened because of thermal vibrations, spin–orbit coupling, and the Jahn–Teller effect (see Paul [7]).

The orbital splitting may be summarized as follows. In a free ion, the d-electrons occupy five d-orbitals (Fig. 19-2): d_{xy}, d_{xz}, d_{yz}, $d_{x^2-y^2}$, and d_{z^2}, with equal probability (i.e., the orbitals are degenerate). In an octahedral symmetry of negatively charged ligands along the $\pm x$, $\pm y$, and $\pm z$ axes, the electrostatic repulsion causes the orbitals to avoid the $d_{x^2-y^2}$ and d_{z^2} orbitals (because these lobes point towards the ligands at the octahedral locations L, see Fig. 19-2). This results in a split in the population to occupy different energy levels, e_g and t_{2g} (Fig. 19-3). (The notation used here is that by Mulliken. Other notations, particularly by Bethe, and Tanabe and Sugano, are also used. See Bates (6) for the various notations.) The energy difference between the two sets of split levels is the ligand field, denoted $10Dq$. (Some authors use Δ_0 or simply Δ.) The group ($d_{x^2-y^2}$ and d_{z^2}) is at the e_g level, which is $6Dq$ above, and the (d_{xy}, d_{xz}, d_{yz}) group is at the t_{2g} level, which is $4Dq$ below the energy level of the unsplit orbitals such that the unsplit level is the energy-weighted centroid. In a tetrahedral ligand field, the situation is opposite: the total energy difference is four-ninths of the octahedral value (or $\Delta_T = -4\Delta_0/9$). In a cubic field, the energy difference would be $-8\Delta_0/9$.

The ligand field strength $10Dq$ is computed by estimating the ligand or crystal field stabilization energy (CFSE), which is a relative measure of the stability of the complex with respect to the hypothetical complex with no crystal field splitting. The CFSE may be obtained directly from thermodynamic data of the heats of formation of the complex. The relationship

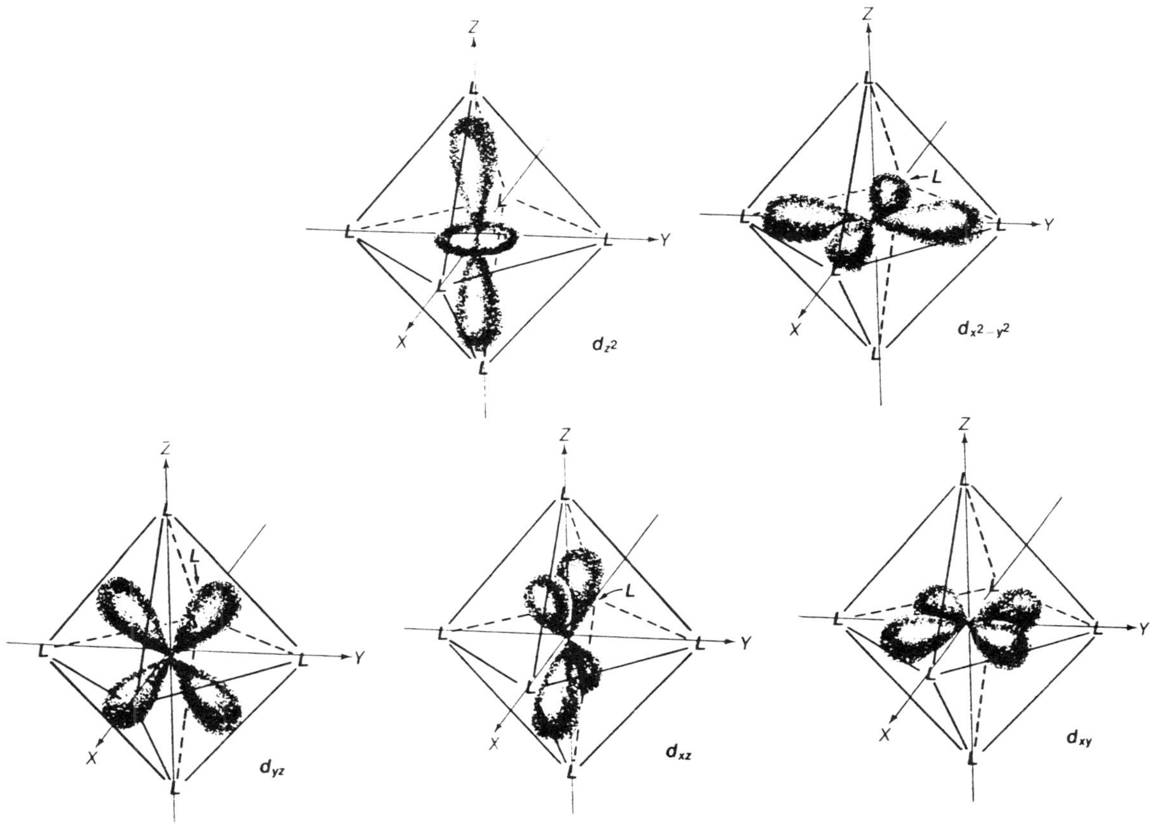

Figure 19-2. The *d*-orbitals. (Reproduced from W. H. Nebergall, H. F. Holtzclaw, and W. R. Robinson, *College Chemistry*, 6th Ed. D. C. Heath & Co., Lexington, Massachusetts, 1980. Reproduced with permission of the publishers.)

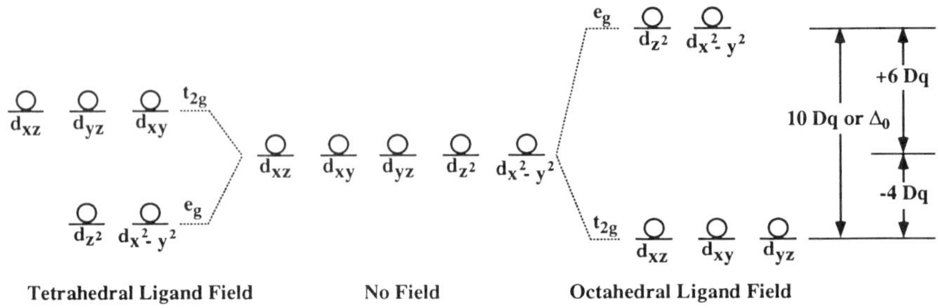

Figure 19-3. Energy levels of the *d*-orbitals in octahedral and tetrahedral fields.

between $10Dq$ and the CFSE may be established, say for Mn^{2+} ion in an octahedral field, as follows:

The Mn^{2+} ion has five *d*-electrons. In the free ion, these electrons are distributed unpaired in the five *d*-orbitals (Hund's Rule of Maximum Multiplicity). In a relatively weak ligand field (Fig. 19-4), the five electrons remain undisturbed; hence, the CFSE $= 2(+6Dq) + 3(-4Dq) = 0$. On the other

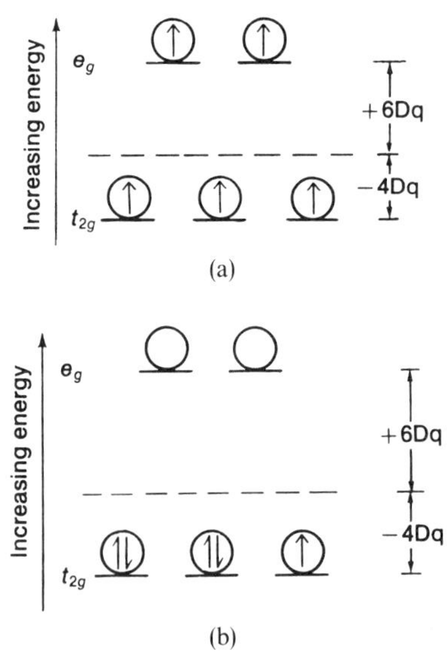

Figure 19-4. Stabilization of the *d*-levels in (a) weak and (b) strong ligand fields. (Reproduced from W. H. Nebergall, H. F. Holtzclaw, and W. R. Robinson, *College Chemistry*, 6th Ed. D. C. Heath & Co., Lexington, Massachusetts, 1980. Reproduced with permission of the publisher.)

hand, in a strong ligand field (Fig. 19-4), the electrons are forced to occupy the lower energy levels (t_{2g}) because spin-pairing costs less energy relative to the excitation to the higher e_g level. This results in the CFSE = $5(-4Dq) = -20Dq$. We would expect a complex of Mn^{2+} with ligands of high field strengths to be considerably more stable than one with low field strengths.

The effect of the ligand field strength on the energy level changes is often represented by an **Orgel diagram** such as that for Ni^{2+}, shown in Fig. 19-5. The ordinate is the dimensionless E/B, where E is the energy difference (written usually in cm^{-1}) and B is the Racah parameter (also in energy units). The abscissa is the ligand field-splitting energy (as measured by Dq, in cm^{-1} units) made dimensionless by dividing by B. The ground state is the abscissa.

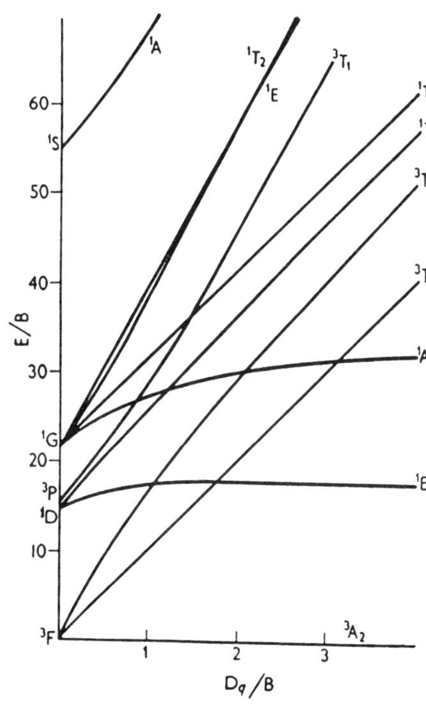

Figure 19-5. Orgel diagram for the energy levels of the Ni^{2+} ion in an octahedral ligand field. $B = 1030\ cm^{-1}$. The field, $10\ Dq$, corresponds to an energy of about $8{,}500\ cm^{-1}$. (After T. M. Dunn in *Modern Coordination Chemistry*, (J. Lewis and R. G. Wilkins, eds.), Fig. 12, p. 250. Interscience Publishers Inc., London, 1960.)

On the extreme left are the Russell–Saunders states of the free ion. Ni^{2+} has a $3d^8$ structure. In solution, Ni^{2+} is known to be present as $[Ni(H_2O)_6]^{2+}$, i.e., in octahedral coordination. The ligand field of the H_2O molecules causes the d-electron energy levels in the central Ni^{2+} ion to split. The allowed transitions among these split levels are from the ground state $^3A_2(F)$ to excited states $^3T_2(F)$, $^3T_1(F)$, and $^3T_1(P)$. The absorption bands (8,500, 13,500, and 25,300 cm^{-1}) shown in Fig. 19-6a occur at roughly the locations predicted by a ligand field strength of $\sim$8,500 cm^{-1}. (Absorption in the near UV and IR gives the solution its green look.) The absorption spectrum of Ni^{2+} in low-alkali borate glass is similar with slightly shifted peaks, which indicates that Ni is surrounded by mostly an octahedral field of oxygens in this glass. The ligand field strength, however, is slightly reduced ($\sim$7,500 cm^{-1}).

The estimation of $10Dq$ from thermodynamic data is only approximate: often the better approach is to use this as a guide and obtain a more accurate value of $10Dq$ by fitting it to one or two observed absorption peaks in known Orgel diagrams. Absorption spectra in other symmetries can then be obtained readily by utilizing the field strength conversion factors as given earlier. For instance, in a high-alkali borosilicate glass (Fig. 19-6b), the absorption bands of Ni^{2+} are consistent with mostly a tetrahedral symmetry of oxygens around the Ni^{2+} in glass [6]. The field strength corresponds to 4,700 cm^{-1}, which is about four-ninths of the octahedral field. Note also the rather high values of the extinction coefficients, which are typical of transition metal ions in tetrahedral symmetry.

Table 19-1 shows the colors generated by some of the transition metal and rare earth ions in glass (8). It may be noted that different valences impart different colors. Now, the higher the valence state of a multivalent metal ion, the higher is its oxidation state. The reduction–oxidation equilibrium, called the **redox equilibrium** in brief, depends upon the composition and the melting conditions. Perhaps the most technologically important redox equilibrium is that of iron. The equilibrium may be stated as follows:

$$4FeO(glass) + O_2(glass) \Longleftrightarrow 2Fe_2O_3(glass). \qquad (19.36)$$

Hence, the thermodynamic equilibrium rate constant K is given by

$$K = \frac{[a_{Fe_2O_3(glass)}]^2}{[a_{FeO(glass)}]^4} \cdot \frac{1}{pO_{2(glass)}}, \qquad (19.37)$$

where a represents the thermodynamic activity, and $pO_{2(glass)}$ is the partial pressure of oxygen in the glass. The former is related, generally nonlinearly, to the concentration of the species, and the latter may be considered equal

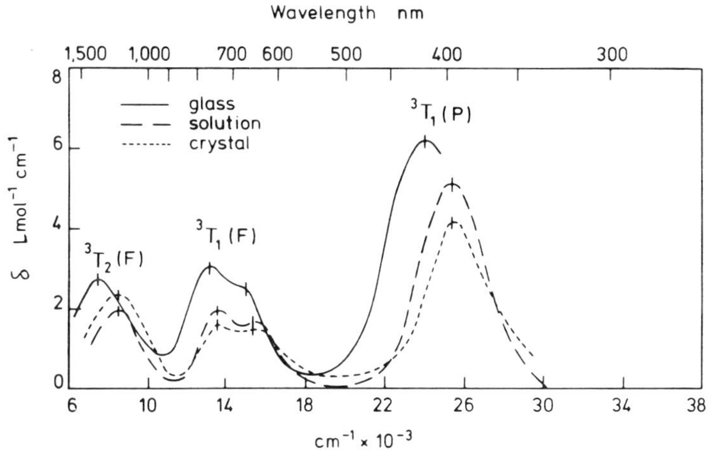

(a)

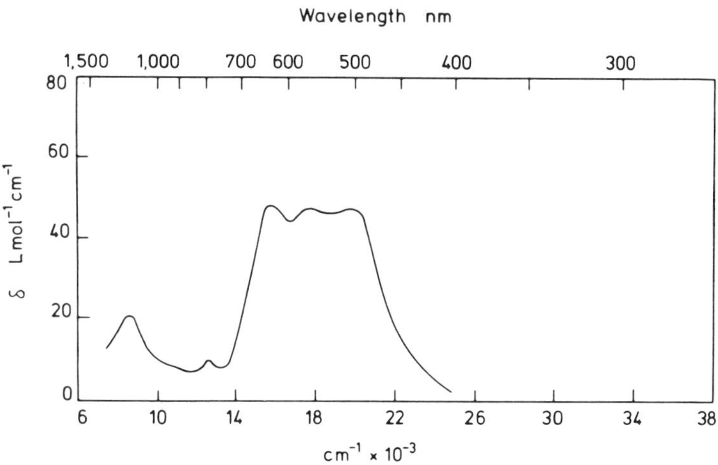

(b)

Figure 19-6. Absorption spectra (a) octahedrally coordinated Ni^{2+} in various media, (b) Ni^{2+} in high-alkali borate glass. (Adapted from Bates [6], Fig. 37, p. 248, and Fig. 38, p. 249. Reproduced with permission of Butterworth-Heinemann Ltd., Oxford.)

Table 19-1. Colors Generated by Transition-Metal and Rare-Earth Ions in Glass

Transition metal ions			Rare earth ions		
Configuration	Ion	Color	Configuration	Ion	Color
d^0	Ti^{4+}	Colorless	$4f^0$	La^{3+}	None
	V^{5+}	Faint yellow to colorless		Ce^{4+}	Weak yellow
			$4f^1$	Ce^{3+}	Weak yellow
	Cr^{6+}	Faint yellow to colorless	$4f^2$	Pr^{3+}	Green
			$4f^4$	Nd^{3+}	Violet-pink
d^1	Ti^{3+}	Violet-purple	$4f^4$	Pm^{3+}	None
	V^{4+}	Blue	$4f^5$	Sm^{3+}	None
	Mn^{6+}	Colorless	$4f^6$	Sm^{2+}	Green
d^2	V^{3+}	Yellow-green		Eu^{3+}	None
d^3	Cr^{3+}	Green	$4f^7$	Eu^{2+}	Brown
d^4	Cr^{2+}	Faint blue		Gd^{3+}	None
	Mn^{3+}	Purple	$4f^8$	Tb^{3+}	None
d^5	Mn^{2+}	Light yellow	$4f^9$	Dy^{3+}	None
	Fe^{3+}	Faint yellow	$4f^{10}$	Dy^{2+}	Brown
d^6	Fe^{2+}	Blue-green		Ho^{3+}	Yellow
	Co^{3+}	Faint yellow	$4f^{11}$	Er^{3+}	Weak pink
d^7	Co^{2+}	Blue-pink	f^{12}	Tm^{3+}	None
d^8	Ni^{2+}	Brown-purple	$4f^{13}$	Tm^{2+}	None
d^9	Cu^{2+}	Blue-green		Yb^{3+}	None
d^{10}	Cu^+	Colorless	$4f^{14}$	Lu^{3+}	None

to the partial pressure of oxygen in the atmosphere above the glass melt under equilibrium conditions. Equation (19.37) shows that if pO_2 is increased, the reaction will produce a higher concentration of the oxidized state. Another way to write Eq. (19.36), more appropriate to glass, is in terms of the complexes of the transition metal ions, for instance for Fe as

$$4Fe^{2+}(\text{glass}) + 14O^{2-}(\text{glass}) + O_2 \Longleftrightarrow 4[Fe^{3+}O_4]^{5-}(\text{glass}), \quad (19.38)$$

such that the rate constant is given by

$$K = \frac{[a_{Fe^{3+}O_4(\text{glass})}]^4}{[a_{Fe^{2+}(\text{glass})}]^4[a_{O^{2-}(\text{glass})}]^{14} pO_{2(\text{glass})}}. \quad (19.39)$$

This equation introduces the concept of the activity of oxygen ions in glass, which may be calculated by measuring the potential developed upon immersing a reversible oxygen electrode in the molten glass. The oxygen ion activity increases with the concentration of the network modifiers in glass, and in the increasing order of the "basicity" of the glass. The latter increases in the order Li < Na < K < Rb < Cs. An increase in the oxygen ion activity will lead to an increase in the oxidation state of the multivalent ion, provided that pO_2 is kept constant. By thermodynamic arguments, it may be shown that the redox equilibrium moves towards the reduced side as the temperature is increased. Figure 19-7, which is a plot of the Fe^{3+}/Fe^{2+} ratio as a function of the temperature for binary silicate glasses, confirms these general arguments [7].

As may be seen from Fig. 19-8, the presence of octahedrally coordinated Fe^{2+} in glass gives a weak absorption band at ~9,000 cm^{-1} (1.1 μm wavelength) which extends into the red. As little as about 0.02–0.04% FeO impurity in the sand and other raw materials is the source of the undesirable bluish-green color observed in many glass products. On the other hand, Fe^{3+} has a very strong charge-transfer band in the UV (discussed below), which tails into the visible. As a result, glasses containing iron in the oxidized state appear straw-yellow, which is complementary to the color of Fe^{2+}. A proper Fe^{3+}/Fe^{2+} ratio, hence, brings about **decolorization** of the glass. Apparently, the addition of (0.25%As_2O_3 or Sb_2O_3) + 0.5%$NaNO_3$ to the batch in large, continuous glass-melting tanks to remove fine bubbles from glass also achieves decolorization by oxidizing part of the Fe^{2+} to Fe^{3+}. Because of the emphasis on safety in the more recent years, the use of As_2O_3 is no longer recommended, and the preferred choices to achieve decolorization by oxidation Fe^{2+} are cerium hydrate (about 1,500 ppm), or nitrates alone. One may also employ about 50 ppm Se to provide the complementary color instead of oxidizing Fe^{2+} to Fe^{3+}.

The amber color of beer bottles, commonly called the carbon–sulfur amber, is obtained by adding 0.2–0.5% Fe_2O_3 plus Na_2SO_4 (to yield ~0.04% SO_3 in the glass). The color is due to two intense charge transfer bands at 0.295 μm and 0.425 μm (Fig. 19-8). The chromophore is believed to be a complex consisting of octahedrally coordinated Fe^{2+}, a tetrahedrally coordinated Fe^{3+} where one of the oxygens has been replaced by S^{2-}, and a sulfate group. Apparently, the redox equilibria of the $Fe^{2+} \Leftrightarrow Fe^{3+}$ and the sulfide ⇔ sulfate both are important in achieving a proper color.

Much of the science of the redox equilibria in glass has been discussed in detail by Paul [7].

When glasses containing a small amount (0.01–0.02 wt. %) of Au, along with a small amount of reducing agents such as SnO_2 and ZnO, are cooled after melting, nothing happens at first, but subsequent heating above T_g causes a ruby-red color to develop. Color developed by such a precipitation scheme is generally termed **striking of a ruby**. A similar phenomenon occurs in copper ruby and silver yellow glasses. The absorption occurs because of plasma resonance of free electrons in the colloidal metal particles. In gold ruby glasses, the gold particles are roughly 200 Å in size, and the color is due to an absorption at around 0.53 μm. (The bulk metal looks bright yellow,

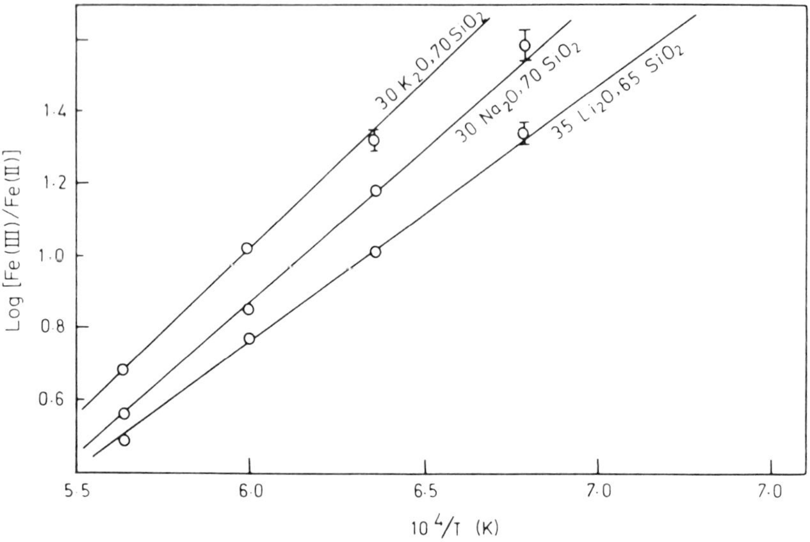

Figure 19-7. Variation of Fe^{3+}–Fe^{2+} equilibrium with temperature in some binary alkali silicate melts. (After Paul [7], Fig. 7.6, p. 233. Reproduced with permission of Chapman & Hall.)

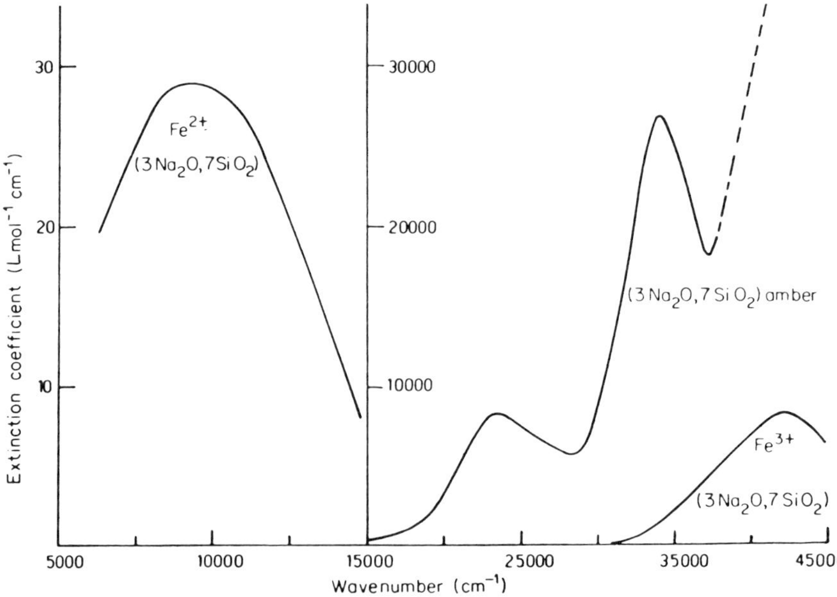

Figure 19-8. Absorption spectra of Fe^{2+}, Fe^{3+}, and "carbon–sulfur amber." (After Paul [7], Fig. 9.31, p. 330. Reproduced with permission of Chapman & Hall.)

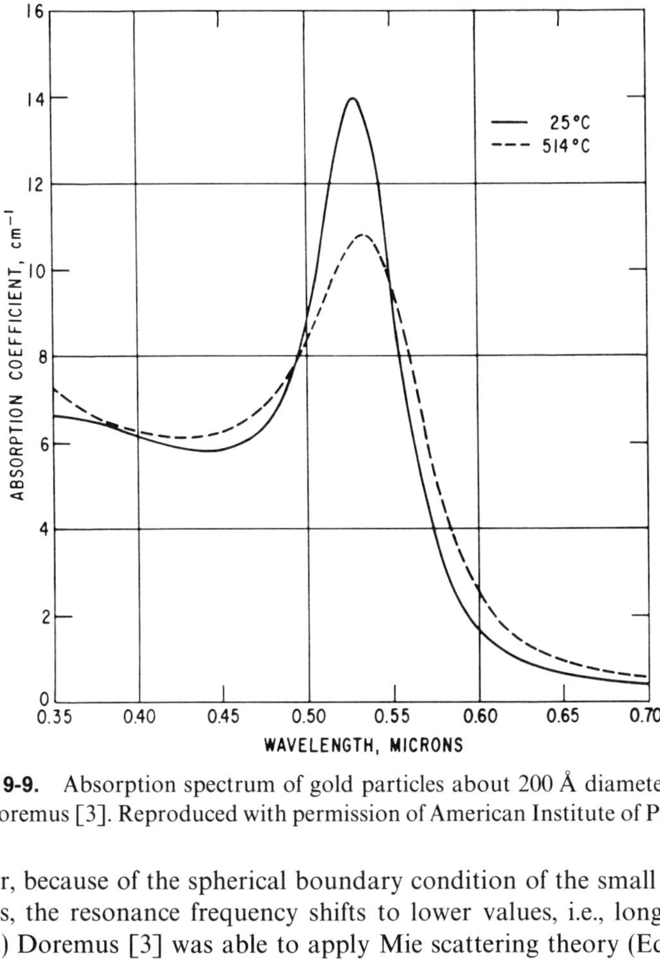

Figure 19-9. Absorption spectrum of gold particles about 200 Å diameter in glass. (After Doremus [3]. Reproduced with permission of American Institute of Physics.)

however, because of the spherical boundary condition of the small colloidal particles, the resonance frequency shifts to lower values, i.e., longer wavelengths.) Doremus [3] was able to apply Mie scattering theory (Eq. (19.34)) to explain the observed absorption coefficient due to the colloidal gold particles in glass (Fig. 19-9).

Electron transitions across the band gap of semiconductor additions to glass can also cause absorption in the visible region of the spectrum. Examples of this type are the yellow to red colors in glasses containing small amounts of CdS–ZnS and CdS–CdSe.

19.2.4. Absorption in the UV

Strong absorption bands arising from electron excitations produce essentially a **UV cutoff**, causing most glasses to appear opaque in the UV. These electron excitations are generally of two types:

1. Intrinsic excitation, where electrons are excited from the valence band to unoccupied states in the exciton or the conduction band (interband transition) levels.

2. Transfer of an electron (charge) between the shells of one ion and the shells of a neighboring ion (interionic transitions). These ions may be major constituents or minor constituents. Electron transfer between neighboring multivalent transition elements, for instance Fe^{2+} to Fe^{3+}, is an example of the latter (Chapter 16).

Intrinsic absorption occurs when photons have energies comparable to the band gap E_g. This means roughly 8.9 eV for SiO_2 and only about 1 to 3 eV for amorphous semiconductors such as the chalcogenides. Thus, the UV cutoff may be estimated [9] by

$$\lambda_c = hc/E_g. \tag{19.40}$$

To be more precise, one should replace E_g in this equation by E_0, the mobility gap measured from the valence band mobility edge to the conduction band mobility edge. For photon energies higher than the mobility gap E_0, the absorption coefficient $\alpha(\omega)$ varies with the angular frequency ω according to

$$\omega\alpha(\omega) = B[\hbar\omega - E_0]^2, \tag{19.41}$$

where B is a constant. A plot (10) of $[\omega\alpha(\omega)]^{1/2}$ vs. $\hbar\omega$ shows a substantial linear region from which a value of E_0 may be extracted by extrapolation (Fig. 19-10). It is hence customary to call E_0 the **optical gap** or the **optical absorption edge**.

Figure 19-10 also shows that, for photon energies less than E_0, a considerable amount of tailing exists. In this region, the absorption coefficient follows the relationship

$$\alpha(\omega) = \alpha_0 \exp[-\Gamma\{E_c - \hbar\omega\}], \tag{19.42}$$

where Γ is a constant and E_c is an energy comparable to E_0. The observed variation of the $\alpha(\omega)$ vs. the photon energy for several chalcogenide glasses is plotted in Fig. 19-11 (also after Ref. 10) and confirms the applicability of Eq. (19.42). This absorption edge well into the gap states is called the **Urbach edge**. Another expression used to describe the absorption coefficient in the Urbach edge is

$$\alpha(\omega) = A_{UV} \exp[E/E_c], \tag{19.43}$$

where A_{UV} is a constant. One may note that many of the amorphous semiconductors are opaque to photons of energies larger than about 1 eV,

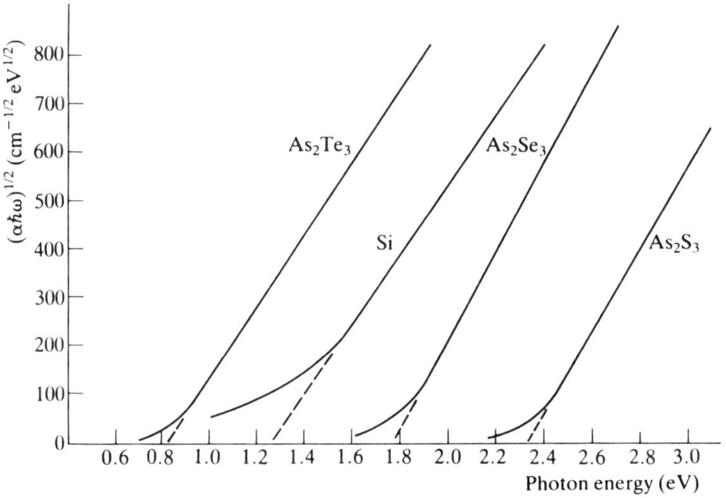

Figure 19-10. Optical absorption edges of some materials. Extrapolation of the linear regime yields the "optical gap." (After N. F. Mott and E. A. Davis, *Electronic Processes in Non-crystalline Materials*, 2nd Ed. Oxford University Press, 1979. Reproduced with permission of Oxford University Press.)

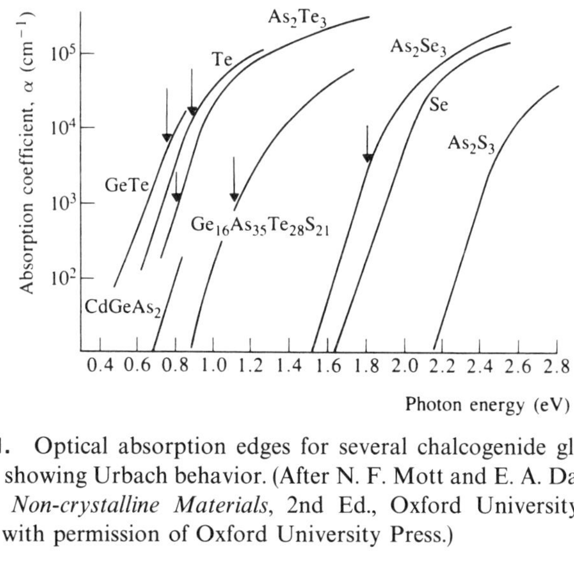

Figure 19-11. Optical absorption edges for several chalcogenide glasses at room temperature, showing Urbach behavior. (After N. F. Mott and E. A. Davis, *Electronic Processes in Non-crystalline Materials*, 2nd Ed., Oxford University Press, 1979. Reproduced with permission of Oxford University Press.)

i.e., wavelengths less than 1,240 nm, which covers essentially all of the visible and the UV spectrum.

The UV spectra of insulator glasses are not easy to understand. They are often interpreted in terms of molecular orbital theory. The fundamental absorption edge in fused silica is at about 10.2 eV and is explained (11) as arising from a transition to an exciton level of an antibonding $2p$ electron level in oxygen lone-pair electron orbitals. The addition of network modifiers shifts this edge to lower energies.

Charge-transfer bands arising from interionic transfer of electrons between transition-metal ion and ligands or another transition-metal ion of a different valence are better understood. Compared to the weak ligand-field-generated absorptions in the visible region, the bands are relatively very strong. As an example, the molar extinction coefficient of the Fe^{3+} charge transfer band in a silicate glass is 7,000 L mol^{-1} cm^{-1}, centered at about 210 nm. This is about 10^4 times stronger than the "weak" absorptions arising from the ligand-field-caused internal electronic transitions (6) (Fig. 19-8).

19.2.5. Absorption in the IR

The IR absorption occurs largely as a result of atomic vibrations in a molecule that has a dipole moment. Thus, homonuclear molecules such as O_2 have no IR vibrational spectra, but an Si–O–Si group may have bond-stretching and bond-bending oscillations that would have characteristic absorption in the IR. Figure 19-12 shows the four fundamental vibrational modes of an XY_4 molecule. Of these, the stretching vibrations are characterized by the higher frequencies, and hence determine the IR cutoff.

For a diatomic molecule having atoms of masses m_1 and m_2, the natural frequency ν_{IR} of the stretching vibration is given by the **Szigeti relation**:

$$\nu_{IR} = [1/2\pi][F/\mu]^{1/2}, \tag{19.44}$$

where $F =$ force constant, and $\mu =$ reduced mass $= m_1 m_2/(m_1 + m_2)$.

Donald and McMillan (9) have suggested the use of Yr_0, where Y is the Young's modulus and r_0 is the bond distance, in place of F as a first approximation to estimating ν_{IR}. The absorption coefficient for energies E ($E < E_0$, where $E_0 = h\nu_{IR}$) is

$$\alpha(\omega) = A_{IR} \exp[-E/E_0]. \tag{19.45}$$

Since the absorption arises from "lattice" vibrations, the term "multiphonon loss" is often used for the losses in this manner.

It may be shown that the cutoff for most oxide glasses lies in the

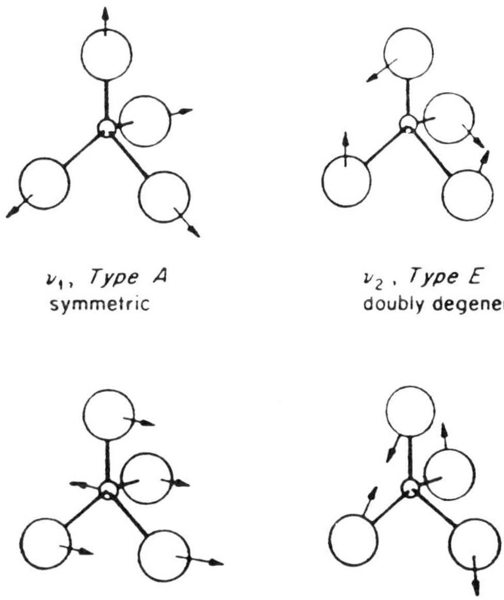

Figure 19-12. Four normal modes of vibration of an XY_4 tetrahedral molecule. Modes ν_1 and ν_2 are only Raman-active; modes ν_3 and ν_4 are both Raman- and IR-active.

4,500–6,500 nm (2.200–1,500 cm^{-1}) range. To increase the IR cutoff, one may note from the Szigeti relation that the higher the masses, the lower the ν. This is very important technologically for optical communication links, where the transmission losses in the visible range due to Rayleigh scattering and electronic excitations limit the choice for a low-loss fiber to IR-transmitting glasses (see Section 19.5.4). Heavy metal fluoride and chalcogenide glasses have therefore attracted much attention in the recent years.

IR absorption bands due to electron excitations are few. Perhaps the most important of these is the 1,100 nm absorption band due to Fe^{2+}, which makes the iron impurity-containing glasses appear blue-green and heat-absorbing.

Complementary to IR-absorption spectroscopy is Raman spectroscopy, where the frequency of an incident photon increases or decreases inelastically in quantized amounts corresponding to the vibrational frequencies of the molecular groups.

19.2.6. Photoelastic Properties

Glass in a well-annealed and homogeneous state is expected to be isotropic because of its random structure in all directions. However, in the presence of nonhydrostatic stresses, the optical isotropy is disturbed. Perpendicular vibrations of an unpolarized light travel at different velocities through the medium. This implies that the glass develops a **double-refraction** or **birefringence** property. This is shown in Fig. 19-13 for a uniaxially applied stress σ_{33}. The optical path difference δ between the vibrations along the x_3- and x_1-axes when viewed along the x_2-axis is given by

$$\delta = C_\lambda \sigma_{33} t, \tag{19.46}$$

where t is the thickness of the specimen. C_λ is called **Brewster's constant** or the **stress-optic coefficient**. Traditionally, when the stress is written in bars ($=10^6$ dyn/cm^2), the thickness is measured in centimeters, and δ is measured in nanometers, then C_λ is written in brewsters ($=10^{-13}$ cm^2/dyn). For conversion to SI units, 1 brewster $= 10^{-12}$ Pa^{-1} ($=1T$Pa^{-1}).

For the setup shown in Fig. 19-13, the vibration along the x_1-axis is called the "ordinary ray," and that along the x_3-axis is called the "extraordinary ray." The x_3-axis is also the optic axis, since glass appears "isotropic" for light propagating along this axis. Glass is said to have developed a positive or a negative birefringence, depending upon whether the refractive index of the e-ray is higher or lower than that of the o-ray. Most glasses tend to display positive birefringence when a tensile stress acts upon them.

The relationship between the applied strains ε_{33} and the observed bi-

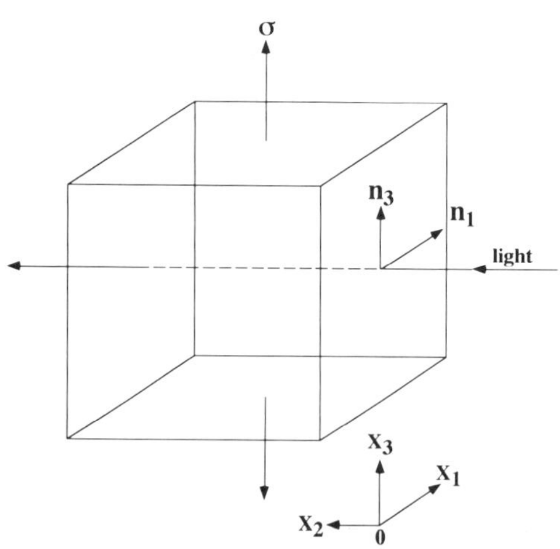

Figure 19-13. Propagation of light in uniaxially stressed glass.

refringence is given by the Neumann relations:

$$\Delta n_3 = n_3 - n = -n^2 Q\varepsilon_{33},$$
$$\Delta n_1 = n_1 - n = -n^2 P\varepsilon_{33}. \tag{19.47}$$

The constants P and Q are called **Neumann's constants**. Note that if both n_3 and n_1 were written as complex numbers, then for double refraction the real parts of the refractive indices are not equal, whereas the imaginary parts remain equal. (Inequality of imaginary parts of the refractive indices would give rise to absorption over different wavelength regions, otherwise called **dichroic behavior**.)

The phenomenology of stress-optical behavior requires a consideration of the propagation of electromagnetic waves in an anisotropic dielectric medium. In an anisotropic medium, the electric displacement vector $\boldsymbol{D}$ and the electric field vector $\mathbf{E}$ are no longer parallel; hence, we need to restate the definitions of the electric permittivity ε, the relative dielectric constant ε_s, and the dielectric susceptibility χ given by Eqs. (15.7) and (15.9):

$$D_i = \varepsilon_{ij}E_j = \varepsilon_0(\varepsilon_s)_{ij}E_j, \tag{19.48}$$

$$P_i = \varepsilon_0\chi_{ij}E_j. \tag{19.49}$$

The ε, ε_s, and χ are second-rank tensors. There may be three principal refractive indices n_1, n_2, and n_3, which are related to the relative dielectric constant by $n_1 = [(\varepsilon_s)_{11}]^{1/2}$, $n_2 = [(\varepsilon_s)_{22}]^{1/2}$, and $n_3 = [(\varepsilon_s)_{33}]^{1/2}$. The three refractive indices of the light waves form an ellipsoid in Cartesian space (principal axes x_1, x_2, x_3) called the indicatrix, defined by

$$\frac{x_1^2}{n_1^2} + \frac{x_2^2}{n_2^2} + \frac{x_3^2}{n_3^2} = 1. \tag{19.50}$$

Defining $B_1 = 1/n_1^2$ = relative dielectric impermeability along x_1, and similarly for B_2 and B_3, this equation can be rewritten as

$$B_1x_1^2 + B_2x_2^2 + B_3x_3^2 = 1. \tag{19.51}$$

More generally, for an arbitrary orientation of axes, this equation to a second-degree surface (quadric) is written as

$$B_{ij}x_ix_j = 1. \tag{19.52}$$

In the presence of stress σ_{ij} or electric fields E_i, the shape, size, and orientation of the indicatrix changes, i.e., the coefficients B_{ij} change. We may write

$$\Delta B_{ij} = z_{ijk}E_k + q_{ijkl}\sigma_{kl}, \tag{19.53}$$

where the third-rank tensors z_{ijk} are the **electro-optical coefficients**, and the fourth-rank tensors q_{ijkl} are called the **stress-optic** or the **piezo-optic coefficients (photoelastic effect)**. The equation may also be rewritten in terms of the strains as

$$\Delta B_{ij} = z_{ijk}E_k + p_{ijkl}e_{kl}, \tag{19.54}$$

where the p's are called the **strain-optic** or the **elasto-optical coefficients**. The p's and q's, together, are called **Pockel's coefficients**. The units of q's are Pa^{-1} (SI) or cm^2/dyn (cgs). The p's are dimensionless. In a manner analogous to the two-suffix notation in linear elasticity, the Pockel's coefficients can be written as p_{ij} or q_{ij}, where $i, j = 1, 6$. There are 81 independent elements to each set of coefficients; however, isotropic symmetry of a glass causes

$$p_{11} = p_{22} = p_{33}; \qquad p_{12} = p_{21} = p_{13} = p_{31} = p_{23} = p_{32}; \qquad \text{and}$$

$$p_{44} = p_{55} = p_{66} = [p_{11} - p_{12}]/2. \tag{19.55}$$

Thus, there are only two independent p's. The relations for q's are similar, except that $q_{44} = q_{55} = q_{66} = (q_{11} - q_{12})$.

The matrix equation

$$\Delta B_i = q_{ij}\sigma_j \tag{19.56}$$

can be readily solved for the case of a uniaxial stress σ_3 to show that

$$\Delta n_3 = -[n^3/2]q_{33}\sigma_3 = -[n^3/2]q_{11}\sigma_3 \tag{19.57}$$

and

$$\Delta n_1 = -[n^3/2]q_{12}\sigma_3. \tag{19.58}$$

This means that for light traveling along the x_2-axis,

$$\Delta n_3 - \Delta n_1 = -[n^3/2][q_{11} - q_{12}]\sigma_3. \tag{19.59}$$

Likewise,

$$\Delta n_3 - \Delta n_1 = -[n^3/2][p_{11} - p_{12}]\varepsilon_3. \tag{19.60}$$

Thus, Neumann's constants are related to Pockel's coefficients through

$$P = [n/2]p_{12} \qquad \text{and } Q = [n/2]p_{11}. \tag{19.61}$$

By comparing (19.59) with (19.46) and recognizing that $\Delta n_3 - \Delta n_1 = \delta/t$, we get the relationship between the Brewster's constant and the Pockel's coefficients:

$$C_\lambda = -[n^3/2][q_{11} - q_{12}]. \tag{19.62}$$

Stress-optical phenomena are very important in glass technology. Because of the usual transparency of glass, it is easy to measure the optical retardation using birefringence compensators (e.g., a quartz wedge, Babinet–Soleil, a Berek or a Senarmont compensator) in a stressed product and relate it to the stresses using Eq. (19.46). While analyzing a glass product for stresses, one should remember thc following:

(i) The absence of optical retardation is not a sufficient proof for the absence of stresses. Likewise, as shown in the section following, stress is not the only source of a nonzero retardation.
(ii) A biaxial stress system (stresses along two perpendicular axes) cannot be separated into its individual stress components using a transmission measurement. Only the difference of the principal stresses integrated over the light path is obtained.

Values of C_λ for some commercial glasses are shown in Table 19-2. These are generally in the range of about 2.5–3 brewsters, although in high PbO-containing glasses, heavy metal fluorides, and chalcogenides the Brewster's constant can be near zero or even negative.

Table 19-2. Stress-Optic Coefficients for Some Commercial Glasses

Glass	Type	Stress-Optic constant in brewsters
Corning 0080 Owens-Illinois R-6	Soda-lime	2.83
Corning 0120 Owens-Illinois KG-12	Lead alkali silicate	3.23
Corning 1723	Aluminosilicate	2.77
PPG 6695	Aluminosilicate	2.65
Corning 7052 Owens-Illinois EN-1	Kovar sealing	3.65
Corning 7570	High lead	0.25
Corning 7740	Borosilicate, low expansion	4.01
Corning 7900	96% silica	3.65
Corning 7940	Fused silica	3.52
F 5795/410	Light flint	3.18
F 689/309	Dense flint	2.18
BaC 5725/574	Light barium crown	2.56
BaC 611/588	Dense barium crown	1.83
BSC 517/645	Borosilicate crown	2.81

Suggested reading: J. F. Nye, *Physical Properties of Crystals*. Oxford Science, Publ., 1987.
T. S. Narasimhamurty, *Photo-elastic & Electro-optic Properties of Crystals*. Plenum Press, 1981.
K. Matusita, C. Ihara, T. Komatsu, and R. Yokota, *J. Am. Ceram. Soc.* **67**(10), 700–714 (1984).

19.2.7. Anomalous Birefringence

Anomalous birefringence refers to the double refraction property observed in phase-separated glasses even in the absence of any internal or external stress. The origin of such double refraction is hardly anomalous, particularly in those glasses that have discrete second-phase particles (separation by a nucleated mechanism). The presence of an anisotropic shape of the particles gives rise to **form birefringence**, and an anisotropic spatial distribution gives rise to **distribution birefringence**. The two types may be observed in stretched specimens of Corning code 7070 glass that have been given a prior phase-separation treatment. The glass is a low-alkali borosilicate glass where discrete borate-rich particles grow in a continuous silica-rich matrix. In Fig. 19-14a, corresponding to the unstretched, phase-separated specimen, the particles are circular and randomly distributed. Birefringence is neither expected nor observed. Stretching at around 10^9–10^8 Pa·s viscosity causes the softer particles to develop anisotropies of both shape and distribution (Fig. 19-14b). Apparently, the form birefringence is strong positive, and the distribution birefringence is weak negative. The specimen therefore shows a net positive birefringence. Upon annealing, the particles spheroidize (Fig. 19.14c), with the result that only the negative distribution birefringence is left. Typical results (12) of the observed birefringence are shown in Fig. 19-15.

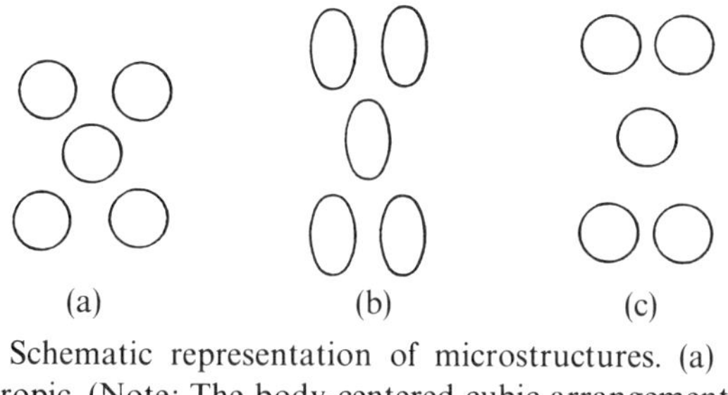

Figure 19-14. Schematic representation of microstructures. (a) The unstretched specimen is isotropic. (Note: The body-centered cubic arrangement shown is merely incidental.) (b) After stretching, the form as well as the distribution become anisotropic. (c) After a post-pulling annealing treatment, the particles spheroidize; however, the distribution remains anisotropic.

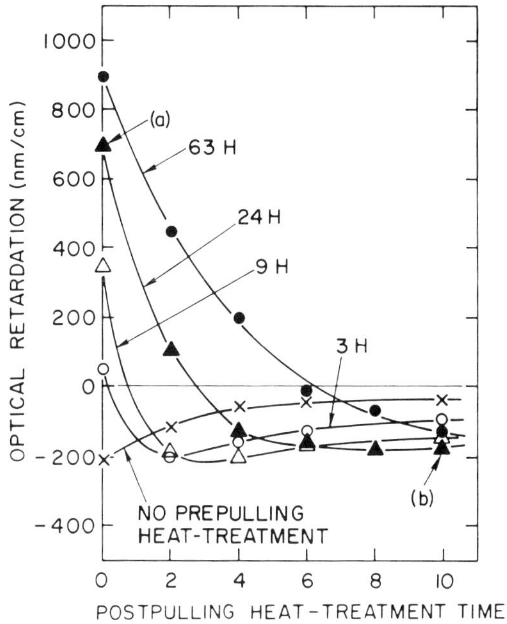

Figure 19-15. Birefringence observed as a function of the post-pulling heat treatment time at 564°C. Times shown on the curves are the pre-pulling heat treatment times at 595°C. (After Takamori and Tomozawa, *J. Am. Ceram. Soc.* **59**, 377 (1976). reproduced with permission of The American Ceramic Society.)

Suggested reading: T. Takamori and M. Tomozawa, *Treatise of Mat. Sci. & Eng.*, Vol. 12, *Glass I* (M. Tomozawa and R. H. Doremus, eds.), pp. 123–155. Academic Press, New York, 1977.

A. K. Varshneya and K. Hosmer, *J. Am. Ceram. Soc.* **72**(11) 2078–2083 (1989).

19.3. Measurement of Optical Properties

Refractive index is generally measured using a commercial refractometer, such as an Abbe or a Pulfrich refractometer. Both the instruments are based upon measuring the critical angle of incidence between the specimen and a standard prism provided with the instruments. In both cases, the required specimen is a plate type with one surface flat polished and an adjoining vertical edge capable of transmitting light. The contact between the polished surface of the specimen and the prism of the instrument is made better by using an oil with a nearly matching refractive index. A split image is made to coincide, and the refractive index is read directly. Refractive index may also be measured using an ellipsometer, where one analyzes the polarization state of the reflected light. Application of the Fresnel equations for the intensities of the perpendicular components of the polarized vibrations yields the refractive index.

An approximate measurement of the refractive index of a transparent glass may be made using the **Becke line technique**. A small chip of the glass having a sharp edge is immersed in a refractive-index oil standard and placed in focus in a transmission microscope. The focal plane is then raised by raising the objective. A bright line contouring the specimen, called the Becke line, moves into the medium of higher refractive index. Successive measurements using different immersion oils of known indices until the movement of the Becke line is hard to distinguish yield the n_d of the glass generally to within ± 0.02.

Absorption/transmission spectra in the near UV, visible, and near IR are measured using a variety of commercial spectrophotometers. One employs a plate, a thin blown film, or a pressed KBr pellet containing fine particles of the specimen, depending upon the absorption characteristics. Specialized equipment are available to measure Raman spectra and mid- to far-IR spectra. In the region where the specimen becomes essentially opaque (UV and far-IR), the refractive index and the absorption coefficient may be calculated using a Kramers–Kronig analysis [13], which involves the integration of the measured reflectivity over a broad range of energies.

19.4. Data on Optical Properties

Suggested reading: I. Fanderlik, *Optical Properties of Glass*. Elsevier, Amsterdam, 1983.
C. R. Bamford, *Colour Generation and Control in Glass*, Elsevier, Amsterdam, 1977.

The electronic polarizabilities of anions in glass can be arranged as

$$F^- < OH^- < Cl^- < O^{2-} < S^{2-} < Se^{2-} < Te^{2-}.$$

As a result, fluoride glasses generally have lower molar refractivities than oxide glasses.

A plot of the refractive index at *d*-line n_d versus the Abbe number ν_d for optical glasses manufactured by Schott is shown in Fig. 19-16. The compositions are classified into "flint" glasses ($\nu_d < 50$–55) having names ending with "F," and "crown" glasses ($\nu_d > 50$–55) having names ending with "K." One may note the general tendency of glasses to fall around a curve going from

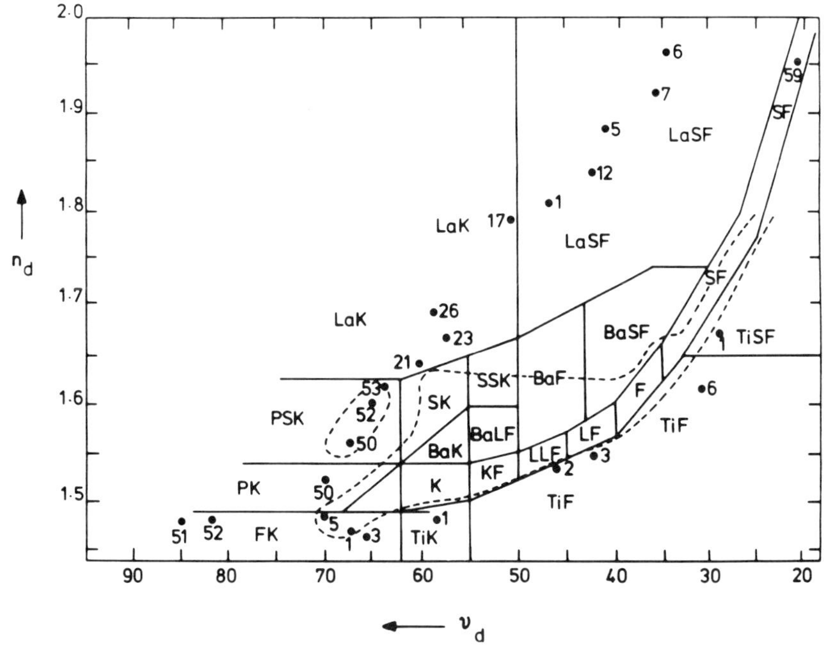

Figure 19-16. Refractive index, n, vs. Abbe number, ν, for several optical glasses. K = crown; F = flint; KF = crown flint; LF = light flint; LLF = extra-light flint; SF = heavy flint; SK = heavy crown; SSK = extra-heavy crown.

lower left to upper right. Glasses at the lower left corner are "fluor-crowns," which are generally borosilicates with high fluorine content and a small amount of Al_2O_3 to inhibit crystallization. P_2O_5 additions also yield glasses in the lower left corner. Glasses at the upper right corner are generally R_2O–PbO–SiO_2 and R_2O–BaO–PbO–SiO_2 glasses. An example of the variation of n_d with PbO in R_2O–PbO–SiO_2 glasses is shown in Fig. 19-17. Some other oxides commonly used to modify the optical properties are La_2O_3 and TiO_2. A description of many of these glasses may be read in the book by Rawson [14] and in a review article by Marker [15].

In order to reduce the chromatic aberrations in lens assemblies, a combination of two lenses with different Abbe numbers is required. The doublet is generally designed to match the focal length at the F and C lines. To obtain a match beyond the F and C lines, a larger number of components with matching of the second derivatives of the refractive index is generally advised. Another criterion often sought in optical component assemblies is the athermal property, i.e., the optical path length W should not change with temperature. Now, the change in the optical path length ΔW can be

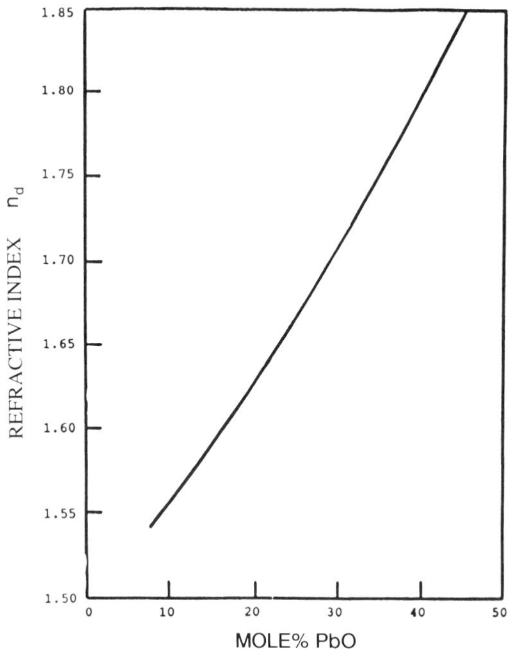

Figure 19-17. Increase of the refractive index with PbO in R_2O–PbO–SiO_2 glasses.

written as

$$\Delta W = s[\alpha(n - 1) + dn/dt] = sG\Delta T, \tag{19.63}$$

where s is the actual distance in glass, α is the linear thermal expansion coefficient, and G may be termed the **thermo-optic** coefficient. It follows that, for $\Delta W = 0$, dn/dt must be negative. Commercially available athermal glasses are also discussed by Marker.

UV absorption edges in glasses shift to higher energies where the electrons participate in stronger chemical bonds. Thus, in B_2O_3, SiO_2, and P_2O_5 glasses, the UV edge moves from 170 to 160 to 145 nm. Presence of nonbridging oxygens reduces the average chemical bond strength, and hence the UV edge shifts to longer wavelengths. In an Na_2O–RO–SiO_2 glass, where R = Ba, Sr, Ca, and Mg, the MgO-containing glasses are the most UV–transmitting.

As seen in Table 19-1, colors produced by the transition metal ions depend upon the oxidation state. Intense colors are often produced by as little as 0.02% colorant. Contamination due to Fe and Cr from the machine tools used in the mineral extractions is thus often a serious problem in the manufacture of glass containers and sheets. Decolorization of glass is carried out either by introducing complementary colors or by changing the valence state that might be "colorless." Thus, the blue-green color due to Fe^{2+} is decolorized by the complementary straw-yellow color due to Fe^{3+} obtained by oxidation. As stated earlier, oxidation in soda lime silica glass melts may be achieved quite effectively by the addition of 0.25 wt. % As_2O_3 + 0.5% $NaNO_3$ to the batch. As_2O_3, however, is toxic and has largely been replaced by Sb_2O_3, cerium hydrate or nitrates alone.

The most technologically important IR absorption in glass is due to water molecules and –OH groups. The IR transmission of a Corning synthetic fused silica containing about 1,200 ppm of OH is shown in Fig. 19-18 (after Sigel [8]). The most intense absorption is at 2.73 μm (3,660 cm^{-1}), with a sharp overtone at 1.38 μm (7,250 cm^{-1}). Not shown is a second overtone at 0.945 μm (10,580 cm^{-1}), of concern to those fiber-optic waveguide systems that employ GaAs light sources. The three bands are identified as arising

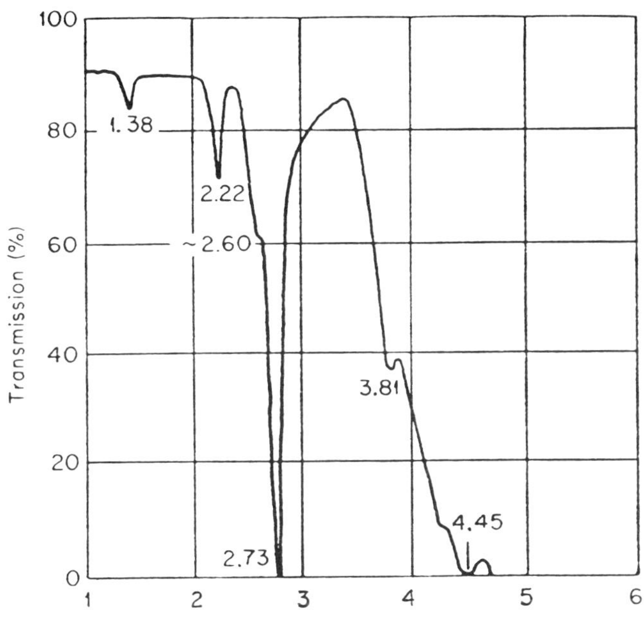

Figure 19-18. Transmission curve of a Corning synthetic fused silica (sample thickness = 5.26 mm) containing about 1,200 ppm OH. The fundamental O–H stretching band lies at 2.73 μm. (After R. W. Adams and R. W. Douglas, *J. Soc. Glass Tech.* **43**, 147T (1959). Reproduced with permission of The Society of Glass Technology.)

from –OH groups. The bands at 2.22 μm and 2.6 μm have been shown to arise from –OH and SiO_4 combination frequencies.

In silica glasses and crystalline forms, one generally observes three primary absorption bands: 1,100 cm^{-1} (9.09 μm), 800 cm^{-1}, and 480 cm^{-1}. The one at 1,100 cm^{-1} is the fundamental Si–O–Si stretching vibration and corresponds to the displacement of the bridging oxygen. The bands at 800 cm^{-1} and 480 cm^{-1} are identified as due to Si–O–Si bond-bending. Upon the addition of alkalis to the structure, another band at around 940–950 cm^{-1} appears and is hence identified as arising from the nonbridging oxygens.

The main IR-absorption peaks in B_2O_3-containing glasses lie at 710, 1260, and 1,420 cm^{-1}. According to Krogh-Moe (16), the 1,420 cm^{-1} peak is due to the ring stretching of the boroxol groups, and the 1,260 cm^{-1} vibration is due to the B–O–B bond constituting the linkage of boroxol groups to neighboring groups (Fig. 19-19). The strong 808 cm^{-1} peak in the Raman spectrum of B_2O_3 glass (Fig. 5-10) is also ascribed to the boroxol groups. The intensities of both the 1,260 cm^{-1} IR absorption and the 808 cm^{-1} Raman line decrease with the addition of alkali, indicating the conversion of BO_3 triangles to BO_4 tetrahedra. The Raman line apparently disappears either by the addition of about 25 mol % alkali, or by bringing the alkali borate glasses into the supercooled liquid range.

For good transmission in the 3–5 μm region of the IR, germanates, calcium aluminates, heavy metal oxides, heavy metal fluorides, chalcogenides, and chalcohalides may be used. Germanates and calcium aluminates transmit out to about 5.5 μm, but are not as durable as the silica-based glasses. Some of the heavy metal oxide glasses involving the use of As_2O_3, Sb_2O_3, Ga_2O_3, Bi_2O_3, Ru_2O_3, MoO_3, WO_3, TeO_2, CdO, and PbO transmit well up to 8 μm; however, most of these glasses have poor chemical and physical properties. The heavy metal fluorides, chalcogenides, and chalcohalides were introduced in Chapter 1. The **Hruby criterion for glass formability** ($= T_x - T_g$,

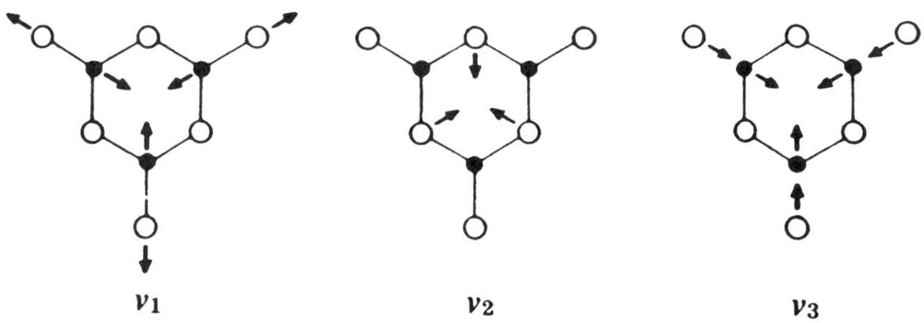

Figure 19-19. The three symmetrical vibrations of the boroxol group. (After J. Krogh-Moe, *Phys. Chem. Glasses* **6**, 46 (1965). Reproduced with permission of The Society of Glass Technology.)

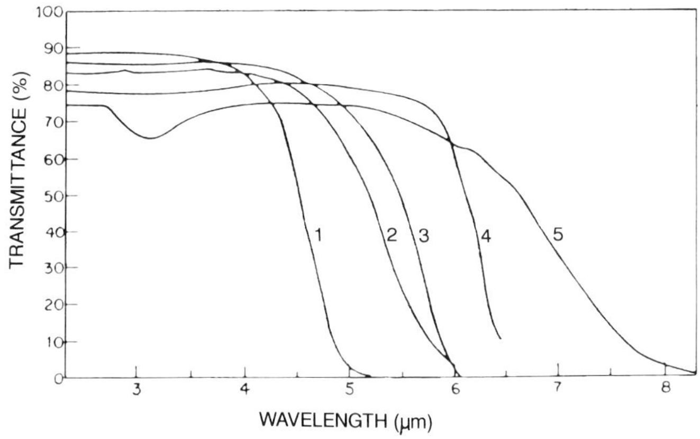

(a)

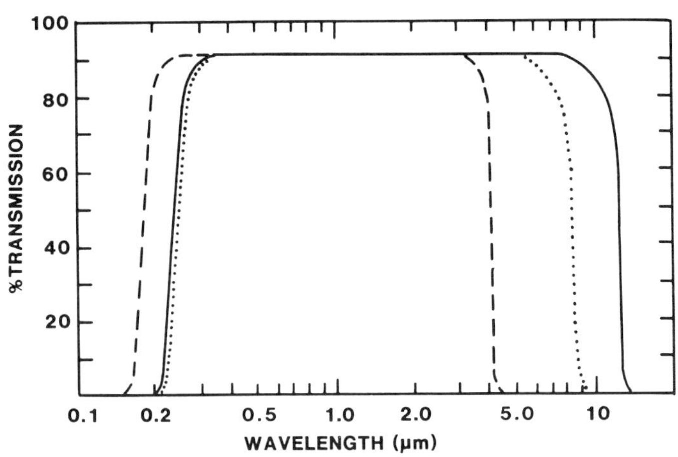

(b)

Figure 19-20. (a) Infrared transmission of some oxide glasses. 1 = 2 mm thick silicate glass; 2 = 2 mm thick germanate glass; 3 = 2 mm thick calcium aluminate glass; 4 = 1 mm thick tellurite glass; 5 = 2 mm thick $PbO–Ga_2O_3–Bi_2O_2$ glass (Corning code EO). (After W. Dumbaugh, *Proc. SPIE* **505**, 101 (1984). Reproduced with permission of the International Society for Optical Engineering.) (b) Infrared transmission of heavy metal fluorides: $57HfF_4·36BaF_2·3LaF_3·4AlF_3$ (——) and $19BaF_2·27ZnF_2·27LuF_3·27ThF_4$ (·····), compared to that of fused silica (-----), each ~5 mm thick. (After M. Drexhage in *Treatise on Mat. Sci. & Tech.*, Vol. 26, *Glass IV* (M. Tomozawa and R. H. Doremus, eds.), pp. 151–245. Academic Press, New York (1985).)

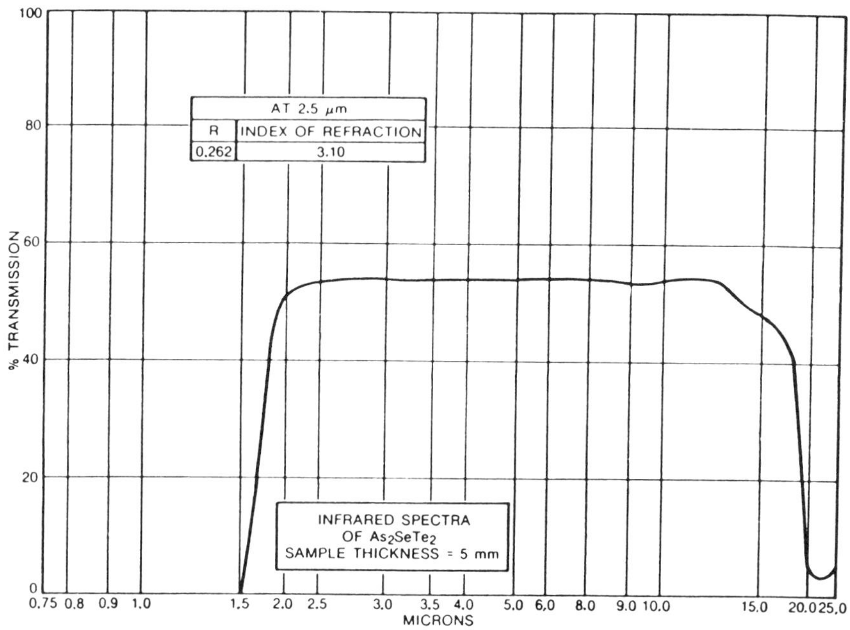

(c)

Figure 19-20. (c) Infrared transmission of As_2SeTe_2 chalcogenide glass (thickness = 5 mm). (After E. Hartouni, F. Hulderman, and T. Guiton, *Proc. SPIE* **505**, 131–140 (1984). Reproduced with permission of the International Society for Optical Engineering.)

where T_x is the crystallization temperature) of most fluorides is less than 100°C, placing them in the "poor" category. In addition, their chemical durability is also quite poor. The chalcogenides and chalcohalides require careful melting procedures involving almost total exclusion of oxygen and water vapor. Relative to the silica-based glasses, their T_g and strengths are also quite low. For good transmission in the 8–12 μm region, the choice is almost exclusively limited to the chalcogenides and chalcohalides. Typical IR transmission spectra of IR-transmitting glasses are shown in Fig. 19-20.

CRT and television faceplate glass often contains significant amounts of BaO/SrO to absorb the x-rays emitted by electron bombardment on the glass. The funnel glass for the television tubes is generally a R_2O–BaO/SrO–PbO–SiO_2 glass. For shielding windows in a nuclear environment, the use of glasses containing a large amount of the highly absorbing PbO is advisable. As it happens, these glasses discolor quite readily by α- and β-radiation. Addition of as little as 0.5% CeO_2 "stabilizes" the PbO glasses such that the modified glass only needs to be half as thick as CeO_2-free glass.

19.5. Special Applications

19.5.1. Photosensitive Glasses

Addition of ~0.5% Ag halides along with a small amount of Cu^+ ions in a suitable host glass causes it to display photochromic properties. During heat treatment the glass phase-separates such that Ag halide particles, roughly 10–30 nm size, are present supersaturated in one phase. Exposure to daylight (mostly higher frequencies) causes the dissociation of Ag^+ and halide, as in the normal photographic process. The two ions are displaced slightly. The Ag^+ is reduced to Ag^0 by the Cu^+, which in turn is oxidized to Cu^{2+}. The metallic Ag causes the darkening of the glass. Upon removal of the light radiation, a recombination between the Ag and halide ions takes place to restore the transmission. In some compositions, the dissociation requires extended exposure to UV. The separated particles then provide suitable nucleation sites for crystallization of Cu by a subsequent heat treatment. An image produced in this manner is not readily reversible.

Suggested reading: R. A. Araujo, *Treatise on Mat. Sci. & Tech.*, Vol. 12, *Glass I* (M. Tomozawa and R. H. Doremus, eds.), pp. 91–122. Academic Press, New York, 1977.

19.5.2. Nonlinear Optics

When a beam of photons travels through a medium, it has been observed in some cases that the frequency-dependent refractive index depends upon the intensity I of the beam inside the medium as

$$n(\omega) = n_0 + \gamma I(\omega) = n_0 + n_2\langle E^2(\omega)\rangle, \tag{19.64}$$

where E is the rms optical electric field strength inside the medium. γ is called the nonlinear index coefficient, and n_2 is called the nonlinear refractive index. In either case, the second term represents nonlinearity of the total refractive index n. It may be negligible at low intensities, but may be significant at high intensities. The **nonlinear refractive index** n_2 may be negative or positive. Negative n_2 is good for bistable optical devices; zero or low n_2 is needed for good high-power laser hosts; and high positive n_2 is needed for optical memories, modulators, and switches. The nonlinear behavior is important in optical-fiber communications technology, since the information-carrying capacity is limited mostly by the electronics used for switching, routing, and signal processing. The best of the modern electronics operate over the picosecond (10^{-12} s) time domain of switching, whereas optical devices can perform the same task over femtoseconds (10^{-15} s). Directional couplers have now been built where the optical coupling from one waveguide to the other is achieved by modulating the intensity of the input beam or having separate control beams. The units of γ are $m^2 \cdot W^{-1}$ (or cm^2/W), and those of n_2 are cm^3/erg (esu) and $m^2 \cdot V^{-2}$ (SI). Conversion between γ and n_2 may be made via

$$n_2(cm^3/erg) = [cn_0/40\pi]\gamma(m^2 \cdot W^{-1}) = 2.387 \times 10^6 n_0 \gamma(m^2 \cdot W^{-1}), \quad (19.65)$$

where c is the velocity of light ($m \cdot s^{-1}$) in vacuum. Also,

$$n_2(cm^3/erg) = 7.162 \times 10^7\, n_2(m^2 \cdot V^{-2}). \quad (19.66)$$

(The reader is cautioned against the unfortunate mix-up of notation and esu/SI units in the literature.)

The origin of the nonlinear refractive index lies in the nonlinear electronic polarization. When a beam of monochromatic waves of angular frequencies $\omega_1, \omega_2\ \omega_3, \ldots, \omega_r$ is incident upon the medium, the ith Cartesian component $\mathbf{P}_i(r)$ of the induced polarization at the frequency $\omega_\sigma = \omega_1 + \omega_2 + \omega_3 + \cdots + \omega_r$ due to the rth-order nonlinear dielectric susceptibility $\chi_{ijkL\ldots r}(r)$ is given by

$$\begin{aligned} \mathbf{P}_i^{(r)}(\omega_\sigma) &= \varepsilon_0 \sum_{jkL\ldots r} g(-\omega_\sigma)\chi_{ijkL\ldots r}{r^{(r)}}(-\omega_\sigma;\ \omega_1, \omega_2, \omega_3 \ldots \omega_r) \\ &\quad \times [\mathbf{E}_j(\omega_1)\mathbf{E}_k(\omega_2)\mathbf{E}_L(\omega_1) \ldots \mathbf{E}_r(\omega_r)]. \end{aligned} \quad (19.67)$$

Here $\mathbf{E}_j(\omega_1)$, $\mathbf{E}_k(\omega_2)$, $\mathbf{E}_1(\omega_3) \ldots$ are separately applied electric fields having their own frequencies $\omega_1, \omega_2, \omega_3, \ldots,$ and $g(-\omega_\sigma)$ is a degeneracy factor arising from the number of distinguishable permutations of the frequencies. The term corresponding to $r = 1$ on the right-hand side is the linear term as in Eq. (15.7) and is the basis of all classical optics. The second term describes second-harmonic generation (SHG) and the electro-optic effect. In the SHG, an optical signal is generated at twice the frequency of the incident frequency ($\omega_1 = \omega_2$). In the electro-optic effect, the optical signal is changed by the application of a dc field. Here the generated signal has the same frequency, i.e., $\omega_2 = 0$. The second term is nonzero only for acentric systems, i.e., crystals that lack a center of symmetry. The third term contains the dielectric susceptibility χ, which, in general, is a fourth-rank tensor possessing 81 possible combinations of three electric field and one polarization vectors, each taking values 1, 2, and 3 corresponding to the three possible spatial directions. This is the term that describes the nonlinearity of the refractive index. For glasses possessing isotropicity, there are only three nonzero components $\chi^{(3)}_{1111}$, $\chi^{(3)}_{1122}$, and $\chi^{(3)}_{1221}$, of which only two are independent because of the isotropy relation

$$\chi^{(3)}_{1111} = 2\chi^{(3)}_{1122} + \chi^{(3)}_{1221}. \tag{19.68}$$

For a linearly polarized beam in an isotropic medium such as glass with $\omega_1 = \omega_2 = \omega_3$, the degeneracy factor $g = \frac{1}{4}$, and it may be shown that

$$n_2 = [12\pi/n_0][\chi^{(3)}_{1111}(-3\omega; \omega, \omega, \omega)] \qquad \text{(in esu)}$$

or

$$n_2 = [3/2n_0][\chi^{(3)}_{1111}(-3\omega; \omega, \omega, \omega)] \qquad \text{(in SI)}. \tag{19.69}$$

Away from absorption edges, the real part of the complex susceptibility may be used without significant error. $\chi(3)$ may be estimated from $\chi(1)$ by the application of the generalized **Miller's rule** [17]:

$$\chi^{(3)}_{ijkL} = M[\chi^{(1)}_{ii}\chi^{(1)}_{ij}\chi^{(1)}_{kk}\chi^{(1)}_{LL}], \tag{19.70}$$

where M is roughly a constant independent of frequencies and roughly equal to 10^{-10} esu. This relation connects the third-order susceptibilities to the first-order "normally measured" susceptibilities. Hence,

$$\chi^{(3)}_{1111} = M[\chi^{(1)}_{11}]^4. \tag{19.71}$$

Near resonance (see the earlier section on dispersion), the imaginary part of the susceptibility becomes important. One may argue that a part of the light beam will be absorbed near the resonance, resulting in the redistribution of electron energy levels depending upon the intensity of the beam. The absorption frequency would then shift toward the frequency of the incident beam. Since the frequency shifts, the refractive index also changes, giving rise to the nonlinearity.

Another method of estimating n_2 or γ is to use the approximation by Boling *et al.* [18], who found that in the long-wavelength region, γ is given approximately by

$$\gamma = \frac{K'[n_d - 1][n_d^2 + 2]^2}{n_d v_d \left\{1.52 + \dfrac{[n_d + 1][n_d^2 + 2]v_d}{6n_d}\right\}^{1/2}} \tag{19.72}$$

where $K' = 2.8 \times 10^{-18}\ \text{m}^2\cdot\text{W}^{-1}$, and v_d is the Abbe number as defined in Eq. (19.9). From Eq. (19.72), it may be seen that n_2 is large for large n_d and low v_d. Chalcogenides, chalcohalides, and heavy-metal rare-earth oxide glasses are thus likely to give high n_2. Table 19-3 lists [19] the estimated n_2 at 588 and 350 nm for some glasses of interest.

As indicated in Section 19.4, the oxide glasses in general have higher

Table 19-3. Nonlinear Refractive Index Coefficient n for Some Glasses

Glass	Calculate n_2 ($\times 10^{-11}$ esu) at 588 nm	at 350 nm
Fused silica	0.02	0.02
EDF 68939	0.11	0.14
$72PbO–28SiO_2$ (mol %)	1.4	—
$90Ta_2O_5$–10CaO (wt. %)	1.4	1.6
$80PbO–20TeO_2$ (mol %)	1.6	—
$90Nb_2O_5$–10CaO (wt. %)	1.7	—
$83Bi_2O_3–7.2Ga_2O_3$–9.8CdO (wt. %)	3.6	—
$42Bi_2O_3–12Ga_2O_3$–46PbO (wt. %)	4.0	—
$66PbO–17Tl_2O–17SiO_2$ (mol %)[a]	9.0	—
$66PbO–17Tl_2O–17TeO_2$ (mol %)[a]	13	—

[a] Designed composition for high n_2.

electronic polarizabilities. It is also known that the nonbridging oxygen is more polarizable than the bridging oxygen. Thus, vitreous silica has a low $\gamma = 3.2 \times 10^{-20}$ m^2/W at $\lambda = 1{,}060$ nm, whereas some glasses, such as those in the $PbO–Bi_2O_3–Ga_2O_3$ system, are reported to have $\gamma = 12 \times 10^{-19}$ m^2/W, attributed to the presence of polarizable heavy-metal cations. In $TiO_2–Nb_2O_3–Na_2O–SiO_2$ glasses, the high γ (values ranging between 0.5 and 4×10^{-19} m^2/W) is related to the concentration of the transition metal ions and not to the nonbridging oxygens.

Three interesting approaches for obtaining high n_2 are being developed: glass-ceramics, organic-impregnated porous glasses, and organics-clathrated gel-produced glasses. In the first approach, CdS_xSe_{1-x} crystallites are precipitated in borosilicate glasses that have sharp cutoffs making the glasses look yellow ($x > 0.8$) to red ($x < 0.2$). As indicated earlier, the cutoffs arise from plasma resonance and Mie scattering. Near the resonance frequencies, the measured n_2 are reported to be 2–3 orders of magnitude higher than in several pure glass candidates, and significantly higher than bulk crystals of CdS–CdSe. The other two approaches utilize the high-n_2 characteristics of some organics, in particular quinizarin. Thus, porous glasses (obtained by leaching of spinodally separated glass) may be impregnated with quinizarin solution in an organic solvent; the resulting composite shows high n_2. Likewise, the starting organometallic sols, e.g., tetramethyl orthosilicate, may be mixed with quinizarin, hydrolyzed, and gelled. Apparently, quinizarin remains unchanged after gelation, and the resulting gel shows strong n_2.

Suggested reading: S. Singh, in *Handbook of Laser Sci. & Tech.*, (M. Weber, ed.), Vol. III, Part 1, p. 3. CRC Press, Boca Raton, Florida, 1986.
W. L. Smith, *ibid.*, p. 259.
M. J. Weber, D. Milam, and W. L. Smith, *Opt. Eng.* **17**(5), 463–469 (1978).
E. M. Vogel, M. J. Weber, and D. M. Krol, *Phys. Chem. Glasses* **32**(6), 231–254 (1991).
N. F. Borelli and D. W. Hall, in *Optical Properties of Glass* (D. R. Uhlmann and N. J. Kreidl, eds.), pp. 87–124. Am. Ceram. Soc., Columbus, Ohio, 1991.

19.5.3. Glass as a Laser Host

The underlying principle of a laser (light amplification by the stimulated emission) is, in fact, quite simple. An atom in a medium may be excited to a higher energy state E_1 by absorption of a photon of energy $h\nu$. Subsequently, the atom may give up the energy as a spontaneous emission to fall back to the ground state E_0. The released photon is emitted in a random direction having no correlation with regard to the direction or the phase of the initially incident photon. In stimulated emission, an external radiation comprising photons possessing resonant energy $h\nu = E_1 - E_0$ passes by an atom in the excited state E_1 and stimulates the atom to release the photon in phase in the same direction as the incident beam — in other words, the emission is coherent. If there is a higher population of atoms in the excited state than in the ground state, and if the photon losses do not exceed the gains, then the incident beam is effectively "amplified." Often what works better is a three energy-level system (or a four energy-level system) where the population is inverted between the ground level E_0 and the excited state E_2 such that a radiationless spontaneous transition occurs between E_2 and an intermediate level E_1, and a stimulated emission occurs between E_1 and E_0 after spending an average fluorescence lifetime τ at E_1 (see Fig. 19-21).

Lasers comprise three basic elements: an external energy source, called the pump, to carry out the population inversion; a host medium; and an optical cavity called the resonator. The host medium is the topic that concerns us. The reader is referred to Pedrotti and Pedrotti [20] for more details on laser basics.

The most important lasing ions are Nd^{3+}, Er^{3+} and Yb^{3+}, and Ho^{3+}. These "activator" ions are incorporated in various host media. Laser host media can be gas, liquid, or solid. Some of the requirements for a good-quality laser host are a high absorption of the pump radiation, high induced emission cross-section σ, high number N of the active lasing ions, and a long fluorescence lifetime τ. For solid hosts, high thermal shock resistance and a

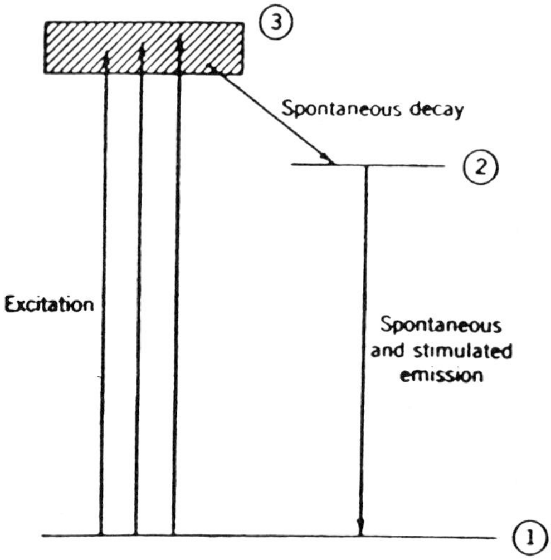

Figure 19-21. Simplified energy level diagram for a three-level lasing system. (After W. D. Kingery, H. K. Bowen, and D. R. Uhlmann, *Introduction to Ceramics*, Fig. 13.26, p. 693. J. Wiley & Sons, New York, 1976. Reproduced with permission of the publishers.)

Table 19-4. Properties of Glasses Suitable for Laser Applications[a]

Glass type	Fluorescence		
	Lifetime τ (10^{-6} s)	Maximum wavelength (nm)	Fluorescence width at half maximum (nm)
Silicate	300–1,000	1,060	30–40
Borate	50–400	1,055	30–40
Borosilicate	200–800	1,058	30
Phosphate	100–500	1,054	20–30
Tellurite	140–240	1,060	25–30
Fluorosilicoborate	300	1,060	30
Fluorophosphate	350–500	1,052	25–31
Fluoroberyllate	460–900	1,048	18–28
Fluorozirconate	430–450	1,049	26
Fluoroaluminate	400–600	1,050	30–33
Chloride	180–220	1,063	20
Chlorophosphate	290–300	1,055	20–23
Sulfide	60–100	1,076	21

[a] After N. Neuroth, *Opt. Eng.* **26**(2), 96 (1987).

low nonlinear refractive index coefficient n_2 are additional important requirements. Table 19-4 shows [21] the effect of matrix in glasses that have 1.5–3 wt. % of Nd^{3+}. Note that the chloride and sulfide glasses have too high an n_2. Fluorophosphates have low n_2 and smaller fluorescence linewidths, but they are expensive to fabricate. Phosphates have moderate n_2 and low linewidths. Silicates, in general, have large τ (as much as 1,000 μs), a moderate n_2 (about 1.5×10^{-13} esu), but large linewidths ($\sim$35 nm).

It is possible to dope the glass with a second ion that would absorb the pump radiation not absorbed by the primary activator, and then transfer the energy to the primary activator. Chromium, for instance, absorbs spectral energy of 450 and 650 nm, which is not absorbed by Nd^{3+}. The Cr^{3+} ions subsequently transfer their energy to the Nd^{3+} with an efficiency of 45%. Likewise, efficient transfer of energy from Nd^{3+} to Yb^{3+} and from Yb^{3+} to Er^{3+} has been observed. The erbium laser with 1.54 μm laser light is important for performing surgery on the eye lens without damaging the retina. The presence of OH^- ions, on the other hand, has a deleterious "quenching" effect: it reduces the fluorescence lifetime. Likewise, other transition metal ions such as Fe^{2+} and Cu^{2+} tend to absorb the lasing wavelength. Thus, for efficient operation of a Nd^{3+} laser, care should be taken during glass melting so as to avoid OH^-, Fe^{2+}, and Cu^{2+} impurities.

19.5.4. Fiber Optics

A beam of light launched at one end of a transparent glass fiber exits at the other end primarily because of a large number of total internal reflections within the fiber. The channeling of light through such a conduit, the **optical waveguide (OWG),** obviously has tremendous significance in terms of telecommunications, computers, optical sensors, and laser-assisted surgery. Long-distance communication, in particular, requires multiplexing, i.e., the simultaneous transmission of several bits of information along the same path. The range of frequencies covered by the ordinary human voice is roughly 200–6,000 Hz. A coaxial cable may carry this conversation amidst carrier frequencies of 1.0002–1.006 MHz. However, a He–Ne laser communication system ($\lambda = 632.8$ nm or 4.738×10^{14} Hz) may carry the same through an optical pipeline at frequencies 473,800,000,000,200 Hz and 473,800,000,006,000 Hz. Thus, the optical communication system occupies only 1.5×10^{-9}% of the available carrier frequency regime, whereas roughly 0.6% of the coaxial channels are occupied. The 10^9 factor in the carrying capacity in favor of the optical communication system speaks for itself.

The propagation of a ray of light along a glass fiber of refractive index n_1 (core) with an encasing medium (clad) of refractive index n_2, and with its ends facing refractive index n_0, is shown in Fig. 19-22. Ray A strikes the

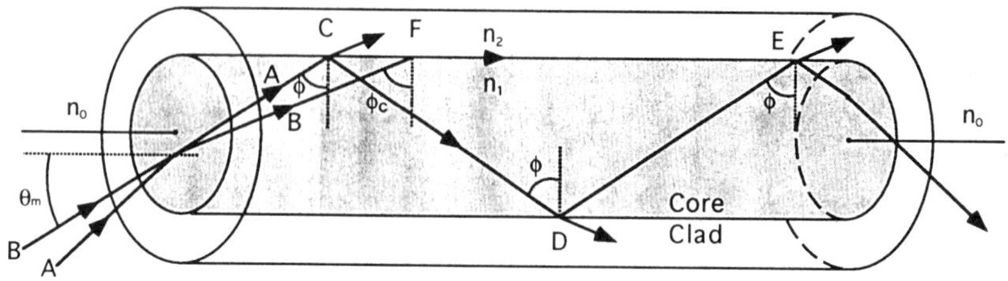

Figure 19-22. Propagation of light rays through an optical fiber. After entering the fiber, a typical ray A undergoes several total internal reflections until exiting. Ray B defines the maximum cone of rays that suffer internal reflections. (After F. L. Pedrotti and L. S. Pedrotti, *Introduction to Optics*, Fig. 10-1, p. 215. Prentice-Hall, 1987.)

inner wall at point C and is partly transmitted into $n_2(n_1 > n_2)$ and partly reflected back into n_1. Ray B, on the other hand, strikes at point F at the Brewster's angle ϕ_c. The angle of incidence of ray B defines a cone of light entering the fiber that is total internally reflected without loss of energy. It can also be shown that within the cone, there are only a discrete number of angles greater than the critical angle that yield a propagating ray. These angles are given by

$$\theta_m = \cos^{-1}[4rn_1/m\lambda], \tag{19.73}$$

where r is the core radius and m is an integer. Each of the m values can be called a **mode**; the total number of modes in a fiber is

$$m \cong 2\pi^2 r^2 [n_1^2 - n_2^2]/\lambda^2. \tag{19.74}$$

The numerical aperture (NA) is defined by

$$\mathrm{NA} = [n_1^2 - n_2^2]^{1/2}. \tag{19.75}$$

The loss of energy as the light rays travel down the length of the fiber can be described in terms of various mechanisms. The total absorption coefficient α_1 may be written as

$$\alpha_1 = A_1/\lambda^4 + B_1 \exp(B_2/\lambda) + C_1 \exp(-C_2/\lambda), \tag{19.76}$$

where the first term is the loss due to Rayleigh scattering arising out of microscopic density and composition fluctuations, the second term is the Urbach edge (electronic absorptions, see Eq. (19.43)), and the third term is the infrared edge (multiphonon absorption, see Eq. (19.45)). The loss is measured in terms of decibels (dB), defined by

$$\mathrm{loss(dB)} = -10 \log_{10}(I/I_0) = -4.343 \ln(I/I_0). \tag{19.77}$$

Thus, a 1 dB loss implies roughly 80% transmission, and a 10 dB loss implies

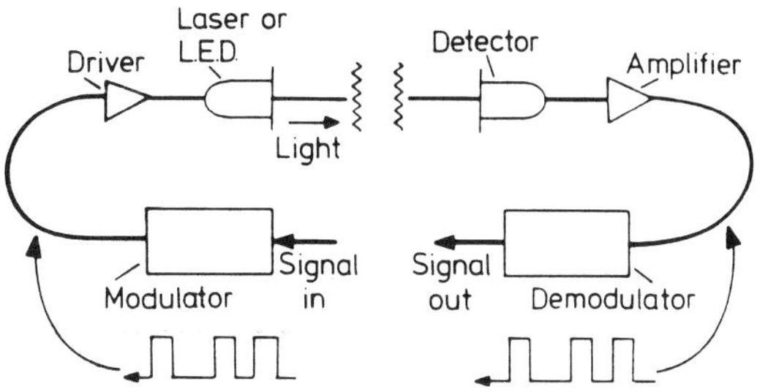

Figure 19-23. Fiber-optic communication system. (After H. Rawson, *Properties and Applications of Glass*, Fig. 103, p. 187. Elsevier, Amsterdam, 1980. Reproduced with permission of the publishers.)

only a 10% transmission. Also, the application of Beer–Lambert's law gives

$$1 \text{ cm}^{-1} \text{ absorption coefficient} = 4.343 \times 10^5 \text{ dB/km}. \qquad (19.78)$$

If the losses during the signal transmission are significant, then for a telecommunication application, a repeater station to boost the signal is required. The schematics of a typical telecommunication system using a laser light source are shown in Fig. 19-23. For repeaterless transcontinental or transatlantic phone lines, one needs to have a total loss no more than 10 dB over the 5,000 km, i.e., <0.002 dB/km. Because of its high transparency over a wide range of wavelengths, silica glass is the most favored material for application in telecommunication optical waveguides.

Figure 19-24 shows the optical attenuation diagram for silica glass. The useful wavelength region for telecommunication is about 600 to 2,500 nm. The common sources of attenuation loss in silica-based fibers are the –OH and the transition metal ion impurities. Figure 19-25 (after Ref. 22, p. 291) shows the variation of the loss per kilometer per ppm –OH impurity as a function of the operating wavelength width. Table 19-5 (after Ref. 22, p. 290) shows the calculated attenuation at 800 nm due to the transition metal ion impurities. The manufacturing technology to produce low-loss fibers has indeed faced formidable challenges over the past three decades. Much of the silica glass or commercial waveguides is made by the vapor deposition techniques discussed in Section 20.4.4. Fibers with losses below 0.1 dB/km have now been reported.

In addition to the attenuation losses, the performance of OWG can be impaired greatly by pulse broadening. A narrow square pulse, for instance, traveling along an OWG will emerge broadened (Fig. 19-26a). The broadening in this case occurred because a large number of light rays (modes) could enter the fiber at a variety of angles such that the total distance traveled by each mode is a broad distribution. The low-order modes (traveling closer

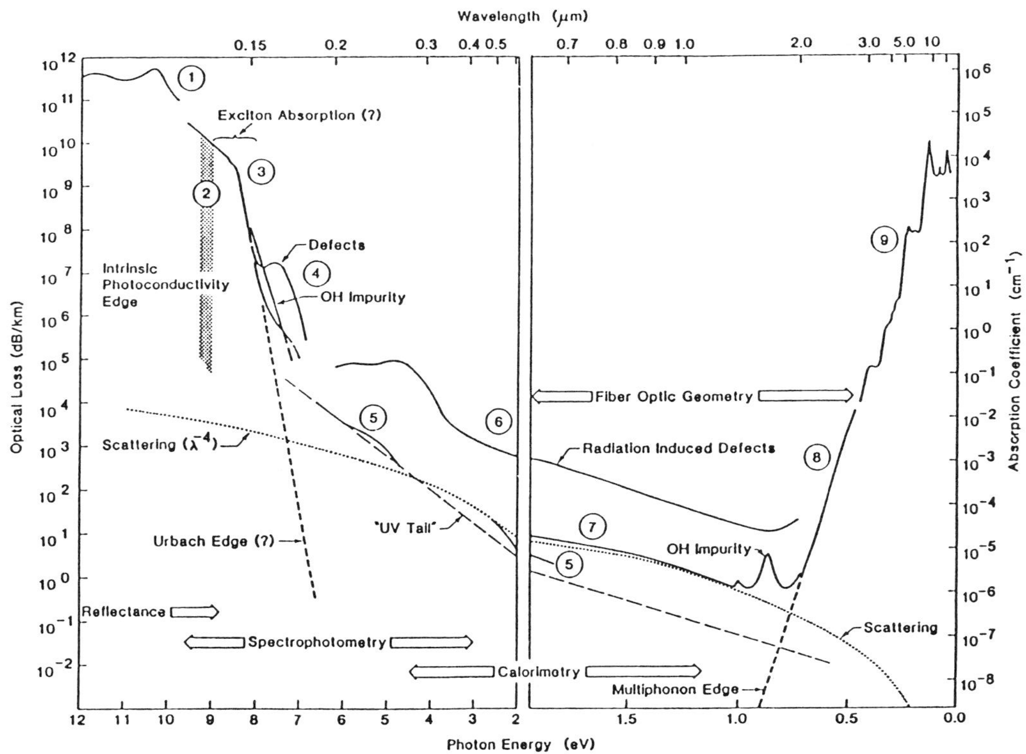

Figure 19-24. Optical attenuation spectra of fused silica [[1]–[6], [9]] and germanium-doped silica fibers [[7], [8]]. (After P. C. Schultz, private communication.)

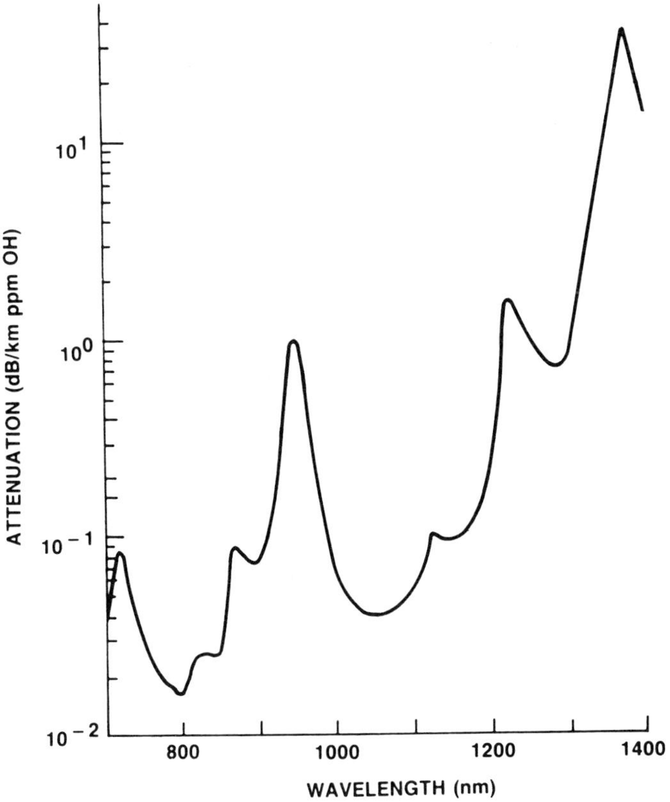

Figure 19-25. Absorption due to hydroxyl impurities in silica-based glasses. (After M. K. Barnoski and S. D. Personick, *Proc. IEEE* **66**, 429 (1978). Reproduced with the permission of the Institute of Electrical and Electronics Engineers.)

Table 19-5. Calculated Attenuation (dB/km/ppm) due to Transition Metal Impurities in Glass Fibers[a]

	Soda–lime–silica	Sodium borosilicate	Fused silica
V	—	39	1,050
Cr	20	50	610
Mn	80	11	20
Fe	100	10	50
Co	—	9	20
Ni	150	130	30
Cu	620	500	10

[a] After Ref. 2.

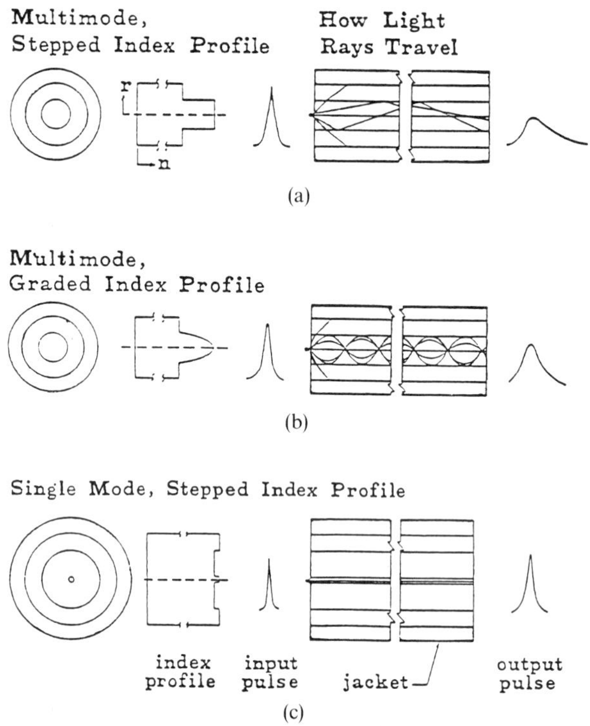

Figure 19-26. Distinctions among (a) single-mode with stepped refractive index, (b) multimode with stepped refractive index, and (c) multimode with graded refractive index optical fibers.

to the fiber axis) arrive faster than the higher modes (traveling zigzag paths at high angles). This **multimode fiber** (Fig. 19-26a), typically 60 to 150 μm in diameter, does, however, allow the transmission of a fairly intense beam. The modal dispersion could be reduced by a **single-mode fiber**—typically <10 μm diameter, with a core-clad refractive index difference reduced to 0.001 from 0.014 for a multimode fiber—which only carries the lower modes. Such a fiber, obviously, can only carry a low-intensity signal (Fig. 19-26c). Both the fiber types have core-clad configurations with a stepped refractive index profile. One of the more advanced configurations is the **parabolic gradient refractive index (GRIN) profile multimode fiber** (Fig. 19-26b), where gradual convergence reduces the delay times of the higher-order modes. GRIN profiles are obtained by mixing vapors of various dopants (in the halide forms) such as Ge, B, and P. The index change Δn per mol % dopant relative to silica is 0.0015 for GeO_2, -0.0005 for B_2O_3, and about 0.0008 for P_2O_5. GRIN profiles may also be obtained by ion exchange and by sol–gel techniques. The former may be accomplished by immersion of porous glass preform in an alkali salt solution (in one or more steps [23]), or by immersion in molten alkali salt baths.

Pulse broadening may also originate from the dispersion characteristics of glass as discussed earlier, and from the waveguide characteristics such as **microbending** introduced by out-of-roundness, local stress gradients, etc. One may readily imagine that an externally applied stress or temperature would cause OWG microbending that could be detected quantitatively and subsequently used as a sensor.

Suggested reading: R. M. Klein, *Glass Sci. & Tech.*, Vol. 2 (D. R. Uhlmann and N. J. Kreidl, eds.), pp. 285–340. Academic Press, New York, 1984.
K. J. Beales and C. R. Day, *Phys. Chem. Glasses* **21**(1), 5–21 (1980).
R. Olshansky, *Rev. Mod. Phys.* **51**(2), 341–367 (1979).
J. Wilson and J. F. B. Hawkes, *Optoelectronics.* Prentice-Hall, 1989.
M. A. Newhouse, in *Optical Properties of Glass* (D. R. Uhlmann and N. J. Kreidl, eds.), pp. 185–204. Am. Ceram. Soc., Columbus, Ohio, 1991.

19.5.5. Coatings on Glass to Improve Optical Quality

The most important uses of coatings on glass are in the area of solar radiation control and antireflection. During summertime, window glass should prevent solar heat from entering the building and necessitating expensive air conditioning. During the winter season, heat should be prevented from escaping through the window. Antireflection coatings increase transmission and are obviously important in optical systems using lenses and filters.

For antireflection, there are two primary conditions to be met:

(i) The refractive index n_1 of the coating is a geometric mean of n_g and n_0, which are the indices of the glass substrate and the air, respectively, i.e.,

$$n_1 = [n_0 n_g]^{1/2}. \tag{19.79}$$

(ii) The thickness t_1 of the coating should be such that optical interference occurs between the waves reflected from the top surface and the glass/coating interface. This requires that

$$n_1 t_1 = \lambda/4,\ 3\lambda/4,\ 5\lambda/4 \ldots. \tag{19.80}$$

A single $\lambda/4$ coating of MgF_2 ($n = 1.38$–1.40) provides effective blooming on a light crown glass (the reflected light is mostly blue, the rest is transmitted). Baking the film after the evaporation imparts good abrasion resistance to the coating. LiF ($n = 1.30$–1.31) and AlF_3 ($n = 1.38$–1.39) have also been found suitable. For an N-layer system with layers of thicknesses $1, 2, 3, \ldots, N \times \lambda_0/4$, the reflectance curve is symmetrical about λ_0 and possesses N zero-reflection points. The refractive indices should alternate between high and low. Thus, a light flint glass ($n = 1.62$) may be bloomed by a three-layer system comprising MgF_2 ($n = 1.38$), ZnS ($n = 2.30$), and Al_2O_3 ($n = 1.76$).

For applications requiring increased reflectivity (e.g., for mirroring and IR reflection), single-layer dielectric films have been found to be quite effective. Some of the useful materials are shown in Table 19-6. Again, the use of high, low, high, low, ... $\lambda/4$ odd-numbered stacks can increase the reflectance to near 1 over a broad range of the spectrum. As an example, an 11-layer stack of Sb_2S_3 ($n = 2.70$) and CaF_2 ($n = 1.28$) has a reflectance of 0.999 at $\lambda = 1$ µm.

For the control of solar radiation, the coatings are designed to reflect or absorb a part of the radiation (300–2,500 nm). As discussed earlier, reflection and absorption of certain wavelengths is controlled by controlling the complex refractive index over desired wavelength regions. Interference effects produced by thin films are utilized additionally to provide good reflection/transmission characteristics with good desired color appearance. Thin coatings (10–20 nm) of metals such as Ag, Au, Cu, alloys such as inconel and stainless steel, and ceramic dielectrics such as TiO_2, Ti_3N_4, CrN, and Zn_2SnO_4 are used in commercial processes. As an example, Table 19-7 shows the appearance of Ag and Au layers in varying thicknesses.

Glass surfaces can be coated using one of many processes: vacuum evaporation/sputtering, solution spray of halides or acetyl acetonates followed by hydrolysis or pyrolysis, chemical vapor deposition (CVD) with concurrent hydrolysis/pyrolysis/reduction, mirroring (wet reduction), and dip-coating using the sol–gel route.

Table 19-6. Materials for Dielectric Films. Reflectance of Surface of Index 1.50 When Covered by Various $\lambda/4$ Films[a]

Film material	Refractive index	Wavelength	Reflectance of single film
ZnS	2.30	5,461 Å	0.31
TiO_2	2.6	5,461 Å	0.40
Sb_2S_3	2.7	1 µm	0.43
Ge	4.0	2 µm	0.69
Te	5.0	4 µm	0.79

[a] After Ref. 24. Table 7-3, p. 217. Reproduced with permission of Butterworth-Heinemann Ltd., Oxford UK.

Table 19-7. The Appearance of Au and Ag Films[a]

Light yellow 10 Å	Ruddy purple 15 Å
Golden yellow 21 Å	Indigo 20 Å
Orange yellow 32 Å	Blue 27 Å
Ruddy orange 43 Å	Green 32 Å
Crimson 52 Å	Yellow-green 40 Å
Indigo 60 Å	(Golden yellow > 40 Å)
(colors disappear at 70 Å)	

[a] After Ref. 24. Table 6-2, p. 169. Reproduced with permission of Butterworth-Heinemann Ltd., Oxford UK.

Suggested reading: O. S. Heavens, *Optical Properties of Thin Solid Films*. Academic Press, New York, 1955.
H. K. Pulker, *Coatings on Glass*. Elsevier, Amsterdam, 1984.
H. Dislich, in *Glass Sci. & Tech.* (D. R. Uhlmann and N. J. Kreidl, eds.), Vol. 2, pp. 251–283. Academic Press, New York, 1984.

Exercises

1. Continuation of Question 3 in Chapter 10: In a semi-infinite plate, where z is the thickness direction and $T = f(z)$ only, the stresses $\sigma_{xx} = \sigma_{yy}$. Hence, when the plate is viewed along z under crossed polarized light, the total optical retardation, which is proportional to $\int_{-L}^{+L} (\sigma_{xx} - \sigma_{yy})\, dz$, is equal to zero. Would you expect net zero optical retardation if the specimen were a rod of diameter $2b$, having a parabolic temperature distribution $T = T_0[1 - (r^2/b^2)]$, which was oriented with its axis at 45° to the cross polarizers and was viewed along the radial direction? (Hint: See *Theory of Thermal Stresses* by B. A. Boley and J. H. Weiner, pp. 287–291, John Wiley & Sons, New York, 1960 for the appropriate stress formulae.)

References

1. M. Garbuny, *Optical Physics*, pp. 278–290. Academic Press, New York, 1965.
2. R. Olshansky, *Rev. Mod. Phys.*, **51**(2), 341 (1979).
3. R. H. Doremus, *J. Chem. Phys.* **40**(8), 2389 (1964).
4. H. C. Van de Hulst, *Light Scattering from Small Particles*. J. Wiley, New York, 1957.

5. C. R. Bamford, *Control of Colour and Generation in Glass*. Elsevier–North Holland Publications, Amsterdam, 1977.
6. T. Bates, in *Modern Aspects of the Vitreous State*, Vol. 2 (J. D. Mackenzie, ed.), pp. 195–254. Butterworths, London, 1962.
7. A. Paul, *Chemistry of Glasses*, 2nd Ed., pp. 276–348. Chapman and Hall, London, 1990.
8. G. H. Sigel, in *Treatise on Mat. Sci. & Tech.*, Vol. 12, *Glass I* (M. Tomozawa and R. H. Doremus, eds.). Academic Press, New York, 1977.
9. I. W. Donald and P. W. McMillan, *J. Mat. Sci.* **13**, 1151–1176 (1978).
10. S. R. Elliott, *Physics of Amorphous Materials*, 2nd Ed. (Figure 5.38). Longman Science and Technical, Harlow, Essex, 1990.
11. G. Sigel, in *Treatise on Mat. Sci. & Tech.*,, Vol. 12, *Glass I* (M. Tomozawa and R. H. Doremus, eds.), pp. 5–90. Academic Press, New York (1977).
12. T. Takamori and M. Tomozawa, in Ref. 11, pp. 123–155.
13. R. J. Powell and W. E. Spicer, *Phys. Rev.* **B2**, 2182 (1970).
14. H. Rawson, *Properties and Applications of Glass*. Elsevier, Amsterdam, 1980.
15. A. Marker, in *Advances in Fusion of Glass* (D. F. Bickford, E. N. Buolos, F. Olix, W. E. Horsfall, J. N. Lingscheit, W. C. LaCourse, F. E. Wooley, F. Harding and L. D. Pye, eds.), pp. 4.1–4.24. Am. Ceram. Soc., 1988.
16. J. Krogh-Moe, *Phys. Chem. Glasses*, **6**, 46 (1965).
17. C. C. Wang, *Phys. Rev.* **B2**, 2045 (1970); also see Y. R. Shen, *The Principles of Nonlinear Optics*, p. 37, J. Wiley, New York, 1984.
18. N. L. Boling, A. J. Glass, and A. Owyoung, *IEEE J. Quant. Electron.* **QE-14**, 601 (1978).
19. H. Nasu and J. D. MacKenzie, *Opt. Eng.* **26**(2), 102–106 (1987).
20. F. L. Pedrotti and L. S. Pedrotti, *Introduction to Optics*. Prentice-Hall, 1987.
21. N. Neuroth, *Opt. Eng.* **26**(2), 96–101 (1987).
22. R. M. Klein, *Glass Sci. and Tech.*, Vol. 2 (D. R. Uhlmann and N. J. Kreidl, eds.), pp. 285–340. Academic Press, New York, 1984.
23. C. R. Bamford and D. G. Loukes, *Glass Technol.* **20**(5), 166–169 (1979).
24. O. S. Heavens, *Optical Properties of Thin Solid Films*, Academic press, New York, 1955.

Chapter 20

Fundamentals of Inorganic Glassmaking

20.1. Laboratory Melting of Oxide Glasses

Oxide glasses are obtained by melting a well-mixed "batch" of appropriately weighed quantities of "raw materials," which are the constituent oxides or compounds that convert to oxides upon heating. Table 20-1 lists commonly used raw materials for melting glass, along with the appropriate conversion factors to the oxide form and their reciprocals (after Tooley [1]). (Also shown in Table 20.1 are the names of the minerals that are used as sources for the various oxides in a commercial operation.) In calculating the batch composition, it is generally assumed that

1. hydrated crystals lose their water completely;
2. organics burn off completely;
3. carbonates, sulfates, sulfites, nitrates, nitrites, etc., change to oxides with no C, S, or N remaining in the structure;
4. fluorides, chlorides, bromides, and iodides convert to the equivalent amount of oxides with little trace of F, Cl, Br, or I left in the glass;
5. a pure element added converts to its equivalent oxide form; and
6. metal borate, aluminate, silicate, phosphate, arsenate, molybdate, etc., yield metal oxides as well as the oxides B, Al, Si, P, As, Mo, etc.

Table 20-1. Raw Materials Used in Glass Melting[a]

Material	Alternative name	Theoretical formula	Oxides supplied	Fraction	Reciprocal[b]
Alumina	Calcined alumina	Al_2O_3	Al_2O_3	1.000	1.000
Aluminum hyd.	Hydrated alumina	$Al_2O_3 \cdot 3H_2O$	Al_2O_3	0.654	1.531
Aplite (typical composition)			Al_2O_3	0.240	4.167
			$Na_2(K_2)O$	0.100	
			SiO_2	0.600	
			CaO	0.060	
Feldspar	Microcline (composition is of typical commercial spar)	$K_2O \cdot Al_2O_3 \cdot 6SiO_2$	Al_2O_3	0.180	5.556
			$K_2(Na_2)O$	0.130	
			SiO_2	0.680	
Nepheline syenite (typical composition)			Al_2O_3	0.250	4.000
			$Na_2(K_2O)$	0.150	
			SiO_2	0.600	
Calumite	Calcium–aluminum Silicate	$2CaO \cdot MgO \cdot 2SiO_2$	SiO_2	0.380	
		$2CaO \cdot Al_2O_3 \cdot SiO_2$	Al_2O_3	0.117	
		$2(CaO \cdot SiO_2)$	CaO	0.400	
			MgO	0.080	
Kyanite (90% concentrate)		$Al_2O_3 \cdot SiO_2$	Al_2O_3	0.567	1.763
			SiO_2	0.433	
Kaolin	China clay	$Al_2O_3 \cdot 2SiO_2 \cdot 2H_2O$	Al_2O_3	0.395	2.57
			SiO_2	0.465	
Cryolite	Kryolith	Na_3AlF_6			
Antimony oxide		Sb_2O_3	Sb_2O_3	1.000	1.000
Arsenious oxide	White arsenic	As_2O_5	As_2O_5	1.160	0.860
Barium carbonate		$BaCO_3$	BaO	0.777	1.288
Barium oxide	Baryta	BaO	BaO	1.000	1.000
Barium sulfate	Barytes	$BaSO_4$	BaO	0.657	1.523
Boric acid	Boracic acid	$B_2O_3 \cdot 3H_2O$	B_2O_3	0.563	1.776
Borax		$Na_2O \cdot 2B_2O_3 \cdot 10H_2O$	B_2O_3	0.365	2.738
			Na_2O	0.163	6.135

Anhydrous borax	("Pyrobor")	$Na_2O{\cdot}2B_2O_3$	B_2O_3	0.692	1.445
			Na_2O	0.308	3.245
Lime, burnt	Quick lime	CaO	CaO	1.000	1.000
Lime, hydrated	Calcium hydrate	$CaO{\cdot}H_2O$	CaO	0.757	1.322
Limestone	Calcium carb.	$CaCO_3$	CaO	0.560	1.786
Calcium carb.	Whiting	$CaCO_3$	CaO	0.560	1.786
Lime, dolomitic	Burnt dolomite	$CaO{\cdot}MgO$	CaO	0.582	1.720
			MgO	0.418	2.390
Dolomite	Raw limestone (dolomitic)	$CaO{\cdot}MgO{\cdot}2CO_2$	CaO	0.304	3.290
			MgO	0.218	4.580
Lime, hydrated, dol.	Finishing lime	$CaO{\cdot}MgO{\cdot}2H_2O$	CaO	0.423	2.363
			MgO	0.304	3.290
Litharge	Lead oxide, yellow	PbO	PbO	1.000	1.000
Red lead	Minium	Pb_3O_4	PbO	0.977	1.024
Bone ash	Calcium phosphate	$3CaO{\cdot}2P_2O_5 + x\ CaCO_3$	CaO	0.372	2.700
			P_2O_5	0.628	1.592
Iron oxide. red	Rouge	Fe_2O_3	Fe_2O_3	1.000	1.000
Potassium hyd.	Caustic potash	KOH	K_2O	0.838	1.194
Potas. nitrate	Saltpeter	KNO_3	K_2O	0.465	2.151
Potassium carb.	Calcined carbonate of potash	K_2CO_3	K_2O	0.681	1.469
Glassmaker's potash	Potassium carbonate, hyd.	$K_2CO_3{\cdot}1.5H_2O$	K_2O	0.570	1.754
Sand	Glass sand, quartz	SiO_2	SiO_2	1.000	1.000
Soda ash	Sod. carb. coml.	Na_2CO_3	Na_2O	0.585	1.709
Sodium nitrate	Saltpeter chili	$NaNO_3$	Na_2O	0.365	2.741
Salt cake	Sodium sulfate	Na_2SO_4	Na_2O	0.437	2.290
Sodium silicofluoride	Sodium fluosilicate	Na_2SiF_6			
Zinc oxide		ZnO	ZnO	1.000	1.000

[a] From Ref. 1.

[b] This column gives pounds of material required to supply one pound of oxide.

As an example, an appropriate batch composition for obtaining 1 kg of $K_2O{\cdot}CaO{\cdot}5SiO_2$ glass may be calculated according to the table.

Moles	wt. %	For 1 kg glass	Obtain from
$K_2O = 1$	20.9 K_2O	209 g K_2O	$209 \times 1.469 = 307$ g K_2CO_3
$CaO = 1$	12.4 CaO	124 gCaO	$124 \times 1.786 = 221.5$ g $CaCO_3$ (limestone)
$SiO_2 = 5$	66.7 SiO_2	667 g SiO_2	$667 \times 1.000 = 667$ g SiO_2 (sand)

A slightly more complex example would be, for instance, to calculate a suitable batch for obtaining 1 kg of $20.0Na_2O{\cdot}25.0B_2O_3{\cdot}54.98SiO_2{\cdot}0.02FeO$ (wt. %) glass. Note that, in this case, the FeO is a trace amount. Such amounts are normally present in the raw materials (in particular, in the sand). One would then proceed as if the FeO were present as part of the sand (and not as a separate raw material). For the sake of argument, let us assume that all the raw materials are reagent-grade, and that the FeO is in fact desired in the final glass. Let us also assume that we are given Na_2CO_3, borax, silica, and Fe_2O_3 as raw materials. The computations proceed according to the table.

wt. %	For 1 kg glass	Obtain from
25.0 B_2O_3	250 g	$250 \times 2.738 = 684.5$ g borax
		(also gives $684.5 \times 0.136 = 111.6$ g Na_2O)
20.0 Na_2O	200 g	$(200 - 111.6) \times 1.709 = 151.1$ g Na_2CO_3
54.98 SiO_2	549.8 g	$549.8 \times 1.000 = 549.8$ g silica
0.02 FeO	0.2 g	$0.2 \times 1.111 = 0.222$ g Fe_2O_3

In the tabulated calculation, the added borax provided part of the soda requirement. The balance had to be secured through the Na_2CO_3 addition. One other assumption we made in these computations is that all the iron in the glass is present in the ferrous (Fe^{2+}) state. In actual fact, regardless of the manner added, we expect an equilibrium to be established between Fe^{2+} and Fe^{3+}; the exact value of the ratio will depend upon the melting conditions (Section 19.2.3).

The **melting** of small amounts of oxide glasses (from a few grams to a few kilograms) is carried out in a laboratory environment by placing the batch inside a crucible and heating the crucible in an electric (or gas-fired) furnace. The melting temperature to be used is determined by the phase diagram. For oxide glasses containing silica, temperatures in the range of 1,200–1,550°C are generally used. Because of the rather high temperature of operations, it is imperative that the operator use body-protective gear, such as heat-insulating gloves, eyeshields (with dark glasses), and a laboratory coat.

Three types of laboratory electric resistance furnaces are common: the nichrome-wire-wound, which achieves a maximum of about 1,100°C; SiC element-heated, which reaches a maximum of about 1575°C; and $MoSi_2$ (molybdenum disilicide, more popularly called by its trade name "Super Kanthal"), which can achieve up to 1,750°C. Clearly, individual needs dictate the type of furnace employed. The furnaces may be obtained as bottom-loading or front-loading. The chief advantages of a bottom-loading furnace relative to a front-loading furnace are that the batch loading can be readily mechanized using an elevator mechanism, and that the operator does not "see" the heat. Its chief disadvantage is that the crucible (with the contained glass) can cool rapidly during lowering of the elevator mechanism, which could make subsequent operations, such as pouring the glass melt, more difficult.

Suitable crucible materials are slip-cast pulverized fused silica, alumina, mullite, and clay. The silica crucibles are most recommended because of their high thermal shock resistance, low reactivity, and generally low cost. If contamination from the crucible material is to be minimized, then one may find pure Pt, Pt–5%Au, or Pt–20%Rh crucibles quite useful. As usual, care must be taken in handling the expensive Pt apparatus. Cl^- and NO_3^- in the batch can readily react with Pt. Metal oxides (e.g., those of Cu or Fe) melted under reducing conditions can yield metals that readily alloy with Pt. Likewise, handling a Pt crucible with lightweight iron tongs at high temperatures should be avoided. The reader may find it safer to first try a melt in ceramic crucibles before proceeding to use noble metal crucibles. One advantage of Au- and Rh-alloyed Pt is that glass can be left in the crucible after the melting is complete, and can be taken out as a solid block by gently tapping the crucible upon cooling. (Au alloying works because of the increased nonwetting behavior, and the Rh alloying works because of the increased crucible hardness and nonwetting: upon cooling, a higher-expansion glass simply shrinks away from the container.)

Depending upon the size of the crucible employed, it may be necessary to recharge the crucible with the batch after some time in the furnace. For this, a suitably sized strong pair of tongs should be employed. During heating, the following reactions in the raw materials may occur:

1. Release of adsorbed moisture around 100 °C.
2. Release of volatiles.
3. Organics burning off around 440°C.

4. Evolution of gases such as CO_2 produced by decomposition of, for instance, the carbonates.
5. Solid-state reactions between the various oxides to form binaries and ternaries.
6. Appearance of the eutectic liquid.
7. Increased rate of reaction to form a molten mass.
8. Continuing sublimation of volatiles.
9. Gradual removal of bubbles (or "seeds") from the molten mass by rising to the surface or by dissolution back into the glass.
10. Homogenization of the melt.
11. Occasional exsolution of the dissolved gases (otherwise called "reboil").
12. Reaction with the container throughout the melting period.

Reactions (9) and (10) together are termed **fining** of glass. Higher melting temperatures generally aid in fining; one occasionally adds fining agents discussed in the next section.

For good **homogenization**, it is important to start with well-mixed, preferably fine particle-sized raw materials. It may be necessary to stir the melt using a Pt-capped refractory or simply a fused-silica rod. Sometimes it may be helpful to break up the glass into small pieces after the first melt and subject the clean pieces to remelting.

Pouring of glass is generally carried out in graphite or clean steel molds. The molds may simply be slabs of the material wired together. Graphite in particular is quite convenient to use because of its nonwetting properties. In case the glass is to be chilled to avoid crystallization, the glass is poured on an aluminum or steel chill plate using a metal roller to roll the glass into thin strip/ribbon form. *Solid blocks of glass should be annealed* before cooling through the glass transformation range to avoid cracking at lower temperatures. For this, another furnace held at around the glass annealing point should be on standby. After the glass is placed in the furnace, the furnace may be shut off, or subjected to a predetermined cooling schedule.

Solid blocks of glass can be core-drilled, sawed, rough-ground, smoothed, and fine-polished using standard ceramographic techniques. For drilling and sawing, it is helpful to use diamond-impregnated tool bits. Rough grinding is customarily performed using abrasive paper or, preferably, a slurry of successively increasing grit number (generally 240, 320, 400, or 600 grit is adequate) on a rotating wheel. (Experience in safe handling of a rotating wheel should be developed.) The final smoothing using a 1,000 grit slurry is recommended to be performed by hand on a glass plate. In between each step of the grinding and smoothing process, the specimen should be thoroughly cleaned by immersion in an ultrasonic cleaner. Fine polishing is accomplished by lapping against a diamond, alumina, cerium oxide, or rouge-laden paper/cloth, again on a rotating wheel. (The reader may find cerium oxide to work quite satisfactorily for most silicate glasses.) For ease of handling of specimens through the ceramographic process, it is convenient to mount the glass specimens, along with dummy specimens for grinding control, on flat plates using a thermoplastic cement (such as Canada balsam). In some cases, it may be necessary to mount glass specimens in standard 1-in. or $1\frac{1}{4}$-in. molds using cold-mounting media or hot-mounting thermosetting resins. Glass specimens should be stored in open air or in a desiccator, depending upon the reactivity of the glass to humidity.

20.2. Continuous Melting of Glass

The single most important glass property that makes high-speed continuous glass-melting and product-forming a reality is the viscosity. A batch can be fed from one end of a melting tank, converted to molten glass having a viscosity roughly 10 Pa·s while in the tank, subsequently delivered to a forming machine at around 10^3 Pa·s (such that it will not flow away), worked on by the machine while cooling gradually, and then transferred to a conveyor belt in a seemingly solid form at 10^9 Pa·s or greater to undergo annealing and subsequent controlled cooling down to packaging. The schematics of a typical commercial glassmaking operation are shown in Fig. 20-1. A variety of glass products can be made continuously at high speeds this way. Notable among these are beverage containers, lamp envelopes, flat plates, and fiberglass. Because of the high efficiencies of modern-day glass machinery, good-quality glass products can be made quite inexpensively, often making glass the material of choice.

The **raw materials** (Table 20-1) utilized commercially are, in general, *not as pure* as those used normally in a laboratory. Calculation of the batch composition for a desired glass composition or *vice versa*, though more involved on account of the impurities, is exactly the same as that outlined earlier. The only significant difference in a commercial operation is the inclusion of broken glass, otherwise called the **cullet**, of the same general composition. The cullet provides an early liquid phase to enhance the reaction rates, and hence helps reduce the overall energy requirements for melting significantly. The usual practice is to add about 20 to 40% cullet by weight. Often the cullet is the in-house production of rejected product. Occasionally, cullet may be bought from recycling centers or the like (provided sufficient care has been taken in the sorting of glass), or else the product of marginally acceptable quality is deliberately broken.

In addition to the cullet, *fining agents are invariably necessary* to make the glass-melting process more efficient. Traditional fining agents, namely,

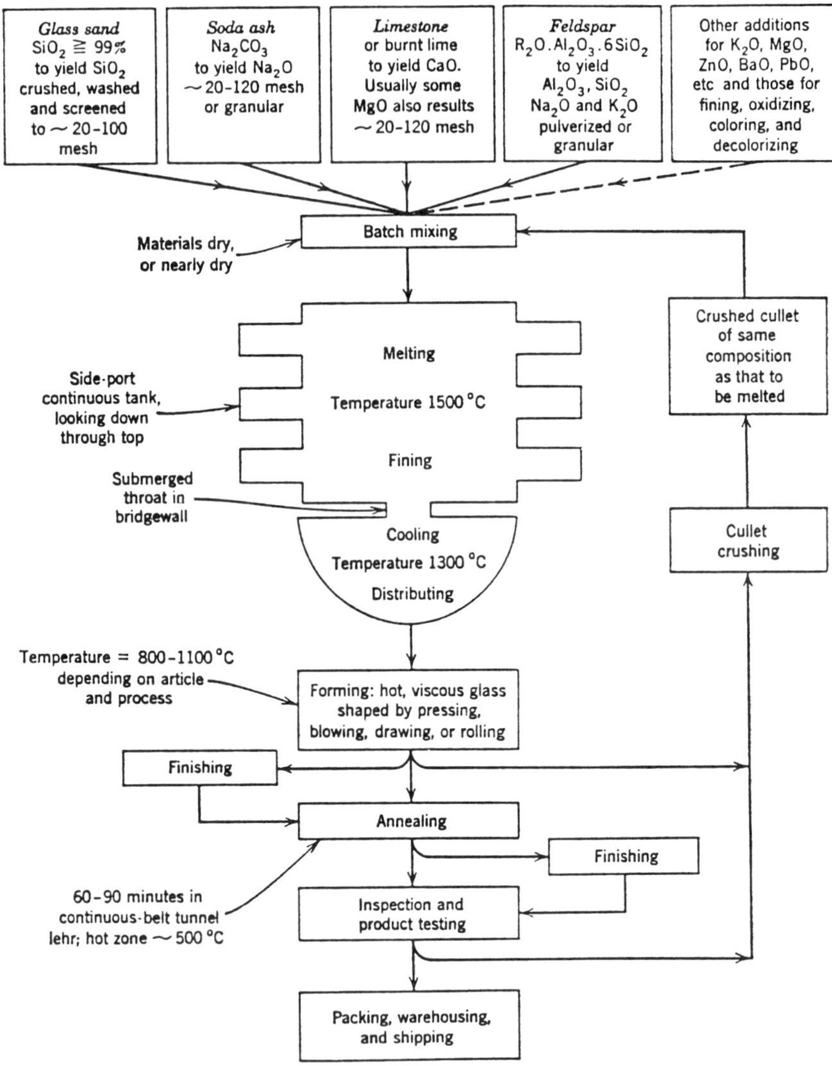

Figure 20-1. Schematics of commercial glassmaking.

0.25–0.5% As_2O_3 plus 0.5–1.0% $NaNO_3$, are seldom used (because of the emphasis on worker safety). Instead, the more common practice is to provide part (about 1 to 2%) of the Na_2O from NaCl or Na_2SO_4 (in the latter case, one has to be concerned about S emissions from the chimney). Where clear glass is the desired product, the batch may also contain **decolorizers**, such as Se or cerium concentrate, to oxidize part of the Fe^{2+} (which imparts a greenish color) to Fe^{3+} (which imparts a complementary straw-yellow color).

The **particle size** of the raw materials is important for determining the total **melting time** and the resultant **homogeneity** of glass. Raw materials are stored in bins/hoppers from which they are taken out as needed, weighed in proper amounts, blended, and charged into the glass tank (the charging end is called the **doghouse**) using feeder mechanisms. Popular feed mechanism are based on a screw device, a shoveling device, or "blanket feeding." Depending upon the distance of travel between the blender and the feeder, the batch is occasionally moistened lightly (using a water spray) to avoid segregation of widely sized batch particles.

Glass-melting tanks are typically quite large, often as large as 35 × 100 ft, and may contain as much as 1,400 tons of molten glass over a tank depth of about 4 to 5 feet (Fig. 20-2). The entire structure is invariably constructed using high-temperature, corrosion-resistant refractories. In particular, premium refractories such as electrocast zirconia/alumina are used to line the glass-contact regions. The tank consists of two portions: the melting chamber and the conditioning chamber, separated by a **bridge wall** below the glass line (**metal line**) and a **shadow wall** above the glass line. A **throat** within the bridge wall connects the two chambers. *The tanks are fired using gas or oil generally;* however, in those regions where hydroelectricity is inexpensive, *melting may also be carried out using electricity*. Electrodes, made from Mo or graphite, are inserted from the tank bottom or sides and provide a part (**boosters**) or all of the energy required to perform the melting of glass. (In the latter case, the glass has to be initially brought to an adequately conductive state by auxiliary means.)

Large glass tanks are invariably equipped with regenerative heating. Firing is carried out from one side of the tank; the combustion products, including batch reaction products and some batch carryover particles, exit from the other side through a **regenerator** chamber, where the hot "flue gases" impart some of their heat to the brickwork (**checkers**) before exiting through the main chimney stack. After some period of firing (generally 10 to 15 minutes), the direction is reversed by the action of gate valves in the gas and air ducts. The gas and air now enter the tank after passing through the preheated regenerators, in effect becoming preheated themselves and thus raising the thermodynamic efficiency of the combustion. Occasionally, the tanks are fitted with energy **recuperators**, which are continuous heat exchangers with the exiting flue gases flowing through metal/ceramic tubes and incoming combustion air/gas going around them (without the need for a reversing action).

Glass is essentially all molten by the time it passes through the throat into the conditioning chamber. Here the glass is "conditioned" for delivery: it is homogenized (primarily by diffusive mixing), the fine bubbles (seeds) are redissolved, decolorization (if needed) is achieved, and the glass is cooled a little. Onwards, the glass travels through narrow tunnels, called the **fore-hearth** (for a container-making tank), leading to forming machines. Depending upon the requirements of the forming machine, the glass may be delivered in the form of a continuous stream or in the form of a discretized piece, called a **gob**. In a popular version of a gob feeder, the glass at the end of the forehearth passes into a rotating refractory bowl in which the pumping action

(a)

(b)

(c)

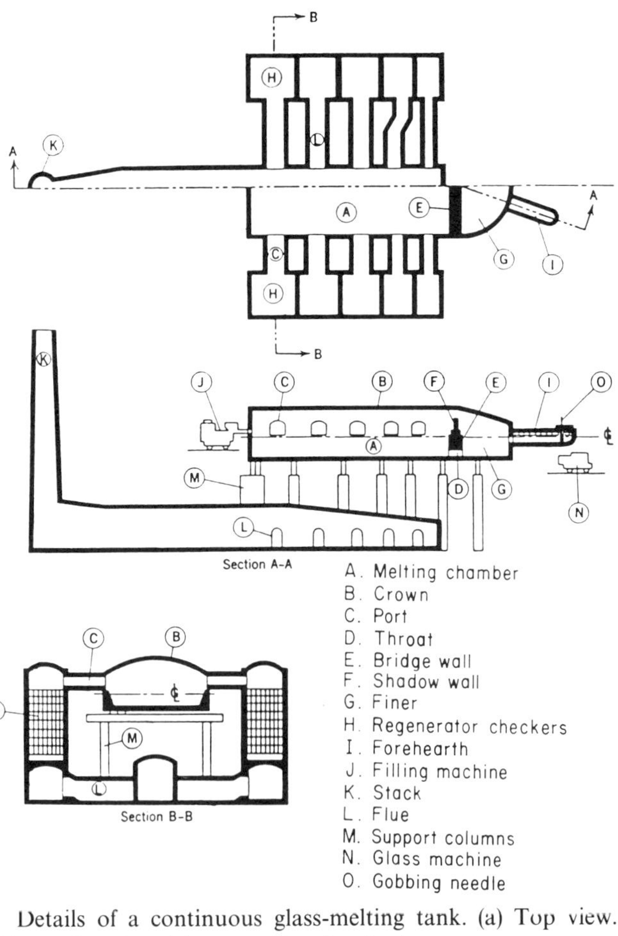

Figure 20-2. Details of a continuous glass-melting tank. (a) Top view. (b) Side elevation. (c) Front elevation. (Courtesy Corning Inc.)

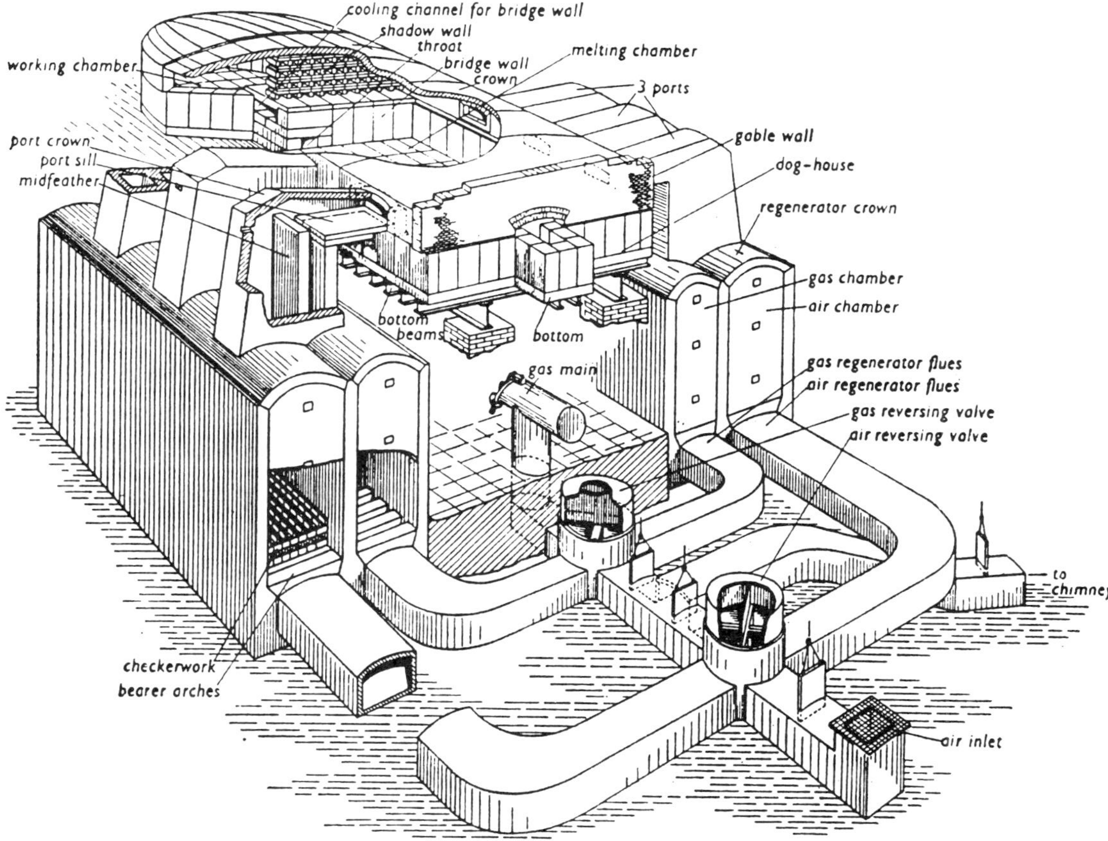

Figure 20-2. (d) Isometric view. (Note: The furnace is different from that in (a)–(c)). (After R. Günther, *Glass-Melting Tank Furnaces*, Fig. 1A, p. 4. Soc. of Glass Tech., Sheffield, 1958. Reproduced with permission of the publisher.)

of a refractory needle pushes the glass out of the orifice in the form of a lumpy stream. The stream is cut into gobs using synchronized metallic shears. The gob then travels along suitably positioned chutes and drops into the forming-machine molds. The exact shape and weight of the gob is often an artistic performance.

As one might expect, the doghouse end of the melting tank is the coolest portion. Piles of raw batch float on semi-molten glass and shield the mass below from the heat of the combustion products. The temperature rapidly rises to a maximum value corresponding to about 10 Pa·s (10^2 poise) viscosity about two-thirds of the distance into the melting chamber. In the conditioning chamber, the glass viscosity is maintained around 10^2–$10^{2.5}$ Pa·s, and is brought to about 10^3–10^4 Pa·s at the gob feeding device, ready for the forming operation.

The applied **pull** (output at the forming end), the temperature distribution, the density gradients (including the rise of bubbles), and the charging of the batch establish currents in the glass. The pull basically causes a slow laminar or "piston" flow of glass. Because it is inherently non-mixing, such a type of flow is undesirable. The temperature and the density gradients, on the other hand, produce convective flow, which greatly helps in the initial homogenization of the glass (**deterministic mixing**). The final stages of homogenization (occurring mostly in the conditioning chamber and the forehearth) are almost entirely controlled by **diffusive mixing**, which, being an atomic process, is quite slow. These secondary flows would, over a period of time, give rise to perfect mixing if the pull were reduced to zero. Since a no-pull condition is also very undesirable, the flow of glass in the tank is generally a combination of the **perfect mixing** and the **piston type**. (In reality, the flow is described closely by a **dead-water** flow, where some portions of the glass contribute more to the flow and others are relatively stagnant.) The **residence time** of the glass in the tank may therefore vary from as little as a few hours to several days. These considerations become quite important in controlling the final homogeneity of the glass product, particularly during glass composition changeover. (See Cooper [2] for a discussion on the flow in glass tanks.)

In a gas- or oil-fired tank, there is a significant amount of batch carryover with the flue gases. Likewise, sulfur concentration in the flue gases is high when either Na_2SO_4 is included in the batch for fining or the firing is carried out using high-sulfur bearing oil. To reduce pollutant release to the atmosphere, it is a virtual necessity to pass the flue gases either through scrubbers or through electrostatic precipitators prior to exiting through the chimney.

All-electric firing, on the other hand, provides heat to the glass from submerged electrodes. As a result, not only is there no need for any regenerators, there also is not much by way of flue gases. There is no batch carryover, and hence there is no need for any precipitators. *The thermal efficiency of all-electric tanks can, indeed, be quite high.*

The melting process is controlled by controlling the raw materials input and the tank temperature. The glass level is monitored by optical reflection, radioisotope shadowgraphy, or simply by a dip stick (generally in a forehearth), and the signal is used to trigger batch charging. The tank temperature is monitored at a number of places by immersed, sheathed thermocouples or optical/radiation pyrometers. The temperature signal is fed to the fuel input control valves. Occasionally, oxygen sensors may also be used to monitor the combustion conditions.

20.3. Non–fusion-Based Techniques of Glassmaking

It was stated in Section 2.1 that *fusion (and subsequent rapid cooling) is not the only means of making glass* (where the word "glass" is intended to include all x-ray amorphous solids). *Glass may be made directly from any of the other states of matter or by growth out of a "low-temperature" liquid solution.* Of these, methods based upon converting a solid into a glass have yet to gain any commercial significance. However, the forming of glass directly from the gaseous state or by growth out of a low-temperature liquid, the **sol–gel** process, has seen a vast expenditure of R&D in the past two decades, and has enjoyed considerable commercialization, to the point that many of these techniques can be made continuous. Indeed, some of the non–fusion-based techniques of glassmaking can no longer be billed as "non-traditional."

20.3.1. Glassmaking Directly from the Solid State

The transformation of a solid directly into a glass may be achieved by high rates of shearing caused, for instance, by a shock wave during an impact or by irradiation with high-energy particles. The former, called **diaplectic** (or **thetomorphic**) glasses by geologists, have been of considerable interest in determining whether glassy fragments in the lunar soil resulted from volcanic activity or from a constant barrage of meteoritic impacts [3]. The diaplectics retain much of the parent mineral contours: they are quite inhomogeneous even to the naked eye. Shock waves in nuclear explosions have been known to convert some of the local soil to diaplectic glasses. Likewise, glass particles associated with shock waves have been discovered near Barringer crater (Arizona), Ries crater (Germany), and Lonar crater (India). From laboratory tests, it has been determined that diaplectics form up to a shock-wave pressure of about 40 GPa. At higher pressures, the cycle of compression and decompression produces enough heat to cause fusion. Such glasses are similar to the normal melt-quenched glasses [4].

Bombardment by high-energy particles such as alpha particles or fast neutrons can often cause atoms to be irreversibly displaced from their regular atomic locations. It is known as the **metamict state** by mineralogists [5]. As an example, crystalline forms of silica, with the exception of coesite (the high-pressure form), can all be disordered by irradiation to neutrons. The end product is generally similar to that obtained after irradiation of vitreous silica (Section 7.4). The density decreases to about 2.205 g/cm^3. A part of the disorder could also be ascribed to fusion caused by thermal spikes that cover perhaps about 40,000 atoms. However, the thermal spikes do not cause the gross disordering: it is the cooperative displacement of atoms that ultimately accumulates to appear as the disorder [6].

20.3.2. Glassmaking Directly from the Gaseous State

Glasses may also be prepared by deposition directly from a gas phase. The deposition may be nonreactive or reactive. In a **nonreactive deposition**, elements such as Si, Ge, and Se and alloys such as As–S may be vacuum-evaporated by the application of heat and condensed onto a cold substrate (chamber pressure generally $<10^{-6}$ torr). The heat may be provided by electrical-resistance heating of a boat containing the substance or by electron bombardment. The atoms deposit in an amorphous structure as thin films and are prevented from rearranging themselves to crystallize by their low mobility. Typical deposition rates are of the order of 0.1–1 μm s^{-1}. Another method of driving the atoms out of a solid into a gas phase is **sputtering**. The solid is made in a "target" form and is impinged upon by high-velocity ions from a low-pressure plasma of an inert gas such as Ar bled into the chamber at 1–20 mtorr. In the dc sputtering process applicable to electrically conducting materials, the target is the cathode and the substrate the anode. Ar^+ ions are accelerated towards the target, sputtering target atoms or atom clusters that are then carried towards the substrate. Insulating materials may be sputtered nonreactively by applying an rf field, typically ~13 MHz, between the substrate and the capacitatively coupled target of the desired solid. Sputtering generally preserves the stoichiometry of the target, which itself may or may not be homogeneous. Typical sputtering rates for insulators in an Ar atmosphere are 1–10 Å/s and somewhat higher for metals.

In a **reactive deposition**, the desired species are formed after chemical reaction/decomposition of one or more vapor-phased chemicals. Glass is formed either as a thin film upon deposition on a cold substrate, or as "soot" particles that are allowed to agglomerate into a bulk, but porous, structure. The porous bulk is sintered subsequently by heating to form dense monoliths.

Because the reactions/decompositions are carried out at temperatures well below the equilibrium melting temperature of the corresponding crystals, glassmaking by these techniques remain classified as non–fusion-based.

There are three common techniques to carry out **reactive thin-film deposition**:

1. **Reactive sputtering**, which is much like the sputtering described earlier, except that in addition to Ar, a reactive gas is also introduced.
2. **Glow-discharge (GD)**, where the vapor-phased material or material mixture itself is the plasma gas. Under appropriate conditions, the gases inside the plasma decompose and react, and deposit on a suitable substrate. This method is used for the production of *a*-Si:H starting with silane (SiH_4) gas with or without Ar, and for producing *n*-type or *p*-type *a*-Si:H by doping the mix with a known amount of phosphine (PH_3) or diborane (B_2H_6). These have already been mentioned in Section 16.3.
3. **Chemical vapor deposition (CVD)**, where the decomposition is pyrolytic using thermal energy or an rf field to heat up the chemical mix.

The CVD technique is extended into "vapor-phase hydrolysis," for instance, of $SiCl_4$ to form SiO_2 "soot" particles of colloidal dimensions that are essentially amorphous. A mixture of $SiCl_4$ and H_2O vapor, with or without dopant gases such as $GeCl_4$ (to vary the refractive index), is introduced into a reaction vessel at high temperature. Bulk forms of doped or undoped "synthetic silica" are made commercially using this technique, described in greater detail in Section 20.4.4.

20.3.3. Glassmaking via the Sol–Gel Process

Suggested reading: C. J. Brinker and G. W. Scherer, *Sol–Gel Science*, Academic Press, New York, 1990.

C. J. Brinker, D. E. Clark and L. L. Hench, *Better Ceramics through Chemistry*, Elsevier Science Publishers, Amsterdam, 1984.

S. Sakka, in *Treatise on Mat. Sci. & Tech.*, Vol. 22, *Glass III*, pp. 129–167. eds. M. Tomozawa and R. H. Doremus, Academic Press, New York, 1982.

20.3.3.1. Introduction. A colloidal suspension of solid particles, less than about 1 μm in size, in a liquid may be called a **sol**. A **gel** is formed when the colloidal particles coalesce to form a continuously linked network of solid with a continuous interstitial liquid phase, both phases having colloidal dimensions. In the sol–gel technique for making glass, one generally uses metal alkoxides ($M(OR)_n$, R = alkyl group), salts, and oxides as precursors. These compounds are gradually dissolved or *hydrolyzed* by reaction with water to form a sol at first. *The colloidal particles in the sol gradually crosslink and polymerize to form a gel instead of growing independently to form a precipitate.* The differences among a sol, a gel, and a precipitate are shown schematically in Fig. 20-3. The wet gel is subsequently *dehydrated* to remove the interstitial liquid phase, and *sintered* by the application of heat to convert it, one hopes, into a glass.

The chief advantage of the process is the use of low processing tempera tures. The only time any significant amount of heat has to be applied is during the sintering stage, where one needs only to be slightly above the presumed T_g of the glass. The use of low processing temperatures allows a lower degree of chemical corrosion with the container. This, coupled with the availability of high-purity reactants, virtually assures a high degree of purity for the end product. Since the reactants are generally stirred well during the hydrolysis, microscale homogeneity is also expected. (Nanoscale homogeneity is not often achieved.) Because sintering may be carried out at temperatures below the maximum nucleation and crystallization rates, it is possible to produce many glasses that may otherwise be difficult (or even "impossible?") to produce by conventional methods, such as melting. For instance, TiO_2–SiO_2 compositions essentially up to 100% TiO_2 can be obtained [7] by sol–gel processing.

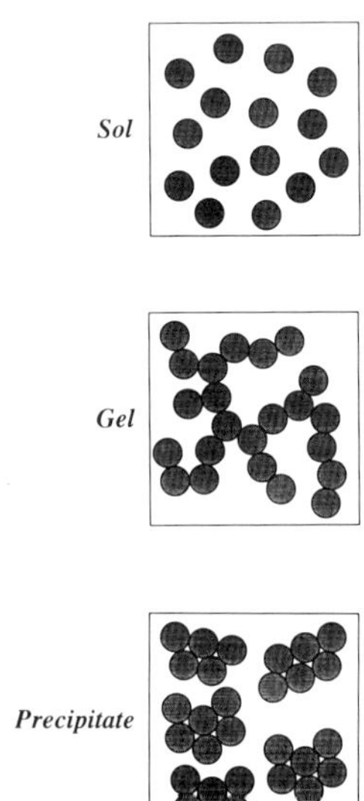

Figure 20-3. Differences among a sol, a gel, and a precipitate.

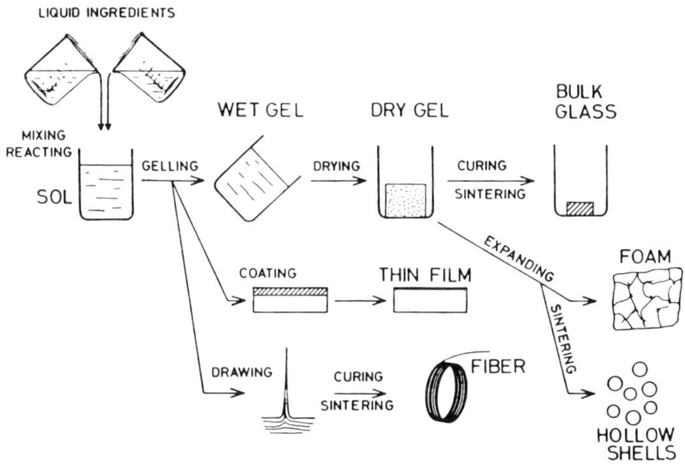

Figure 20-4. Schematic diagram of the sol–gel route to glass making. (After J. Zarzycki, Proc. III Int Conf Adv in Fusion & Processing of Glass, *Ceram. Trans.* **29** (A. K. Varshneya, D. F. Bickford and P. P. Bihuniak, eds.). pp. 519–590, 1993. Reproduced with permission of The American Ceramic Society.)

There are three major disadvantages associated with the sol–gel technique:

1. production of stresses during the drying stage makes preparation of monolithic pieces extremely difficult;
2. the total cycle is often very slow for many commercial applications; and
3. the raw materials' cost definitely outweighs any energy savings in lower-temperature processing.

Figure 20-4 is a schematic of the sol–gel process, outlining the formation of some of the glass products that can be made. The most important application of the sol–gel technique is in the area of thin-film deposition because of the ease of application (e.g., dipping or spinning) and because dense thin films (<1 μm thick) can be readily obtained crack-free. Antireflective, conductive, and protective (corrosion-, scratch-, and wear-resistant and electronic passivating) coatings by the sol–gel process are already well commercialized. As an example, TiO_2–SiO_2 compositions are used to obtain optical films of refractive indices ranging between 1.4 and 2.3, and for the production of ultralow-expansion (ULE) glass (Section 10.4). Other commercial applications of the sol–gel process include high-purity optical components, monosized powders, abrasive grains (particularly alumina, by the hydrolysis of aluminum alkoxides), whiskers for composites, and hollow ceramic or glass spheres. High-purity optical fibers for sensors and telecommunication are an example of potential applications. In nature, the formation of opal is an acknowledged sol–gel process occurring over geologic times.

20.3.3.2. Raw Materials. Silica additions are generally made using either tetraethyl orthosilicate, $Si(OCH_2CH_3)_4$ (abbreviated as TEOS, also called tetraethyloxysilane), or tetramethyl orthosilicate, $Si(OCH_3)_4$, because of their excellent hydrolyzing and polymerizing properties. Higher-chain monomers and their hybrids have also been used, the need dictated by their solubility, their thermal stability, and the desired rate of hydrolysis, which is lowered by higher-branched alkoxide ligands (steric hindrance). Because most metal-organics are hard to dissolve in water, anhydrous alcohol is commonly used as a mutual solvent for the metalorganics and water for the hydrolysis stage. Gels having alcohols and water mixtures as the interstitial liquid are called **alcogels**. Silica gels may also be prepared by the hydrolysis of salts such as sodium silicate, in which case water is the only interstitial fluid, and the gel is termed a **hydrogel**.

Like silicon, boron and aluminum are also added often as alkoxides. Alkalis and alkaline earths are generally introduced as methylates or ethylates. Calcium may be introduced conveniently by dissolving the metal in perchloric acid (60% in water) and adding the solution to the reactants slowly. Aluminum may be added, in many instances, as a nitrate or a chloride.

20.3.3.3. Typical Experimental Procedure. The first stage is the mixing of proper quantities of reactants dissolved in anhydrous alcohol. Mixing may be carried out at 0°C with magnetic stirring for several hours (days!) or refluxing at elevated ($\sim$80°C) temperatures. Mixing of components under dry nitrogen is encouraged because it eliminates any premature hydrolysis of the reactants with atmospheric humidity. The next stage is hydrolysis and gelation, where a known amount of water is added gradually during continued stirring. The pH of the sol may be adjusted by the use of mineral acids (generally HCl) or bases (generally NH_4OH). These additives also act to catalyze the hydrolysis and influence the microstructure of the particles (see the following). For multicomponent systems, an important aim here is to minimize the differences between the hydrolysis rates of the different reactants, and hence to avoid premature precipitation of an unwanted phase. By the action of suitable catalysts, crosslinking and polymerization or condensation proceed, with the result that the viscosity of the solution increases with time. At the **gel point**, the viscosity of the system rises steeply and gelation is achieved. Presumably, this is the rigidity percolation threshold. (In practical terms, the stirrer no longer stirs at this point!) The gelled mass is slowly dried by ordinary means at, say, $\sim$40°C or by hypercritical

Table 20-2. Formulations of TEOS, Alcohol, and Water Used for Making Glass by the Sol–Gel Route[a]

	mol %					
SiO_2 Gel Types	TEOS	EtOH	H_2O	HCl	NH_3	H_2O/Si
Bulk						
1-step acid	6.7	25.8	67.3	0.2	—	10
1-step base	6.7	25.8	67.3	—	0.2	10
2-step acid–base						
1st step acid	19.6	59.4	21.0	0.01	—	1.1
2nd step acid	10.9	32.8	55.7	0.6	—	5.1
2nd step base	12.9	39.2	47.9	0.01	0.016	3.7
fibers	11.31	77.26	11.31	0.11	—	1.0
films	5.32	36.23	58.09	0.35	—	10.9
Monodisperse spheres	0.83	33.9	44.5	—	20.75	53.61

[a] After C. J. Brinker and G. W. Scherer, *Sol-Gel Science*, Table 2, p. 110. Academic Press, New York, 1990.

drying techniques (explained later). Since the evaporation of the fluid from the capillaries creates layers with large strain incompatibility, resulting in the generation of large stresses, cracking is inevitable if extreme care is not used during the drying stage. Following the drying stage, the mass is sintered (viscous sintering) above the T_g of the glass to achieve a high density. Some typical formulations using TEOS, ethanol, and water are shown in Table 20-2 (after Klein [8]).

20.3.3.4. Chemical Reactions. The hydrolysis reaction for $Si(OR)_4$ results in the production of silanol ($\equiv$SiOH) groups as

$$Si(OR)_4 + H_2O \longrightarrow (RO)_3SiOH + ROH.$$

The partial hydrolysis may go to completion as

$$Si(OR)_3OH + 3H_2O \longrightarrow Si(OH)_4 + 3ROH.$$

In most cases, condensation or polymerization of the silanol groups begins to occur after partial hydrolysis. The condensation reactions may be represented by the following two reactions:

$$2(RO)_3SiOH \longrightarrow (RO)_3Si\text{–}O\text{–}Si(OR)_3 + H_2O \quad \text{(water condensation)},$$

$$(RO)_3SiOH + ROSi(RO)_3 \longrightarrow (RO)_3Si\text{–}O\text{–}Si(OR)_3 + ROH \quad \text{(alcohol condensation)}.$$

It may be recognized that further hydrolysis of the alkyl group and subsequent condensation leads to the growth of the $n\text{SiO}_2$ molecule. The overall reaction is

$$n\text{Si(OR)}_4 + 2n\text{H}_2\text{O} \longrightarrow n\text{SiO}_2 + 4n\text{ROH}.$$

The hydrolysis reactions imply that, in principle, a completion of hydrolysis would require four moles of water for each mole of Si(OR)_4. However, water is a by-product of the condensation reaction, and thus the theoretical mole ratio of H_2O to Si(OR)_4 from the overall reaction is 2. Researchers have varied this ratio from <1 to as much as 50, depending upon a variety of factors such as the use of higher-chain alkoxides and acids or bases to catalyze the hydrolysis.

The catalytic action of mineral acids or bases is seen in terms of influencing the rate of hydrolysis and the size of the particle aggregate before the onset of gelation. At low pH and a low water/alkoxide ratio, particles are generally about 2–5 nm and mostly chainlike or linear polymers at the gel point. On the other hand, particles grow to 5–10 nm and more branched, and hence, more spherical at high pH and large water/alkoxide ratio before gelation. Under acidic conditions, where hydronium ions (H_3O^+) are in excess, an alkoxide group is rapidly protonated by H^+, causing electron density to be withdrawn from the silicon atom. The water molecule is able to attack the protonated group, generally from the opposite end, making a transition state that subsequently decays, releasing the alcohol. Those substituents that lower the steric hindrance around the silicon are thus expected to increase the rate of hydrolysis. Under basic conditions, water dissociates to produce OH^- ions, which displace OR^- through a five-coordinated intermediate in which the silicon atom acquires a negative charge. Electron-withdrawing substituents thus accelerate the hydrolysis [9]. The rate of hydrolysis is, in general, expected to be affected by the type (strength) and the concentration of the acid/base. Effects of some catalysts on the pH and gel times for TEOS hydrolyzed with four equivalents of water are shown in Table 20-3 (after Pope and Mackenzie [10]).

The choice of solvents also affects the hydrolysis rate, particularly when the solvent molecules are able to hydrogen-bond to either the OH^- or the H_3O^+. Gel times as a function of water:TEOS and ethyl alcohol:TEOS ratios are shown in Fig. 20-5.

20.3.3.5. Drying. Wet gels may be dried by the application of gentle heat. Dried gels are called **xerogels**. Differential pressures developed in the capillaries ($\Delta p = 2\gamma \cos \theta / r$; γ = surface tension at the air–liquid interface, θ = contact angle at solid–liquid–air interface, r = radius of the capillary), in addition

Table 20-3. Effect of Catalysts on Gel Times[a]

Catalyst	Concentration (mol.: TEOS)	Initial pH of solution	Gelation time (h)
HF	0.05	1.90	12
HCl	0.05	0.05[b]	92
HNO_3	0.05	0.05[b]	100
H_2SO_4	0.05	0.05[b]	106
HOAc	0.05	3.70	72
NH_4OH	0.05	9.95	107
No catalyst	—	5.00	1000

[a] Source: Pope and Mackenzie [10].
[b] Between 0.01 and 0.05.

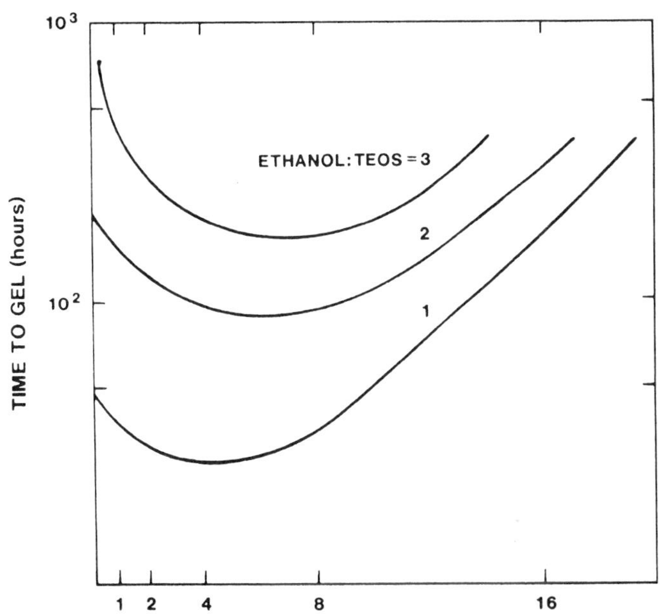

Figure 20-5. Variation of gel times for various water:TEOS and EtOH:TEOS ratios. (After Klein [8], Fig. 6, p. 237. Reproduced with permission of Annual Reviews, Inc.)

to the differential shrinkage, tend to develop high stresses in the gels during the drying stage. In general, thin gel films may dry without cracking when heated at atmospheric pressure; however, gel bodies of any significant physical dimensions are almost certain to develop cracks because of the stresses exceeding the tensile strength. Two methods are used effectively to eliminate or reduce cracking in order to obtain monolithic specimens from alcogels: use of drying-control chemical additives [11] (DCCA), and hypercritical drying [12]. The most successful of the DCCAs is formamide, which appears to favor the formation of uniformly distributed fine pores, thus reducing the development of differential capillary stresses. In the second method, the interstitial liquid is first flushed with alcohol. The gel is then placed (starting point "P_1T_1" in Fig. 20-6) inside an autoclave which is filled to 100% of its free volume (=empty autoclave volume minus the volume of solids, where the gel volume is counted at full density) with ethyl alcohol. The pressure is then raised in small increments to a pressure p_2, which is above the desired critical pressure. The temperature is then gradually raised to T_2, which is above T_c, while maintaining constant p_2. Under hypercritical conditions at p_2T_2, there is no interface between the alcohol liquid and its vapor. At this point, this critical (single-phased) mixture is repeatedly flushed

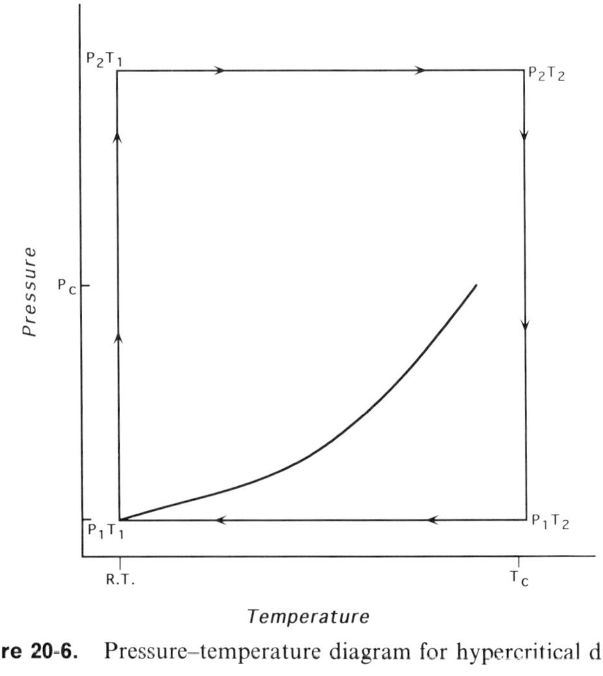

Figure 20-6. Pressure–temperature diagram for hypercritical drying.

with an inert gas such as argon and brought gradually to p_1T_2, and finally to room temperature (p_1T_1). Since no interface existed at the point p_2T_2, the exchange with the inert gas does not generate capillary stresses. Dried gels produced in this manner could have extremely high, as much as 99%, porosity with fine pore size, which makes them quite suitable as thermal insulators and catalytic supports. Dried gels with large porosity are called **aerogels**.

20.3.3.6. Sintering. To obtain higher-density structures, xerogels or aerogels may be heated gradually to slightly above the T_g of the glass, where viscous flow-assisted sintering causes the gradual elimination of the porosity. As may be expected, the removal of open porosity is much easier than that of closed porosity. Upon gradual heating, say at 4°C/min, trapped water and alcohols are first given off, followed by the oxidation of residual organics. The silica skeleton then begins to come closer together by viscous sintering. The sintering process may also bring unreacted hydroxyls and cause them to condense and create Si–O–Si bonds. In this manner, gels are able to achieve densities close to those of the glass at the T_g. During sintering, the temperature excursion must be such that one stays to the left of the nose in the *T–T–T* diagram.

20.3.3.7. Micro- and Nanostructures. The nanostructures of the sol particles during various stages of gelation, and of the gels during various stages of drying and sintering, have been studied by several methods such as IR, Raman, ^{1}H and ^{29}Si NMR, capillary gas chromatography, small angle x-ray scattering (SAXS), and transmission electron microscopy. For instance, Raman spectroscopy of TMOS hydrolysis shows ([13] a gradual formation of the hydrolyzed molecules HO–Si–$(OCH_3)_3$, $(HO)_2$–Si–$(OCH_3)_2$, and $(HO)_3$–Si–(OCH_3), and subsequent condensation to form ≡Si–O–Si–$(OH)_n$–$(OCH_3)_{3-n}$ end groups, and tri- and tetrasubstituted chains and rings. Schematic representations of gel desiccation under acid- and base-catalyzed conditions are shown in Fig. 20-7. *The acid-catalyzed gels are much more linear (chain-like), whereas the base-catalyzed structures tend to be more branched.* During drying and sintering, water and alcohols gradually disappear to yield *a structure that is essentially the same as that of the melted glasses.*

It is possible for gels to phase-separate. In particular, hydrated surfaces of alkali silicate glasses, which are essentially silica-rich gel layers, do show phase separation. During leaching of the glass, the gel structure breaks down and repolymerizes into silica (with few silanol groups) and aqueous silicate species [14].

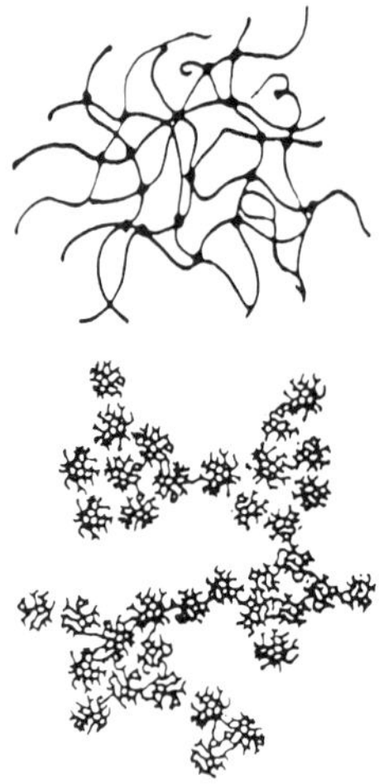

Figure 20-7. Schematic representation of microstructures upon gel desiccation: (a) acid-catalyzed (b) base-catalyzed.

20.4. The Forming of Glass Products

Suggested reading: *Handbook of Glass Manufacture*, by F. V. Tooley, Vols. 1 and 2, 3rd Ed. Books for Glass Industry, Inc., New York, 1984.
Glass Engineering Handbook, by G. W. McLellan and E. B. Shand. McGraw-Hill, New York, 1984.
Glass Science and Technology, Vol. 2, *Processing*, D. R. Uhlmann and N. J. Kreidl, eds. Academic Press, New York, 1984.

Having gone through the various processes of making glass, we turn our attention to the continuous forming of a few of the more common glass products, bearing in mind that many of the technological details are out of the scope of this book.

20.4.1. Glass Containers

As indicated in Chapter 1, *glass for beverage-container applications is a soda lime silicate. The glass usually has about 2%* Al_2O_3 *to improve its chemical durability*. Despite the low strength-to-density ratio, the relatively inexpensive raw materials making up this glass system and high-speed, highly automated processing techniques enable the glass to continue to enjoy a high market share in the container industry. The low permeability of glass to atmosphere is desirable particularly for carbonated beverages.

Many years of experience has indicated that the glass container is best formed by a two-step process. In the first step, a **parison** (or a "blank") is made either by blowing air into or by pressing the glass. In the second step, the parison is blown (by air) into the final shape. Narrow-neck containers are typically made by the **"blow-and-blow" process** (blowing of the parison and blowing of the final shape), whereas wide-mouth jars are made more often by the **"press-and-blow" process**.

The most popular "blow-and-blow" automatic machine is the Hartford **I.S. (Independent Section) machine**, shown schematically in Fig. 20-8. The machine typically consists of five to seven "separate" mini-machines operating in a certain sequence, each mini-machine composed of the parison-making and the final shape-making sections. The gob is dropped into the parison (blank) side of a section and fills the cavity below, making the neck of the container. Air is then blown into the glass, making it into the parison shape. The parison molds then open, and the arm holding the parison by the neck swings onto the finishing side, where the finishing molds close around the parison. Air is blown into it, making the glass into its final shape. The molds upon, the bottom plate drops, and the finished container is cooled a little bit before being set on the conveyor belt for transportation to a quality-check point and on to the annealer. From the instant that the gob drops to the point of transfer onto the conveyor belt, a typical beverage container takes hardly 10 to 12 seconds. During the period that one section works on the glass, the gob is dropped into another section, and so on in sequence. Occasionally, the molds may be double or even triple molds.

In the press-and-blow process (e.g., a Miller machine), the gob is dropped into the parison mold, where a needle plunger pushes into the glass to force it to fill the mold and make the parison. In the step that follows, the parison molds open and drop, and the parison is massaged (mostly reheated). The parison is then blown to the finished form, usually in more than one operation, and finally taken out.

Perhaps a true marvel of glass-blowing technology is the **ribbon machine** (Fig. 20-9), used to make glass shells for lamps. A stream of molten glass is made into a ribbon by passing it through a pair of rollers. The ribbon passes under the puffing orifices of a fast-running blow head turret. Gradually, the glass begins to take form of a blown shell, which is then enveloped by the rotating molds of the lower (synchronized) turret. In the straight portion of the turrets, the shells are blown further and smoothed to finished form. Finished shells are cracked away from the ribbon, collected, and carried through the annealer. The most striking thing is the speed at which the entire machinery operates. Light bulb shells (in single molds) are made at speeds exceeding 50 per second! One has to adjust the eyes to be able to watch the running ribbon.

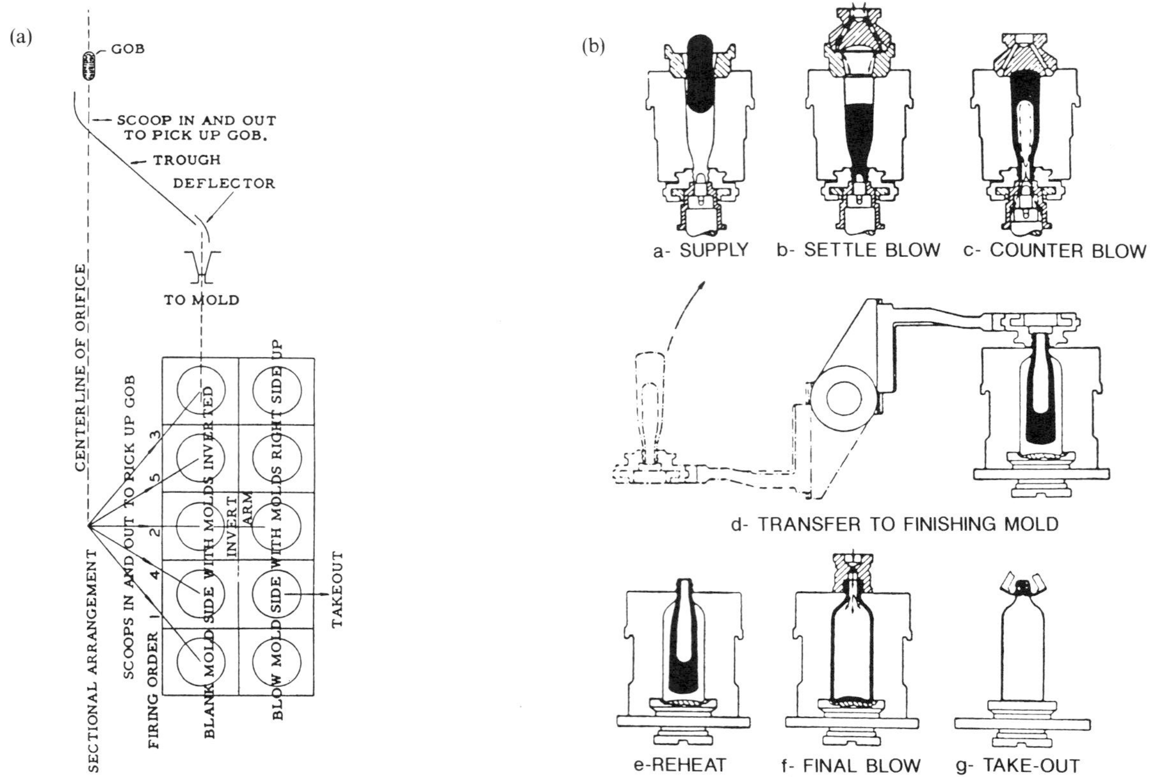

Figure 20-8. Functioning of a "blow-and-blow" IS machine. (a) The gob is delivered to the parison side of one of the machine "sections." The various sections function in rotation on a cam-driven schedule. (b) The details of the parison blowing, transfer to the finishing side, and the finishing process. (After F. V. Tooley, *Handbook of Glass Manufacture*, 3rd ed. Fig. 19, pp. 610–611, Ashlee Publishing Co., 1984. Reproduced with permission of the publisher.)

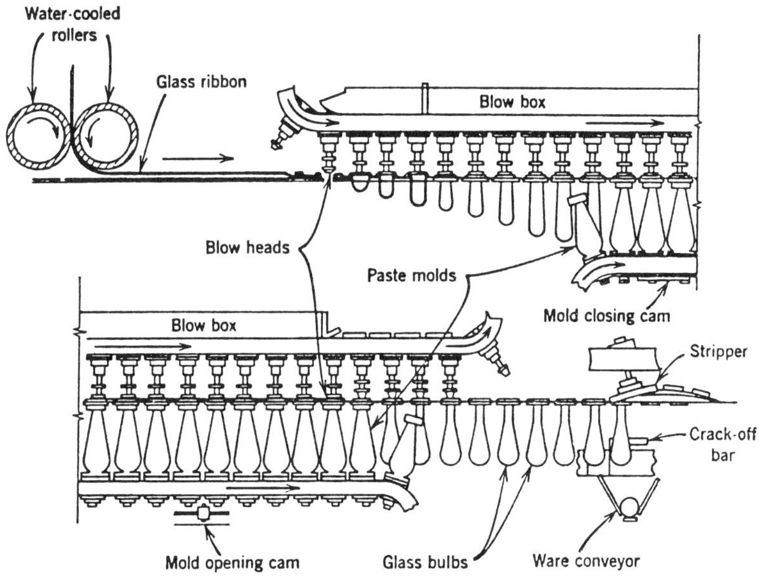

Figure 20-9. Functioning of a ribbon machine. (Courtesy Corning Inc.)

Glass-forming molds are made out of iron and steels. The required properties are a high melting point, high oxidation resistance, good thermal conductivity, and nonwetting to hot glass. Often the molds have vents for air or water flow to control the temperature distribution over the mold body. The purpose is not only to extract the heat from the glass at a controlled rate, but also to assure that the exterior of the glass has as uniform a temperature as possible (otherwise, warping could result upon cooling to room temperature).

The containers are often coated either at the entrance of the annealing lehr (**hot-end coatings**) or at the exit (**cold-end coatings**) *to protect them from abrasion against other hard surfaces* such as those of the metals and ceramics encountered during the manufacture, or simply against each other. The reduced risk of abrasion allows higher packaging speeds. The hot-end coating is most often carried out by spraying an atomized solution of tin chloride, which decomposes to tin oxide as the container travels through the lehr. The tin oxide coatings render the glass surface harder than the parent glass, thus developing a more scratch-resistant surface. There is, however, some debate as to whether the hard coating of tin oxide itself isn't more damaging to the containers when they impinge against each other, at least during travel through the annealing lehr. The cold-end coatings are generally stearates (soaps) to provide a lubricious surface. The lesser friction during contact with other surfaces creates a lesser degree of abrasive damage. All of the coatings are applied at temperatures that will not shock the ware.

Because of the emphasis on ecology and resource conservation, there has been a great deal of effort to recycle beverage containers. The most important cost hurdle is the cost of transportation, which would clearly be lower if the container were lighter for the same strength. This "lightweighting" could be achieved, for instance, by flame-spraying compression coatings of a lower-expansion glass on the exterior of a thinner-wall container. Likewise, a desired modification of the container surface could also be achieved by the application of air plasma coatings or by allowing the surface to crystallize. However, the technology for making a thinner-wall container needs a great deal of refinement. The control of the wall thickness in a typical blow-and-blow machine currently is far too inadequate to allow the high-speed manufacture of thin-wall containers. Indeed, the challenges to the glass scientist/technologist for "lightweighting" of glass are immense, and much work needs to be done.

20.4.2. Flat Glass

Suggested reading: W. C. Hynd, in *Glass Science and Technology*, Vol. 2: *Processing* (D. R. Uhlmann and N. J. Kreidl, eds.), pp. 46–106. Academic Press, New York, 1984.

Clear flat glass is also a simple soda lime silicate system similar to the one used for containers, except for *minor additions such as about 4% MgO that help lower the liquidus temperature* of the glass melt in the glass drawing chamber to reduce the risk of devitrification.

In days gone by, flat glasses used to be made by gathering a gob of glass at the end of a spindle and spinning it till the centrifugal force made it into a flat disk shape. This kind of "flat" glass may still be seen in some old European churches. The next development in this direction was to blow a huge cylinder, cut axially when cold, and then reheat it and unroll it gradually while soft. In another development, the glass was melted in a pot and poured or ladled onto a clean cast-iron table, where a heavy roller was employed to roll the glass into a flat uniform thickness. The first important continuous flat glass process was the **Fourcault process** (Fig. 20-10). In this process, the glass from the conditioning chamber is allowed to enter the Fourcault canal (basically a forehearth with a compartmentalized top opening at the end). A three-meter long clay block, called the "debiteuse," with a slot in the center floats on the molten glass in this compartment. The glass sheet is picked up through the slot using an iron "bait" and is pulled upward through a system of cast-iron or steel rollers about seven meters high. The sheet at the mouth

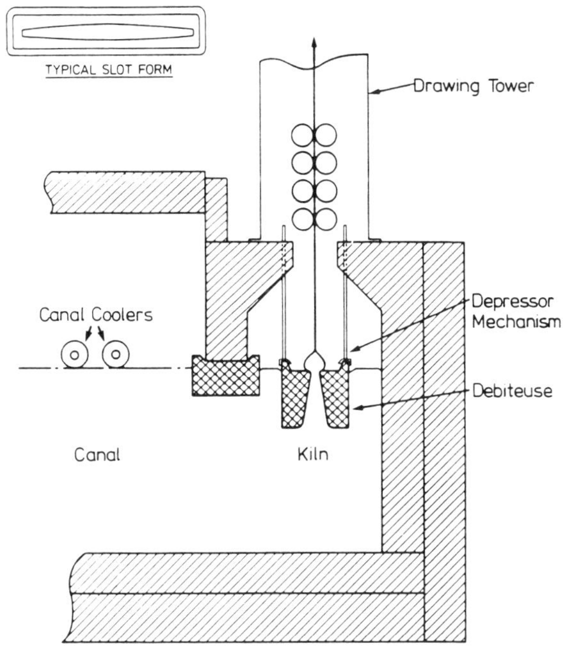

Figure 20-10. Fourcault flat glass drawing process. (After W. C. Hynd, in *Glass Science & Tech.*, Vol. 2, *Processing I* (D. R. Uhlmann, and N. J. Kreidl, eds.), Fig. 8, p. 72. Academic Press, New York, 1984.)

of the debiteuse is cooled at its edges in order to retain the width. The glass traveling upward is annealed along the travel, cut using a scorer, and picked up using suction cups. In a different version (without the need for the debiteuse), the glass is first drawn vertically and then turned around a polished nickel alloy roller horizontally (more or less). This is called the **Colburn–Libbey–Owens process**, Fig. 20-11. The thickness of the glass in both processes is determined by the glass viscosity and the drawing rate: thicker sheets are produced by slower drawing rates. Sheet glass so produced is clearly strong, because of its direct exposure to furnace heat (fire-polishing), although there do exist flaws due to the roller contact. The main disadvantage of this type of glass is its nonuniform thickness, which clearly makes the view through such glass quite distorted. Because of poor optics, it has been unsuitable for mirror manufacture. (Circus mirrors with funny-face images are made out of such types of glass.) Many improvements have been made in terms of the controls and the glass quality to obtain at least tolerable optics for architectural and automotive applications. An additional disadvantage of the sheet glass process is the inevitable presence of defects

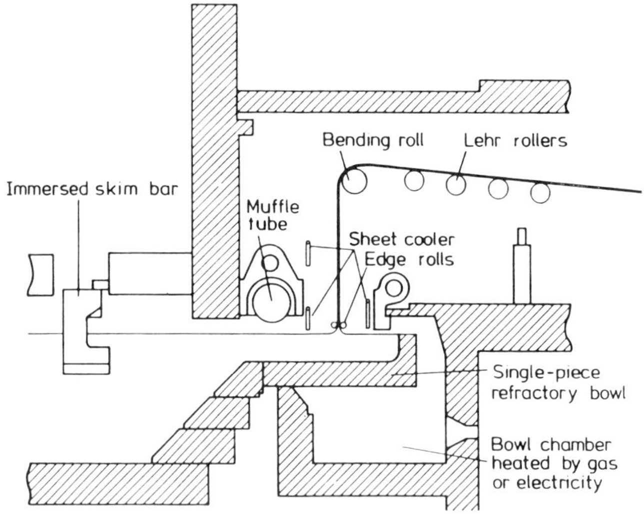

Figure 20-11. Colburn–Libbey–Owens flat draw process. (After W. C. Hynd, in *Glass Science & Tech.*, Vol. 2, *Processing I* (D. R. Uhlmann, and N. J. Kreidl, eds.), Fig. 9, p. 75. Academic Press, New York, 1984.)

such as lengthwise stretch marks, **striations**, and drawn bubbles (sometimes called **airlines**).

Flat glass of better optical quality and superior dimensional control, suitable for the manufacture of automobile windshields, microscope cover slips, and optical filters, can be made using a **downdraw process**. In the Corning downdraw process, the glass is allowed to fill a trough and overflow from both sides (Fig. 20-12). Both the streams are then recombined below to form a single sheet, which is drawn downwards with the help of roller tractors. A modified version of the downdraw process is used to make the Corning Corelle ware, where the two streams are brought to sandwich a centrally located third stream of a higher-expansion opal glass. The composite ribbon so formed is allowed to wrap over a rotating wheel, where circular patties are sucked out of the ribbon to form the desired dinnerware shape. The difference of thermal contraction coefficient between the two glasses causes the outer clear glass layers to develop a high degree of compression (and a balancing tension in the middle layer of the opal glass). An additional tempering process down the line adds further compression to the outer layers (the total compression is generally of the order of 300–400 MPa).

Instead of drawing upwards or downwards, continuous flat glass has also been produced by allowing the molten glass to come out of the forehearth over a close-fitting refractory lip and flow through a set of water-cooled

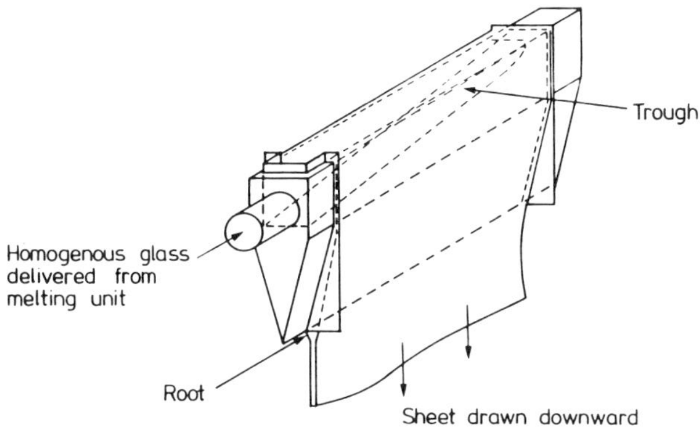

Figure 20-12. The Corning fusion downdraw process. (After W. C. Hynd, in *Glass Science & Tech.*, Vol. 2, *Processing I* (D. R. Uhlmann and N. J. Kreidl, eds.), Fig. 13, p. 82. Academic Press, New York, 1984.)

double rollers. An example of such a method is the **Boudin process**, shown schematically in Fig. 20-13. The rotational speed of the rollers, the spacing between them, and the glass pull rate control the glass thickness. The double-roller processes yield glass that has significantly less distortion than that produced by typical updraw processes and is quite suitable for automotive windshield applications. It has also been suitable for producing wire-meshed glass.

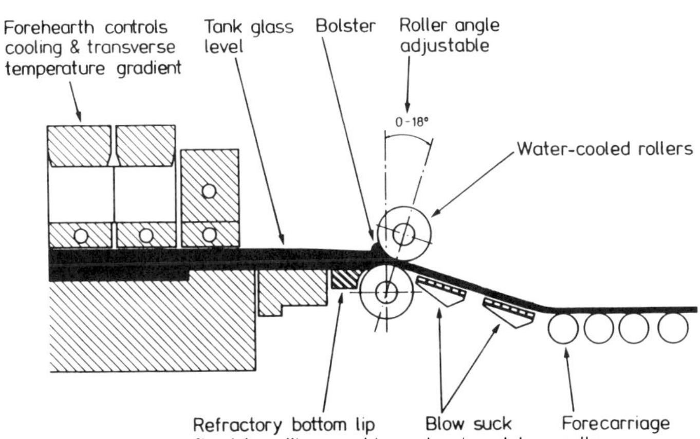

Figure 20-13. Boudin continuous double-roll process. (After W. C. Hynd, in *Glass Science & Tech.*, Vol. 2, *Processing I* (D. R. Uhlmann and N. J. Kreidl, eds.), Fig. 3, p. 61. Academic Press, New York, 1984.)

Plates for mirror applications have been made by grinding/polishing the sheet glass on both sides. In the most automated form, the sheet in the Libbey–Owens process, or in the double rollers process, is allowed to pass through large, slurry-fed grinding and polishing disks contacting the glass on both sides (called the **twin grinding process**). Flat glass so obtained has the desired quality of optics, but lacks strength because of the surface flaws generated by the grinding and polishing.

By far the most advanced process for making flat glass is the **float process** by Pilkington plc of the U.K. In this process, shown schematically in Fig. 20-14, the glass is brought out from a horizontal lip and allowed to pass between rollers and float over a bath of molten tin. The melting point of tin is 232°C, and it can be kept to a metallic state by maintaining a lightly reducing atmosphere. Gravity and surface forces spread the fluid glass evenly over the heavier fluid tin. At the exit, the temperature of the glass is brought down to about 600°C, at which the tin is still fluid, but the glass can be safely removed in a "rigid" condition. The glass is allowed to enter the annealing chambers and on to controlled cooling and cutting. It can be shown that this process readily yields glass plate thicknesses between

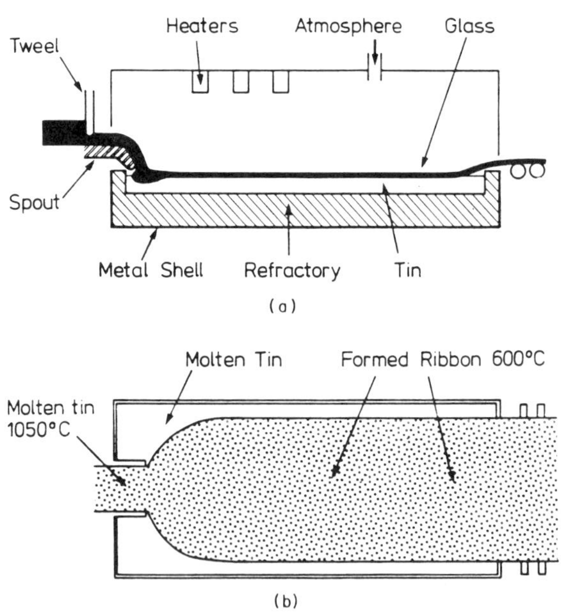

Figure 20-14. Float glass process. (a) Float bath. (b) Equilibrium mode of operation. (After W. C. Hynd, in *Glass Science & Tech.*, Vol. 2, *Processing 1* (D. R. Uhlmann and N. J. Kreidl, eds.), Fig. 14, p. 86. Academic Press, New York, 1984.)

about 2 mm and 20 mm, depending upon the glass viscosity and the drawing speed.

The superior quality of the float glass, combined with the high efficiency and the economics of the process, have virtually rendered all other flat glassmaking processes obsolete (except for a few specialized applications).

Concerns of ecology relating to flat glass have, in fact, farther-reaching ramifications than those relating to glass containers. The student will perhaps appreciate that a significant portion of the energy consumed by the Western world is simply in the area of heating and cooling offices and residences, and that much of this energy is fossil-fuel produced. This conversion generates CO_2 faster than the plant life on this planet could possibly consume — a scenario that could ultimately lead to the "greenhouse effect." A reduction in the rate of our consumption of fossil-fuel energy and exploitation of other cleaner sources of energy are mandatory if we are to avoid the global warming trends now predicted by some. Usage of radiation-controlled flat glass can help.

Three areas of radiation control are important: (i) Solar radiation, where the ratio of the transmitted visible light to the transmitted total radiation energy is improved relative to that of a clear glass window. This would reduce the amount of energy required to air-condition a room in warm season. (ii) Infrared reflective or lower emissivity windows to effect higher retention of expensive room heat during cold seasons. (iii) Antireflection coatings to reduce the otherwise about 4% reflection loss of solar radiation energy per surface for improved efficiency of flat-plate solar energy collectors.

Several processes are used for producing radiation-controlled glass. The conventional one for the heat absorption application is the full body tint obtained by adding about 0.5% iron to the glass. The ferrous portion of the iron generates a green color, which can readily absorb solar radiation around 1.1 micron. A 6-mm thick glass window may absorb about half the energy, but only about a quarter of the visible light in the solar radiation. As a result of the energy absorption, the full body tinted glass does have a tendency to become quite hot, which is a disadvantage by itself. In addition, the temperature gradients produced can result in high temporary stresses that could initiate glass fracture.

Copper, gold, silver, and several other metal films can be deposited on the glass by a mirroring type of cold process (reduction of metallic salt solutions). These films are not very adherent and are quite prone to external abrasion. Such films are thus suitable only for internal surface applications. More adherent wet process films are produced by the sol–gel route described in more detail earlier. Pyrolysis of organometallic compounds in contact with hot glass surface around 500°C can readily yield adherent coatings of metal oxides. Single coatings and controlled thickness, multiple-paired coatings of high- and low-refractive-index oxides (such as titania and silica) to enhance near-infrared reflectivity can be obtained in this manner.

Coatings of controlled thicknesses are more often obtained by vacuum evaporation, sputtering, or dc magnetron plasma sputtering. Again, coatings derived by simple vacuum evaporation are too soft for an external surface application. The magnetron-assisted coatings, on the other hand, can be quite hard and hermetic.

Another interesting technique that has been tried to obtain radiation-controlled float glass is to install copper electrode plates along the full width of the ribbon about halfway down the tin bath, a dragging contact with the ribbon being provided by a 2% Cu–Pb fluid alloy. When a dc voltage is applied such that the copper electrode is the anode, and the tin bath the cathode, then the copper ions from the molten alloy migrate into the upper surface of the glass, and sodium ions from the bottom surface are rejected into the tin bath. Copper in the molten alloy is continuously replenished by the copper plate. The copper ions in the glass surface are reduced to copper metal by the reducing atmosphere of the tin bath. These copper atoms then coagulate and form clusters, which imparts a metallic bronze reflection to the glass. Many modern buildings utilize these bronze-appearing windows for energy conservation.

The most modern method of radiation control is the use of **electrochromics**. Compounds such as WO_3 are coated on the glass using a variety of the processes mentioned earlier. When a dc voltage is applied, using a household battery in most cases, the optical absorption characteristics of the compound change. These changes can be custom-tailored. For instance, it is possible to adjust the level of greyness achieved by merely adjusting the voltage. This would have application in automobile and aircraft windshields. Likewise, it is possible to think that the same coating would act as an effective energy transmission barrier during the warm season, as it would to retain room heat during the cold season.

One of the important concerns in coatings technology is the development of stresses in the glass, due primarily to a difference in expansion between the coating and the substrate glass from the coating temperature to the ambient. Of course, there are other sources of stress generation that apply to specific technologies. For instances, in magnetron sputtering, the plasma ions bombarding the glass can produce stresses not only by getting incorporated into the interstices, but also through an "atomic-peening" effect [15]. The presence of tensile stresses in flat glass can severely limit its application because of increased risk of fracture. Likewise, for architectural applications, the windows and their mounting have to be properly designed to account for wind pressures.

20.4.3. Glass Tubes and Rods

Glass tubes and rods can be formed much more easily compared to the forming of containers and flat glass. There are four basic processes that are used for tube/rod forming: the Danner process, updraw, downdraw, and the Vello process.

In the **Danner process**, the glass from the forehearth is allowed to flow in ribbon form over a downward-inclined (about 20° from the horizontal), rotating, and hollow refractory mandrel placed inside a temperature-controlled muffle. The continuous rotation (roughly 10 rpm) makes the glass wrap around the mandrel while flowing gradually downstream. The glass ultimately flows off the mandrel, where it is picked up, cooled gradually, and placed between horizontal tractors. (See Fig. 20-15.) The tractors are located at a far distance (20 to 40 m) away from the mandrel at a level about 2–3 m below. Thus, the glass is now in the form of a catenary suspended freely between the mandrel and the tractors. The upper tractor mechanism applies traction at a slight angle to the lower tractor, enabling the glass to rotate in synchronism with the rotation of the mandrel. To make tubes, air is blown gently through the mandrel, whereas for rod-forming, a gentle vacuum is applied (which causes the glass to collapse on the inner core as it leaves the mandrel).

There are several issues one must watch for in the Danner process. Firstly, the mandrels themselves can be quite large (as much as 2 ft in diameter and

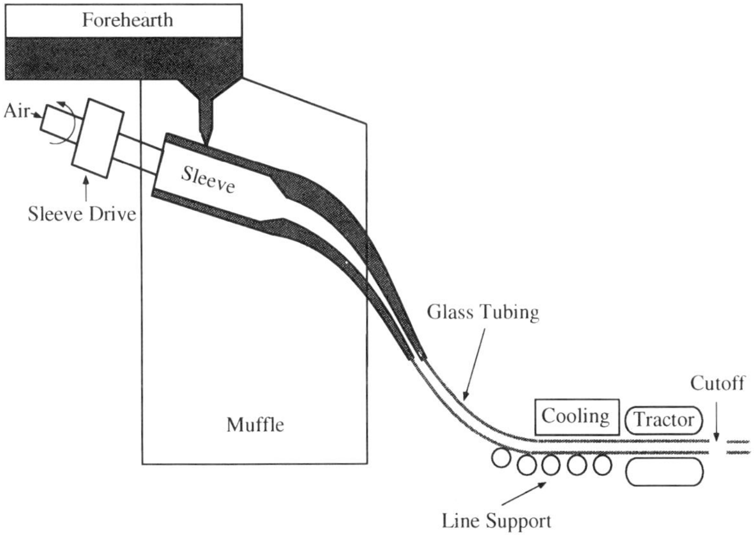

Figure 20-15. The Danner tube/rod drawing process.

about 8 ft long). When a new mandrel is installed, it should be heated very, very slowly to allow any condensation inside the mandrel (formed during storage) to escape, or else the mandrel could rupture. Also, the muffle chamber should have good uniformity of temperature over the region of the mandrel; otherwise, glass leaves the mandrel with varying viscosities depending upon the rotation speed. This can be witnessed in the form of the catenary swinging up and down with each rotation of the mandrel. Clearly, the result of the swing is variation in the wall thickness ("siding" variation for a tube) and ovality of the section.

In the **updraw process**, the glass is drawn vertically upwards off a hollow refractory cone located within a surrounding, rotating refractory sleeve partly immersed in the glass. Again, air is blown through the hollow cone to allow the formation of the tube. The updraw process has the advantage that a variety of tube sizes and wall thicknesses can be made by relatively simple variations applied in the location of the refractory cone with respect to the glass level. Large, thick-walled tubes could be made using this process. However, it may also be realized that the process is indeed overly sensitive to fluctuations in glass level, which itself is a serious disadvantage. For several such reasons, the updraw process has now become obsolete in favor of the downdraw process.

In the **downdraw process**, the glass flows through a carefully controlled orifice ring and is pulled vertically downwards by a tractor mechanism. If a tubing is to be drawn, the glass flows over a "bell" located concentrically within the orifice ring (Fig. 20-16). Air pressure (or a slight vacuum) is applied

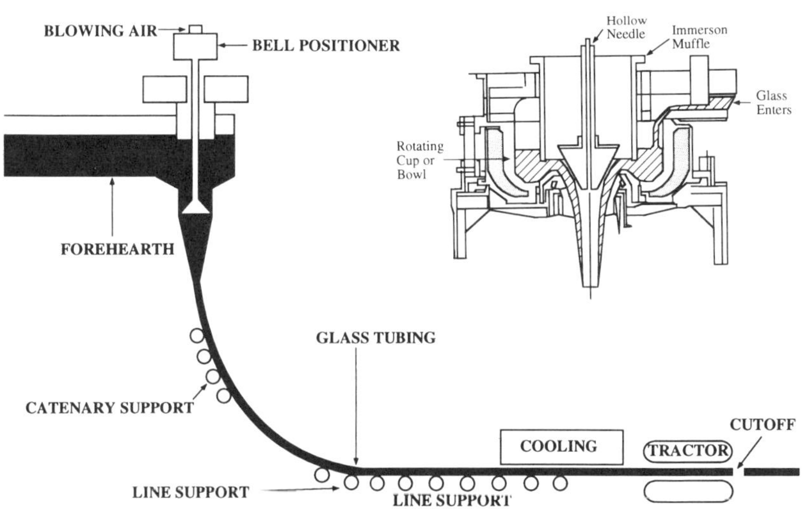

Figure 20-16. The Vello process.

to vary the inner diameter (and the wall thickness) of the tube. A wide range of tube diameters and wall thicknesses can be made by adjusting the air pressure and the glass flow, etc., without the need for changing the bell itself. In the **Vello process** (Fig. 20-16), the tube or the rod is turned horizontally along support rollers. Unlike the Danner process, the tube or the rod is not rotated. Because of this, and because of the orifice/bell controls, both the downdraw and the Vello processes give an excellent-quality product in terms of OD and ID control. The bell and the orifice shapes can be varied to give various cross-sectional shapes to the inside as well as the outside surfaces. For instance, thermometer tubing with a magnifying cylindrical lens shape on the outside and an elliptical bore on the inside is made using the Vello process. In addition, colored stripes can also be introduced via channels carried into the Vello bowl just before the glass passes through the orifice.

20.4.4. Glass Fibers

Suggested reading: R. M. Klein, in *Glass Science & Technology*, Vol. 2, *Processing* (D. R. Uhlmann and N. J. Kreidl, eds.), pp. 285–339. Academic Press, New York, 1984.

Glass fibers can be readily formed by rapidly stretching a softened mass of glass at just about any angle (horizontally, upwards, or downwards). As in the case of tube drawing, a downdraw process allows the possibility of custom-tailoring the cross-section and gives better control of size. To draw glass fibers in the laboratory, a small piece of glass is placed inside a platinum crucible that has a small hole drilled in the bottom. The crucible is mounted between two electrodes and acts as an electric resistance heater. By gradually increasing the voltage supplied across the crucible, glass can be brought to melting and flow through the hole and through a surface-coating device, before being wrapped on a high-speed rotating drum.

The manufacture of optical waveguide requires more care. The starting point is the formation of a glass blank or "preform" (much like the "parison" for a container). Instead of melting the glass in the crucible (and risking raw material inclusions, seeds, and other inhomogeneities in the drawn fiber), a rod-shaped preform is first prepared and examined for its quality. Acceptable preforms are then softened at the end of the rod in a furnace or by laser heating. A fiber is drawn much like that described earlier. The drawing equipment is synchronized to advance the preform rod into the softening zone to keep up with the rate of the draw. Single refractive-index fibers, both single-mode and multimode, can be readily manufactured using this technique. Yet another method to make single refractive-index fibers, particularly silica fibers, is to use the sol–gel process (Section 20.3.3).

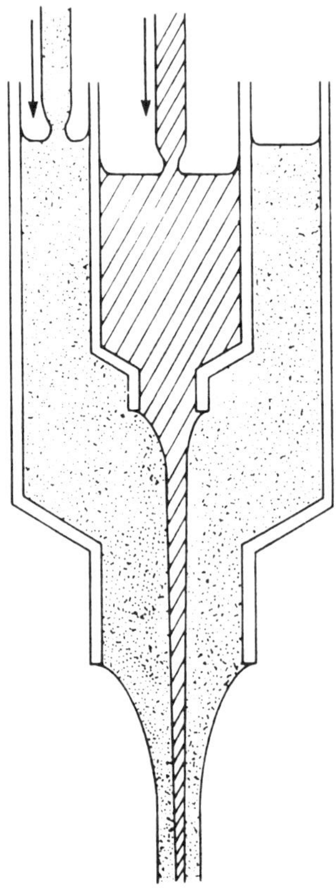

Figure 20-17. Double crucible technique for clad-core fiber drawing. (After R. M. Klein, in *Glass Science & Tech.*, Vol. 2, *Processing* I (D. R. Uhlmann and N. J. Kreidl , eds.), Fig. 20, p. 331, Academic Press, New York 1984.)

There are two techniques to make core-clad waveguides. The first is the **double-crucible technique** (16, 17), applicable only to a step-index fiber and shown schematically in Fig. 20-17. Preform rods of the two desired composition glasses are gradually advanced into a double-wall crucible where the two glasses melt. Two concentric orifices located at the bottom of each chamber allow the glasses to flow out in a core-clad form and be drawn together. A second and more advanced technique, applicable to the manufacture of graded index fibers as well as step-index fibers, is to prepare a compound preform and draw it much like a single refractive-index fiber.

The compound preform itself is prepared by **vapor or plasma deposition** in a variety of ways, depending upon the shape of the refractive-index profile desired. To perform vapor deposition, vapors of $SiCl_4$ mixed with small amounts of another halide such as $GeCl_4$, $POCl_3$, BCl_3, SiF_4, or CCl_2F_2 are hydrolyzed by passing through a high-temperature flame or oxidized by reaction with oxygen, again at a high temperature. The hydrolysis reaction

$$SiCl_4 + 2H_2O = SiO_2(\text{glass}) + 4HCl$$

or the oxidation reaction

$$SiCl_4 + O_2 = SiO_2(\text{glass}) + 2Cl_2$$

produces silica glass with Ge, P, B, and F incorporated into the network. Ge and P additions to silica raise the refractive index, whereas additions of B and F decrease the refractive index (Section 19.5.4). By precisely controlling the composition of the halide liquid mix, the refractive-index profile can be accurately shaped at will. The vapor deposition is carried out in one of three ways: radial outside vapor deposition (simply called the **OVD process**), radial inside vapor deposition (**IVD process**, also called the internal thermal oxidation or the **ITO process**), and the axial vapor deposition (**AVD process**). These processes are shown schematically in Fig. 20-18. In the OVD, the

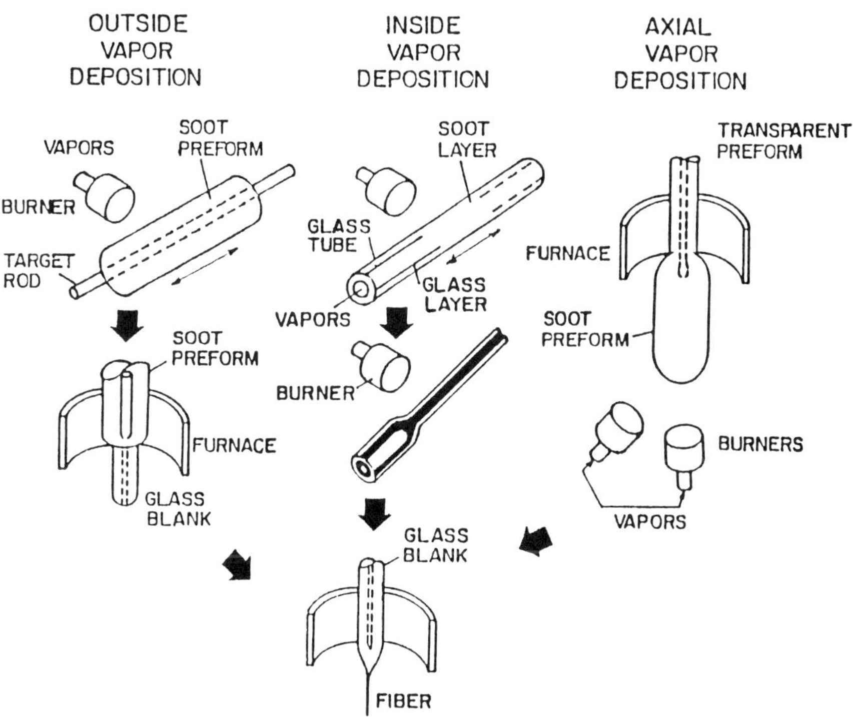

Figure 20-18. Summary of the various CVD processes to make a fiber preform.

mixed halide vapors are introduced through a suitably shaped burner, and the hydrolyzed product is deposited in soot form on the outside of a (generally) fused-silica target rod or boule, building the soot layer gradually in the radial direction. The target rod is rotated as well as translated with respect to the burner. The deposited porous soot layer is dehydrated at around 1,200°C and sintered to full density subsequently by heating to 1,500–1,700°C. In IVD, the deposition is carried out on the inside of a tube. Higher temperatures than the OVD process are needed to carry out the oxidation in a restricted space, and the sintering of a soot layer is carried out before the deposition of the next soot layer. The chlorine that is evolved itself acts as a strong dehydrant; as a result, the IVD process yields significantly lower –OH impurity preforms than does the OVD process. In the AVD process, the deposition is carried out by hydrolysis of the halide mixture at the end of the seed rod in an axial direction. The distance between the burner and the end of the preform is kept constant by gradually retracting the preform away as it grows (see Fig. 20-19). The radial concentration distribution of the various elements is varied by controlling the radial temperature distribution during the deposition. Continuing retraction of the porous preform causes it to pass through a combined dehydration/sintering chamber held at 1,200–1,500°C (Fig. 20-19). For dehydration, gases such as $SOCl_2$ or Cl_2 mixed with He are used. The sintered preform is then jacketed by vacuum-collapsing a fused-silica tube over it.

The **plasma deposition process** resembles the IVD process, except that the oxidation reaction is carried out by plasma. Unlike the IVD process, where the glass is formed in the vapor phase, the glass in the plasma process is formed in a sintered form directly upon deposition on the cooler walls of the tube.

Some research has also gone into making a compound preform by partial leaching of a phase-separated glass, much like the Vycor process carried out to partial completion only. A phase-separable alkali borosilicate glass rod is heat-treated to effect phase separation, and immersed subsequently in an acid bath to dissolve the soluble alkali borate phase partially from the surface inwards, such that the core is still unleached. In one version, this partially leached rod is sintered as-is to become a preform where the inside is alkali borosilicate glass clad with a gradually reducing alkali borate, till the surface is essentially 95% silica. In another version, the partially leached rod is immersed in a solution containing a heavier alkali (such as Cs) that is allowed to soak through its interconnected capillaries. Upon drying and sintering, a cladding of graded refractive index higher than the parent is obtained.

As indicated earlier, fibers are then drawn from these preforms in a draw tower arrangement such as that shown in Fig. 20-20.

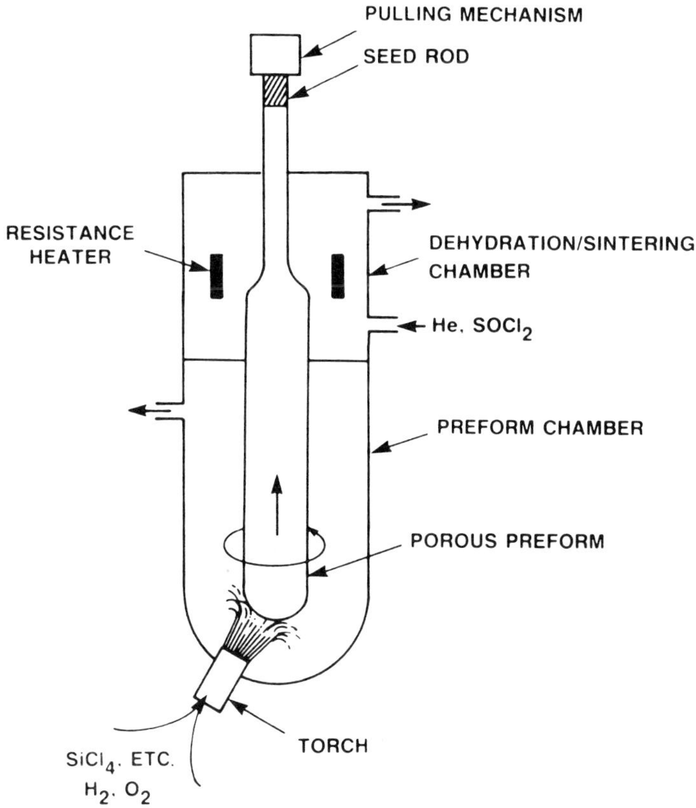

Figure 20-19. Axial flame hydrolysis technique to make a fiber preform. (After R. M. Klein, in *Glass Science & Tech.*, Vol. 2, *Processing I* (D. R. Uhlmann and N. J. Kreidl, eds.), Fig. 14, p. 310, Academic Press, New York, 1984.)

Prior to spooling, *the fibers must be coated* (at high speeds, continuously) to protect them from abrasion and subsequent weakening. It is also generally desirable to minimize the penetration of humidity, which would eventually corrode the glass. Coatings have to be of the type that would not induce stresses in the fiber; otherwise, such stresses are likely to change the radial distribution of the refractive index and cause loss of optical transmission. (Such losses are called **microbending losses**). The common procedure is to coat the fibers with a thermally cured or UV-cured organic polymer overcoated with a harder thermoplastic resin. Polyimides are probably the more popular thermally cured coatings, whereas epoxy or urethane acrylates

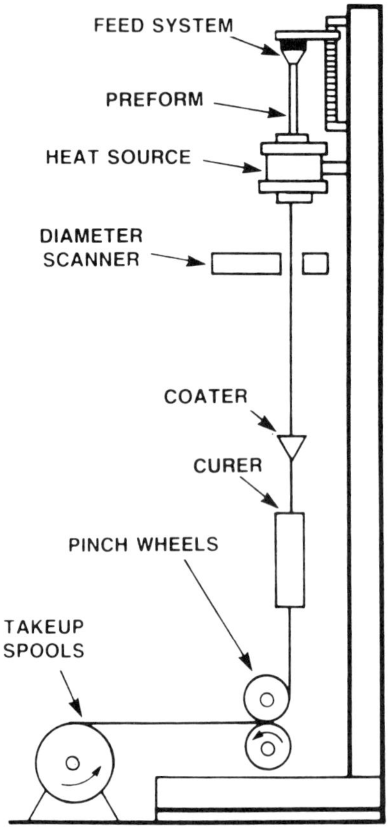

Figure 20-20. Typical fiber drawing tower. (After R. M. Klein, in *Glass Science & Tech.*, Vol. 2, *Processing I.* (D. R. Uhlmann, N. J. Kreidl, eds.), Fig. 16, p. 315, Academic Press, New York, 1984.)

are the more common UV-cured coatings. Extruded nylons or Teflon serve as common overcoats. Where the fibers are more readily corroded by water, e.g., the proposed HMFG glass fibers, it would almost be necessary to coat them hermetically. The long-term hermeticity requirement virtually eliminates the use of organic coatings. Research is being carried out currently to apply inorganic coatings of metals and metal oxide combinations using a dc plasma magnetron sputtering process [18, 19].

The word "research" is a natural reminder to the author of the backlog of things to do that have piled up. It is time to quit writing this book!

Exercise

1. Calculate a suitable batch composition to give 1 kg of $11Na_2O{\cdot}9CaO{\cdot}2MgO{\cdot}2B_2O_3{\cdot}75.5SiO_2{\cdot}0.5SO_3$ (wt. %) glass. Use sand, soda ash, Na_2SO_4, dolomite, limestone, and borax as starting materials. Assume that 100% of the water, 75% of the S, 5% B_2O_3, and 2% Na_2O from the batch are lost during the melting.

References

1. F. V. Tooley, *Handbook of Glass Manufacture*. Ogden Publishing Co., New York, 1953.
2. A. R. Cooper, *J. Am. Ceram. Soc.* **42**(2), 93 (1959). *Glass Technol.* **7**, 2 (1966); *Proc. Adv. in the Fusion of Glass*, Paper No. 23, *Am. Ceram. Soc.*, 1988.
3. J. G. O'Keefe, in *Natural Glasses* (L. D. Pye, J. G. O'Keefe, and V. D. Frechette, eds.), pp. 1–17. North-Holland, Amsterdam, 1984.

4a. D. Stöffler, in Ref. 3, pp. 465–502.

4b. J. Arndt, U. Hoernemann, and W. F. Müller, *Phys. Chem. Glasses* **12**(1), 1–7 (1971).

5. A. Pabst, *Amer. Minerol.* **37**(3&4), 137 (1952).
6. W. Primak, *J. Phys. Chem. Sol.* **13**, 279–286 (1960).
7. W. C. LaCourse and S. Kim, in *Science of Ceramic Chemical Processing* (L. L. Hench and D. R. Ulrich, eds.), pp. 304–310. J. Wiley, New York, 1986.
8. L. Klein, *Ann. Rev. Mater. Sci.* **15**, 227–248 (1985).
9. C. J. Brinker, *J. Non-cryst. Sol.* **100**, 31–50 (1988).
10. E. J. A. Pope and J. D. Mackenzie, *J. Non-cryst. Sol.* **87**, 185 (1986).
11. S. Wallace and L. L. Hench, in *Better Ceramics Through Chemistry, Proc. MRS Symp. 32* (C. J. Brinker, D. E. Clark, and L. L. Hench, eds.), pp. 47–52. Elsevier Science, Amsterdam, 1984.
12. R. A. Laudise and D. W. Johnson, *J. Non-cryst. Sol.* **79**, 155–164 (1986).
13. J. L. Lippert, S. B. Melpolder, and L. W. Kelts, *J. Non-cryst. Sol.* **104**, 139–147 (1988).
14. B. C. Bunker, T. J. Headley, and S. C. Douglas, in Ref. 11, pp. 41–46.
15. D. W. Hoffman and M. R. Gaerttner, *J. Vac. Sci. Technol.* **17**, 425 (1980).
16. N. S. Kapany and R. J. Simms, *Infrared Phys.* **5**, 69–80 (1965).

17. T. Kanamori, Y. Terunuma, S. Takahashi, and T. Miyashita, *J. Non-cryst. Sol.* **69**, 231–242 (1985).
18. D. W. Hoffman and J. A. Thornton, *J. Vac. Sci. Technol.* **20**(3), 355–358 (1982).
19. K. Yasui, K. Komaki, and S. Kaneda, *J. Non-cryst. Sol.* **127**, 1–7 (1991).

Appendix I

Elements of Linear Elasticity

A force vector **F**, applied in an arbitrarily chosen direction, has three components F_x, F_y, and F_z acting along the three axes x, y, and z of the Cartesian coordinate system. Since stress is defined as force per unit area, and the area can be represented by a vector normal that can itself be represented by its three components, the stress σ_{ij} is a second-rank tensor having nine components written in a matrix form as (see Fig. A1-1)

$$\sigma_{ij} = \begin{vmatrix} \sigma_{xx} & \sigma_{xy} & \sigma_{xz} \\ \sigma_{yx} & \sigma_{yy} & \sigma_{yz} \\ \sigma_{zx} & \sigma_{zy} & \sigma_{zz} \end{vmatrix}. \tag{A1.1}$$

σ_{ij} is the stress component acting on the i plane (i.e., whose normal is the i direction) in the j direction. The nine stress components are not all independent, since $\sigma_{ij} = \sigma_{ji}$. This then yields six independent components of stresses due to an arbitrarily applied external force. These components are often written in abbreviated form as

$$\sigma_{xx} = \sigma_1; \quad \sigma_{yy} = \sigma_2; \quad \sigma_{zz} = \sigma_3; \quad \sigma_{yz} = \sigma_4; \quad \sigma_{xz} = \sigma_5; \quad \sigma_{xy} = \sigma_6 \tag{A1.2}$$

(The best way to remember this abbreviation is to consider x, y, z as 1, 2, 3; replace ii with i and, for mixed indices, add 3 to the missing index.) Note that σ_{ii} are the normal components, whereas $\sigma_{ij} (i \neq j)$ are the shear components.

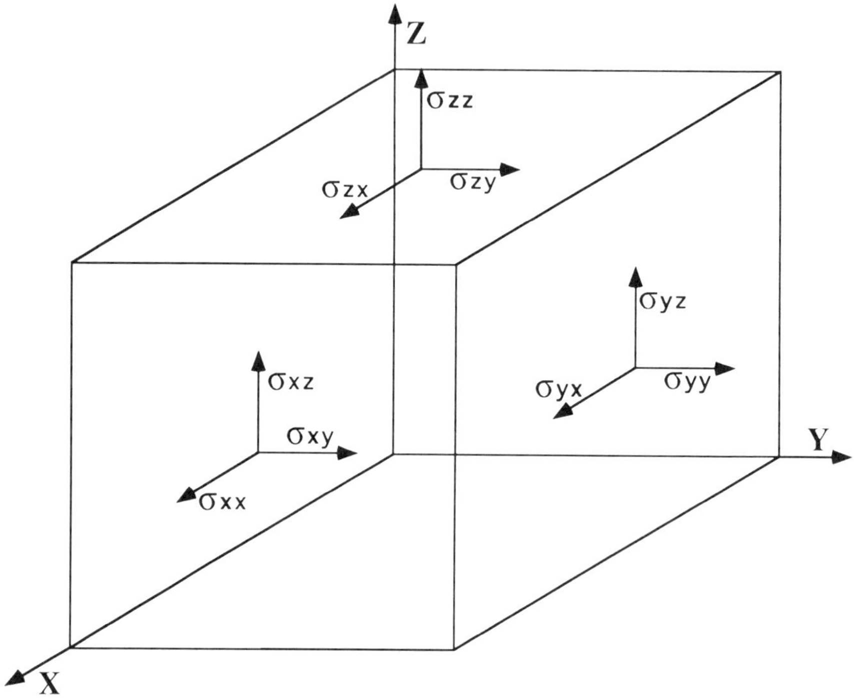

Figure A1-1. Components of stress tensor in Cartesian coordinates.

Equilibrium of forces at the point of application requires that the following three equations be obeyed:

$$\frac{\partial\sigma_{xx}}{\partial x} + \frac{\partial\sigma_{xy}}{\partial y} + \frac{\partial\sigma_{xz}}{\partial z} + f_x = 0,$$

$$\frac{\partial\sigma_{yy}}{\partial y} + \frac{\partial\sigma_{xy}}{\partial x} + \frac{\partial\sigma_{yz}}{\partial z} + f_y = 0,$$

$$\frac{\partial\sigma_{zz}}{\partial z} + \frac{\partial\sigma_{xz}}{\partial x} + \frac{\partial\sigma_{yz}}{\partial y} + f_z = 0, \tag{A1.3}$$

where f_i denotes the intensity of the body forces (body force per unit volume) such as gravitational force, magnetic force, or inertial force for a body in motion. Equations (A1.3) are called the **equilibrium equations.**

If the displacements of a point Q in the body with respect to another reference point P are described by a vector having components u, v, and w

along the three axes, then the **engineering strain** tensor e_{ij} is a second-rank tensor having nine components of strain. The matrix of e_{ij} is written as

$$e_{ij} = \begin{vmatrix} e_{xx} & e_{xy} & e_{xz} \\ e_{yx} & e_{yy} & e_{yz} \\ e_{zx} & e_{zy} & e_{zz} \end{vmatrix}, \tag{A1.4}$$

where

$$e_{xx} = \partial u/\partial x; \qquad e_{yy} = \partial v/\partial y; \qquad e_{zz} = \partial w/\partial z;$$
$$e_{xy} = \partial u/\partial y + \partial v/\partial x; \qquad e_{xz} = \partial u/\partial z + \partial w/\partial x;$$
$$e_{yz} = \partial v/\partial z + \partial w/\partial y;$$

and

$$e_{xy} = e_{yx}; \qquad e_{xz} = e_{zx}; \qquad e_{yz} = e_{zy}. \tag{A1.5}$$

Thus, only six of the strain components need be independent.

The e_{ii} are the extensional stains that change both the shape and the volume of the body. The $e_{ij} (i \neq j)$ are shear strains that change only the shape, not the volume, of the body.

The abbreviation of the various e_{ij} $(i, j = 1 \text{ to } 3)$ to e_i $(i = 1 \text{ to } 6)$ is given by

$$e_{xx} = e_1; \quad e_{yy} = e_2; \quad e_{zz} = e_3; \quad e_{yz} = e_4; \quad e_{xz} = e_5; \quad e_{xy} = e_6. \tag{A1.6}$$

It may be shown that each of the shear components is composed of a symmetric deformation part and an antisymmetric rotational part. The **mathematical strain** components, denoted by ε_{ij}, ignore the rotational part and are given by

$$\varepsilon_{xx} = \partial u/\partial x; \qquad \varepsilon_{yy} = \partial v/\partial y; \qquad \varepsilon_{zz} = \partial w/\partial z;$$

and

$$\varepsilon_{xy} = \tfrac{1}{2}[\partial u/\partial y + \partial v/\partial x]; \qquad \varepsilon_{xz} = \tfrac{1}{2}[\partial u/\partial z + \partial w/\partial x];$$
$$\varepsilon_{yz} = \tfrac{1}{2}[\partial v/\partial z + \partial w/\partial y],$$

i.e.,

$$\varepsilon_{ii} = e_{ii} \qquad \text{and} \qquad \varepsilon_{ij} = \tfrac{1}{2} e_{ij}. \tag{A1.7}$$

Hence, the mathematical strain tensor may be written as

$$\varepsilon_{ij} = \begin{vmatrix} \varepsilon_{xx} & \varepsilon_{xy} & \varepsilon_{xz} \\ \varepsilon_{yx} & \varepsilon_{yy} & \varepsilon_{yz} \\ \varepsilon_{zx} & \varepsilon_{zy} & \varepsilon_{zz} \end{vmatrix} = \begin{vmatrix} e_1 & \frac{1}{2}e_6 & \frac{1}{2}e_5 \\ \frac{1}{2}e_6 & e_2 & \frac{1}{2}e_4 \\ \frac{1}{2}e_5 & \frac{1}{2}e_4 & e_3 \end{vmatrix}. \tag{A1.8}$$

The generalized Hooke's law relationship between the stress and the strain is written as

$$\sigma_1 = c_{11}e_1 + c_{12}e_2 + c_{13}e_3 + c_{14}e_4 + c_{15}e_5 + c_{16}e_6, \qquad \text{(A1.9)}$$

and so on for σ_2, σ_3, σ_4, σ_5, and σ_6, where the constants c_{ij} are called the **elastic stiffness coefficients**. The six equations are abbreviated simply as

$$\sigma_i = c_{ij}e_j \qquad (i = 1 \text{ to } 6) \qquad \text{(A1.10)}$$

(summation over the dummy index $j = 1$ to 6 is automatically implied). Likewise,

$$e_i = s_{ij}\sigma_j \qquad (i = 1 \text{ to } 6), \qquad \text{(A1.11)}$$

where the constants s_{ij} are called the **elastic compliance coefficients**.

The generalized Hooke's law can also be written in the fully expanded version involving mathematical strains, which is

$$\begin{aligned}\sigma_{11} = {} & c_{1111}\varepsilon_{11} + c_{1112}\varepsilon_{12} + c_{1113}\varepsilon_{13} + c_{1121}\varepsilon_{21} + c_{1122}\varepsilon_{22} \\ & + c_{1123}\varepsilon_{23} + c_{1131}\varepsilon_{31} + c_{1132}\varepsilon_{32} + c_{1133}\varepsilon_{33},\end{aligned} \qquad \text{(A1.12)}$$

and so on for the eight other components of σ_{ij}. Note that the elastic stiffness coefficients are fourth-rank tensors. In a similar manner, the inverse relationship can also be written in terms of the compliance coefficients, i.e.,

$$\varepsilon_{ij} = s_{ijkl}\sigma_{kl} \qquad (i, j, k, l, = 1, 2, 3). \qquad \text{(A1.13)}$$

Some useful relationships to note are

$$\sum c_{ij}s_{ij} = 1 \qquad (j = 1 \text{ to } 6); \qquad \sum c_{ij}s_{jk} = 0 \qquad (i \neq j); \qquad \text{(A1.14)}$$

$$c_{ijkl} = c_{mn} \qquad (i, j, k, l = 1, 2, 3), \qquad \text{(A1.15)}$$

and the indices m and n are to substitute for ij and kl, respectively, in accordance with the rule stated earlier:

$$s_{ijkl} = s_{mn} \qquad \text{when } m \text{ and } n \text{ are 1, 2, or 3;}$$

$$2s_{ijkl} = s_{mn} \qquad \text{when either } m \text{ or } n \text{ is 4, 5, or 6;}$$

$$4s_{ijkl} = s_{mn} \qquad \text{when both } m \text{ and } n \text{ are 4, 5, or 6.} \qquad \text{(A1.16)}$$

(The reader may note that relationship between the compliance coefficients is not as simple as that between the stiffness coefficients, simply because of the relationship (A1.7) between ε_{ij} and e_i.)

For a cubic crystal,

$$c_{11} = c_{22} = c_{33},$$

$$c_{12} = c_{13} = c_{21} = c_{23} = c_{31} = c_{32},$$

$$c_{44} = c_{55} = c_{66}, \tag{A1.17}$$

and the remaining 24 out of the 36 stiffness coefficients are zero. In other words, there are three independent moduli of elasticity.

For isotropic solids such as glass,

$$c_{44} = (c_{11} - c_{12})/2. \tag{A1.18}$$

Writing $c_{12} = \lambda$, and $c_{44} = G$ (same as the shear modulus), it is possible to show that the Young's modulus

$$E = G(3\lambda + 2G)/(\lambda + G), \tag{A1.19}$$

and the Poisson's ratio

$$v = \lambda/2(\lambda + G). \tag{A1.20}$$

Alternatively,

$$\lambda = vE/(1 + v)(1 - 2v). \tag{A1.21}$$

In this context, λ and G are called the **Lame's constants**. (The reader should not confuse E with c_{11}.) Hence, there are only two independent elastic moduli for isotropic solids.

Returning to considering stresses and strains, it is useful to understand the effect of rotating the axes. We have noted that, given an arbitrary orthogonal system of axes x, y, and z, a complete description of the stress state at the point of application requires six independent components of the stress tensor. We could rotate the axes in space. It is possible to show that in a specific orientation of the new orthogonal axes 1, 2, and 3, the shear stress components on planes perpendicular to these axes vanish. In other words, only σ_{11}, σ_{22}, and σ_{33} are sufficient to describe the stress state: σ_{12}, σ_{13}, and σ_{23} are zero. This, however, is not a case of six independent variables somehow being reduced to three (that is impossible). What is required is the specification of three independent direction cosines a_{ij}, which the 1, 2, and 3 axes make with the initially chosen x, y, z axes. The planes along which the shear components vanish are called the **principal planes**, the normals (which are the 1, 2, and 3 axes) are called the **principal axes**, and the stresses along these 1, 2, and 3 axes are called the **principal stresses** (often denoted by σ_1, σ_2, and σ_3 and written such that $\sigma_1 > \sigma_2 > \sigma_3$). The magnitude of these principal stresses may be determined from the relation

$$\begin{vmatrix} \sigma_{xx} - \sigma_i & \sigma_{xy} & \sigma_{xz} \\ \sigma_{xy} & \sigma_{yy} - \sigma_i & \sigma_{yz} \\ \sigma_{xz} & \sigma_{yz} & \sigma_{zz} - \sigma_i \end{vmatrix} = 0, \qquad \text{(A1.22)}$$

or, in other words,

$$\begin{aligned} &\sigma_i^3 - (\sigma_{xx} + \sigma_{yy} + \sigma_{zz})\sigma_i^2 + (\sigma_{xx}\sigma_{yy} + \sigma_{yy}\sigma_{zz} + \sigma_{xx}\sigma_{zz} - \sigma_{xy}^2 - \sigma_{xz}^2 - \sigma_{yz}^2)\sigma_i \\ &\quad -(\sigma_{xx}\sigma_{yy}\sigma_{zz} + 2\sigma_{xy}\sigma_{xz}\sigma_{zz} - \sigma_{xx}\sigma_{yz}^2 - \sigma_{yy}\sigma_{xz}^2 - \sigma_{zz}\sigma_{yz}^2) = 0 \qquad \text{(A1.23)} \end{aligned}$$

There are three real roots of this equation that yield the three principal stresses. The principal axes may be determined from any two of the relations

$$a_{ix}(\sigma_{xx} - \sigma_i) + a_{iy}\sigma_{xy} + a_{iz}\sigma_{xz} = 0,$$

$$a_{ix}\sigma_{xy} + a_{iy}(\sigma_{yy} - \sigma_i) + a_{iz}\sigma_{yz} = 0,$$

$$a_{ix}\sigma_{xz} + a_{iy}\sigma_{yz} + a_{iz}(\sigma_{zz} - \sigma_i) = 0,$$

combined with the relation

$$a_{ix}^2 + a_{iy}^2 + a_{iz}^2 = 1 \qquad \text{for each } i. \qquad \text{(A1.24)}$$

The axes could be rotated by 45° to obtain a new set of orthogonal axes 1′, 2′, and 3′ such that the perpendicular planes have only shear stresses acting upon them and no normal stresses. These shear stresses, called principal shears, may be calculated from

$$\tau_{1'} = \tfrac{1}{2}(\sigma_2 - \sigma_3); \qquad \tau_{2'} = \tfrac{1}{2}(\sigma_3 - \sigma_1); \qquad \tau_{3'} = \tfrac{1}{2}(\sigma_1 - \sigma_2). \quad \text{(A1.25)}$$

By substituting ε_{ij} in place of σ_{ij} in Eq. (1.24), the principal strains ε_i can also be calculated.

The most important property of principal stresses is that they represent the maximum and minimum normal stress at a point. Thus, σ_1 is a maximum normal stress and σ_3 a minimum. It follows that the maximum shear stress is $\tau_{2'}$. The consideration of principal axes, principal stresses, principal strains, and principal shears is important with respect to criteria for failure in materials. Most materials fail either because of the magnitude of the maximum principal stress (tension or compression), or because the maximum principal shear exceeds some intrinsic strength. Most brittle materials such as glasses fail because of a principal tensile stress exceeding an engineering strength (see chapter on the strength of glass), and the fracture plane is perpendicular to the direction of the principal tension, i.e., the fracture plane is one of the principal planes. The measurement of stresses in glass by photoelastic techniques and the estimation of failure probability thus require a knowledge of principal axes, etc. In ductile materials, such as metals, plastic flow occurs because of excessive shears. A useful criterion to determine the extent of yielding due to application of a three-dimensional stress is the **von Mises criterion**, where yielding occurs when the effective stress σ_{eff}, defined by

$$\sigma_{eff} = [\tfrac{1}{2}\{(\sigma_1 - \sigma_2)^2 + (\sigma_2 - \sigma_3)^2 + (\sigma_3 - \sigma_1)^2\}]^{1/2}, \qquad (A1.26)$$

exceeds the yield stress in simple tension.

Since Eq. (A1.23) applies to any arbitrary orientation of the x, y, z axes, yet still yields the same value of σ_i, Eq. (A1.23) must be invariant, and hence the quantities in the parentheses in (A1.23) do not change. An important consequence of this is that

$$\sigma_{xx} + \sigma_{yy} + \sigma_{zz} = \sigma_1 + \sigma_2 + \sigma_3 = 3\sigma_m = \text{a constant.} \qquad (A1.27)$$

Likewise,

$$\varepsilon_{xx} + \varepsilon_{yy} + \varepsilon_{zz} = \varepsilon_1 + \varepsilon_2 + \varepsilon_3 = \varepsilon_m = \text{a constant.} \qquad (A1.28)$$

The quantity σ_m is called the **hydrostatic (or the spherical) component** of the stress, and ε_m is identically equal to e_m, the volumetric strain or the dilatation defined in Eq. (8.6). It may be shown that

$$\sigma_m = K\varepsilon_m. \qquad (A1.29)$$

In problems involving viscous fluids, it is useful to separate the total stress into the volume-conserving **deviatoric component** S_{ij}, which is the pure shear part producing shape change only, and the hydrostatic component producing dilatation only. In a glass near its transition range, the hydrostatic component of the stress may relax with time by what is called the **volume viscosity**, which is a structural relaxation mechanism. Once an equilibrium structure has been achieved, no further relaxation may occur. In purely viscoelastic bodies, the structure does not change and viscous relaxation occurs by shape change. It is the S_{ij} which causes true viscous flow. The S_{ij} is given by

$$S_{ij} = \sigma_{ij} - \delta_{ij}\sigma_m, \qquad (A1.30)$$

where δ_{ij} is the Kronecker delta, given by

$$\begin{aligned} \delta_{ij} &= 0 \qquad \text{for } i \neq j, \\ &= 1 \qquad \text{for } i = j. \end{aligned} \qquad (A1.31)$$

Likewise the deviatoric components of strain ϵ_{ij} is given by

$$\varepsilon_{ij} = \varepsilon_{ij} - \delta_{ij}(\varepsilon_m/3). \qquad (A1.32)$$

Note that the traces of the S_{ij} and ϵ_{ij} matrices are each zero, i.e.,

$$S_{11} + S_{22} + S_{33} = 0, \quad \text{and} \quad \epsilon_{11} + \epsilon_{22} + \epsilon_{33} = 0, \tag{A1.33}$$

confirming the premise that the deviatoric components of the stress do not produce any dilation. Also,

$$S_{ij} = \sigma_{ij}, \quad \text{and} \quad \epsilon_{ij} = \varepsilon_{ij}, \, (i \neq j), \tag{A1.34}$$

so that from (8.4) and (A1.7) we get

$$S_{ij} = \sigma_{ij} = Ge_{ij} = 2G\varepsilon_{ij} = 2G\epsilon_{ij}, \, (i \neq j). \tag{A1.35}$$

It may be shown that the relationship between the deviatorics, i.e., $S_{ij} = 2G\epsilon_{ij}$, holds true even for $i = j$, having five independent equations. These, along with (A1.29), are another way of expressing the constitutive equations (8.3) and (8.4).

One may realize that a uniaxial stress σ_x may be thought to be composed of a hydrostatic component equal to $\sigma_x/3$ and a deviatoric component equal to $2\sigma_x/3$.

Calculation of stresses at various points in a body due to application of external forces involves calculation of strains. The six strain components arise from only three displacements. All six strain components cannot be arbitrarily prescribed if we are to maintain single-valued continuous displacement functions (without incurring fracture of the body). This implies that the strain components must have three additional equations of interdependence. This interdependence is called the **Saint-Venant compatibility criterion** and is generally written as a set of six second-order equations (which are equivalent to three independent fourth-order equations):

$$\begin{gathered}
\frac{\partial^2 e_{xx}}{\partial y^2} + \frac{\partial^2 e_{yy}}{\partial x^2} = \frac{\partial^2 e_{xy}}{\partial y\,\partial x}, \\
\frac{\partial^2 e_{yy}}{\partial z^2} + \frac{\partial^2 e_{zz}}{\partial y^2} = \frac{\partial^2 e_{yz}}{\partial y\,\partial z}, \\
\frac{\partial^2 e_{xx}}{\partial z^2} + \frac{\partial^2 e_{zz}}{\partial x^2} = \frac{\partial^2 e_{xz}}{\partial z\,\partial x}, \\
2\frac{\partial^2 e_{xx}}{\partial y\,\partial z} = \frac{\partial}{\partial x}\left\{-\frac{\partial e_{yz}}{\partial x} + \frac{\partial e_{xz}}{\partial y} + \frac{\partial e_{xy}}{\partial z}\right\}, \\
2\frac{\partial^2 e_{yy}}{\partial z\,\partial x} = \frac{\partial}{\partial y}\left\{\frac{\partial e_{yz}}{\partial x} - \frac{\partial e_{xz}}{\partial y} + \frac{\partial e_{xy}}{\partial z}\right\}, \\
2\frac{\partial^2 e_{zz}}{\partial x\,\partial y} = \frac{\partial}{\partial z}\left\{\frac{\partial e_{yz}}{\partial x} - \frac{\partial e_{xz}}{\partial y} - \frac{\partial e_{xy}}{\partial z}\right\}.
\end{gathered} \tag{A1.36}$$

One way of imagining the enforcement of the compatibility criterion is to imagine a body cut up into a large number of pieces upon which strains are prescribed arbitrarily. The various pieces could not fit together to fill the entire space unless the strains were adjusted such that the displacements at the ends of each piece meshed with those of the neighboring pieces.

Solutions of problems in elasticity require solving the three equilibrium equations (A1.3), the six definitions of strain (A1.5), and the six constitutive equations (8.3) and (8.4) with the prescribed boundary conditions. (The 15 equations give the 15 variables: six stress components, six strain components, and three displacement components. Note that the compatibility criterion would be satisfied automatically.) Often, the differential equations are not easy to solve, and so one turns to approximate methods. The most important of these methods is the minimization of the potential energy, which in this case is the mechanical strain energy.

The mechanical strain energy per unit volume (or simply the strain energy density) U_0 due to a uniaxial stress is given by

$$U_0 = \tfrac{1}{2}\sigma_x e_x = \tfrac{1}{2}E e_x^2. \tag{A1.37}$$

In the presence of a pure shear σ_{xy}, the strain energy density is

$$U_0 = \tfrac{1}{2}\sigma_{xy} e_{xy} = \tfrac{1}{2}G e_{xy}^2. \tag{A1.38}$$

Under the general stress condition, the strain energy density is

$$U_0 = \tfrac{1}{2}[\sigma_{xx}e_{xx} + \sigma_{yy}e_{yy} + \sigma_{zz}e_{zz} + \sigma_{xy}e_{xy} + \sigma_{yz}e_{yz} + \sigma_{xz}e_{xz}]. \tag{A1.39}$$

Simple situations of boundary conditions are the **plane stress** and the **plane strain** conditions. The plane stress state approximates a two-dimensional stress condition, i.e.,

$$\sigma_x = \sigma_x(x, y); \quad \sigma_y = \sigma_y(x, y); \quad \sigma_{xy} = \sigma_{xy}(x, y); \quad \text{and } \sigma_{zz} = \sigma_{xz} = \sigma_{yz} = 0. \tag{A1.40}$$

This condition is fulfilled approximately by thin slices where the thickness dimension is small compared to the length and the width.

A state of plane strain exists if the displacements in the body take the form

$$u = u(x, y); \qquad v = v(x, y); \qquad \text{and } w = 0. \tag{A1.41}$$

These conditions imply that $e_{zz} = e_{xz} = e_{yz} = 0$ and $\sigma_{xz} = \sigma_{yz} = 0$. In turn, these lead to a body force component in the z-direction equal to zero; however, a finite $\sigma_{zz} = \nu(\sigma_{xx} + \sigma_{yy})$. (Note, again, that the strain and the body force component in the z-direction are zero, but not the z-stress.) Most long cylindrical bodies or prismatic bodies often approximate the plane strain condition.

In closing, there are two other principles that are quite useful in linear elasticity problems. The first of these is **Boltzmann's principle of superposition**, which states that the dependent variables (strain, stress, and displacement components) resulting from each set of separately acting external loads (body forces, surface tractions, or prescribed surface displacements) may be superposed to give the total value of the variables due to a combined set of external loads. The order in which a given external load may be applied is also immaterial. The reader may acknowledge that this is a common property of all linear equations. In linear elasticity, the displacements are so small that the principle of superposition is not very difficult to obey. The second is **Saint–Venant's principle**, which states that the stresses due to statistically equivalent loadings applied over a small area differ only over the one to two smallest dimensions in the vicinity of the area of the application. Away from these "ends," the overall stresses are essentially identical. This principle is quite useful in approximating elasticity situations in that a certain set of boundary forces may be replaced by an equivalent loading for which the solution may be easier to obtain. It also means that **many of the analytical solutions obtained for semi-infinite geometries (e.g., large plates, long rods) do not apply at the ends**. This **end problem** is of great concern to the success of many products where finite-sized components are subjected to high stresses at the ends (for instance, gripped by or sealed to other components).

For additional reading, the following references are recommended:

1. *Theory of Elasticity*, by Timoshenko and J. N. Goodier. McGraw-Hill, New York, 1970.
2. *Elasticity: Tensor, Dyadic, and Engineering Approaches*, by P. C. Chou and N. J. Pagano. Van Nostrand, New York, 1967.
3. *Elasticity, Fracture, and Flow*, by J. C. Jaeger. John Wiley & Sons, New York, 1964.
4. *Photoelastic and Electro-optic Properties of Crystals*, by T. S. Narasimhamurty. Plenum Press, New York, 1981.
5. *Elements of Elasticity*, by H. T. Smyth and L. C. Klein. *J. Matls. Ed.* **8**(1–2), 239–265 (1986).

Appendix II

Units, Conversions and General Data

A. Basic and Supplementary SI Units

Quantity	*Unit*	*Symbol*
Length	meter	m
Mass	kilogram	kg
Time	second	s
Electric current	Ampere	A
Thermodynamic temperature	kelvin	K
Amount of substance	mole	mol
Luminous intensity	candela	cd
Plane angle	radian	rad
Solid angle	steradian	sr

B. Prefixes for Fractions and Multiples

10^{12}	10^{9}	10^{6}	10^{3}	10^{-1}	10^{-2}	10^{-3}	10^{-6}	10^{-9}	10^{-12}
tera	giga	mega	kilo	deci	centi	milli	micro	nano	pico
T	G	M	k	d	c	m	μ	n	p

C. Some Derived SI Units

Area	square meter	m^2
Volume	cubic meter	m^3
Density	kilogram per cubic meter	$kg{\cdot}m^{-3}$
Force	newton	$N = kg{\cdot}m{\cdot}s^{-2} = J{\cdot}m^{-1}$
Pressure (stress)	Pascal = newton per square meter	$Pa = N{\cdot}m^{-2}$
Energy	Joule	$J = kg{\cdot}m^2{\cdot}s^{-2} = N{\cdot}m$
Power	watt	$W = kg{\cdot}m^2{\cdot}s^{-3} = J{\cdot}s^{-1}$
Electric charge	coulomb	$C = A{\cdot}s$
Electric potential difference	volt	$V = kg{\cdot}m^2{\cdot}s^{-3}{\cdot}A^{-1} = J{\cdot}A^{-1}{\cdot}s^{-1}$
Electric resistance	ohm	$\Omega = V{\cdot}A^{-1}$
Electrical conductance	sieman	$S = \Omega^{-1}$
Electric capacitance	farad	$F = A{\cdot}s{\cdot}V^{-1}$
Magnetic flux	weber	$Wb = V{\cdot}S$
Magnetic flux density	tesla	$T = Wb{\cdot}m^{-2}$
Magnetic inductance	henry	$H = Wb{\cdot}A^{-1}$

D. Some Non-SI Units and Conversions

Quantity	*Name*	*SI equivalent*
Length	angstrom (Å)	10^{-10} m
	inch	0.0254 m
Volume	liter (L)	10^{-3} m^3
Mass	pound (lb)	0.454 kg
Force	dyne (dyn)	10^{-5} N
	kilogram force (kgf = kp)	9.807 N
	pound force (lbf)	4.448 N
	poundal	0.138 N
Pressure	atmosphere (atm)	1.013×10^5 Pa
	torr (mm Hg)	133.3 Pa
	bar	10^6 dyn/cm^2 = 10^5 Pa
	pounds per square inch (psi)	6895 Pa

Energy	erg	10^{-7} J
	calorie (cal)	4.187 J
	electronvolt (eV)	0.1602×10^{-18} J
	kilowatt-hour (kWh)	3.600×10^{6} J
Power	horsepower (hp), metric	735.5 W
Frequency	hertz (Hz)	s^{-1}
Magnetic field	Gauss (G)	10^{-4} T
Viscosity	poise (P)	10^{-1} Pa·s
Stress-optical coefficient	Brewster (B)	10^{-12} Pa^{-1} ($=1$ TPa^{-1})

E. Constants

Constant	*Usual notation*	*Value*
Speed of light	c	2.998×10^{8} $m{\cdot}s^{-1}$
Charge of proton	e	1.602×10^{-19} C
Mass of electron	m_e	9.11×10^{-31} kg
Mass of proton	m_p	1.673×10^{-27} kg
Acceleration due to gravity	g	9.806 $m{\cdot}s^{-2}$
Avogadro's number	N	6.022×10^{23} mol^{-1}
Loschmidt number	N_L	2.687×10^{25} m^{-3}
Boltzmann constant	k_B	1.381×10^{-23} $J{\cdot}K^{-1}$
Gas constant	R $(=k_BN)$	8.314 $J{\cdot}K^{-1}{\cdot}mol^{-1}$ 1.987 $cal{\cdot}K^{-1}{\cdot}mol^{-1}$
Faraday constant	F $(=eN)$	96,487 $C{\cdot}mol^{-1}$
Planck constant	h	6.626×10^{-34} J·s
Stefan-Boltzmann constant	σ	5.67×10^{-8} $W{\cdot}m^{-2}{\cdot}K^{-4}$
Vacuum permittivity	ε_0	8.854×10^{-12} $J^{-1}{\cdot}C^{2}{\cdot}m^{-1}$ ($= F{\cdot}m^{-1}$)
Vacuum permeability (Note: $\varepsilon_0\mu_0 = 1/c^2$)	μ_0	$4\pi \times 10^{-7}$ $J{\cdot}C^{-2}{\cdot}s^{2}{\cdot}m^{-1}$ ($= H{\cdot}m^{-1}$)

F. Other Useful Data

$$\text{Energy of wavelength } \lambda \text{ (nm)} = \frac{1239.8}{\lambda} \text{ eV}$$

$$\text{Wavenumber for } \lambda \text{ (nm)} = \frac{10^7}{\lambda} \text{ cm}^{-1}$$

Absorption coefficient $1\ \text{cm}^{-1} = 4.343 \times 10^5$ dB/km (Note: dB = decibels)
$\pi \cong 3.14159$
Base of the natural logarithm: e = 2.71828
$\log_e(a) = \log_e(10) \times \log_{10}(a) = 2.3026 \log_{10}(a)$
[Usually, ln = $\log_e(a)$; and log = $\log_{10}(a)$]
1 MPa $\cong$ 145 psi
1 year $\cong 3.15 \times 10^7$ s .... ticking away!

Subject Index

D

E

F

G

H

I

J

K

R

S

T

U

V

W

X

Y

Z

ISBN 0-12-714970-8

90051

9 780127 149707